D1238659

The chemistry of

organophosphorus compounds

Volume 3

THE CHEMISTRY OF FUNCTIONAL GROUPS

A series of advanced treatises under the general editorship of Professor Saul Patai

The chemistry of alkenes (2 volumes)
The chemistry of the carbonyl group (2 volumes)
The chemistry of the ether linkage
The chemistry of the amino group
The chemistry of the nitro and nitroso groups (2 parts)
The chemistry of carboxylic acids and esters
The chemistry of the carbon–nitrogen double bond
The chemistry of amides
The chemistry of the cyano group
The chemistry of the hydroxyl group (2 parts)
The chemistry of the azido group
The chemistry of acyl halides
The chemistry of the carbon–halogen bond (2 parts)
The chemistry of the quinonoid compounds (2 volumes, 4 parts)
The chemistry of the thiol group (2 parts)
The chemistry of the hydrazo, azo and azoxy groups (2 parts)
The chemistry of amidines and imidates (2 volumes)
The chemistry of cyanates and their thio derivatives (2 parts)
The chemistry of diazonium and diazo groups (2 parts)
The chemistry of the carbon–carbon triple bond (2 parts)
The chemistry of ketenes, allenes and related compounds (2 parts)
The chemistry of the sulphonium group (2 parts)
Supplement A: The chemistry of double-bonded functional groups (2 volumes, 4 parts)
Supplement B: The chemistry of acid derivatives (2 volumes, 4 parts)
Supplement C: The chemistry of triple-bonded functional groups (2 parts)
Supplement D: The chemistry of halides, pseudo-halides and azides (2 parts)
Supplement E: The chemistry of ethers, crown ethers, hydroxyl groups and their sulphur analogues (2 volumes, 3 parts)
Supplement F: The chemistry of amino, nitroso and nitro compounds and their derivatives (2 parts)
The chemistry of the metal–carbon bond (5 volumes)
The chemistry of peroxides
The chemistry of organic selenium and tellurium compounds (2 volumes)
The chemistry of the cyclopropyl group (2 parts)
The chemistry of sulphones and sulphoxides
The chemistry of organic silicon compounds (2 parts)
The chemistry of enones (2 parts)
The chemistry of sulphinic acids, esters and their derivatives
The chemistry of sulphenic acids and their derivatives
The chemistry of enols
The chemistry of organophosphorus compounds (3 volumes)
The chemistry of sulphonic acids, esters and their derivatives
The chemistry of alkanes and cycloalkanes
Supplement S: The chemistry of sulphur-containing functional groups

UPDATES

The chemistry of α-haloketones, α-haloaldehydes and α-haloimines
Nitrones, nitronates and nitroxides
Crown ethers and analogs
Cyclopropane-derived reactive intermediates
Synthesis of carboxylic acids, esters and their derivatives
The silicon–heteroatom bond
The chemistry of lactones and lactams

Patai's 1992 guide to the chemistry of functional groups—*Saul Patai*

—P

The chemistry of **organophosphorus compounds**

Volume 3

Phosphonium salts, ylides and phosphoranes

Edited by

FRANK R. HARTLEY

Cranfield University
Cranfield, UK

1994

JOHN WILEY & SONS
CHICHESTER—NEW YORK—BRISBANE—TORONTO—SINGAPORE

An Interscience® Publication

Other Wiley Editorial Offices

John Wiley & Sons, Inc., 605 Third Avenue,
New York, NY 10158-0012, USA

Jacaranda Wiley Ltd, 33 Park Road, Milton,
Queensland 4064, Australia

John Wiley & Sons (Canada) Ltd, 22 Worcester Rd,
Rexdale, Ontario M9W 1L1, Canada

John Wiley & Sons (SEA) Pte Ltd, 37 Jalan Pemimpin #05-04,
Block B, Union Industrial Building, Singapore 2057

Library of Congress Cataloging-in-Publication Data

(Revised for vol. 3)

The Chemistry of organophosphorus compounds.

(The Chemistry of functional groups)
'An Interscience publication.'
Includes bibliographical references and indexes.
Contents: v. 1. Primary, secondary, and tertiary
phosphines, polyphosphines, and heterocyclic organo-
phosphorus(III) compounds—v. 2. Phosphine oxides,
sulphides, selenides, and tellurides—v. 3. Phosphonium
salts, ylides, and phosphoranes.
1. Organophosphorus compounds. I. Hartley, F. R.
II. Series.
QD412.P1C444 1990 547'.07 89-22591
ISBN 0 471 92607 8 (v. 1)
ISBN 0 471 93056 3 (v. 2)
ISBN 0 471 93057 1 (v. 3)

British Library Cataloguing in Publication Data

A catalogue record for this book is available from the British Library

ISBN 0 471 93057 1

Typeset in Times 9/10 pt by Thomson Press (India) Ltd, New Delhi and
Printed and bound in Great Britain by Biddles Ltd, Guildford, Surrey

To

V. H., S. M. H., J. A. H. and E. J. H.

Contributing authors

Steven M. Bachrach	The Michael Faraday Laboratories, Department of Chemistry, Northern Illinois University, DeKalb, Illinois 60115, USA
Ramon Burgada	Université Pierre et Marie Curie, Laboratoire de Chimie des Organoéléments, CNRS, 4 Place Jussieu, F-75252 Paris Cedex 05, France
Henri-Jean Cristau	Laboratoire de Chimie Organique, URA 458, Ecole Nationale Supérieure de Chimie, Université de Montpellier, 8 rue de l'Ecole Normale, 34053 Montpellier Cedex 1, France
Manfred Dankowski	Degussa AG, ZN Wolfgang, AC-PT-PS, Postfach 13 45, 63403 Hanau, Germany
Hannah Feilchenfeld	Department of Organic Chemistry, The Hebrew University of Jerusalem, Jerusalem 91904, Israel
Declan G. Gilheany	Department of Chemistry, St Patrick's College, Maynooth, Co. Kildare, Ireland
Carmen I. Nitsche	The Michael Faraday Laboratories, Department of Chemistry, Northern Illinois University, DeKalb, Illinois 60115, USA
Francoise Plénat	Laboratoire de Chimie Organique, URA 458, Ecole Nationale Supérieure de Chimie, Université de Montpellier, 8 rue de l'Ecole Normale, 34053 Montpellier Cedex 1, France
Kalathur S. V. Santhanam	Chemical Physics Group, Tata Institute of Fundamental Research, Colaba, Bombay 400 005, India
Ralph Setton	CNRS, Centre de Recherche sur la Matière Divisée, 1B rue de la Férollerie, F-45071 Orléans Cedex 2, France

Foreword

The Chemistry of Organophosphorus Compounds is a multi-volume work within the well established series of books covering *The Chemistry of Functional Groups*. It is proposed to cover the extensive subject matter in four volumes.

Volume 1 covers primary, secondary and tertiary phosphines (PR_3H_{3-n}, $n = 1$–3), polyphosphines (both P—$(C)_n$—P and $R(P)_nR'$, $n > 1$) and heterocyclic compounds containing phosphorus.
Volume 2 covers phosphine oxides, sulphides, selenides and tellurides.
Volume 3 cover phosphonium salts, phosphonium ylides and phosphoranes.
Volume 4 will cover phosphinous, phosphonous, phosphinic and phosphonic acid compounds and their halogen derivatives R_2PY, RPY_2 and $R_2P(X)Y_2$, where Y = halogen and X = O, S or Se.

For many years the nomenclature used in organophosphorus chemistry was extremely frustrating, with different compounds being given the same name by different authors. The nomenclature has, however, now been rationalized and is summarized in Volume 1, Chapter 1, Section IV.

In common with other volumes in *The Chemistry of the Functional Groups* series, the emphasis is laid on the functional group treated and on the effects which it exerts on the chemical and physical properties, primarily in the immediate vicinity of the group in question, and secondarily on the behaviour of the whole molecule. The coverage is restricted in that material included in easily and generally available secondary or tertiary sources, such as *Chemical Reviews* and various 'Advances' and 'Progress' series, as well as textbooks (i.e. in books which are usually found in the chemical libraries of universities and research institutes) is not as a rule repeated in detail, unless it is necessary for the balanced treatment of the subject. Therefore, each of the authors has been asked *not* to give an encyclopaedic coverage of his or her subject, but to concentrate on the most important recent developments and mainly on material that has not been adequately covered by reviews or other secondary sources by the time of writing of the chapter, and to address himself or herself to a reader who is assumed to be at a fairly advanced postgraduate level. With these restrictions, it is realized that no plan can be devised for a volume that would give a complete coverage of the subject with no overlap between the chapters, while at the same time preserving the readability of the text.

The publication of the Organophosphorus Series would never have started without the support of many people. This volume would never have reached fruition without the help of Mr Mitchell and Mrs Perkins with typing and the efficient and patient cooperation of several staff members of the Publisher. Many of my colleagues in England, Israel and elsewhere gave help in solving many problems, especially Professor Saul Patai, without whose continual support and encouragement this work would never have been attempted.

Finally, that the project ever reached completion is due to the essential support and partnership of my wife and family, amongst whom my eldest daughter provided both moral support and chemical understanding in the more difficult areas of the subject.

Cranfield, England FRANK HARTLEY

Contents

List of abbreviations used

abd	azobisisobutyl diacetate
Ac	acetyl (MeCO)
acac	acetylacetone
Ad	adamantyl
aibn	azobisisobutyronitrile
all	allyl
an	acetonitrile
An	anisyl
Ar	aryl
ATP	adenosine triphosphate
bipy	2,2′-bipyridine
bpr	Berry pseudorotation
BSA	bovine serum albumin
btsa	*N,O*-bis(trimethylsilyl)acetamide
Bu	butyl (also *t*-Bu or Bu^t)
Bz	benzyl
CD	circular dichroism
CI	chemical ionization
cod	cycloocta-1,5-diene
cp	cyclopentadienyl
mCPBA	*m*-chloroperoxybenzoic acid
CP-MAS	cross-polarization magic angle spinning
Cy	cyclohexyl
dbn	1,5-diazabicyclo[5.4.0]non-5-ene
dbso	dibenzoyl sulphoxide
dbu	1,8-diazabicyclo[5.4.0]undec-7-ene
$DDPN^+$	deamino diphosphopyridine nucleotide
diop	2,3-*o*-isopropylidene-2,3-dihydroxy-1,4-bis(diphenylphosphino) butane
dme	1,2-dimethoxyethane
dmf	dimethylformamide
dmg	dimethylglyoximate
dmpe	bis(1,2-dimethylphosphino)ethane
dmso	dimethyl sulphoxide
DNA	deoxyribonucleic acid

$dpbO_2$, $dpbS_2$, $dpbSe_2$, $dpeO_2$, $dpeS_2$, $dpeSe_2$, $dpmO_2$, $dpmS_2$, $dpmSe_2$, $dppO_2$, $dppS_2$, $dppSe_2$	$Ph_2P(E)(CH_2)_nP(E)Ph_2$ b, $n = 4$ e, $n = 2$ m, $n = 1$ p, $n = 3$ $E = O, S, Se$
dpmPS	$Ph_2P(S)CH_2PPh_2$
dpmPSe	$Ph_2P(Se)CH_2PPh_2$
DPN^+	diphosphopyridine nucleotide
DPNH	dihydronicotinamide adenine dinucleotide
dppb	bis(1,4-diphenylphosphino)butane
dppe	bis(1,2-diphenylphosphino)ethane
dppm	bis(1,1-diphenylphosphino)methane
dppp	bis(1,3-diphenylphosphino)propane
dpso	diphenyl sulphoxide
DTG	differential thermal gravimetry
ECE	electron transfer followed by chemical reaction followed by further electron transfer
edta	ethylenediaminetetraacetic acid
ee	enantiomeric excess
EI	electron impact
EPR	electron paramagnetic resonance
ESR	electron spin resonance
FAB	fast atom bombardment
FAD	flavine adenine dinucleotide
FDMS	field desorption mass spectrometry
FMN	flavine mononucleotide
FT	Fourier transform
GLC	gas–liquid chromatography
Hba	benzoylacetone
Hbfa	benzoyltrifluoroacetone
Hdbm	dibenzoylmethane
H_2dehp	di(2-ethylhexyl)phosphoric acid
H_2dmg	dimethylglyoxime
H_2dz	dithizone (3-mercapto-1,5-diphenylformazan)
Hex	hexyl
Hhfa	hexafluoroacetylacetone
hmde	hanging mercury drop electrode
hmpa	hexamethylphosphoramide
hmpt	hexamethylphosphorotriamide
HOMO	highest occupied molecular orbital
Hox	8-hydroxyquinoline
HPLC	high-performance liquid chromatography

Hpmap	1-phenyl-3-methyl-4-acylpyrazol-5-one
Hpmbp	1-phenyl-3-methyl-4-benzoylpyrazol-5-one
Hpmbup	1-phenyl-3-methyl-4-butyrylpyrazol-5-one
Hpmdbp	1-phenyl-3-methyl-4-(3,5-dinitrobenzoyl)-pyrazol-5-one
Hpmop	1-phenyl-3-methyl-4-octanoylpyrazol-5-one
Hpmsp	1-phenyl-3-methyl-4-stearoylpyrazol-5-one
Hpmtfp	1-phenyl-3-methyl-4-trifluoroacetylpyrazol-5-one
Hpva	pivaloyltrifluoroacetone
Hpvta	dipivaloylacetone
Htbfa	thiobenzoyltrifluoroacetone
Htfa	trifluoroacetylacetone
Htfma	1,1,1-trifluoro-5-methylhexane-2,4-dione
Htta	1,1,1-trifluoro-3-(2-thenoyl)acetone
IP	ionization potential
LC_{50}	concentration causing lethality to 50% of the population
LD_{50}	dose causing lethality to 50% of the population
lda	lithium diisopropylamide
lp	lone pair of electrons
LUMO	lowest unoccupied molecular orbital
M	metal
Me	methyl
mibk	methyl isobutyl ketone
MIS	metal–insulator semiconductor
MNDO	modified neglect of diatomic overlap
MS	mass spectrometry
NADP	nicotinamide adenine dinucleotide phosphate
nba	*N*-bromoacetamide
NCI	negative ion chemical ionization
NHN	nicotinamide ribose monophosphate
Np	naphthyl
OAc	acetate
ORD	optical rotatory dispersion
PCI	positive ion chemical ionization
Pe	pentenyl
Pen	pentyl (C_5H_{11})
PES	photoelectron spectroscopy
Ph	phenyl
phen	1,10-phenanthroline
ppa	polyphosphoric acid
ppm	parts per million
Pr	propyl (also *i*-Pr or Pr^i)
R	any radical
RNA	ribonucleic acid
SCE	standard calomel electrode

SCF	self-consistent field
SIMS	secondary ion mass spectrometry
tbap	tetra-*n*-butylammonium perchlorate
tbp	trigonal bipyramid (when referring to a structure) or tertiary butyl peroxide (when referring to a chemical)
tbpo	tri-*n*-butyl phosphate
TCNQ	tetracyanoquinone
tfa	trifluoroacetic acid
tfb	tetrafluorobenzobicyclo[2.2.2]octatriene
TG	thermogravimetric
thf	tetrahydrofuran
tht	tetrahydrothiophene
TLC	thin-layer chromatography
tmeda	*N,N,N',N'*-tetramethylethylenediamine
tmpo	2,2,6,6-tetramethylpiperidine-1-oxyl
Tol	tolyl ($CH_3C_6H_4$)
topo	tri-*n*-octyl phosphate
tos	tosyl
tp	tetragonal pyramid
TPN	triphosphopyridine nucleotide
tr	turnstile rotation
t_r	adjusted retention time (in GLC)
TSP	thermal spraying
VSEPR	valence shell electron pair repulsion
X	halide
XRD	X-ray diffraction

CHAPTER **1**

Structure and bonding in phosphonium ylides, salts and phosphoranes

D. G. GILHEANY

Department of Chemistry, St. Patricks College, Maynooth, Co-Kildare, Ireland

The chemistry of organophosphorus compounds, Vol. 3
Edited by F. R. Hartley

I. INTRODUCTION

The general features of bonding to phosphorus have been discussed in the first two volumes of this series[1,2], including likely bonding schemes[1], the strengths of single and double bonds to phosphorus[1] and the non-involvement of virtual d orbitals in any of the three possible current descriptions of the phosphoryl bond[2]. This latter point is also relevant for the systems discussed here and, once again, as we shall see, there are more powerful alternative descriptions of the bonding than those involving d orbitals. Because there is a strong similarity between the structure and bonding in tertiary phosphine oxides and phosphonium ylides, the latter are discussed first here, followed by the less controversial salts and finally the pentacoordinate phosphoranes.

II. PHOSPHONIUM YLIDES

In phosphonium ylides, as we shall see in Section II.A, the phosphorus atom is tetracoordinate and the unique (anionic) carbon is tricoordinate. For exactly the same reasons as in the oxides[2], all previous discussions[3–14] of the P—C bond have been in terms of a resonance hybrid between a dipolar form **1A** and a double bond form **1B**:

$$\underset{\textbf{(1A)}}{R_3\overset{+}{P}-\overset{-}{C}R_2} \longleftrightarrow \underset{\textbf{(1B)}}{R_3P{=}CR_2}$$

For the purposes of chemical reactivity, the dipolar form **1A** is considered the more important[3]. Structure **1B** is meant to indicate $d\pi-p\pi$ bonding involving back-donation of electron density from a doubly occupied 2p orbital of the ylidic (anionic) carbon into vacant phosphorus 3d orbitals in an overlap scheme such as that in Figure 1. This $d\pi-p\pi$ bonding has been invoked to explain a number of the properties of phosphonium ylides (see Section II.B.1), particularly the fact that they are more stable than their nitrogen analogues, for which such a stabilizing interaction is not possible, nitrogen not having the requisite low-energy vacant orbitals[3,15]. Contrariwise, many authors have also taken these properties as evidence for $d\pi-p\pi$ bonding so that the structure and bonding of phosphonium ylides are part of the general controversy about $d\pi-p\pi$ bonding (see ref. 2 for a more general discussion). For these reasons, many studies in the past 20 years have addressed the two fundamental problems associated with phosphonium ylides, namely the extent (if any) of the contribution of structure **1B** and the related question of the geometry (i.e. configuration) at phosphorus and the anionic carbon.

As discussed in detail in Volume 2[2], it is now known that d orbitals are not involved in bonding in Main Group compounds. In particular, it is now clear that the requirement for

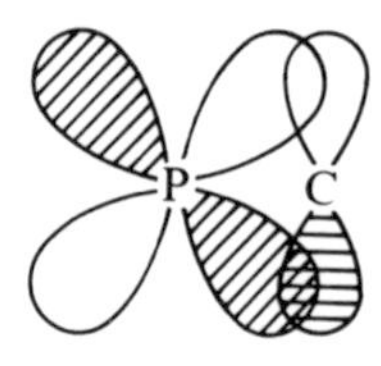

FIGURE 1. The obsolete view of the p-type overlap in phosphonium ylides

d functions in *ab initio* LCAO treatments of the bonding is as polarization functions and that they are unrelated to atomic d orbitals[16]. There is a greater need for polarization functions in these systems because there is a large charge variation within the molecule, e.g. the high electric field across the P—C bond as a result of the opposite charges on P and C. Further, as pointed out earlier[2], structures **1A** and **1B** are not mutually exclusive because any back-bonding that occurs will still lead to a highly unsymmetrical charge distribution.

The prototypical ylide $H_3P{=}CH_2$ has been detected by neutralization–reionization mass spectrometry[17] as predicted by calculation[18], but no data are yet available for it.

A. Structure

Structure in organophosphorus chemistry was last reviewed comprehensively in 1974 by Corbridge[19]. However, work on phosphorus ylides was not well developed at that time and so the material presented below is mostly from the intervening period. A useful overview of structures determined up to early 1972 is given in a paper by Howells *et al.*[20] and there is a comprehensive list of references up to 1981 in a paper by Vincent *et al.*[21]. Table 1 gives structural data on most of the phosphorus ylides that have been reported since 1973, along with a few earlier structures. Structure numbers in Table 1 refer to compounds presented in Scheme 1. Other ylides whose structures have been determined since 1973 but which are not included in Table 1 are given in Scheme 2. The literature has been surveyed up to the end of 1991 and a determined effort has been made to include a high proportion of available structures in Table 1 and Schemes 1 and 2. However, they should not be regarded as complete and readers requiring data on particular compounds, particularly metal-containing systems, should also refer to *Chemical Abstracts*. Readers should also refer to the original publications for details of the structural determinations, error limits and the other molecular dimensions of the compounds quoted.

It will be seen from Table 1 and Scheme 1 that, although there is a fairly large number of structure studies on phosphorus ylides, only some of these[22,23,25–27,29] have been systematic studies of the factors influencing structure. Also there are only three electron diffraction studies, none of which can be directly compared with an X-ray study. A further difficulty is the high R value of many structure determinations, particularly those pre-1973, many of which have $R > 10$. This is partly because many have been done purely to determine the constitution of unusual molecules containing the $Ph_3P{=}C$ moiety where data refinement was not critical. As shown by a recent study[23] discussed below, it would be unwise to use the results of studies with $R > 6$ for detailed structural comparisons and only a few of these are included in Table 1. Also, comparisons with other members of Group 5 are not well developed[80] so only two are included in Table 1. Further, irreversible isomerization processes involving rapid proton transfer[81,82] mean that the triisopropylphosphonium case[25] is the simplest available trialkyl-substituted isopropylide, further limiting the comparisons which can be made.

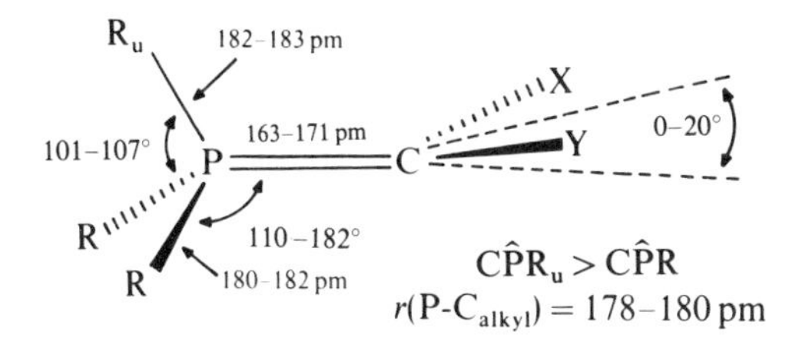

FIGURE 2. Structure of a typical phosphonium ylide

TABLE 1. Structural data[a] for some phosphonium ylides $R_3P{=}CXY$

Molecule[b]	r(P=C)	r(P—C)	∠CPR	∠RPR	∠PCX/Y	∠XCY	Σ(C)[c]	R[d]	Ref.
$Me_3P{=}CH_2$[e]	164.0	181.5	116.5	101.6	—	—	—	10	22
$Ph_3P{=}CH_2$[f]	169.3	182.0	112.2	103.8	114.8	119.5	349	4.3	23
		183.3u	116.3u	105.8u					
$Ph_3P{=}CH_2$	166.1	182.3	113.5	105.0	117.0	124.0	358	6.4	24
$Fc_3P{=}CH_2$	162.9	180.4	115.3	103.1	—	—	—	3.4	23
$Pr^i_3P{=}CMe_2$	173.1	183.4	108.1	114.9	120.1	114.0	354	4.9	25
			117.5u	104.2u					
$Ph_3P{=}C(CH_2)_2$	169.6	181.0	110.0	106.5	117.5	58.8	294	6.3	26
		181.6u	117.0u						
$Ph_3P{=}C(CH_2)_3$	166.8	181.2	108.7	108.7	131.3	91.4	354	5.0	27
		184.6u	123.5u	103.4u					
$Ph_3P{=}C(CF_2)_3$	171.3	179.8	110.7	108.2	—	93.4	—	5.2	20
$Bu^n_3N{=}C(CF_2)_3$	146.4	152.8	108.5	110.5	—	92.6	—	8.6	28
$Me_3P{=}CHSiH_3$[e]	165.3	180.7	115.0	—	123.4	—	—	7	29
2	168.6	181.5	116.6C	100.1C	121.5	117.0	360	3.9	30
			118.7P	100.9P					
3	168.6	178.3C	115.8C	106.7C	119.4	121.4	360	3.4	31
		205.0Cl	113.4Cl	101.7Cl					
4	164.6	179.3C	118.4C	105.4C	120.4	117.0	358	4.2	31
		236.6Br	116.4Br	97.2Br					
5	169.6	—	—	—	123.3B	118.8	360	4.7	32
					117.9Si				
$Me_3P{=}CHPPh_2$	168.4	179.2	113.8	104.8	122.2P	128	360	5.3	23
					110H				
$Ph_3P{=}CHPMe_2$	161.8	182.1	113.5	103.1	122.3P	—	—	4.9	23
$Ph_3P{=}C(PPh_2)_2$	172.0	181.6	113.6	104.8	118.1[g]	119.4	356	7.1	33
$Ph_3P{=}C(AsPh_2)_2$	169.8	181.8	113.8	104.6	118.2[h]	116.4	353	4.6	34
$Ph_3P{=}C(SbPh_2)_2$	169.2	182.8	114.5	104.0	121.0[i]	116.2	358	3.0	35
$Ph_3P{=}C(SePh)SPh$	170.7	181.8	113.1	105.2	118.5	122.2	359	5.9	36
6	172.6	180.4	114.5	104.0	119.1	121.7	360	6.1	37
$PhMe_2P{=}CHU(Cp)_3$	169	180Ph	—	—	—	—	—	7.1	38
		183Me							
$Ph_3P{=}CHCH{=}CMe_2$	167.5	182.0	113.2	105.5	126.7C	—	—	6.7	39
$Bu^t_2ClP{=}CPh_2$	166.9	188.0C	116.6Cl	—	—	—	—	6.4	40
		221.9Cl							
$Ph_3P{=}CHCOPh$	171	181	—	—	121.1	—	—	5.4	41
$Ph_3As{=}CHCOPh$	187	194	—	—	114.6	—	—	5.8	41
7	173.2	181.7	111.7	107.1	120.8	117.6	359	5.4	42

8	171.6	182.1	112.0	106.8	121.6	116.7	360	7.6	42
9	171.8	180.7	112.7	106.1	118.5	121	358	8.9	43
10	174.6	181.1	112.1	106.7	115.9C 117.2Se	123.9	357	5.3	44
11	168.1	180.2	112.4	106.4	—	—	—	4.0	45
12	171.8	179.0	112.2	106.7	118.2	—	—	3.6	46
13	175	181	—	—	120.3	—	—	4.9	47
$PhPr^{n}_{2}P{=}C(CN)_2$	174.3	179.9Pr 181.0Ph							48
14	176.1	179.5	110.2	108.7	120.0	118.3	358	6.1	49
15	176.8	179.8	—	—	—	—	—	5.5	50
$Ph_3P{=}C(CH)_4$	171.8	180.6	111.4	107.5	125.3	107.4	358	4.2	51
$Me_3P{=}C{=}PMe_3$ [e]	159.4	181.4	116.7	101.4	147.6	—	—	6.6	52
$Ph_3P{=}C{=}PPh_3$ [j]	161.0	185.3	—	—	134.4	—	—	5.9	53
$Ph_3P{=}C{=}PPh_3$ [k]	163.5	183.1	—	—	131.7	—	—	4.1	53
$Ph_3P{=}C{=}PPh_3$ [l]	162.9	183.2	114.9	103.9	143.8	—	—	8.9	54
$Ph_3P{=}C{=}PPh_3$ [l]	163.3	183.7	114.5	104.0	130.1	—	—	8.9	54
$MePh_2P{=}C{=}PPh_2Me$	164.8	182.1Me 182.5Ph	114.1Me 117.2Ph	103.7Me 102.3Ph	121.8	—	—	6.9	55, 56
16	164.9	182.3Me 183.0Ph	117.2Me 114.1Ph	103.6Me 102.7Ph	116.7	—	—	7.2	56, 57
$Ph_3P{=}C{=}C{=}O$	164.8	180.5	111.5	107.3	145.5	—	—	9.2	58
$Ph_3P{=}C{=}C{=}S$	167.7	179.5	110.8	108.1	168.0	—	—	7.1	59
$Ph_3P{=}C{=}C{=}NPh$	167.7	180.0	112.4	107.2	134.0	—	—	5.9	60
$Ph_3P{=}C{=}C(OEt)_2$	168.2	183.2	113.9	104.5	125.6	—	—	7.9	61
17	170.3	179.2	110.4	108.4	136.4	—	—	5.1	62
18	167.9	177.8	110.5	108.5	164.0	—	—	6.5	63
$(Ph_3P)_2C^+Br^-$	170.3	180.8	112.5	106.2	116PCH	128.2	360	3.6	64
$(Ph_3P)_3C^{2+}2I^-$	175	179	113.5	105.1	120.0	—	360	5.1	65, 66

[a] r(P—C) = bond length in pm; ∠XYZ = bond angle in degrees: mean values are quoted, unless indicated otherwise; C, P, Cl, Br, etc., identify R, X and Y groups; u refers to the unique substituent on phosphorus shown in Figure 2.
[b] All studies by single-crystal X-ray crystallography at room temperature, unless indicated otherwise.
[c] Sum of angles about the anionic carbon.
[d] Reliability index = 100 × conventional R for both X-ray and electron diffraction studies; see original papers for definitions.
[e] Electron diffraction study.
[f] Performed at 100 K with high-order data refinement.
[g] Mean of 107.4° and 128.8°.
[h] Mean of 108.4° and 128.0°.
[i] Mean of 130.5° and 111.5°.
[j] Orthorhombic.
[k] Performed at −160 °C.
[l] Monoclinic; two independent molecules in the unit cell, both reported.

(**2**)[30]

(**3**)[31]

(**4**)[31]

(**5**)[32]

(**6**)[37]

(**7**)[42]

(**8**)[42]

(**9**)[43]

(**10**)[44]

(**11**)[45]

(**12**)[46]

(**13**)[47]

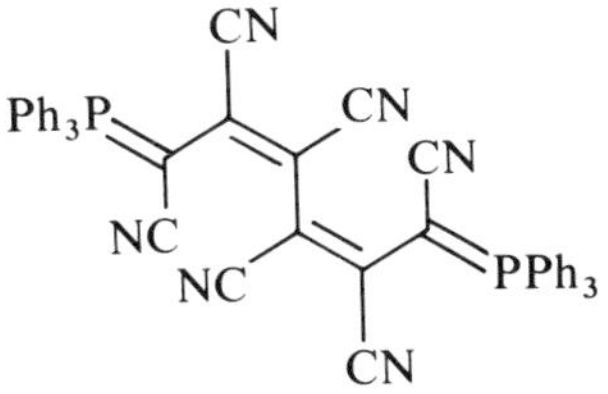

(**14**)[49]

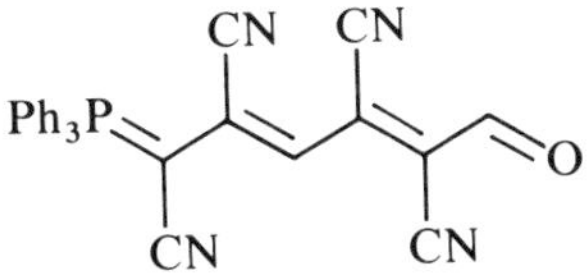

(**15**)[50]

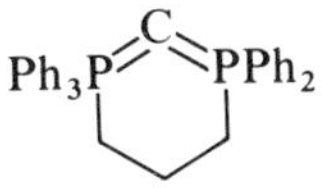

(**16**)[56,57]

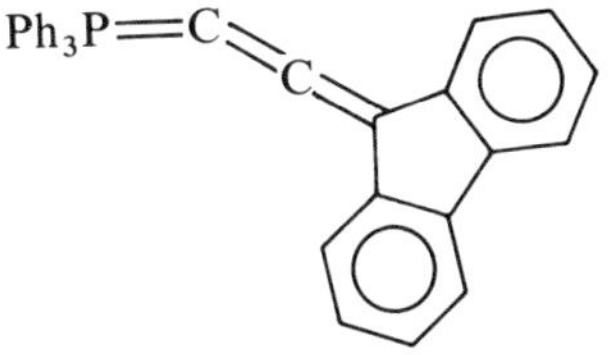

(**17**)[62]

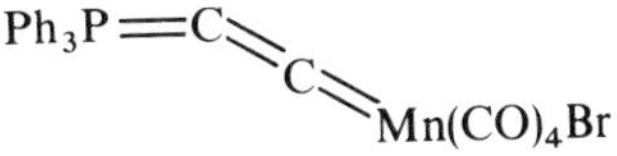

(**18**)[63]

SCHEME 1

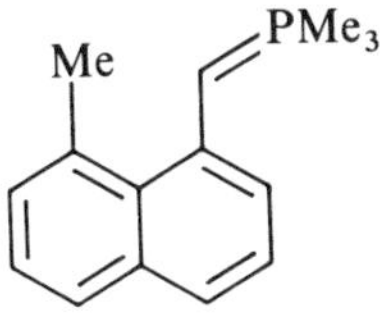

Ref. 37

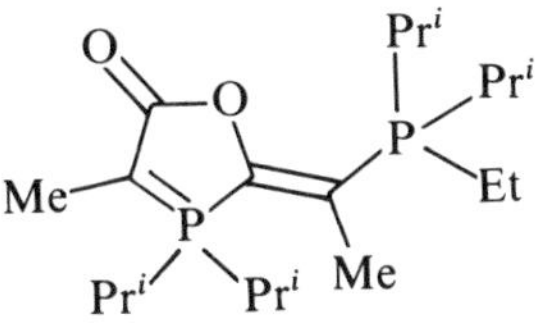

Ref. 67

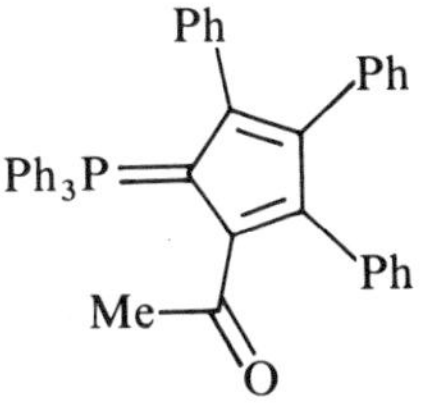

Ref. 68

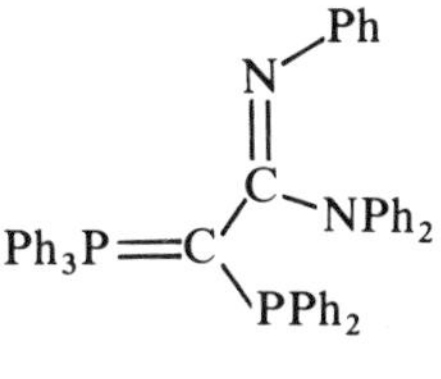

Ref. 69

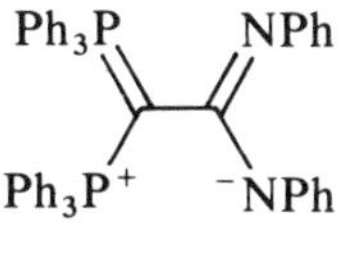

Ref. 70

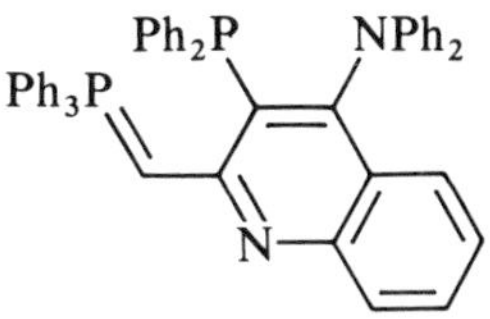

Ref. 71

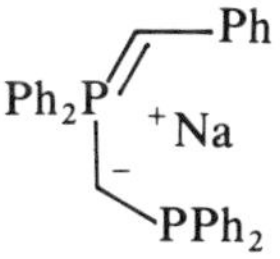

Ref. 72

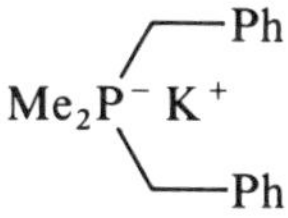

Ref. 73

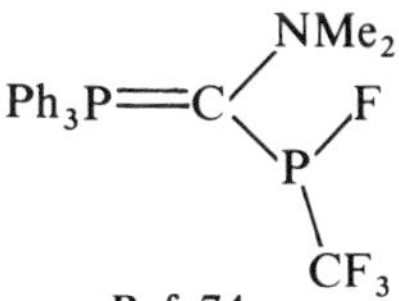

Ref. 74

SCHEME 2

(continued)

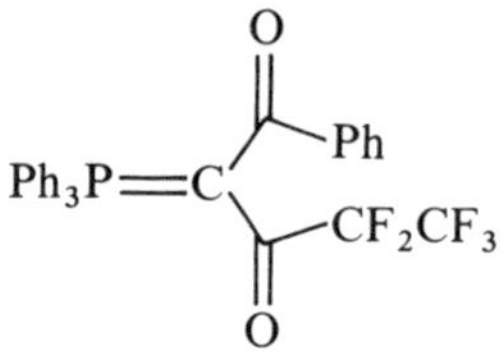

Ref. 75 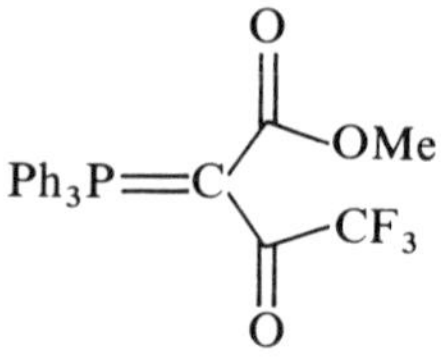 Ref. 76  Ref. 77

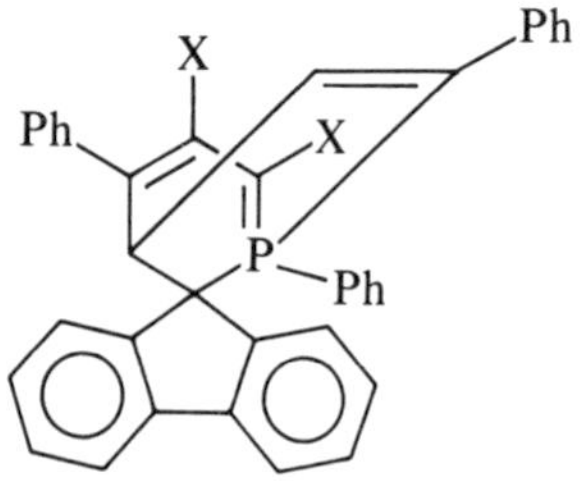

Ref. 78 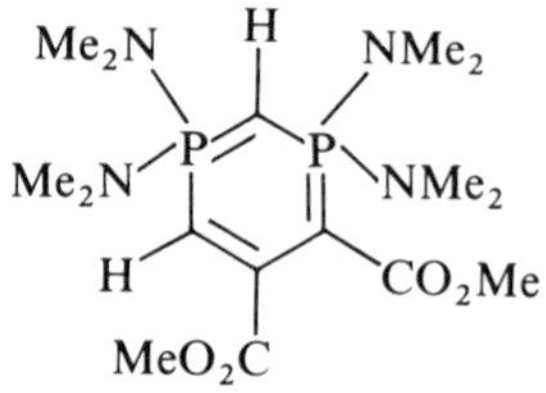 Ref. 79

SCHEME 2

It has been the practice[3] to divide ylides into stabilized, semi-stabilized and non-stabilized. It can be seen that this division breaks down somewhat for structural studies, as shown by the variation in P=C bond lengths and the more varied nature of the substituents in more recent studies. However, an effort has been made in Table 1 to quote non-stabilized ylides first, followed roughly by semi-stabilized, then stabilized and finally sp ylides.

1. Bond lengths

Given the comments above, it is to be expected that there will be uncertainties in some of the structural data for phosphorus ylides. This is very well illustrated by the length of the P=C ylidic bond for which the data are particularly unsatisfactory. From Table 1, it can be seen that it varies from 163 pm to 173 pm for unstabilized ylides, so that it is not possible to give a typical value for it. Our best guess must be the value of about 169 pm reported recently by Schmidbaur *et al.*[23] for $PH_3P{=}CH_2$. However, they also reported a value of 163 pm for the trisferrocenyl derivative[23] while 164 pm was found for $Me_3P{=}CH_2$ by electron diffraction[22], although the latter study had a high R value and C_3 symmetry was imposed in the analysis. A similar 4–5 pm difference in r(P=C) between trimethyl and triphenyl derivatives was found for carbodiphosphoranes (see Section II.A.4), so this may be a *bona fide* effect, but there are not enough data to be sure. The only isopropylidene ylide reported has a P=C length of 173 pm but this is probably anomalous because of the hindered nature of the substituents (see discussion of bond angles in Section II.A.2) which cannot easily be varied because of the proton transfer isomerization problems already referred to. The difficulties associated with high R values are well illustrated by the recent study on $Ph_3P{=}CH_2$ which was done at $-100\,^{\circ}C$ and with a high-order data refinement[23]. The previous study on this molecule by Bart[24] had a reasonable R value (6.4) but the P═C distance was 3 pm shorter. It is interesting that very similar observations have been made for the analogous phosphine oxide[2] where a decrease in R from 7.8 to 3.5 in a low-temperature study also caused an increase in P═O distance of 3 pm. Despite these difficulties, we may still say that r(P═C) is considerably shortened with respect to r(P—C) (180–183 pm, see below) and is in a range expected for a double bond between phosphorus and carbon. Thus, for example, the P—C bond length in phosphaalkenes[1,83] can vary between 161 and 171 pm with an average of 167 pm while P≡C in phosphaalkynes[1] is < 154 pm. Pauling[84] predicts 166.5 pm for a P═C double bond which includes the Schomaker–Stevenson correction[85].

One observation which has been known[20] from the 1960s, and which has stood the test of time as shown in Table 1, is that stabilized ylides have a longer P═C bond length (> 171 pm) than non-stabilized ylides. This can be rationalized by noting that stabilizing groups reduce the negative charge at the ylide carbon thereby weakening P═C. Another related and well known observation[20] is that in conjugatively stabilized ylides the conjugated multiple bond (C═O, C≡N or C═C) is also lengthened whereas the intervening C—C bond is shortened. This observation is also confirmed by the more recent data shown in Table 2, although the effect is not as marked as has been previously claimed[20]. Thus there are increases of 3–4 pm over a typical non-conjugated C═O length[86] of 122 pm and decreases of up to 7 pm relative to the 146 pm expected for an sp^2–sp^2 single bond[86]. These observations are easily explained by simple resonance arguments[20]. A further example shown in Table 2 is the small difference in C—C and C═C bond lengths around the five-membered ring in cyclopentadienylidenetriphenylphosphorane.

Since most studies have been on triphenyl-substituted ylides, there are plenty of data on the P—C(phenyl) bond length, which is in the range 181–182 pm with very little variation,

TABLE 2. Bond lengths[a] in some conjugatively stabilized phosphonium ylides R_3P═CX—C(═Y)Z

Molecule[b]	r(P═C)	r(C—C)	r(C═O)	r(C≡N)	r(C═C)	Ref.
Ph_3P═CHCOPh	171	139	126			41
Ph_3As═CHCOPh	187	138	124			41
7	173.2	139.2	124.3			42
8	171.6	141.7	122.1			42
9	171.8	141.7	121.9			43
10	174.6	140.6	125.2			44
12	171.8	139.1	123.4			46
13	175	136	127			47
$PhPr^n_2P$═C(CN)$_2$	174.3					48
14	176.1	141.7		115.3		49
15	176.8	141.9		114.8		50
Ph_3P═C(CH)$_4$	171.8	142.5			138.4	51
Ph_3P═CHCH═CMe$_2$	167.5	143.3			134.9	39
Ph_3P═C═C═O	164.8	121.0	118.5			58
Ph_3P═C═C═S	167.7	120.9	159.5S			59
Ph_3P═C═C═NPh	167.7	124.8	125.2N			60
17	170.3	124.3			135.8	62
Ph_3P═C═C(OEt)$_2$	168.2	131.7	—			61
18	167.9	121.6	198.1Mn			63

[a] In pm.
[b] All studies by single-crystal X-ray crystallography at room temperature.

as can be seen from Table 1. This is definitely longer than in phosphine oxides (179–180 pm)[2], about the same as in the few phosphine sulphides and selenides[2] and shorter than in phosphines (183–184 pm)[1]. As in the phosphine chalcogenides[2], it may be that the P—C(alkyl) bond is slightly shorter (178–180 pm) than the P—C(phenyl) bond, but there are not enough data to be sure.

2. Bond angles

Regarding the bond angles reported in Table 1, there is the expected tetrahedral arrangement about phosphorus. Again, just as in the phosphine chalcogenides[2], the bond angles to the ylide carbon are normally wider than those to the other carbons, which can be rationalized as a bond length–bond angle relationship[2] in that the closer a ligand approaches the central atom, the more space it takes up around that atom. Consistent with this is that the bond angles at phosphorus in a stabilized ylide are nearer to regular tetrahedral than in unstabilized ylides. Thus as the P=C bond length in stabilized ylides is longer, the carbanion centre does not take up as much space near the phosphorus atom, allowing the angles to the other ligands to widen. Note that in trisisopropylphosphonium isopropylide this normal disposition of angles around phosphorus is inverted just as in highly hindered phosphine chalcogenides, e.g. tri-*tert*-butylphosphine oxide[2], which are also anomalous in having a much larger P=O bond than normal.

It has been commonplace in the past[3–11] to state that the ylide carbanionic carbon is planar and that, since this is counter to VSEPR expectations for an isolated carbanion in an sp^3 orbital on carbon, this is evidence for back-bonding from carbon to phosphorus. It can be seen from Table 1 that there are small but definite deviations from planarity for many ylides. This is strikingly so for the most careful study[23] on $Ph_3P{=}CH_2$ where the angle sum at carbon is only 349°. Obviously, when the carbanion is part of a ring the deviation from planarity can be large, as in the cyclopropylides. Note, however, that even where there is significant carbanion pyramidalization, the P=C distance remains short[23,26,27].

3. Conformation

Liu and Schlosser[87] have recently shown convincingly by ^{13}C NMR that in solution there is free rotation about the P=C bond in unstabilized triphenylphosphonium ylides. This was because they failed to find any evidence for non-equivalence of the *ipso*-carbon signals down to temperatures as low as −105 °C.

However, there does appear to be a preferred conformation in the solid state, which is shown in Figure 2. Thus it is frequently observed in X-ray studies that the carbanion substituents tend to take up an orientation at right-angles to the plane of one of the P—C bonds, referred to here as the unique substituent and labelled u in Table 1 and Figure 2, and the conformation is usually referred to as *perpendicular*. The conformation resulting from a 90° rotation of the carbanion substituents is called *parallel* or *eclipsed*. The perpendicular conformation has also been found by electron diffraction[29]. Also, where a deviation from carbanion planarity is found, it is commonly towards this unique substituent and, in those cases where a difference can be discerned, the bond angle to the unique substituent is increased compared with that to the other substituents and the P—C bond length is increased. These generalizations are still tentative because the data set is not large, but there does appear to be a definite effect, so that we may give the overall solid-state structure of a typical phosphonium ylide as that shown in Figure 2.

The lengthening of the P—R(unique) distance detailed above provides evidence relevant to the qualitative discussion on bonding discussed in Section II.B.3.a. This is because it appears to be strong support for the existence of negative hyperconjugation.

This model envisages back-bonding as being from a p orbital on the anionic carbon into an antibonding orbital on the R_3P moiety. The putative p orbital on carbon would be eclipsed by the unique substituent (R_u) and so would be best placed to back-bond into a p-type antibonding orbital lobe on phosphorus colinear with P—R_u (see Section II.B.3.a), thus lengthening P—R_u to a greater extent than the other P—R bonds. Further support comes from a recent, very elegant, study reported by Grützmacher and Pritzkow[40] on $Bu^t_2ClP{=}CPh_2$. The bond to a halogen has σ^* lower in energy than to carbon so we may expect halogen to be the unique ligand. Grützmacher and Pritzkow reported that this was indeed the case but, better still, there were three independent molecules in the unit cell, all with slightly different dispositions of the CPh_2 group and as this group becomes more nearly perpendicular to the P—Cl bond that bond is lengthened the most[40].

In those cases where there is a deviation from carbanion planarity the resulting conformation, shown in Figure 2, is called the *trans*-bent conformation. This is a known phenomenon for other double bonds and cumulenes containing second or higher row atoms and has been treated theoretically[88,89]. A simple rule has been devised to predict when this distortion will occur in any given system based on the singlet-triplet separation of the constituent carbenoid fragments[89]. However, in this case it might be argued that it is a simple steric effect which operates once negative hyperconjugation is invoked. In those cases where there is significant pyramidalization of the carbanion, there is also rapid carbanion inversion at room temperature which can be detected by NMR[26] and by the relatively large displacement parameters for the substituents in X-ray crystal studies, e.g. the hydrogen atoms of $Ph_3{=}CH_2$[33].

The *trans*-bent conformation is not the preferred conformation in those ylides where the carbanion bears a substituent with a lone pair of electrons. Schmidbaur and coworkers have made a study of phosphino-substituted ylides and related species[23,33–35,90–93]. In the case of mono(phosphino)methylides there are a number of possible conformations, shown in Figure 3, all with the lone pair of electrons at the phosphino group perpendi-

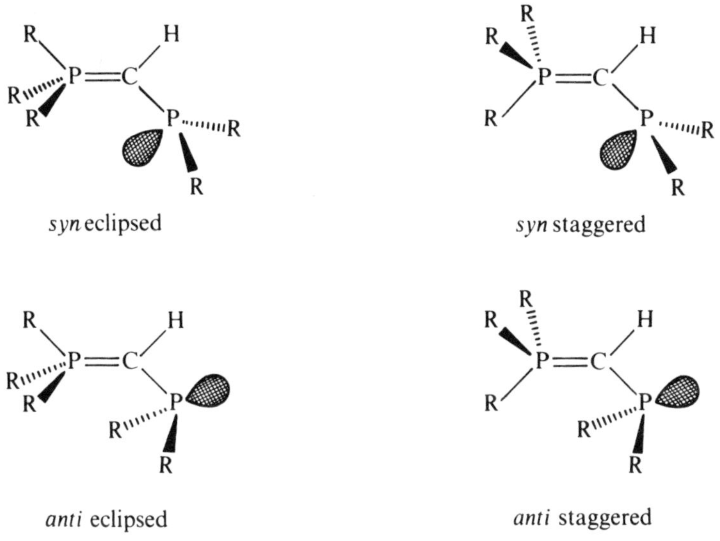

syn eclipsed

syn staggered

anti eclipsed

anti staggered

FIGURE 3. Possible conformations of monophosphino-substituted phosphonium ylides. The preferred conformations is *syn*

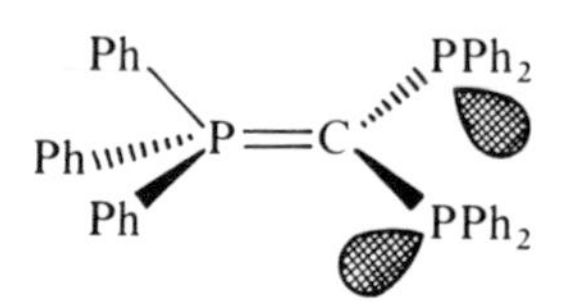

FIGURE 4. The preferred conformation of triphenylphosphonium bis(diphenylphosphino)methylide

cular to the p orbital at the ylidic carbon and none of which corresponds to a *trans*-bent conformation. Of those shown in Figure 3, the *syn* conformation is the favoured one with both eclipsed and staggered forms occurring[23]. In the case of bisphosphino-substituted methylides (and other Group 5 analogues) the *anti* conformation is more common[33–35,90–93] and the deviation from a *trans*-bent arrangements is less pronounced. Also in the latter case the relative orientations of the lone pairs of electrons on the substituents have to be taken into account[33,35,36] and it is found that the preferred conformation is the unsymmetrical rotamer shown for the heptaphenyl case in Figure 4. In this compound both lone pairs are in the $P{=}CP_2$ plane and the barrier to rotation about the P—C(ylide) single bond is about 50 kJ mol^{-1}.

It is well known[94–96] that in conjugatively stabilized ylides, especially ester-stabilized ylides, there is restricted rotation about the C_α—C_β bond with rotation barriers of up to about 75 kJ mol^{-1}. Barluenga *et. al.*[97] have reported some further examples of this phenomenon with barriers in the range 45–62 kJ mol^{-1}. Similarly, as part of the study mentioned above, Liu and Schlosser[87] found that there was a small barrier (35 kJ mol^{-1}) to rotation of the phenyl group of benzylides[87].

4. *sp ylides*

There are ylides in which the ylidic carbon is also double bonded to another group which includes carbodiphosphoranes and cumulated ylides. Since they tend to be more stable than analogous sp^2 ylides, they are well represented in Tables 1 and 2. Once again it can be seen in the studies of hexaphenylcarbodiphosphorane that reducing R can lengthen $r(P{=}C)$ by up to 3 pm. Two structural observations on sp ylides in general are noteworthy; they have reduced P=C bond lengths and the angle at carbon may be significantly less than 180°.

It can be seen from Table 1 that, for comparable systems, there is a reduction of about 5–6 pm in $r(P{=}C)$ in the sp ylide relative to the sp^2 ylide. However, the distances reported for most sp ylides are still within the low end of the range reported for sp^2 ylides. Note that in hexamethylcarbodiphosphorane $r(P{=}C)$ is significantly less than in the hexaphenyl derivative, consistent with the same observation for the sp^2 analogues. The shorter P=C distance may be a steric effect in that the sterically less encumbered carbanion can approach closer to the phosphorus atom.

It is more difficult to explain the fact that the angle between the substituents on the carbanion is often significantly less than 180°, as can be seen in Table 1. Some of these difficulties may explainable by the high R value of some of the studies, e.g. the early reports on the monoclinic modification of hexaphenylcarbodiphosphorane, but there is no doubt that the effect is real. It is consistent with the slight pyramidalization observed in many sp^2 ylides and it is tempting to try to explain it on this basis[98], although the effect appears much larger. Also, it is related to the occurrence of *trans*-bent conformations and has been treated theoretically in an exactly analogous manner[88,89]. In what may amount to the same thing, Ebsworth *et al.*[52] argued cogently that the observed angle may be due a large-amplitude bending vibration for these molecules where the central carbon can move easily relative to the P atoms. Thus deformation of the molecule does not cost much energy, which leads to the wide variation in observed angles, especially between different crystal modifications, and to the observed triboluminescence[53]. This would be consistent with the more rapid inversion which is observed in those sp^2 ylides which are pyramidalized. A good comparison which is made[52] is with molecules such as carbon suboxide which is a linear molecule with a low frequency for the CCC bending vibration and has an apparent CCC angle by electron diffraction of 158°. On this basis Ebsworth *et al.*[52] estimate the PCP bending frequency to be about $80\,cm^{-1}$.

B. Bonding

For reasons similar to those for the oxides[2], studies of the bonding in phosphonium ylides have concentrated on the nature of the P=C bond regarding the following issues: (i) the strength of the bond, although in reality this has meant its reactivity (see below); (ii) the exact distribution of the electron density with discussions very similar to those for the oxides; and (ii) the difference between phosphorus, nitrogen and (to a lesser extent) arsenic ylides.

1. Other experimental observations

The bond length data, although not satisfactory, do suggest that the P=C bond has multiple character with a bond order of about 2. Other physical measurements which support this and provide further insight are detailed below. Note that many of the original studies which reported these data discuss them in terms of $d\pi$–$p\pi$ overlap on the basis that it is evidence for such overlap. Since the d orbital concept is now discarded, these discussions are obsolete, but the experimental results still give useful information.

a. Bond energies and reactivity. There are no experimental data on the bond dissociation energies of phosphorus ylides. The only guide we have is the series of σ and π 'increments' for various multiple bonds developed by Kutzelnigg[99], which were derived by a combination of calculation and limited experimental results. His value for the π increment of the PO bond, $155 \pm 20\,kJ\,mol^{-1}$, was in rough agreement with other estimates[2] of $200\,kJ\,mol^{-1}$, so we may have some confidence in his value of $200 \pm 20\,kJ\,mol^{-1}$ derived for the π increment for the P=C bond with a σ increment of $268\,kJ\,mol^{-1}$. This total value of $468\,kJ\,mol^{-1}$ contrasts markedly with the binding energy of H_3PCH_2 (with respect to H_3P and CH_2), which has been calculated to be in the range 222–$286\,kJ\,mol^{-1}$ at various theory levels[98,100,101]. There has been one heat of combustion study of phosphorus ylides[102] which found that, relatively, $Ph_3P{=}CHC(O)OMe$ had a significantly higher energy content than $Ph_3P{=}CHC(O)Ar$.

It is often loosely stated that phosphorus ylides are more stable than nitrogen ylides. However, in comparing stabilities we must be particularly careful to define the sort of stability referred to. Thus it has been well established[3,15,103] that ylides become less reactive (owing to decreased basicity) as more powerfully electron-withdrawing groups are substituted on the ylide carbanion. The electron withdrawal may be by inductive and/or resonance effects. Thus $Ph_3P{=}CH_2$ is very reactive whereas $Ph_3P{=}CHC(O)OMe$ and cyclopentadienylidenetriphenylphosphorane are fairly stable. However, in this context stable means both capable of isolation and handling in the atmosphere and slow rates of Wittig reaction. If precautions are taken to exclude air, water and light, the reactive ylides are just as long-lived as the stable ylides . Similarly, the presence of such electron-withdrawing groups on a potential ylide carbon facilitates the formation of that ylide from its conjugate acid (the phosphonium salt), and this in turn correlates with the strength of base required in the Wittig reaction[104]. Likewise, a comparison[3] of the acidity of fluorene ($pK_a = 25$), fluorenylammonium salts ($pK_a \approx 25$) and fluorenylphosphonium salts- ($pK_a < 10$) shows that the effect of an ammonium group does not alter the stability of the fluorenyl anion very much, whereas the presence of a phosphonium group stabilizes it considerably. Since a purely electrostatic stabilization has been estimated[105] to be 30% stronger for nitrogen, some additional stabilization in the case of phosphorus is suggested. Extension to other Main Group onium centres gives the stability order Se $\approx$ S > P > As > Sb[106,107]. However, this order does not parallel the reactivity of these ylides with, for example, aldehydes and nitrosobenzene, since the phosphonium ylide is less reactive than the others[107] and anyway, rather than showing a special stabilization in second-row ylides, these results may suggest a special destabilizing effect for nitrogen ylides.

It is worth recording that for no meaning of the term stability have phosphorus and nitrogen ylides been directly compared experimentally. This is because their reactivities are rarely comparable since there are few reactions which both nitrogen and phosphorus ylides undergo to give analogous products. The main difficulty is that nitrogen ylides readily undergo Stevens rearrangement[108,109]. Now in fact, contrary to what was believed until fairly recently, phosphorus ylides are known to undergo a Stevens rearrangement process[110,111] and indeed this requires temperatures in the range 100–200 °C whereas the normal Stevens rearrangement occurs at room temperature. However, this qualitative observation may be merely a reflection of the relative strengths of PC and NC single bonds.

b. Infrared spectroscopy. The IR spectroscopy of ylides is not well developed and there is considerable confusion over the frequency assignments[112]. The only unequivocal assignment[113], based on solvent studies, is a value of 1190 cm^{-1} for ν(P=C) in carboethyoxytriphenylphosphorane. It appears[112–114] that stabilized ylides have a value of ν(P=C) equal to or greater than this, the precise value depending on the extent of stabilization. Two attempts have been made to determine the force constants and bond orders in non-stabilized ylides. Luttke and Wilhelm[115] assigned frequencies by H/D and ^{13}C/^{12}C substitution in methylenetriphenylphosphorane. They found a force constant of 4.9 mdyn Å^{-1} and hence derived a bond order of 1.3 from Siebert's rules[116]. Sawodny[117] investigated methylenetrimethylphosphorane using IR and Raman spectroscopy and found a force constant of 5.59 mdyn Å^{-1} with a bond order of 1.65. This is consistent with the difference in P=C bond length between these compounds.

c. Dipole moments. Phosphonium ylides are very polar substances consistent with the charge separation across the P=C bond. A number of dipole moment measurements and studies have been reported[118–125]. Typical values are 7.0 D for cyclopentadienylidenetriphenylphosphorane[119] and 5.5 D for Ph_3P=CHC(O)OMe[120]. Some authors have attempted to analyse these values for an estimate of the contribution of back-bonding, then thought to be dπ–pπ bonding, e.g. Ramirez and Levy[119] estimated it at 50% for cyclopentadienylidenetriphenylphosphorane. Such calculations are fraught with difficulties[126], some of which are (1) the σ-bond moment may be in the opposite direction to the π-bond moment, (2) stabilizing groups shift the negative charge and (3) the overlap in a such a bond is highly unsymmetrical (cf. P=O bond[2]), so that it is expected to have a high moment anyway. Two detailed and careful studies[123,125] have attempted to derive a value for the dipole moment of methylentriphenylphosphorane from other, experimentally determined, values. The latter of these[125] was more extensive and a value of 4.5 D was deduced from which an estimate of 50–55% ionic character was derived. However, even in these studies the unsymmetrical nature of bonding was not explicitly taken into account.

d. Ultraviolet–visible spectroscopy. Phosphonium ylides are often coloured substances, sometimes intensely so, and the colour is probably associated[8] with tailing into the visible of a broad UV absorption which usually occurs between 300 and 400 nm[127].

This bond, occurring as it does in most ylides and located away from other possible interfering absorptions, offers a good opportunity for the study of bonding. However, although there are many reports of the UV–visible spectra of ylides[12,49,119,121,127–134], only a few of these are concerned in detail with bonding[122,127–130]. This may be partly because of assignment difficulties; the ylide absorption has been variously described as a $\pi \rightarrow \pi^*$ transition[121], a $\pi \rightarrow d$ transition[128], an intramolecular charge-transfer transition from anionic to cationic moiety[127] and as a transition from the ground state of the ylide to that of the ylene[129]. The intensities of the ylide absorption are referred to less often and this is often due to difficulties in their evaluation because of air sensitivity[128]. One result from UV spectroscopy which may be of use in the study of the bonding in ylides has been summarized by Ramsey[127]. In a number of cases there is a close correspondence between the transition energies of ylides and those of alkali metal ion intimate ion pairs and he noted that in many cases the best model for predicting the transition energy of an ylide $R_3P{=}CR_2$ is not the hydrocarbon $R_2C{=}CH_2$ or the free carbanion R_2CH^-, but the intimate ion pair $R_2CH^-M^+$.

e. Nuclear magnetic resonance spectroscopy. Protons attached to the anionic carbon in unstabilized ylides are strongly shielded, for example, $Ph_3P{=}CH_2$ and $Me_3P{=}CH_2$ have $\delta_H + 0.13$ and -0.78 ppm, respectively[82,135]. This may be taken as evidence for a high negative charge load at the ylidic carbon. For comparison, these shifts are close to those observed[136] for alkali metal alkyls R^-M^+ where δ_H is in the range -1 to -2 ppm and where a distinct polarity is beyond doubt. The shielding is less in stabilized ylides e.g. $\delta_H + 4.40$ ppm in $Ph_3P{=}CHC(O)Ph$[102], which refleclts the removal of charge density from the ylidic carbon. Methylides[82,135] tend to have values of $^2J_{PH}$ slightly less than 10 Hz whereas those in stabilized ylides[102] are usually greater than 20 Hz. However, other substitution patterns have values[81,137] which are less readily systematized and attempts by some workers[138,139] to rationalize $^2J_{PH}$ in terms of the percentage of s character on carbon have not been very successful. For alkyl groups attached to phosphorus there is another $^2J_{PH}$. For example, in methylenetriphenylphosphorane $^2J_{P-CH_3} = -12.5$ Hz and a camparison[102] of this with values for phosphonium salts is consistent with the assumption of sp^3-hybridized phosphorus.

The phosphorus atom in an ylide is usually shielded relative its conjugated acid, the

TABLE 3. Core ionization potentials (eV) for some phosphonium ylides and salts

Ylides	P(2p)	C(1s)	Ref.	Salts[a]	P(2p)	Ref.
$Me_3P{=}CH_2$[b]	137.0	287.8	156	Bu^n_4P	132.3	160
$Ph_3P{=}CH_2$	132.6	283.9	159	Ph_3PCH_3	132.6	159
					130.2	161
				Ph_3PCH_2Ph	130.5	161
					132.5	160
$Ph_3P{=}CHC(O)Ph$	131.2		102	$Ph_3PCHC(O)Ph$	132.0	102
$Ph_3P{=}CHC(O)C_6H_4NO_2$	130.8		102	$Ph_3PCHC(O)C_6H_4NO_2$	131.8	102
$Ph_3P{=}CHC(O)C_6H_4OMe$	131.2		102	$Ph_3PCHC(O)C_6H_4OMe$	132.0	102
$Ph_3P{=}CHC(O)OMe$	131.7		102	$Ph_3PCHC(O)OMe$	132.5	102
	134.7		158	Ph_4P	133.4	162
$Ph_3P{=}CHCN$	134.7		158			
$Ph_3P{=}O$	131.5		102			
	132.8		2			

[a] Counter ion and charge on phosphorus omitted.
[b] In the gas phase. All others in the solid phase.

TABLE 4. Results[a] of *ab initio* calculations on $H_3P{=}CH_2$

Calculation[b]	Energy[c]	r(P=C)[d]	r(P—H)[d]	∠CPH[e]	∠PCH[e]	∠HCH[e]	ΣC[f]	μ[g]	Ref.
STO-3G	−377.0505	189.7	138.7u	132.1u	100.0				163, 164
			137.3	110.9					
STO-3G*	−377.1701	160.7	139.9u	127.4u	119.7				163
			137.6	113.7					
3–21G	−379.3986	172.8	141.5	128.7		119.8		2.71	165
3–21G*[h]	−379.5667	166.6	140.5	128.7		117.4		2.34	165
3–21G* + D[h]	−379.5769	167.7	140.2	128.7		117.5		2.9	165
3–21G*		164.6					360	2.36	166
3–21G*	−379.5494	164.9	141.4u	126.6u	120.3		360		167
			139.1	113.5					
3–21G*	−379.5608	165.8	141.0u			125.0			98
			139.0						
4–31G	−380.8750	172.1	146.1u	129.1u	117.2				163
			141.0	112.2					
4–31G*	−380.9968	165.2	142.8u	128.8u	118.3				163
			139.8	112.8					
6–31G*	−381.3923	166.7					354	2.48	168
6–31G* + D	−381.3975	167.2	141.4u	127.5u	118.5	117.5	354		98
			138.9						
DZ	−381.1349	171.7	144.8u	126u	119				
			142.4	114					
DZ + P[i]	−381.2738	164.0	140.1		120.7	118.5	360		169
DZ + P[i]		167.2	140.2	118	119.0	121	359	3.18	170
DZ + P[j]		168.2	142.1		120.9	118.1	360		171
DZ + P + D	−381.4025	167.5	141.4u	127.7u	118.4	117.4	354	2.86	101, 172
			138.9	112.1					

GVB–SOPP[k]	−381.4351	166.8	140.1	118.6u 117.7	119.2	120.4	359	173
GVB–POL–CI	−381.5029							173
6.31G* + MP2	−381.6470	167.4	144.0u 140.4	127.2	116.7	116.7	350	100
6–31G* + MP4[l]	−381.7070							167
MNDO		162.0	136.4u 135.5	114.5u 115.4	119.3	120.7	359	174
EXPT[m]		163–171					350–360	

[a] All results are for the more stable perpendicular conformation (see Section II.A.3). Selected geometrical data only are shown; see footnotes *d* and *e* for other data.
[b] All self-consistent field, contracted Gaussian-type basis sets, geometry optimized by the gradient method unless indicated otherwise. STO-3G, DZ, 3–21G, 4–31G, 6–31G and MNDO have their usual meanings[175]; * = a set of six d-type polarization functions added to the basis set; P = other combinations of d functions added; D = diffuse functions added to the basis set of the anionic carbon; MPx = electron correlation by xth order Møller–Plesset perturbation theory; GVB–SOPP = generalized valence bond calculation employing the strong orthogonality and perfect pairing approximations; POL–CI = polarization configuration interaction.
[c] Total electron energy in hartree; 1 hartree = 27.2 eV = 2625 kJ mol^{-1}.
[d] Bond length in pm; u refers to the unique hydrogen on P (see Figure 2). r(C—H) = 107–109 pm.
[e] Bond angle in degrees. ∠HPH ≈ 105–105.5° and ∠HPHu ≈ 98.5°.
[f] Sum of bond angles at carbon.
[g] Dipole moment in debye.
[h] Unoptimized exponent of d polarization functions.
[i] C_{3v} symmetry forced at phosphorus.
[j] Pseudopotential method.
[k] Using HF geometry determined at the DZ + P + D level.
[l] Using geometry determined at the 3–21G* level.
[m] Range of observed values for unstabilized ylides, from Section II.A and Figure 2.

TABLE 5. Comparison of results of calculations at the same levels on the perpendicular (per) *trans*-bent and parallel (par) (eclipsed) conformations[a] of $H_3P{=}CH_2$

Calculation[b]	Conformation	Energy[c]	$r(P{=}C)$[d]	$r(P{-}H)$[d]	∠CPH[e]	∠PCH[e]	∠HCH[e]	ΣC[f]	μ[g]	Ref.
6–31G*	per	−381.3923	166.7					354	2.48	168
6–31G*	par	−381.3908	165.5					360	2.43	168
6–31G* + D	per	−381.3975	167.2	141.4u 138.9	127.5u	118.5	117.5	354		98
6–31G* + D	par	−381.3967	166.2	138.5u 140.3	110.3u		119.7			98

[a] For explanation of conformational terms, see Section II.A.3 and Figure 2.
[b] See footnote *b* in Table 4.
[c] Total electron energy in hartree; 1 hartree = 27.2 eV = 2625 kJ mol^{-1}.
[d] Bond length in pm; u refers to the unique hydrogen on phosphorus shown in Figure 2.
[e] Bond angle in degrees.
[f] Sum of bond angles at carbon.
[g] Dipole moment in debye.

phosphonium salt[140–143]. The effect is greater for unstabilized than for stabilized ylides, e.g. $\delta_P = 23$ ppm for $Me_3P{=}CH_2$, falling to 5.4 ppm for $Ph_3P{=}CHC(O)Ph$, presumably reflecting the lesser shielding ability of the anionic carbon in stabilized ylides as a result of its reduced charge[144].

The ylidic carbon in unstabilized phosphonium ylides is strongly shielded. Thus $Ph_3P{=}CH_2$ and $Me_3P{=}CH_2$ have $\delta_C - 5.3$ and -2.3 ppm, respectively[145,146]. For comparison, the anionic carbon in methyl lithium falls at $\delta_C = -16$ ppm in thf[136]. Again, this is evidence of a high charge load at carbon. As in the ^1H NMR spectra, the shielding is less in the case of stabilized ylides.

The magnitude of one-bond ^{31}P–^{13}C coupling has been successfully related to the percent s character in the carbon hybrid orbital comprising the P—C bond for a variety of tetracoordinate compounds[139,140,143,147,148]. On this basis the fact that $^1J_{PC}$ couplings in phosphorus ylides are high (90–130 Hz) has been confidently interpreted[142,145,149,150] as indicating sp^2 hybridization at the ylidic carbon. Similarly, in unsubstituted ylides bearing a hydrogen on the ylidic carbon, $^1J_{CH}$ is large, e.g. + 149 Hz in $Me_3P{=}CH_2$[135,151] and 153 Hz in $Ph_3{=}CH_2$[146,152]. These high values have also been interpreted[152] as indicating sp^2 hybridization at the ylidic carbon.

f. Photoelectron spectroscopy. Phosphonium ylides have been studied by both UV PES[153–155] and X-ray PES (ESCA)[51,102,156–159]. The most comprehensive study was a UV PES investigation[155] of a large series of unstabilized ylides in the gas phase. It was found that the first ionization usually occurs at between 6 and 7 eV and this low value was assigned to carbanion ionization, indicating a high degree of negative charge on carbon. Core ionization potentials for ylides are given in Table 3 along with those for analogous phosphonium salts. Although there are a number of inconsistencies among different workers, it can be seen that the high positive charge at phosphorus is confirmed. This can be seen most clearly in the study of carbonyl-stabilized ylides by Seno *et al.*[102] where the phosphorus 2p binding energy was usually found near 131 eV, lower than in the corresponding salt and similar to the oxide, indicating a large charge separation as implied by structure **1A**.

2. Survey of theoretical calculations on phosphonium ylides

Tables 4–7 show the results of most calculations on methylidenephosphorane and its derivatives reported from 1977 until the end of 1991. There was one earlier *ab initio* calculation[178] which was not geometry optimized. A number of earlier empirical calculations had also been reported[64,126,129,130,142,146,179–187] but these are now only of historical interest since they were mainly concerned with the influence of d orbitals. Exceptions are (i) the extended Hückel calculation of Boyd and Hoffmann[187], which predicted some interesting hyperconjugative effects in cyclopropylides; (ii) correlations of ^{13}C chemical shifts with atomic charges[142,146]; (iii) the correct prediction of the UV spectrum of cyclopentadienylidenetriphenylphosphorane[129]; and (iv) an MO study[64] of the bonding in carbodiphosphoranes by a number of semiempirical methods which successfully predicted the PCP bond angle of 130°. The latter observation was also predicted in *ab initio* calculations[188].

As we have previosly noted for the phosphine chalcogenides[2], we seek, as a minimum, two related objectives from calculations such as these: an understanding of the bonding in these unusual systems and prediction of the structure and properties of unknown species. Unfortunately, as regards the prediction of unknown species, few data are available on any of the molecules studied so that we cannot calibrate these calculations against experiment. We have also noted previously[1,2] that basis sets of at least double zeta quality with polarization and diffuse terms, full geometry optimization and some form of electron

TABLE 6. Valence molecular orbital energies[a] and Mulliken populations in H_3PCH_2 determined by calculation at various theory levels[b]

MO	DZ + P[c]	DZ + P[d]	DZ + P[e]	DZ + P + D[f]	6–31G*[g]	3–21G*[h]	MNDO[i]
10*a*′	−7.39	−7.54	−7.40	−7.71			−7.17
3*a*″	−13.94	−13.90	−14.02				
9*a*′	−14.49	−14.61	−14.44				
8*a*′	−15.47	−15.59	−15.46				
2*a*″	−16.14	−16.27	−16.20				
7*a*′	−21.86	−21.85	−21.88				
6*a*′	−26.51	−26.67	−26.51				
qP		0.47	1.26	0.48	0.56	0.58	
qC		−0.85	−1.10	−0.90	−0.78	−0.92	

[a] In eV.
[b] See footnote *b* in Table 4.
[c] Pseudopotential method from ref. 171.
[d] C_{3v} symmetry forced at P from ref. 169.
[e] C_{3v} symmetry forced at P from ref. 170.
[f] From ref. 101.
[g] From ref. 168.
[h] From ref. 166.
[i] From ref. 174.

correlation will be necessary for the detailed *ab initio* description of compounds of third-row elements. It can be seen from Tables 4 and 7 that only a few of the reported studies are near to this quality[100,101,167,172] and, not surprisingly, therefore, there is a fairly wide variation in the reported parameters, especially the important *r*(P=C) and the degree of planarity at the carbanion. Hence we may only tentatively say that if the structure of H_3PCH_2 is ever determined it will be similar to the results obtained at the DZ + P + D[101,172] or 6–31G* + MP2[100] levels shown in Table 4.

Many of the reported calculations, especially those from the 1970s and early 1980s, were concerned with the involvement or not of valence d orbitals but, as with phosphine oxides[2], their significance changed during this time. Undoubtedly, as can be seen in Table 4, d functions are required in the computation of accurate energies and geometries for the ylide bond. However, it is now clear[2,16,189] that they are required in relatively large amounts in this case to describe the severe polarization caused by the strong electric field across the bond[163,170]. This is nicely illustrated in the calculations of Mitchell *et al.*[163] in both STO-3G/STO-3G* and 4–31G/4–31G* comparisons where the P=C bond is too long without d functions because electron density is not being directed efficiently into the internuclear region. The detrimental effects of any geometry restraints were shown explicitly in these calculations[163].

Some general observations may be made about the body of results reported in Table 4. In the best calculations the geometry at phosphorus is found to be tetrahedral and that at carbon approximately planar, but almost always with slight pyramidalization towards the *trans*-bent conformation (see Figure 2). Almost free rotation is found about the ylide bond[98,163,168] with barriers in the range 4–5 kJ mol^{-1}, as can be seen in Table 5, which compares the perpendicular (*trans*-bent) and parallel (eclipsed) conformations of methylenephosphorane. A useful viewpoint, developed in detail in studies by Trinquier and Malrieu[171] and Bestmann *et al.*[98], is to consider the construction of the ylide bond by interaction of a phosphine and a carbene. This is allowed if a non-least motion pathway is adopted[171] and the triplet carbene would give rise to a pyramidal carbanion while the

TABLE 7. Comparison of results[a] of other calculations on phosphonium ylides $R_3P{=}CXY$

Molecule	Calculation[b]	$r(P{=}C)$[c]	$r(P{-}R)$[c]	∠CPR[d]	∠RPR[d]	∠PCX/Y[d]	∠HCH[d]	ΣC[e]	E_{homo}[f]	qP[g]	qC[g]	μ[h]	Ref.
H_3PCH_2	3–21G*	165.8	141.0u 139.0				125.0						98
	3–21G* + D	167.7	140.2	128.7			117.5					2.9	165
	DZ + P + D	167.5	141.4u 138.9	127.7u 112.1	98.5u 105.1	118.4	117.4	354	−7.71	0.48	−0.90	2.86	101
	MNDO	162.0	136.4u 135.5	114.5u 115.4	99.1u 110.9	119.3	120.7	359	−7.17				174
H_3PCHF	3–21G*	167.8	141.0u 138.4										98
	DZ + P + D	172.3	141.4u 138.4	131.7u 108.0	100.1u 106.9	115.4H 113.0F	109.7	338	−8.12	0.42	−0.35	8.12	101
H_3PCHLi[i]	3–21G* + D	167	140u 142	112.9u	95.4	114H 131Li							176
H_3PCHMe	3–21G*	165.8	141.5u 139.1										98
H_3PCF_2	DZ + P + D	354	140.3u 140.1	165.8u 94.1	95.9u 95.2	104.2	104.8	313	−10.52	−0.05	0.23	0.88	101
$H_3PC(CH_2)_2$	DZ + P	168.2	141.3u 138.1	129.0u 110.9	99.3u 104.6	127.6	61.1	316		0.54	−0.51	3.34	21
$H_3PC(CF_3)_2$	DZ + P + D	170.7	139.2u 138.1	118.9u 112.4	103.3u 105.3	120.8	118.2	360	−10.62	0.34	−0.57	8.30	101
$MePH_2CH_2$	MNDO	163.3	136.2H 178.2C	114.5C 113.6H	108.7	119.1	120.4	359	−6.91				174
Me_2PHCH_2	MNDO	163.3	138.4H 177.9C	113.8C 111.7H	113.3C 101.5H	119.8	120.3	360	−6.70				174
Me_3PCH_2	MNDO	154	182	117.1		124.8							177
$Pr^i_3PCH_2$	MNDO	179	182	106.2		122.1							177
Ph_3PCH_2	MNDO	177	177	109.1		120.8							177

[a] All results are for the more stable perpendicular conformation (see Section II.A.3), unless indicated otherwise.
[b] See footnote *b* in Table 4.
[c] Bond length in pm; u refers to the unique H on P (see Figure 2). H, C identify R group.
[d] Bond angle in degrees. H, C, F, Li identify R/X/Y.
[e] Sum of bond angles at carbon.
[f] Energy of HOMO in eV.
[g] Mulliken atomic population.
[h] Dipole moment in debye.
[i] The parallel conformation reported here was found more stable.

singlet gives the planar. This is related to the general discussion of the *trans*-bent conformation of multiple bonds[88,89], the likely occurrence of which may be predicted by a consideration of the singlet–triplet separation in the carbene components of the bond concerned[89]. The binding energy of $H_3P{=}CH_2$ with respect to H_3P and CH_2 is found to be about $225\,kJ\,mol^{-1}$ at the 6–31G*/MP2[100] and DZ + P + D[101] levels, rising to $256\,kJ\,mol^{-1}$ at higher levels[100], while the use of isodesmic equations at the 6–31G*/MP2 level[98] gives a value of $286\,kJ\,mol^{-1}$. The H_2PCH_3 isomer was found to be $236\,kJ\,mol^{-1}$ more stable at the 6–31G*/MP2 level[100], falling slightly at higher levels[100].

The valence MO energies and Mulliken populations for H_3PCH_2 at various theory levels are shown in Table 6. The low HOMO energy is consistent with the first ionization potentials determined for substituted ylides which are in the range 6–7 eV (Section II.B.1.f). The Mulliken populations show that there is a build-up of electron density on carbon, also reflected in the integrated spatial electron population (ISEP) value of 9.2 at the 3–21G* level[165], with a corresponding decrease in electron density of phosphorus. The semiquantitative approach of Jardine *et al.*[190] gives a similar result.

Table 7 shows the results of the few reported calculations on substituted ylides along with some of the results for methylenephosphorane. Despite the paucity of the data, some useful insights have been derived. It has already been mentioned in Section II.B.1.a that conjugatively electron-withdrawing substituents stabilize P=C as do inductively withdrawing β-substituents. The effects of other types of substituents are harder to quantify experimentally, especially if they destabilize the P=C bond. Bestmann *et al.*[98] have shown that σ-donor, π-acceptor substituents stabilize the P=C bond, e.g. Li, BeH and BH_2, whereas π-donor substituents destabilize it, e.g. OH and NH_2. Thus the following binding energies ($kJ\,mol^{-1}$ with respect to $H_3P + CX_2$) were found *via* isodesmic equations at the 6–31G* + MP2 level: H_3PCH_2 286.2, H_3PCHLi 313.0, $H_3PCHBeH$ 374.9, H_3PCHBH_2 264.0, H_3PCHNH_2 33.9, H_3PCHOH 54.4 and H_3PCHF 99.2. Similarly, Dixon and Smart[101], in calculations at the DZ + P + D level, found H_3PCH_2 222.6, H_3PCHF 69.5, H_3PCF_2 5.0 and $H_3PC(CF_3)_2$ 323.4 $kJ\,mol^{-1}$. In the case of difluoromethylenephosphorane the effect is so strong that the system is really just weakly interacting H_3P and CF_2[101]. Vincent *et al.*[21] found that the rotation barrier in cyclopropylidenephosphorane, $H_3PC(CH_2)_2$ was $24.3\,kJ\,mol^{-1}$ at the DZ + P level, significantly higher than the $4–5\,kJ\,mol^{-1}$ noted above for the unsubstituted molecule. However, the inversion barrier for $H_3PC(CH_2)_2$ was found[21] to be $26.4\,kJ\,mol^{-1}$, which is very similar to the experimental value in $Ph_3PC(CH_2)_2$. The rotation barrier in H_2PCHF was also found[98] to be raised to $11.7\,kJ\,mol^{-1}$ at the 6–31G* + D level. Finally, McDowell and Streitwieser[176] found that the eclipsed conformation was the more stable in H_3PCHLi at the 3–21G* + D level and considered that the species was best described as a contact ion pair whose chemistry is that of the free ion[176].

3. *The two alternative views of the bonding in methylidenephosphorane*

In agreement with experimental observations, all of the detailed theoretical studies of the P=C bond show that it has multiple character and is highly polarized[21,80,98,100,101,163–173]. As mentioned in Section II.A, the standard description of the bonding in ylides has been in terms of a resonance hybride between structures **1A** and **1B** with the latter indicating d-orbital participation in back-bonding. This is similar to the situation in phosphine oxides and we have previously discussed in detail the difficulties with this model and various relevant qualitative considerations[2]. A particular difficulty is that structures **1A** and **1B** are not mutually exclusive because any back-bonding overlap would be highly unsymmetric with most of the electron density near carbon. Since structure **1B** breaks the octet rule, an alternative possibility is that **1A** is the best description with the strengthening of the P=C link being due to an ionic interaction. In support of this Whangbo *et al.*[191], in an MO study, found that the C—X bond length in the CH_3X(X = O, S) radicals, cations and anions could not be correlated with overlap populations but instead correlated linearly with ionic bond order. The other major difficulty is the use of d orbitals but this is easily overcome as detailed below. Note that Glidewell[192] has rationalized the planarity of the ylide carbon purely on the basis of non-bonded interactions.

Again, just as in the phosphine oxides[2], there is disagreement about the exact electron distribution in the P=C bond with one view corresponding to the σ/π description of multiple bonds and the other a bent (banana) bond description.

a. Backbonding/negative hyperconjugation. This is the updated version of the older descriptions based on structures **1A** and **1B**. The lone pair of electrons from phosphorus forms a σ bond to carbon which completes its octet. The extra charge density on carbon is in a p orbital which forms a back-bond by overlap with suitable acceptor orbitals on phosphorus. These are not d orbitals but the σ* LUMO of the phosphine moiety as shown in Figure 5. Thus the back-bonding in phosphorus ylides can be considered exactly analogous to that in phosphine oxides[2] and in transition metal phosphine complexes[1]. Actually, phosphorus ylides may have been the first system whose bonding was explicitly described in this way in a paper by Bernardi *et al.*[80] although , of course, it is essentially the same as the concept of hyperconjugation or double bond/no bond resonance[193,194]. This description of phosphorus ylides is now fairly commonplace[98,163,165,195]. The term negative hyperconjugation is used to distinguish this effect from cases where the interaction is between filled σ orbitals and empty π* orbitals[189,196]. Population of an antibonding orbital in the perpendicular conformation would be expected to lead to a lengthening of the P—C (unique) bond as discussed in Section II.A.3.

Good discussions of this topic are to be found in papers by Bestmann *et al.*[98], Mitchell *et al.*[163] and in a book by Albright *et al.*[195]. Mitchell *et al.*[163] in particular provide a very full and clear discussion showing, for example, that there can be back-bonding to antibonding orbitals in both the perpendicular and parallel conformations, hence explaining the low barrier to rotation in phosphorus ylides[163]. Also, secondary interactions with the other lobes of the σ* orbitals favour the perpendicular conformation, there being one extra antibonding interaction in the eclipsed form[163]. Similarly, minimization of unfavourable secondary interactions in the perpendicular conformation would also lead to a lengthening of the P—C(unique) bond and a wider $C_{ylide}PC_{unique}$ angle, thus explaining the overall *trans*-bent conformation shown in Figure 2[163].

b. Two Ω bonds (τ bonds/banana bonds). This concept has been discussed comprehensively in Volume 2[2]. It was noted that very often in *ab initio* MO calculations of multiple bond systems, if σ–π separation is not forced, localization procedures lead to the description of multiple bonds in terms of equivalent curved regions of electron density called banana bonds[175]. Similar bonds are also obtained in certain generalized valence bond (GVB) calculations, particularly with the strong orthogonality and perfect pairing (SOPP) restrictions removed[197], except that in this case they are the only result which minimizes the energy of the system and they have been referred to as Ω bonds[197]. In the

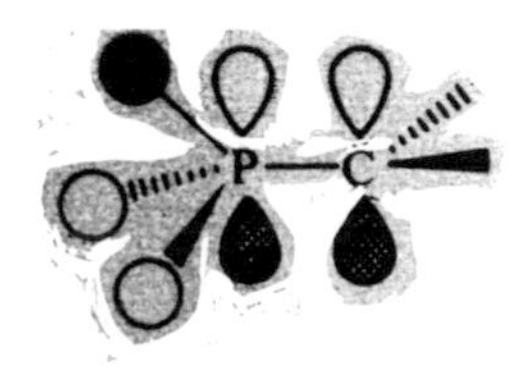

FIGURE 5. View of the P=C bond as a σ bond and a π bond formed by back-bonding into an antibonding orbital of *e* symmetry on the phosphine moiety

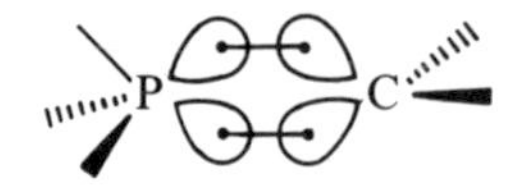

FIGURE 6. View of the P=C bond as two Ω bonds

case of phosphine oxides, the Boys localization procedure gave three banana bonds connecting P and O in all cases where it was applied[2] and GVB–SOPP calculations gave three Ω bonds in the limited number of cases examined[2,197].

Exactly parallel observations have been made in calculations on $H_3P{=}CH_2$, except that now only two banana bonds are involved as depicted in Figure 6. This was found using the Boys procedure by both Molina *et al.*[168] and Lischka[170] and in a single GVB calculation by Dixon *et al.*[173], who used the term τ bond for these regions of electron density. Further, in both the MO and VB calculations the two banana bonds are not exactly equivalent. The Boys localized MOs have both bond pairs near carbon but one is slightly closer to phosphorus than the other[168]. Similarly, the GVB bond pair which is in the plane of the unique PH is more diffuse[173]. The GVB calculation also compared the phosphorus ylide with H_3NCH_2, which was found to have an anionic carbon with no back-bonding[173]. Finally, we have previously noted[2] that often when a localization procedure leads to banana bonds the result is not taken wholly seriously. A similar comment can be made about the results of Lischka[170].

C. Summary

Phosphorus ylides have been succinctly described[23] as an easily pyramidalized carbanion stabilized by an adjacent tetrahedral phosphonium centre. They usually adopt the *trans*-bent conformation (Figure 2) but the barrier to rotation about P=C is very low (4–5 $kJ\,mol^{-1}$).

The P=C bond is highly polar and multiple, as implied by the representation. It is significantly stronger and shorter than a PC single bond, so it may be referred to as a double bond, although the actual order may be less and anyway such terms may have a different meaning in second than in first-row compounds[2]. The actual electron density distribution in the P=C bond is still a matter of debate, with two possible descriptions, in both of which it is strongly skewed towards carbon:

(i) there are two electron pairs, composed of a σ bond between P and C and a π back-bond between a filled p orbital on C and one of two possible acceptor orbitals on P which are antibonding in character with respect to the other ligands on P as shown in Figure 5;

(ii) there are two electron pairs, composed of two Ω bonds (banana bonds) from P to C as shown in Figure 6.

Finally, again as in the phosphine oxides[2], we note that there is an analogy between the bonding in ylides and the three-centre, four-electron bonding in phosphoranes (discussed in Section V). This has been formally stated by Musher[198] thus: 'an ylide is a hypervalent molecule in which a three-centre bond is reduced to a two-centre bond using a single orbital from the hypervalent atom and formally transferring one electron from the main group atom to the ylide carbon'.

III. PHOSPHONIUM SALTS

Structure and bonding in phosphonium salts has not been a controversial subject[4,5,7–11,199]. Approximately regular tetrahedral geometry is expected from four nearly equivalent ligands with bonding by four two-electron bonds, the phosphorus atom necessarily bearing a positive charge. Detailed studies confirm this expectation.

A. Structure

There are a large number of reports of the structures of phosphonium salts. However, many of these are tetraphenyl or methyltriphenyl cases where the anion is of the greater interest. Table 8 shows the results from a very restricted selection of the available reports. Only one or two representative reports of the simplest salts, mostly with $R < 5$ and bond length standard deviations no greater than 0.4 pm, have been included. Notwithstanding the large number of reports, these restrictions reveal that there are some gaps in the coverage available in the literature, especially for alkyl-substituted salts. Nevertheless some generalizations can be made with confidence from Table 8.

The PC bond length in phosphonium salts is usually in the range 179–181 pm, similar to that in phosphine oxides[2] and shorter than in phosphonium ylides (see Section II.A.1). For simple salts, there is a definite effect that the P—C(alkyl) bond is longer than the P—C(phenyl) bond (180–181 vs 179–180 pm). There are exceptions to this generalization in more complex cases, some of which are shown in Table 8, and there are a number of others, in particular three bisphosphonium salts[212,213]. The P—C(alkynyl) bond length is remarkably short at 169.9 pm in $Ph_2MePC{\equiv}CBPh_3$[214]. As expected. bond angles at phosphorus are very nearly tetrahedral with only a slight variation, as shown in Table 8. However, crowded salts break regular tetrahedral symmetry usually by having elongated PC bond lengths and compressed CCC bond angles, e.g. 192.4 pm and 106.5°, respectively in the tetratertbutyl salt[201], the tetraisopropyl[125] and trimesityl[209] cases being similar (see Table 8).

Archer *et al.*[206] surveyed all phosphonium salts known in 1981 in a correlation analysis of the positioning of cation and anion. Exclusive face orientation of the anion with respect to the cation centre was found, i.e. the anion was positioned along the direction of approach leading to the apical position in a trigonal bipyramidal intermediate. The positioning of the anion did not correlate with nucleophilic attack at the phosphorus centre but did correlate with α-hydrogen abstraction (ylide formation)[206].

The conformational arrangements of species with four equivalent groups attached to one centre can be interesting[25,215]. Thus, the tetraisopropyl cation[25] approaches point group S_4 symmetry, in agreement with theoretical predictions[215] for tetrahedral species with four ligands of C_s symmetry. The isopropyl groups are rotated pairwise in opposite directions by 110° away from the fully staggered conformation[25]. Similarly, the tetratert-butyl cation[201] has the predicted[215] T symmetry with a conrotatory distortion of all four threefold rotors (CC_3) by 14° and a concomitant rotation of all CH_3 rotors by 11°. In methyltrimesityl salts, the rate of rotation of the mesityl groups could be measured by ^2H dynamic NMR[209] and the conformation of methyltriphenyl anion is different in the low- and high-temperature modifications of the tetracyanoquinodimethanide salt[216].

Strong S—C—P anomeric effects have been previously reported in 2-phosphinoyl-substituted 1,3–dithianes[2]. This is shown by a strong preference for the conformer which places the phosphinoyl group in the axial position. Mikolajczyk and coworkers[217,218] have also studied 1,3-dithianes with various phosphonium substituents in the 2-position and found that the preference for the axial conformer is not as great as for the phosphinoyl cases. The explanation is presumably related to the greater bulk of the phosphonium groups[217,218].

B. Bonding

The description of the bonding in phosphonium salts is unexceptional. Four approximately equivalent two-electron bonds are made by σ overlap of appropriate atomic

TABLE 8. Structural data[a] for some phosphonium salts

Cation	Anion	$r(P—C_{Alk})$	$r(P—C_{Aryl})$	∠CPC	Δ∠CPC[b]	R	Ref.
$Et_3PC(OH)Me_2$	Br^-	180.0 Et 187.7OH		—		4.2	200
Pr^i_4P	Ph_4B^-	183.0		108.8	107.2–110.4	4.5	25
Bu^t_4P	F_4B^-	192.4		112.3		6.5	201
Ph_4P	Cl^-		178.7	109.5	108.4–111.1	5.8	202
Ph_4P[c]	Br^-		178.4	109.5	107.7–112.0	5.8	202
Ph_4P[d]	Br^-		180.0	109.5	107.8–110.3	6.6	203
Ph_4P	I^-		179.9	109.5	107.6–110.4	3.7	202
Ph_3PCH_3	Cl_4Fe^-	180.5	179.1	109.2Al 109.7Ar	108.7–110.3	3.9	204
$Ph_3PCH_2CH_3$	TCNQ	180.7	179.3	109.1Al 109.6Ar	108.8–110.0	3.9	205
Ph_3PCH_2Ph	I^-	181.2	179.2	109.5	108.0–110.9	5.2	206
$Ph_3PCH_2C_6H_5NO_2$	Br^-	181.3	179.2	109.5	108.6–110.5	4.8	207
$Ph_3PCH_2C(O)CH_2Cl$	Cl^-	179.5	178.8	109.5	105.3–112.5	5.2	208
Mes_3PCH_2CN	I_4Hg^{2-}	185.7	181.7	105.1Al 113.4Ar	103.7–117.0	3.1	209
$Ph_2MePFerrocenyl$	Cl^-	181.9	180.4 176.8F	109.5	106.9–112.1	4.1	210
BzMeNaphPhP	Br^-	177.7Me 182.4Bz	180.2Ph 178.9Nap	109.5	106.0–113.0	5.8	211

[a] All studies by single-crystal X-ray crystallography at room temperature. $r(P—C)$ = mean bond length in pm; ∠CPC = bond angle in degrees; R = reliability index = 100 × conventional R, see original papers for definitions.
[b] Variation in ∠CPC.
[c] Trigonal modification.
[d] Tetragonal modification.

TABLE 9. Results of calculations on phosphonium ions $X_nPH_n{}^+$

Cation	Calculation[a]	Energy[b]	r(P—H)[c]	r(P—X)[c]	∠HPX[d]	∠XPX[d]	Ref.
$PH_4{}^+$	STO-3G	−339.0250	138.2				221, 222
	STO-3G#	−339.0846	137.6				223
	STO-3G*	−339.1048	138.0				221, 222
	3–21G	−340.9982	139.2				224, 225
	3–21G#	−341.0620	137.7				223
	3–21G*	−341.0620	139.1				226
	4–31G	−342.3244	140.3				221
	4–31G#	−342.3881	138.3				223
	4–31G*	−342.4051	139.2				221, 226
	6–31G*	−342.76158	138.0				227–229
	6–31G + P + D + MP2		138.4				230
	DZ + P[e]		139.7				231
	TZ + P		138.9				232
	TZ + P + CI		141.5				232
	EXPT[f]		141.4				233
	MNDO		135.7				226, 234
	PRDDO		141.8				235
$MePH_3{}^+$	STO-3G	−377.6316	138.1	183.1			221
	STO-3G*	−377.7198	137.9	179.8			221
	STO-3G#	−377.6995	137.8	179.4	111.5		223
	3–21G	−379.8491	139.3	186.9	111.4		224, 225
	3–21G#	−379.9156	137.8	179.8	111.7		223
	4–31G	−381.3330	140.0	190.0			221
	4–31G#	−381.4000	138.3	180.4	111.8		223
	6–31G*[g]	−381.8240					224
	6–31G*		138.0	181.1	111.7		236

(continued)

TABLE 9. (*continued*)

Cation	Calculation[a]	Energy[b]	r(P—H)[c]	r(P—X)[c]	∠HPX[d]	∠XPX[d]	Ref.
$Me_2PH_2^+$	3–21G	−418.6977	139.5	186.2	109.3	113.2	224, 225
Me_3PH^+	STO-3G#	−454.9099	137.7	181.3	107.0	111.8	223
	3–21G#	−457.6034	137.8	181.6	107.0	111.9	223
	4–31G#	−459.4042	138.5	182.4	107.1	111.7	223
FPH_3^+	STO-3G#	−436.6088	138.5	150.5	109.6		223
	3–21G#	−439.4110	137.1	152.8	108.8		223
	4–31G#	−441.1302	137.7	153.2	108.2		223
	6–31G*	−441.6275	137.5	152.3	108.3		227, 229
$F_2PH_2^+$	6–31G*	−540.5085	137.0	150.6	108.4	108.0	227, 229
F_3PH^+	STO-3G#	−631.6910	138.5	149.4	109.3		223
	3–21G#	−636.1498	135.5	149.7	111.3		223
	4–31G#	−638.6537	136.3	149.2	111.1		223
	6–31G*	−639.3941	136.4	149.1	110.9	108.0	227, 229

[a] All self-consistent field, contracted Gaussian-type basis sets, geometry optimized by the gradient method. STO-3G, DZ, TZ, 3–21G, 4–31G, 6–31G, MNDO and PRDDO have their usual meanings[175]; * = a set of six d-type polarization functions added to basis set; # = a set of five d-type polarization functions added to basis set; P = other combinations of d functions added; MPx = electron correlation by xth order Møller–Plesset perturbation theory; CI = with electron correlation by configuration interaction.
[b] Total electron energy in hartree; 1 hartree = 27.2 eV = 2625 kJ mol^{-1}.
[c] Bond length in pm; u refers to the unique hydrogen on P (see Figure 2). r(C—H) = 107–109 pm.
[d] Bond angle in degrees. ∠HPH ≈ 105–105.5° and ∠HPHu ≈ 98.5°.
[e] Pseudopotential method.
[f] By neutron diffraction.
[g] At 3–21G geometry.

orbitals. By the valence shell electron pair repulsion (VSEPR) analysis these will be disposed tetrahedrally and phosphorus will carry a unit of positive charge. This latter point is borne out by strong downfield shifts in the ^{31}P NMR[219,220] and the photoelectron spectroscopic data (see Table 3). These expectations are confirmed by some of the calculations which have been done.

Table 9 shows the results of most published calculations on phosphonium ions. Almost all of these were performed either as tests of calculation methods or in relation to the protonation energies and basicity of phosphines and so do not contain much discussion of the nature of bonding in the salts. An exception is an early study[222] at the STO-3G* level that examined which substituents would be likely to stabilize the planar form. As mentioned in Section II.B.2, we have previously noted[1,2] that basis sets of at least double zeta quality with polarization and diffuse terms, full geometry optimization and some form of electron correlation will be necessary for the detailed *ab initio* description of compounds of third-row elements. It can be seen from Table 9 that only two calculations meet this criterion and, not surprisingly, the better of these[232] is the only *ab initio* calculation so far to give the correct value of the PH bond length. As expected, population analyses place a high positive charge on phosphorus[222,225,226,229,231] and approximately sp^3 hybridization was found for PH_4 and various derivatives by Magnusson[237].

IV. PHOSPHONIUM SALT–PHOSPHORANE ISOMERISM

It is well known[238–241] that phosphoranes, especially halophosphoranes, can exist in an ionic phosphonium salt form. This ionic–molecular isomerism is finely balanced in many cases and often accompanies phase change, e.g. organochlorophosphoranes R_nPCl_{5-n} are often molecular (trigonal bipyramidal) in the gas phase or non-polar solvents[238,242] and ionic, $[R_nPCl_{4-n}]^+[Cl]^-$, in the solid[11,238]. The substitution pattern can also affect the relative energies of the isomers, e.g. $PhPCl_4$ is molecular whereas $MePCl_4$ is ionic[238]. More recent work has provided more detail and shown that the situation is more complicated in some cases.

Dillon and Lincoln[243] showed by NMR that the interhalogen derivatives Ph_3PICl and Ph_3PIBr have the ionic structures $Ph_3PI^+X^-$ in the solid state and Dillon and Straw[244] found an ionic species with both anionic and cationic phosphorus. Brown *et al.*[245] studied various tetraarylfluorophosphoranes and showed that while the ionic form was the more stable as expected, there were strong P^+/F^- interactions in it. Similar observations were made in the case of tetraphenylphosphonium nitrite[246]. More interestingly, Al-Juboori *et al.*[247] investigated the three chlorophenylphosphoranes Ph_nPCl_{5-n} ($n = 1-3$) and isolated both ionic and molecular modifications in the solid state.

More interesting still is the discovery that there is another (possibly common) type of isomer in these systems[248–250]. Thus, in the solid state, X-ray analysis shows that dihalotriorganophosphoranes are molecular four-coordinate compounds, Ph_3P—Hal—Hal, where the dihalogen is bound to the P atom as a linear 'spoke'. It seems likely that there is a continuum of structures from this spoke structure to the fully ionic species because in $Bu^t{}_3I_2$, which also has the spoke structure, the I—I bond is considerably longer than in typical iodine charge-transfer complexes[250].

V. PHOSPHORANES

The recent review literature on pentacoordinate phosphorus is dominated by two books by Holmes[240,241]. These compounds have occupied an important place in the development of molecular chemistry, having generated much discussion in the past concerning their structure, dynamic stereochemistry and their non-octet bonding. All of these important issues have been treated in detail by Holmes[240,241] and not much significant material has been published since. One issue which has gradually been clarified in the meantime is that d orbitals are not required for a satisfactory description of the bonding (see Section V.B). Another useful summary of the chemistry of these compounds is that of Sheldrick[251].

A. Structure

The structures of phosphoranes were reviewed exhaustively by Holmes[240] in 1980 and only a small amount of work has appeared since then. Of the two possible ligand arrangements for pentacoordination, the trigonal bipyramid and the square pyramid, the former is by far the most common[240] but there are examples of the latter particularly in bicyclic systems[240]. The two most important structural principles for the trigonal bipyramid concern the relative *apicophilicity* (tendency to occupy the axial site) of ligands and the orientation of small-membered rings. Thus the more electronegative ligands tend to occupy the axial site[240] (have high apicophilicity) and small rings prefer to bridge an axial and an equatorial site[240]. The former observation is well accommodated by the theories of bonding in these systems as is the fact that axial bonds to any ligand are always longer than equatorial bonds to the same ligand (see Section V.B) The 'small-ring rule' is explainable on the basis of relief of ring strain[240]. The electronegativity rule has had to be modified as more complex systems have been investigated to take account of steric factors and π donor/acceptor capabilities of ligands[240]. Thus bulky ligands prefer to be equatorial having only two near neighbours at 90° rather than three, and π donors may also prefer the equatorial site since it allows conjugation[240,241]. Also, these compounds are not rigid, being able to undergo a stereomutation process called *pseudorotation* wherein axial and equatorial ligands exchange places. A number of mechanisms have been suggested for this process, of which the *Berry*[252] and *turnstile*[253] cases are the best known[241,251]. Holmes has shown that the geometries of known phosphoranes fall along the reaction coordinate for the Berry process and in particular the square pyramidal isomer is the transition state for the process[240]. On the other hand, the same exercise has been carried out for the turnstile process[254]. Recent work has shed some more light on these issues.

Structural data are now available for the complete series of chlorofluorophosphoranes and are presented in Table 10. Phosphorus pentafluoride has been particularly well studied by electron diffraction, X-ray crystallography and microwave spectroscopy (Table 10), many studies being reinvestigations. As noted by Macho *et al.*[255] and as can be seen from Table 10, the effect of successive substitution of fluorine in PF_5 by chlorine

TABLE 10. Structural data[a] for the chlorofluorophosphoranes

Molecule	Method[b]	$r(P—F_e)$	$r(P—F_a)$	$r(P—Cl_e)$	$r(P—Cl_a)$	$\angle X_ePX_e$	$\angle X_ePY_a$	Ref.
PF_5	E	153.4	157.7			120	90	255
	E	153.2	158.0					256
	M	153.4	157.5					257
	X	152.2	158.0					258
$PClF_4$	E	153.5	158.1	200.0		117.8	90.3	255
PCl_2F_3	E	153.8	159.3	200.2		121.8	90.0	255
	E	154.6	159.3	200.4		122.0	89.3	259
PCl_3F_2	E		159.6	200.5		120	90	255
PCl_4F	E		159.7	201.1	210.7	120.0	90.9	255
PCl_5	E			202.3	212.7	120	90	255, 260

[a] $r(P—C)$ = mean bond length in pm; $\angle CPC$ = bond angle in degrees; e = equatorial; a = axial.
[b] E = electron diffraction, r_g values are given; M = microwave spectroscopy; X = X-ray crystallography.

causes lengthening of all bonds, both equatorial and axial, although the latter are affected to a larger extent. The equatorial P—F and P—Cl bonds lengthen only very little if equatorial fluorines are substituted by chlorines. On the other hand, replacement of axial fluorine by chlorine has a large effect on the equatorial P—Cl bonds. The axial P—F distances increase monotonically with successive substitution of equatorial fluorines by chlorine, but the first axial chlorine substitution leaves the distance to the remaining fluorine unaffected. Finally, the biggest change observed is in the axial chlorine distance when the opposite fluorine is substituted[255]. As also can be seen from Table 10, there is not much variation in the bond angles across the series.

The structure and conformation of a series of trifluoromethylfluorophosphoranes, $(CF_3)_nPF_{5-n}$ ($n = 1-3$), has also been clarified[261]. For CF_3PF_4 a mixture of two conformers with equatorial and axial CF_3 groups was found. In $(CF_3)_2PF_3$ both CF_3 groups are axial but in $(CF_3)_3PF_2$ all three CF_3 groups occupy equatorial positions. Typical P–C distances found were in the region 188–189 pm with not much variation between axial and equatorial values. Trends similar to those enumerated above for the chlorofluorophosphoranes were found for the variation in P–F distances. Analogous observations have been made for phosphoranes containing oxaphosphole rings[262].

The structure of phenyltetrafluorophosphorane was studied and compared with the analogous methyl and ethynyl derivatives[263]. The respective P—C distances are 179.6, 178.0 and 174.7 pm. This trend is similar to that in other organophosphorus compounds (see refs 1 and 2 and Sections II.A and III.A), where the alkyl and aryl distances are similar, differing only by 1–2 pm either way, while the alkynyl distance is significantly shorter. The structure of $C_6F_5PCl_4$ was studied by NQR spectroscopy[264] and confirmed as trigonal bipyramidal with an axial pentafluorophenyl group. An electron diffraction study[265] of diaminodifluorophosphorane found a relatively undistorted trigonal bipyramidal structure with the fluorines axial as expected and the following bond distances: $r(\text{P—N}) = 164.0$ pm and $r(\text{P—F}) = 164.3$ pm. Two groups[266,267] simultaneously reported the structures of HPF_4 and H_2PF_3 determined by electron diffraction and these results are included in Table 11, where they are compared with the results of *ab initio* calculations.

Comparison of solid and solution ^{31}P NMR data for several phosphoranes[268] confirmed that the structures do not differ substantially between the two phases. Similarly, a vibrational spectroscopic study of $(Cl_3C)_2PCl_3$ showed that the structure was the same in benzene solution[269] as in the solid state[270]. The synthesis, structure and reactions of alkoxy- and aryloxy-phosphoranes have been reviewed[271,272]. The relative apicophilicities of aryl and methyl groups in such compounds can be assessed from $^1J_{PC}$ values[273].

B. Bonding

The issues relevant to bonding in phosphoranes have been well rehearsed in previous reviews of the subject[99,241,251]. Briefly, any bonding theory must explain (i) the preference for trigonal bipyramidal structures over square pyramids, (ii) the correlation of apicophilicity with electronegativity with the deviations that occur in the case of π donor/acceptor substituents and also the similar preference of electronegative ligands for the axial site in the square pyramidal cases and (iii) the greater lengths of the axial bonds in the tirgonal bipyramidal structures. Initially it must be emphasized that the VSEPR theory[274] can explain these observations without explicit consideration of the nature of the bond pairs. Thus, if the repulsion between electron pairs is represented as an inverse force law, then the trigonal bipyramidal arrangement minimizes repulsions[251]. Then, because an axial bonding pair has three nearest neighbouring pairs at 90°, whereas the equatorial pairs have only two such neighbours, minimization of repulsion will require the axial pairs to be at a greater distance from the central atom than the equatorial pairs. Finally, the

(a)

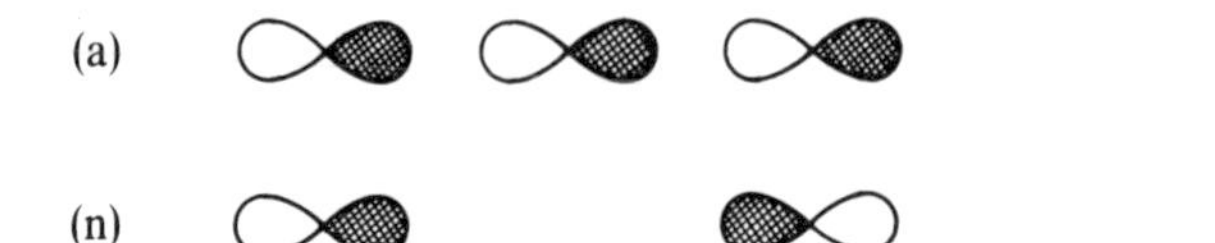

(n)

(b)

FIGURE 7. Molecular orbitals for a three-centre, four electron bond: (a) antibonding; (n) non-bonding; (b) bonding

more electronegative the ligand, the further away from the central atom is its bond pair, taking up less space so that minimization of repulsion will tend to favour occupation of the crowded axial sites by these small electron pairs[251,274]. Other secondary observations can also be explained. For example, substitution of an equatorial fluorine by a methyl group leads to a decrease in the effective electronegativity of the phosphorus and allows all bonding pairs to move away, increasing all bond lengths. The size of the bond pair to carbon will lead to increased axial–equatorial repulsion, lengthening the axial bonds and bending then away from the methyl group as observed[241,251]. Aside from the VSEPR analysis, two bonding models have been used to explain the bonding in phosphoranes, as follows.

The first is the directed valence model, which invokes the involvement of phosphorus d orbitals in sp^3d hybrids. If the d orbital is taken to be the d_{z^2} then the increased axial bond length is reasonable and the preference for axial electronegative ligands can be explained by the requirements for contraction and energy lowering of the d orbital so that its extent and energy are commensurate with those of the s and p orbitals[2]. As explained in Volume 2[2], the concept of virtual d orbital involvement in bonding is now redundant. Since there is a perfectly reasonably alternative bonding description of these systems described below, we may discard this model[16,99,189,275].

The reasonable alternative referred to is the three-centre, four-electron bond model[99] which was developed in detail by Hach and Rundle[276] and later by Musher[277]. The equatorial bonds are considered as normal two-centre, two-electron bonds formed by, for example, overlap of three sp^2 hybrids on the central atom with suitable orbitals on the ligands. That leaves one p orbital on the central atom to bond with two orbitals from axial ligands. There is no particular problem with that, one simply makes linear combinations of the three atomic orbitals to give three molecular orbitals as shown in Figure 7 for the case of ligand p orbitals. The result is one bonding orbital, one antibonding orbital and one non-bonding orbital[99]. Since there are four electrons, both the bonding and non-bonding levels are occupied, explaining why axial bond lengths are longer. The non-bonding orbital places electron density on the axial atoms at the expense of the central atom, explaining why electronegative substituents are preferred in the axial position, i.e. the bonding is partially ionic[99,276,277]. Thus the axial ligands are expected to have lower binding energies for their electrons which is reflected to some extent in the photoelectron spectrum of PF_5[278–282].

The full set of qualitative valence MOs for PH_5 is shown on the left of Figure 8, which is essentially the same as that given Albright *et al.*[195], who presented a detailed account of the construction of these MOs. Also shown in Figure 8 is the correlation between the orbitals for the trigonal bipyramidal geometry and those for the square pyramidal case, showing the Berry pseudorotation process to be orbitally allowed[195]. The arrangement shown in Figure 8 is confirmed by more recent *ab initio* calculations, the results of a selec-

FIGURE 8. Correlation diagram for the trigonal bipyramidal (D_{3h}) and square pyramidal (C_{4v}) forms of an AH_5 molecule, via a C_{2v} structure

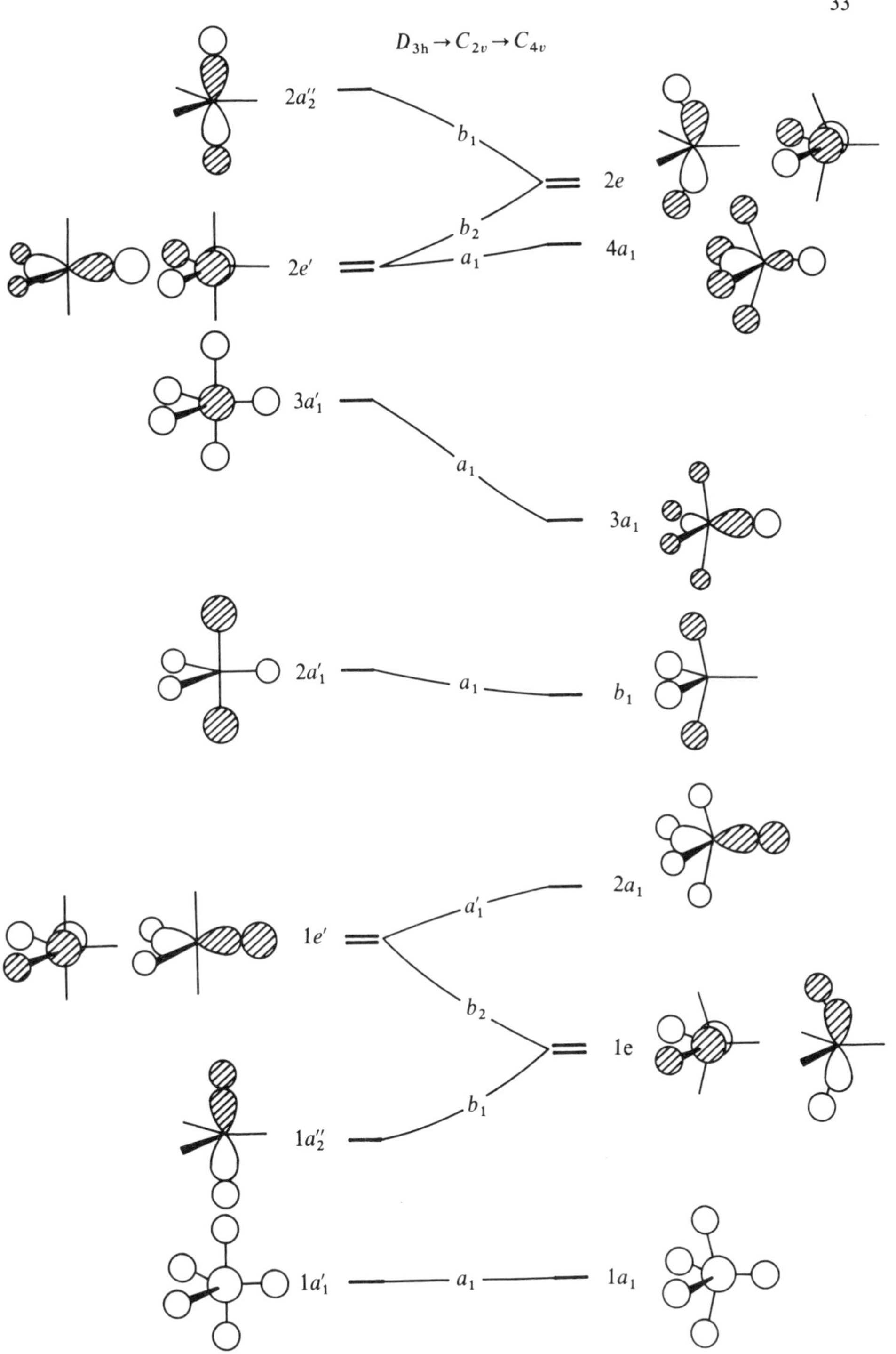

$D_{3h} \rightarrow C_{2v} \rightarrow C_{4v}$
$2a_2''$
b_1
$2e$
b_2
$2e'$
a_1
$4a_1$
$3a_1'$
a_1
$3a_1$
$2a_1'$
a_1
b_1
$2a_1$
a_1'
$1e'$
b_2
1e
b_1
$1a_2''$
$1a_1'$
a_1
$1a_1$

TABLE 11. Results of *ab initio* calculations on selected phosphoranes

Molecule	Calculation[a]	Energy[b]	$r(P—H_e)$[c]	$r(P—H_a)$[c]	$r(P—X_e)$[c]	$r(P—X_a)$[c]	Ref.
PH_5	3–21G*	−341.8545	141.1	147.3			283, 284
	6–31G*	−343.4999	140.7	146.4			283, 285
	DZ + P[d]		141.8	147.8			231
	TZ + P	−343.5286	141.5	147.7			286–288
	MP4/6–31G*	−343.6459					283
	MP4/TZ + P	−343.7258					286
	MP4/66–311G	−343.7631	141.4	146.3			289
CH_3PH_4	3–21G*	−380.7203	140.9	147.8	184.9		283, 284
	6–31G*	−382.5472	140.5	147.1[e]	184.7		283
	MP4/6–31G*	−382.8416					283
FPH_4	3–21G*	−440.2966	139.5	145.7		161.1	283, 284
	6–31G*	−442.4302	138.6	144.0		165.7	229, 283, 290
	6–31G**		138.9	144.4		165.9	288
	MP4/6–31G*	−442.7474					283
F_2PH_3	6–31G*[f]		137.5			163.8	229, 290
	6–31G**		137.8			163.9	288
F_3PH_2	6–31G*[f]		137.1		155.3	161.0	229, 290
	6–31G**		137.3		155.3	161.1	288
	EXPT[g]		137.5		153.9	161.3	266, 267
F_4PH	6–31G*[f]		136.6		154.2	158.8	229, 290
	6–31G**		136.9		154.2	158.8	288
	EXPT[g]		136.3		153.9	159.6	266, 267
PF_5	6–31G*	−838.0507			153.5	156.8	229, 291, 292
	DZ + P	−838.1841			153.7	157.7	286
	6–31G**				153.5	156.8	288
	MP2/6–31G*	−839.0575			156.6	159.5	291
	MP4/DZ + P	−839.1833					286
	EXPT[h]				153.4	157.7	255

[a] All self-consistent field, contracted Gaussian-type basis sets, geometry optimized by the gradient method unless indicated otherwise. DZ, TZ, 3–21G and 6–31G have their usual meanings[175]; * = a set of six d-type polarization functions added to basis set; P = other combinations of d functions added; MPx = electron correlation by xth order Møller–Plesset perturbation theory.

[b] Total electron energy in hartree; 1 hartree = 27.2 eV = 2625 kJ mol^{-1}.

[c] Bond length in pm; e and a refer to atoms in the equatorial and axial positions, respectively. Bond angles mostly approximate regular trigonal bipyramidal values.

[d] By pseudopotential method.

[e] Average of 147.3 and 146.8 pm.

[f] Regular trigonal bipyramidal geometry forced.

[g] By electron diffraction.

[h] From Table 10.

tion of which are shown in Table 11. Results at higher levels of theory only are included in Table 11 and omitted are calculations on systems containing chlorine[231,290] and two sets of calculations, discussed below, on PH_5 monosubstituted with a full series of first- and second-row substituents[283,284]. It can be seen from Table 11 that there is fairly good agreement between theory and experiment where a comparison can be made, so that we may be reasonably confident of the predictions for unknown species in Table 11. Note that bond lengths in phosphoranes are generally longer than those in phosphonium salts[229].

The decomposition pathways of PH_5 were studied by Reed and Schleyer[285], who found that the barrier to dissociation into PH_3 and H_2 was approximately 128 kJ mol^{-1} at the MP4/6–31 + +G** level with correction for zero-point energy. This is significantly lower than earlier values[293] and similar to other calculations[289].

Recent detailed discussions[229,283,284,286,289,290] of the bonding in phosphoranes has usually taken for granted the three-centre, four-electron bond model described above and have concentrated on other related issues of importance in these systems, viz. apicophilicity, pseudorotation and the oxaphosphetanes involved in the Wittig reaction.

1. *Detailed studies of bonding and apicophilicity*

Apicophilicity is usually correlateld with high electronegativity[240,251] but it is also affected by other factors. Thus, simple qualitative MO considerations suggest that π donors should prefer the equatorial rather than the apical position[241,251,294,295] and the introduction of a ring introduces extra strain effects[250]. Some of these issues have been addressed in recent theoretical studies.

McDowell and Streitwieser[284] studied apicophilicity in a wide series of monosubstituted derivatives of PH_5 at the 3–21G* level. The order was found to be Cl > CN > F > CCH > H > CH_3 > OH > O^- > S^- > NH_2 > BH_2. Apicophilicity is enhanced by ligand electronegativity and diminished by π donation. Dieters *et al.*[290] studied fluorine and chlorine apicophilicities in the series Cl_nPF_{5-n}. They found that chlorine is in general more apicophilic than fluorine. However this is not so when the collective electronegativity of the equatiorial atoms is increased sufficiently. They found that in nucleophilic substitution reactions, the apicophilicity of the leaving group is diminished by the composition of equatorial ligands in the order Cl > F > H but is enhanced by the opposite apical atom in the same order.

Recently, a very comprehensive study of substituent effects in phosphoranes was published[283]. The effect of substituting PH_5 was studied with a full range of first- and second-row substituent groups, in both the apical and equatorial positions in the trigonal bipyramidal conformation and in the apically substituted square pyramidal conformation. Geometries were determined at the 6–31* level and energies were obtained both at this level and with correlation up to the MP4 level with population analysis using the natural bond orbital (NBO) method. Most interestingly, it was found[283] that apically substituted square pyramidal structures were the more stable for substituents from Main Groups 1 and 2. The lithium derivative can be considered as a lithium–phosphoranide ion pair. As expected, the trigonal bipyramidal conformation was the most stable for substituents from Main Groups 3–7, but some important other observations were made in these cases also, especially in the disection of apicophilicities into inductive and π-bonding contributions and in new insights into ring systems.

Conjugative effects on apicophilicity can outweigh purely inductive effects and give total apicophilicities not correlated with substituent electronegativity[240]. Thus it was found[283] that there is relatively little preference for CH_3 equatorial (apicophilicity −3.8 kJ mol^{-1}) whereas NH_2 is much more apicophobic (−30.1 kJ mol^{-1}). This latter effect is due to π-bonding effects shown by the fact that the NH_2 group is planar with the nitrogen lone pair in the equatorial plane. In contrast, a PH_2 substituent prefers the pyramidal conformation, again, of course, in the equatorial position. The purely inductive contribution to the apicophilicity of NH_2 was measured[283] as approximately $+46$kJ mol^{-1} by comparison of the apical conformer and the equatorial conformer with the NH_2 group rotated by 90° to eliminate the π-contribution.

For oxygen (OH) and sulphur (SH) substituents, new light was also shed[283] on the 'small-ring rule' (see Section V.A), which states that four- and five-membered rings preferentially span axial–equatorial positions[296]. This is because in the equatorial conformations the POH/PSH plane can be either perpendicular or coplanar with respect to the equatorial plane. The perpendicular conformation is the more stable due to hyperconjugation of O and S lone pairs with P—H antibonding orbitals. It turns out that if ring strain and steric effects are omitted, OH and SH groups have little positional preference (apicophilicity $= +1.7$ kJ mol^{-1} for OH) but in a small-membered diequatorial ring the OR and SR groups have to assume the unfavourable coplanar conformation. This is estimated as adding about 53 kJ mol^{-1} for each O and 38 kJ mol^{-1} for each S, comparable to the effect of ring strain in some cases[283]. Apicophilicities were also divided into to inductive and π contributions for these substituents[283].

In the case of halogen substituents, the equatorial conformation is not a minimum on the potential energy surface but a transition state between the two apical conformations and the P—Hal bond in it is significantly lengthened[283], making it like an edge-associated ion pair, consistent with the observed ionic–covalent isomerism in similar compounds (see Section IV). It is suggested that the term apicophilicity may be less useful in this context because of this high degree of ion-pair character in the equatorial conformer[283].

In a recent study[286], the term *equatoriphilicity* was introduced. This is because it was found, in a wide-ranging study at the MP4/DZ + P level, that the stabilization of the equatorial plane part of the molecule determines the stability of the whole molecule[286]. The same study confirmed the three-centre, four-electron character of the bonds to the axial ligands[286].

In a study which is relevant to the mechanism of hydrolysis of phosphonium salts, Glaser and Streitwieser[297] studied the ions H_4PO^- and H_3PFO^- and their derivatives with Li^+, NH_4^+ and HF at the 6–31G* level augmented by diffuse functions. They found that the structures of the anions are those of a hydride or fluoride ion 'solvated' by or complexed with phosphine oxide, rather than phosphoranes[297]. A very important point is that earlier studies with diffuse functions yielded the pentacoordinated phosphoranes which they judged[297] to be computational artifacts of the small basis set.

2. *Other studies*

A number of theoretical studies have been directly concerned with the mechanism of pseudorotation[231,286,298,299]. It is found[231,299] that the Berry process is more favoured but the turnstile process is not ruled out. Note that these two processes are topologically equivalent but proceed along different pathways on the potential energy surface[293].

Because of their importance as intermediates in the Wittig reaction, oxaphosphetanes have been fairly common targets for theoretical study. This is usually in connection with (i) the mode of their formation from ylide and carbonyl and whether it is via a betaine or not and (ii) the relative energies of the diastereomeric oxaphosphetanes resulting from the reaction of substituted ylides and carbonyls in attempts to explain *E–Z* stereoselectivities. The former concerns are usually addressed by fairly high levels of theory[164,300–303] whereas the latter have used lower levels on the more realistic molecules involved[177,303–306]. Wittig reactions are found to be symmetry allowed[164] but highly asynchronous reactions bordering on two-step processes, with little P—O covalent bonding in the transition state[304]. Despite a lot of work, a satisfactory explanation of the high *Z* stereoselectivity of certain Wittig reactions remains elusive[304].

Bestmann *et al.*[298] showed that there was a fairly small energy difference between the pseudorotational isomers of the oxaphosphetane from $H_3P{=}CH_2$ and $H_2C{=}O$.

VI. ADDENDUM

There have been two recent X-ray crystal structure determinations of stabilized phosphorus ylides.[307,308] Aitken *et al.*[307] determined the structure of an ester-stabilized tributylphosphonium ylide which provides further confirmation of the delocalization of the anionic charge at the ylidic carbon in these systems with bond lengths as follows P=C:172.6, C—C:140.2 and C=O:123.4 pm (R = 5.5, see Table 2 for comparisons). Abell *et al.*[308] studied a triphenylphosphonium ylide stabilized by both an ester and a keto group providing an opportunity to compare the effects of these two groups and indeed the keto group showed the more pronounced effects, the relevant bonds lengths being C—C:141.6 vs 145.6 and C=O:125.3 vs 120.5 pm (R = 4.7). However the molecule studied also contained a carboxylic acid function which was hydrogen bonded to the keto group which complicated this analysis because hydrogen bonding would lead to lengthening of the C=O bond while, on the other hand, the greater the negative charge on O the more likely is hydrogen bonding to it.

VII. REFERENCES

1. D. G. Gilheany, in *The Chemistry of Organophosphorus Compounds* (Ed. F. R. Hartley), Vol. 1, Wiley, Chichester, 1990, Chapter 2, pp. 9–49.
2. D. G. Gilheany, in *The Chemistry of Organophosphorus Compounds* (Ed. F. R. Hartley), Vol. 2, Wiley, Chichester, 1992, Chapter 1, pp. 1–52.
3. A. W. Johnson, *Ylid Chemistry*, Academic Press, New York, 1966.
4. R. F. Hudson, *Structure and Mechanism in Organophosphorus Chemistry*, Academic Press, New York, 1965.
5. A. J. Kirby and S. G. Warren, *The Organic Chemistry of Phosphorus*, Elsevier, Amsterdam, 1967.
6. H. J. Bestmann and R. Zimmermann, in *Organic Phosphorus Compounds* (Eds G. M. Kosolapoff and L. Maier), Vol. 3, Wiley, London, Chichester, 1972, Chapter 5A.
7. B. J. Walker, *Organophosphorus Chemistry*, Penguin, London, 1972.
8. J. Emsley and D. Hall, *The Chemistry of Phosphorus*, Wiley, Chichester, 1976.
9. H. Goldwhite, *Introduction to Phosphorus Chemistry*, Cambridge University Press, Cambridge, 1981.
10. D. J. H. Smith, in *Comprehensive Organic Chemistry* (Eds D. Barton and W. D. Ollis), Vol. 2 (Ed I. O. Sutherland), Pergamon Press, Oxford, 1979, pp. 1313–1328.
11. D. E. C. Corbridge, *Phosphorus, an Outline of its Chemistry, Biochemistry and Technology*, 3rd ed., Elsevier, Amsterdam, 1985.
12. A. Van Dormael, *Bull. Soc. Chim. Fr.*, 2701 (1969).
13. M. Schlosser, in *Methodicum Chimicum* (Ed. F. Korte), Vol. 7B (Eds H. Zimmer and K. Niedenzu), Academic Press, New York, and Georg Thieme, Stuttgart, 1978, Chapter 28.
14. D. G. Morris, *Surv. Prog. Chem.*, **10**, 189 (1983).
15. I. Zugravescu and M. Petrovanu, *Nitrogen Ylid Chemistry*, McGraw-Hill, New York, 1976.
16. E. Magnusson, *J. Am. Chem. Soc.*, **112**, 7940 (1990).
17. H. Keck, W. Kuchen, P. Tommes, J. K. Terlouw and T. Wong, *Angew. Chem., Int. Ed. Engl.*, **31**, 86 (1992).
18. B. F. Yates, W. J. Bouma and L. Radom, *J. Am. Chem. Soc.*, **106**, 5805 (1984).
19. D. E. C. Corbridge, *The Structural Chemistry of Phosphorus*, Elsevier, Amsterdam, 1974.
20. M. A. Howells, R. D. Howells, N. C. Baenziger and D. J. Burton, *J. Am. Chem. Soc.*, **95**, 5366 (1973).
21. M. A. Vincent, H. F. Schaefer, A. Schier and H. Schmidbaur, *J. Am. Chem. Soc.*, **105**, 3806 (1983).
22. E. A. V. Ebsworth, T. E. Fraser and D. W. H. Rankin, *Chem. Ber.*, **110**, 3494 (1977).
23. H. Schmidbaur, J. Jeong, A. Schier, W. Graf, D. L. Wilkinson and G. Müller, *New J. Chem.*, **13**, 341 (1989).

24. J. C. J. Bart, *J. Chem. Soc. B*, 350 (1969).
25. H. Schmidbaur, A. Schier, C. M. F. Frazão and G. Müller, *J. Am. Chem. Soc.*, **108**, 976 (1986).
26. H. Schmidbaur, A. Schier, B. Milewski-Mahrla and U. Schubert, *Chem. Ber.*, **115**, 722 (1982).
27. H. Schmidbaur, A. Schier and D. Neugebauer, *Chem. Ber.*, **116**, 2173 (1983).
28. N. C. Baenziger, B. A. Foster, M. Howells, R. Howells, P. Vander Valk and D. J. Burton, *Acta Crystallogr., Sect. B*, **27**, 2327 (1977).
29. E. A. V. Ebsworth, D. W. H. Rankin, B. Zimmer-Gasser and H. Schmidbaur, *Chem. Ber.*, **113**, 1637 (1980).
30. H. H. Karsch, B. Deubelly and G. Müller, *J. Chem. Soc. Chem. Commun.*, 517 (1988).
31. G. Fritz, W. Schick, W. Hönle and H. G. von Schnering, *Z. Anorg. Allg. Chem.*, **511**, 95 (1984).
32. K. Horchler von Locquenghien, A. Baceiredo, R. Boese and G. Bertrand, *J. Am. Chem. Soc.*, **113**, 5062 (1991).
33. H. Schmidbaur, U. Deschler and B. Milewski-Mahrla, *Chem. Ber.*, **116**, 1393 (1983).
34. H. Schmidbaur, P. Nussstein and G. Müller, *Z. Naturforsch., Teil B*, **39**, 1456 (1984).
35. H. Schmidbaur, B. Milewski-Mahrla, G. Müller and C. Krüger, *Organometallics*, **3**, 38 (1984).
36. H. Schmidbaur, C. Zybill, C. Krüger and H.-J. Kraus, *Chem. Ber.*, **116**, 1955 (1983).
37. L. M. Engelhardt, R. I. Papasergio, C. L. Raston, G. Salem, C. R. Whitaker and A. H. White, *J. Chem. Soc., Dalton Trans.*, 1647 (1987).
38. R. E. Cramer, R. B. Maynard, J. C. Paw and J. W. Gilje, *J. Am. Chem. Soc.*, **103**, 3589 (1981).
39. B. L. Barnet and C. Krüger, *Cryst. Struct. Commun.*, **3**, 427 (1973).
40. H. Grützmacher and H. Pritzkow, *Angew. Chem., Int. Ed. Engl.*, **31**, 99 (1992).
41. M. Shao, X. Jin, Y. Tang, Q. Huang and Y. Huang, *Tetrahedron Lett.*, **23**, 5343 (1982).
42. A. F. Cameron, F. D. Duncanson, A. A. Freer, V. W. Armstrong and R. Ramage, *J. Chem. Soc., Perkin Trans. 2*, 1030 (1975).
43. U. Lingner and H. Burzlaff, *Acta Crystallogr., Sect. B*, **30**, 1715 (1974).
44. S. Husbeye, E. A. Meyers, R. A. Zingaro, A. A. Braga, J. V. Comasseto and N. Petragnani, *Acta Crystallogr., Sect. C*, **42**, 90 (1986).
45. S. Z. Goldberg and K. N. Raymond, *Inorg. Chem.*, **12**, 2923 (1973).
46. H. Blau, W. Malisch, S. Voran, K. Blank and C. Krüger, *J. Organomet. Chem.*, **202**, C33 (1980).
47. J. Buckle, P. G. Harrison, T. J. King and J. A. Richards, *J. Chem. Soc., Chem. Commun.*, 1104 (1972).
48. W. Dreissig, H. J. Hecht and K. Pleith, *Z. Kristallogr. S*, 132 (1973).
49. P. J. Butterfield, J. C. Tebby and T. J. King, *J. Chem. Soc., Perkin Trans. 1*, 1237 (1978).
50. G. Ríhs, C. D. Weis and P. Wieland, *Phosphorus Sulfur Silicon Relat. Elem.*, **48**, 27 (1990).
51. H. L. Ammon, G. L. Wheeler and P. H. Watts, *J. Am. Chem. Soc.*, **95**, 6158 (1973).
52. E. A. V. Ebsworth, Th. E. Fraser, D. W. H. Rankin, O. Gasser and H. Schmidbaur, *Chem. Ber.*, **110**, 3508 (1977).
53. G. E. Hardy, J. I. Zink, W. C. Kaska and J. C. Baldwin, *J. Am. Chem. Soc.*, **100**, 8001 (1978).
54. A. T. Vincent and P. J. Wheatley, *J. Chem. Soc., Dalton Trans.*, 617 (1972).
55. H. Schmidbaur, G. Hasslberger, U. Deschler, U. Schubert, C. Kappenstein and A. Frank, *Angew. Chem., Int. Ed. Engl.*, **18**, 408 (1979).
56. U. Schubert, C. Kappenstein, B. Milewski-Mahrla and H. Schmidbaur, *Chem. Ber.*, **114**, 3070 (1981).
57. H. Schmidbaur, T. Costa, B. Milewski-Mahrla and U. Schubert, *Angew. Chem., Int. Ed. Engl.*, **19**, 555 (1980).
58. J. J. Daly and P. J. Wheatley, *J. Chem. Soc. A*, 1703 (1966).
59. J. J. Daly, *J. Chem. Soc. A*, 1913 (1967).
60. H. Burzlaff, E. Wilhelm and H.-J. Bestmann, *Chem. Ber.*, **110**, 3168 (1977).
61. H. Burzlaff, U. Voll and H.-J. Bestmann, *Chem. Ber.*, **107**, 1949 (1974).
62. H. Burzlaff, R. Hagg, E. Wilhelm and H.-J. Bestmann, *Chem. Ber.*, **118**, 1720 (1985).
63. S. Z. Goldberg, E. N. Duesler and K. N. Raymond, *Inorg. Chem.*, **11**, 1397 (1972).
64. P. J. Carrol and D. D. Titus, *J. Chem. Soc., Dalton Trans.*, 824 (1977).
65. H. H. Karsch, B. Zimmer-Gasser, D. Neugebauer and U. Schubert, *Angew. Chem., Int. Ed. Engl.*, **18**, 484 (1979).
66. B. Zimmer-Gasser, D. Neugebauer, U. Schubert and H. H. Karsch, *Z. Naturforsch., Teil B*, **34**, 1267 (1979).
67. M. A. Kazankova, I. L. Rodionov, I. F. Lutsenko, A. N. Chernega, M. Yu. Antipin and Yu. T.

Struchkov, *Zh. Obshch. Khim.*, **57**, 2482 (1987); Engl. Transl., *J. Gen. Chem. USSR*, **57**, 2211 (1987).
68. D. F. Rendle, *Diss. Abstr. Int. B*, **33**, 5178 (1973).
69. F. K. Ross, L. Manojlovic-Muir, W. C. Hamilton, R. Ramirez and J. F. Pilot, *J. Am. Chem. Soc.*, **94**, 8738 (1972).
70. F. K. Ross, W. C. Hamilton and F. Ramirez, *Acta Crystallogr., Sect. B*, **27**, 2331 (1971).
71. H. -J. Bestmann, G. Schmid, R. Böhme, E. Wilhelm and H. Burzlaff, *Chem. Ber.*, **113**, 3937 (1980).
72. H. Schmidbaur, U. Deschler, B. Zimmer-Gasser, D. Neugebauer and U. Schubert, *Chem. Ber.*, **113**, 902 (1980).
73. H. Schmidbaur, U. Deschler, B. Milewski-Mahrla and B. Zimmer-Gasser, *Chem. Ber.*, **114**, 608 (1981).
74. J. Grobe, D. LeVan and J. Nientiedt, *New J. Chem.*, **13**, 363 (1989).
75. Z. Xia, Z. Zhang and Y. Shen, *Huaxue Xuebao*, **42**, 1223 (1984); *Chem. Abstr.*, **102**, 158435 (1985).
76. C. He, Q. Zheng, F. Shen, S. Dou and G. Lin, *Wuli Xuebao*, **31**, 825 (1982); *Chem. Abstr.*, **97**, 206012 (1982).
77. H. Schmidbaur, R. Pichl and G. Müller, *Chem. Ber.*, **120**, 789 (1987).
78. D. G. Gilheany, D. A. Kennedy, J. F. Malone and B. J. Walker, *Tetrahedron Lett.*, **26**, 513 (1985).
79. E. Fluck, B. Neumüller, G. Heckmann, W. Plass and P. G. Jones, *New J. Chem.*, **13**, 383 (1989).
80. F. Bernardi, H. B. Schlegel, M.-H. Whangbo and S. Wolfe, *J. Am. Chem. Soc.*, **99**, 5633 (1977).
81. H. Schmidbaur and W. Tronich, *Chem. Ber.*, **101**, 595, 604 (1968).
82. R. Köster, D. Simic and M. Grassberger, *Justus Liebigs Ann. Chem.*, **739**, 211 (1970).
83. R. Appel, *Pure Appl. Chem.*, **59**, 977 (1987).
84. L. Pauling, *The Nature of the Chemical Bond*, 3rd ed., Cornell University Press, Ithaca, NY, 1960.
85. V. Schomaker and D. P. Stevenson, *J. Am. Chem. Soc.*, **63**, 37 (1941).
86. F. A. Carey and R. J. Sundberg, *Advanced Organic Chemistry, Part A*, 3rd ed., Planum Press, New York, 1990.
87. Z. Liu and M. Schlosser, *Tetrahedron Lett.*, **31**, 5753 (1990).
88. G. Trinquier and J.-P. Malrieu, *J. Am. Chem. Soc.*, **109**, 5303 (1987).
89. J. -P. Malrieu and G. Trinquier, *J. Am. Chem. Soc.*, **111**, 5916 (1989).
90. H. Schmidbaur, A. Schier, S. Lauteschläger, J. Riede and G. Müller, *Organometallics*, **3**, 1906 (1984).
91. H. Schmidbaur, C. E. Zybill, D. Neugebauer and G. Müller, *Z. Naturforsch., Teil B*, **40**, 1293 (1985).
92. H. Schmidbaur and P. Nussstein, *Chem. Ber.*, **120**, 1281 (1987).
93. C. Schade and P. v. R. Schleyer, *J. Chem. Soc., Chem. Commun.*, 1399 (1988).
94. I. O. Sutherland, *Annu. Rep. NMR Spectrosc.*, **4**, 211 (1971).
95. L. M. Jackman and F. A. Cotton, *Dynamic Nuclear Magnetic Resonance Spectroscopy*, Academic Press, New York, 1975.
96. C. J. Devlin and B. J. Walker, *Tetrahedron*, **28**, 3501 (1972), and references cited therein.
97. J. Barluenga, F. López, F. Palacios and F. Sánchez-Ferrando, *J. Chem. Soc., Perkin Trans. 2*, 903 (1988).
98. H. -J. Bestmann, A. J. Kos, K. Witzgall and P. v. R. Schleyer, *Chem. Ber.*, **119**, 1331 (1986).
99. W. Kutzelnigg, *Angew. Chem., Int. Ed. Engl.*, **23**, 272 (1984).
100. B. F. Yates, W. J. Bouma and L. Radom, *J. Am. Chem. Soc.*, **109**, 2250 (1987).
101. D. A. Dixon and B. E. Smart, *J. Am. Chem. Soc.*, **108**, 7172 (1986).
102. M. Seno, S. Tsuchiya, H. Kise and T. Asahara, *Bull. Chem. Soc. Jpn.*, **48**, 2001 (1975).
103. B. M. Trost and L. S. Melvin, *Sulfur Ylides—Emerging Synthetic Intermediates*, Academic Press, New York, 1975.
104. I. Gosney and A. G. Rowley, in *Organophosphorus Reagents in Organic Synthesis* (Ed. J. I. G. Cadogan), Academic Press, London, 1979, Chapter 2.
105. M. L. Huggins, *J. Am. Chem. Soc.*, **75**, 4126 (1953).
106. A. W. Johnson and R. T. Amel, *Can. J. Chem.*, **46**, 461 (1968).
107. D. Lloyd and M. I. C. Singer, *Chem. Ind. (London)*, 1277 (1968).
108. T. S. Stevens, *Prog. Org. Chem.*, **7**, 48 (1968).
109. W. D. Ollis, M. Rey and I. O. Sutherland, *J. Chem. Soc., Perkin Trans. 1*, 1009 (1983), and references cited therein.

110. D. G. Gilheany, D. A. Kennedy, J. F. Malone and B. J. Walker, *J. Chem. Soc., Chem. Commun.*, 1217 (1984).
111. A. Maercker, B. Basta and R. Jung, *Main Group Met. Chem.*, **10**, 11 (1987).
112. L. C. Thomas, *Interpretation of the Infrared Spectra of Organophosphorus Compounds*, Heyden, London, 1974.
113. P. J. Taylor, *Spectrochim. Acta, Part A*, **34**, 115, (1978).
114. M. I. Shevchuk, M. V. Khalaturnik and A. V. Dombrovskii, *Zh. Obshch. Khim.*, **42**, 2630 (1972); Engl. Transl., *J. Gen. Chem. USSR*, **42**, 2621 (1972).
115. W. Luttke and K. Wilhelm, *Angew. Chem., Int. Ed. Engl.*, **4**, 875 (1965).
116. H. Siebert, *Z. Anorg. Allg. Chem.*, **273**, 170 (1953).
117. W. Sawodny, *Z. Anorg. Allg. Chem.*, **368**, 284 (1969).
118. K. Dimroth, S. Berger and H. Kaletsch, *Phosphorus Sulfur*, **10**, 305 (1981).
119. F. Ramirez and S. Levy, *J. Am. Chem. Soc.*, **79**, 67, 6167 (1957).
120. F. Ramirez and S. Dershowitz, *J. Org. Chem.*, **22**, 41 (1957).
121. A. W. Johnson, *J. Org. Chem.*, **24**, 282 (1959).
122. H. Goetz and B. Klabuhn, *Justus Liebigs Ann. Chem.*, **724**, 1 (1969).
123. H. Lumbroso, C. Pigenet, A. Arcoria and G. Scarlata, *Bull. Soc. Chim. Fr.*, 3838 (1971).
124. H. Lumbroso, D. Lloyd and G. Harris, *C.R. Acad. Sci., Ser. C*, **278**, 219 (1974).
125. H. Lumbroso, J. Cure and H.-J. Bestmann, *J. Organomet. Chem.*, **161**, 347 (1978).
126. Z. Yoshida, S. Yoneda and Y. Murata, *J. Org. Chem.*, **38**, 3537 (1973).
127. B. G. Ramsey, *Electronic Transitions in Organometalloids*, Academic Press, New York, 1969.
128. S. O. Grim and J. H. Ambrus, *J. Org. Chem.*, **33**, 2993 (1968).
129. K. Iwata, S. Yoneda and Z. Yoshida, *J. Am. Chem. Soc.*, **93**, 6745 (1971).
130. Z. Yoshida, K. Iwata and S. Yoneda, *Tetrahedron Lett.*, 1519 (1971).
131. R. I. Yurchenko, O. M. Voitsekhovskaya, I. N. Zhmurova and V. G. Yurchenko, *Zh. Obshch. Khim.*, **46**, 255 (1976); Engl. Transl., *J. Gen. Chem. USSR*, **46**, 251 (1976).
132. B. Freeman, D. Lloyd and M. Singer, *Tetrahedron*, **28**, 343 (1972).
133. B. Freeman and D. Lloyd, *Tetrahedron*, **30**, 2257 (1974).
134. R. A. Loktionova, I. E. Boldeskul and V. P. Lysenko, *Teor. Eksp. Khim.*, **19**, 30 (1983); Engl. Transl., *Theor. Exp. Chem.*, **19**, 23 (1983).
135. H. Schmidbaur, W. Buchner and D. Scheutzow, *Chem. Ber.*, **106**, 1251 (1973).
136. L. D. MeKeever, R. Waak, M. A. Doran and E. B. Baker, *J. Am. Chem. Soc.*, **91**, 1057 (1969).
137. H. Schmidbaur and W. Tronich, *Chem. Ber.*, **100**, 1032 (1967).
138. M. J. Gallagher, *Aust. J. Chem.*, **31**, 1197 (1968).
139. T. A. Albright, *Org. Magn. Reson.*, **8**, 489 (1976).
140. T. A. Albright, W. J. Freeman and E. E. Schweiser, *J. Am. Chem. Soc.*, **97**, 2946 (1975).
141. T. A. Albright and E. E. Schweiser, *J. Org. Chem.*, **41**, 1168 (1976).
142. T. A. Albright, W. J. Freeman and E. E. Schweiser, *J. Am. Chem. Soc.*, **98**, 6249 (1976).
143. T. A. Albright, W. J. Freeman and E. E. Schweiser, *J. Am. Chem. Soc.*, **97**, 2942 (1975).
144. S. O. Grim, W. McFarlane and T. J. Marks, *J. Chem. Soc., Chem. Commun.*, 1191 (1967).
145. H. Schmidbaur, W. Richter, W. Wolf and F. H. Kohler, *Chem. Ber.*, **108**, 2649 (1975).
146. K. A. Ostoja Starzewskii and H. tomDieck, *Phosphorus*, **6**, 177 (1976).
147. T. A. Albright, W. J. Freeman and E. E. Schweiser, *J. Org. Chem.*, **40**, 3437 (1975).
148. G. A. Gray, *J. Am. Chem. Soc.*, **93**, 2132 (1971).
149. H. Schmidbaur, *Adv. Organomet. Chem.*, **14**, 205 (1976).
150. G. A. Gray, *J. Am. Chem. Soc.*, **95**, 5092, 7736 (1973).
151. K. Hildebrand and H. Dreeskamp, *Z. Naturforsch., Teil B*, **28**, 226 (1973).
152. K. A. Ostoja Starzewskii and M. Fiegel, *J. Organomet. Chem.*, **93**, C20 (1975).
153. K. A. Ostoja Starzewskii, H. Bock and H. tomDieck, *Angew. Chem., Int. Ed. Engl.*, **14**, 173 (1975).
154. K. A. Ostoja Starzewskii and H. Bock, *J. Organomet. Chem.*, **65**, 311 (1974).
155. K. A. Ostoja Starzewskii and H. Bock, *J. Am. Chem. Soc.*, **98**, 8486 (1976).
156. W. B. Perry, T. F. Schaaf and W. L. Jolly, *J. Am. Chem. Soc.*, **97**, 4899 (1975).
157. S. C. Avanzino, W. L. Jolly, M. S. Lazarus, W. B. Perry, R. R. Rietz and T. F. Schaaf, *Inorg. Chem.*, **14**, 1595 (1975).
158. E. Fluck, *Pure. Appl. Chem.*, **44**, 373 (1975).
159. Y. Yamamoto and H. Konno, *Bull. Chem. Soc. Jpn.*, **59**, 1327 (1986).
160. M. Pelavin, D. N. Hendrickson, J. M. Hollander and W. L. Jolly, *J. Phys. Chem.*, **74**, 1116 (1970).

161. W. E. Swartz and D. M. Hercules, *Anal. Chem.*, **43**, 1066 (1971).
162. B. Folkesson, *Chem. Scr.*, **20**, 108 (1982).
163. D. J. Mitchell, S. Wolfe and H. B. Schlegel, *Can. J. Chem.*, **59**, 3280 (1981).
164. R. Höller and H. Lischka, *J. Am. Chem. Soc.*, **102**, 4632 (1980).
165. A. Streitwieser, A. Rajca, R. S. McDowell and R. Glaser, *J. Am. Chem. Soc.*, **109**, 4184 (1987).
166. M. M. Francl, R. C. Pellow and L. C. Allen, *J. Am. Chem. Soc.*, **110**, 3723 (1988).
167. M. T. Nguyen and A. F. Hegarty, *J. Chem. Soc., Perkin Trans. 2*, 47 (1987).
168. P. Molina, M. Alajarin, C. L. Leonardo, R. M. Claramunt, M. C. Foces-Foces, F. Hernandez Cano, J. Catalan, J. L. G. de Paz and J. Elguero, *J. Am. Chem. Soc.*, **111**, 355 (1989).
169. A. Strich, *New J. Chem.*, **3**, 105 (1979).
170. H. Lischka, *J. Am. Chem. Soc.*, **99**, 353 (1977).
171. G. Trinquier and J.-P. Malrieu, *J. Am. Chem. Soc.*, **101**, 7169 (1979).
172. R. A. Eades, P. G. Gassman and D. A. Dixon, *J. Am. Chem. Soc.*, **103**, 1066 (1981).
173. D. A. Dixon, T. H. Dunning, R. A. Eades and P. G. Gassman, *J. Am. Chem. Soc.*, **105**, 7011 (1983).
174. J. R. Bews and C. Glidewell, *J. Mol. Struct. THEOCHEM*, **104**, 105 (1983).
175. I. N. Levine, *Quantum Chemistry*, 3rd ed., Allyn and Bacon, Boston, 1983.
176. R. S. McDowell and A. Streitwieser, *J. Am. Chem. Soc.*, **106**, 4047 (1984).
177. F. Mari, P. M. Lahti and W. E. McEwan, *Heteroatom Chem.*, **2**, 265 (1991).
178. I. Absar and J. R. van Wazer, *J. Am. Chem. Soc.*, **94**, 2382 (1972).
179. Y. Ito, M. Okano and R. Oda, *Tetrahedron*, **22**, 2615 (1966).
180. R. Vilceanu, A. Balint and Z. Simon, *Nature, (London)*, **217**, 61 (1968).
181. H. Goetz and F. Marschner, *Tetrahedron*, **27**, 3581 (1971).
182. J. M. F. van Dijk and H. M. Buck, *Recl. Trav. Chim. Pays-Bas*, **93** 155 (1974).
183. B. Klabuhn, *Tetrahedron*, **30**, 2327 (1974).
184. G. Frenking, H. Goetz and F. Marschner, *J. Am. Chem. Soc.*, **100**, 5295 (1978).
185. M. J. S. Dewar, M. L. McKee and H. S. Rzepa, *J. Am. Chem. Soc.*, **100**, 3607 (1978).
186. R. Hoffmann, D. B. Boyd and S. Z. Goldberg, *J. Am. Chem. Soc.*, **92**, 3929 (1970).
187. D. B. Boyd and R. Hoffmann, *J. Am. Chem. Soc.*, **93**, 1064 (1971).
188. T. A. Albright, P. Hofmann and A. R. Rossi, *Z. Naturforsch., Teil B*, **35**, 343 (1980).
189. A. E. Reed and P. v. R. Schleyer, *J. Am. Chem. Soc.*, **112**, 1434 (1990).
190. W. K. Jardine, R. F. Langer and J. A. MacGregor, *Can. J. Chem.*, **60**, 2069 (1982).
191. M. W. Whangbo, S. Wolfe and F. Bernardi, *Can. J. Chem.*, **53**, 3040 (1975).
192. G. Glidewell, *J. Inorg. Nucl. Chem.*, **38**, 669 (1976).
193. L. S. Bartell, L. S. Su and H. Yow, *Inorg. Chem.*, **9**, 1903 (1970).
194. W. B. Perry, T. F. Schaaf and W. L. Jolly, *J. Am. Chem. Soc.*, **97**, 4899 (1975).
195. T. A. Albright, J. K. Burdett and M. H. Whangbo, *Orbital Interactions in Chemistry*, Wiley–Interscience, New York, 1985.
196. P. v. R. Schleyer and A. J. Kos, *Tetrahedron*, **39**, 1141 (1983).
197. R. P. Messmer, *J. Am. Chem. Soc.*, **113**, 433 (1991), and references cited therein.
198. J. I. Musher, *Tetrahedron*, **30**, 1747 (1974).
199. P. Beck in *Organic Phosphorus compounds* (Eds G. M. Kosolapoff and L. Maier), Vol. 2, Wiley, London, 1972, Chapter 4.
200. S. W. Lee and W. C. Trogler, *J. Org. Chem.*, **55**, 2644 (1990).
201. H. Schmidbaur, B. Blaschke, B. Zimmer-Gasser and U. Schubert, *Chem. Ber.*, **113**, 1612 (1980).
202. E. E. Schweizer, C. J. Baldacchini and A. L. Rheingold, *Acta Crystallogr., Sect. C*, **45**, 1236 (1989).
203. N. W. Alcock, M. Pennington and G. R. Willey, *Acta Crystallogr., Sect. C*, **41**, 1549 (1985).
204. T. Glowiak, E. Durcanska, I. Ondrejkovicova and G. Ondrejovic, *Acta Crystallogr., Sect. C*, **42**, 1331 (1986).
205. R. J. Fleming, M. A. Shaikh, B. W. Skelton and A. H. White, *Aust. J. Chem.*, **32**, 2187 (1979).
206. S. J. Archer, T. A. Modro and L. R. Nassimbeni, *Phosphorus Sulfur*, **11**, 101 (1981).
207. P. vander Sluis, K. vander Vlist and H. Krabbendam, *Acta Crystallogr., Sect. C*, **46**, 1297 (1990).
208. J. P. Henichart, R. Houssin, C. Vaccher, M. Foulon and F. Baert, *J. Mol. Struct.*, **99**, 283 (1983).
209. A. J. Bellamy, R. O. Gould and M. D. Walkinshaw, *J. Chem. Soc., Perkin Trans. 2*, 1099 (1981).
210. W. E. McEwen, C. E. Sullivan and R. O. Day, *Organometallics*, **2**, 420 (1983).
211. R. Böhme, H. Burzlaff, M. Gomm, H.-J. Bestmann and R. Luckenbach, *Chem. Ber.*, **108**, 3525 (1975).

212. H. Schmidbaur, R. Herr, T. Pollok, A. Schier, G. Müller and J. Riede, *Chem. Ber.*, **118**, 3105 (1985).
213. P. Batail, L. Ouahab, J.-F. Halet, J. Padiou, M. Lequan and R. M. Lequan, *Synth. Met.*, **10**, 415 (1985).
214. H.-J. Bestmann, H. Behl and M. Bremer, *Angew. Chem., Int. Ed. Engl.*, **28**, 1219 (1989).
215. L. D. Iroff and K. Mislow, *J. Am. Chem. Soc.*, **100**, 2121 (1978).
216. M. Konno and Y. Saito, *Acta Crystallogr., Sect. B*, **29**, 2815 (1973).
217. M. Milolajczyk, P. Graczyk, M. W. Wieczorek and G. Bujacz, *Angew. Chem., Int. Ed. Engl.*, **30**, 578 (1991).
218. P. Graczyk and M. Mikolajczyk, *Phosphorus Sulfur Silicon Relat. Elem.*, **59**, 211 (1991).
219. J. C. Tebby, in *Phosphorus-31 NMR Spectroscopy in Stereochemical Analysis* (Eds J. G. Verkade and L. D. Quin), VCH, Deerfield Beach, FL, 1987, Chapter 1.
220. G. P. Schiemenz, *Phosphorus*, **3**, 125 (1973).
221. S. Ikuta and P. Kebarle, *Can. J. Chem.*, **61**, 97 (1983).
222. M.-B. Krogh-Jespersen, J. Chandrasekhar, E.-U. Würthwein, J. B. Collins and P. v. R. Schleyer, *J. Am. Chem. Soc.*, **102**, 2263 (1980).
223. E. Magnusson, *J. Comput. Chem.*, **5**, 612 (1984).
224. S. F. Smith, J. Chandrasekhar and W. L. Jorgensen, *J. Phys. Chem.*, **86**, 3308 (1982).
225. C. Glidewell and C. Thomson, *J. Comput. Chem.*, **3**, 495 (1982).
226. A. A. Korkin and E. N. Tsvetkov, *Bull. Soc. Chim. Fr.*, 335 (1988).
227. A. E. Reed and P. v. R. Schleyer, *J. Am. Chem. Soc.*, **109**, 7362 (1987).
228. S. C. Choi, R. J. Boyd and O. Knop, *Can. J. Chem.*, **66**, 2465 (1988).
229. J. A. Dieters and R. R. Holmes, *J. Am. Chem. Soc.*, **112**, 7179 (1990).
230. J. E. Del Bene and I. Shavitt, *J. Phys. Chem.*, **94**, 5514 (1990).
231. G. Trinquier, J.-P. Daudey, G. Caruana and Y. Madaule, *J. Am. Chem. Soc.*, **106**, 4794 (1984).
232. S. A. Pope, I. H. Hillier and M. F. Guest, *Faraday Symp. Chem. Soc.*, **19**, 109 (1984).
233. A. Sequeira and W. C. Hamilton, *J. Chem. Phys.*, **47**, 1818 (1967).
234. J. R. Bews and C. Glidewell, *J. Mol. Struct. THEOCHEM*, **94**, 305 (1983).
235. D. S. Marynick and W. N. Lipscomb, *Proc. Natl. Acad. Sci. USA*, **79**, 1341 (1982).
236. L. L. Lohr, H. B. Schlegel and K. Morokuma, *J. Phys. Chem.*, **88**, 1981 (1984).
237. E. Magnusson, *J. Am. Chem. Soc.*, **106**, 1177, 1185 (1984).
238. N. N. Greenwood and A. Earnshaw, *Chemistry of the Elements*, Pergamon Press, Oxford, 1984.
239. P. C. Crofts in *Organic Phosphorus Compounds* (Eds G. M. Kosolapoff and L. Maier), Vol. 6, Wiley, London, 1973.
240. R. R. Holmes, *Pentacoordinated Phosphorus, Vol. 1, Structure and Spectroscopy*, ACS Monograph 175, American Chemical Society, Washington, DC, 1980.
241. R. R. Holmes, *Pentacoordinated Phosphorus, Vol. 2, Reaction Mechanisms*, ACS Monograph 176, American Chemical Society, Washington, DC, 1980.
242. L. M. Sergienko, G. V. Ratovskii, V. I. Dmitriev and B. V. Timokhin, *Zh. Obshch. Khim.*, **49**, 317 (1979); Engl. Transl., *J. Gen. Chem. USSR*, **49**, 275 (1979).
243. K. B. Dillon and J. Lincoln, *Polyhedron*, **8**, 1445 (1989).
244. K. B. Dillon and T. A. Straw, *J. Chem. Soc., Chem. Commun.*, 234 (1991).
245. S. J. Brown, J. H. Clark and D. J. Macquarrie, *J. Chem. Soc., Dalton Trans.*, 277 (1988).
246. J. H. Clark and D. J. Macquarrie, *J. Chem. Soc., Chem. Commun.*, 229 (1988).
247. M. A. H. A. Al-Juboori, P. Gates and A. S. Muir, *J. Chem. Soc., Chem. Commun.*, 1270 (1991).
248. S. M. Godfrey, D. G. Kelly, C. A. McAuliffe, A. G. Mackie, R. G. Pritchard and S. M. Watson, *J. Chem. Soc., Chem. Commun.*, 1163 (1991).
249. N. Bricklebank, S. M. Godfrey, A. G. Mackie, C. A. McAuliffe and R. G. Pritchard, *J. Chem. Soc., Chem. Commun.*, 355 (1992).
250. W.-W. duMont, M. Bätcher, S. Pohl and W. Saak, *Angew. Chem., Int. Ed. Engl.*, **26**, 912 (1987).
251. W. S. Sheldrick, *Top Curr. Chem.*, **73**, 1, (1978).
252. R. S. Berry, *J. Chem. Phys.*, **32**, 933 (1960).
253. F. Ramirez and I. Ugi, *Adv. Phys. Org. Chem.*, **9**, 256 (1971).
254. P. Lemmen, R. Baumgartner, I. Ugi and F. Ramirez, *Chem. Scr.*, **28**, 451 (1988).
255. C. Macho, R. Minkwitz, J. Rohmann, B. Steger, V. Wölfel and H. Oberhammer, *Inorg. Chem.*, **25**, 2828 (1986).
256. H. Kurimura, S. Yumamoto, T. Egawa and K. Kuchitsu, *J. Mol. Struct.*, **140**, 79 (1986).
257. C. Styger and A. Bauder, *J. Mol. Spectrosc.*, **148**, 479 (1991).

258. D. Mootz and M. Wiebcke, *Z. Anorg. Allg. Chem.*, **545**, 39 (1987).
259. R. J. French, K. Hedberg, J. M. Shreeve and K. D. Gupta, *Inorg. Chem.*, **24**, 2774 (1985).
260. B. W. McClelland, L. Hedberg and K. Hedberg, *J. Mol. Struct.*, **99**, 309 (1983).
261. H. Oberhammer, J. Grobe and D. Le Van, *Inorg. Chem.*, **21**, 275 (1982).
262. N. K. DeVries and H. M. Buck, *Phosphorus Sulfur*, **31**, 267 (1987).
263. C. Dittebrandt and H. Oberhammer, *J. Mol. Struct.*, **63**, 227 (1980).
264. K. B. Dillon and J. Lincoln, *Polyhedron*, **4**, 1333 (1985).
265. D. E. J. Arnold, D. W. H. Rankin and G. Robinet, *J. Chem. Soc., Dalton Trans.*, 585 (1977).
266. D. Christen, J. Kadel, A. Liedtke, R. Minkwitz and H. Oberhammer, *J. Phys. Chem.*, **93**, 6672 (1989).
267. A. J. Downs, G. S. McGrady, E. A. Barnfield and D. W. H. Rankin, *J. Chem. Soc., Dalton Trans.*, 545 (1989).
268. L. W. Dennis, V. J. Bartuska and G. E. Maciel, *J. Am. Chem. Soc.*, **104**, 230 (1982).
269. E. S. Kozlov, I. E. Boldeskul, V. I. Karmanov, A. T. Kozulin, I. A. Kyuntsel, V. A. Mokeeva and G. B. Soifer, *Zh. Obshch. Khim.*, **52**, 2513 (1982); Engl. Transl., *J. Gen. Chem. USSR*, **52**, 2219 (1982).
270. M. Yu. Antipin, A. N. Chernega, Yu. T. Struchkov, E. S. Kozlov and I. E. Boldeskul, *Zh. Strukt. Khim.*, **28**, 105 (1987); Engl. Transl., *J. Struct. Chem.*, **28**, 723 (1987).
271. L. N. Markovskii, N. P. Kolesnik and Yu. G. Shermolovich, *Ups. Khim.*, **56**, 1564 (1987); Engl. Transl., *Russ. Chem. Rev.*, **56**, 894 (1987).
272. L. N. Markovskii, E. S. Kozlov and Yu. G. Shermolovich, *Heteroatom Chem.*, **2**, 87 (1991).
273. P. B. Kay and S. Trippett, *J. Chem. Res. (S)*, 62 (1986).
274. R. J. Gillespie, *Molecular Geometry*, Van Nostrand Rheinhold, London, 1972.
275. A. E. Reed and F. Weinhold, *J. Am. Chem. Soc.*, **108**, 3586 (1986).
276. R. J. Hach and R. E. Rundle, *J. Am. Chem. Soc.*, **73**, 4321 (1951); R. E. Rundle, *Surv. Prog. Chem.*, **1**, 81 (1963).
277. J. I. Musher, *Science*, **141**, 736 (1963); *Angew. Chem., Int. Ed. Engl.*, **8**, 54 (1969); *J. Am. Chem. Soc.*, **94**, 1370 (1972).
278. M. Grodzicki and S. Elbel, in *Modelling of Structure and Properties of Molecules* (Ed. Z. B. Maksic), Ellis Horwood, Chichester, 1987, pp. 239–250.
279. R. N. S. Sodhi and C. E. Brion, *J. Electron Spectrosc. Relat. Phenom.*, **37**, 145 (1985).
280. P. A. Cox, S. Evans, A. F. Orchard, N. V. Richardson and P. J. Roberts, *Disscuss. Faraday Soc.*, **54**, 26 (1972).
281. D. W. Goodman, M. J. S. Dewar, J. R. Schweiger and A. H. Cowley, *Chem. Phys. Lett.*, **21**, 474 (1973).
282. R. W. Shaw, T. X. Carroll and T. D. Thomas, *J. Am. Chem. Soc.*, **95**, 2033 (1973).
283. P. Wang, Y. Zhang, R. Glaser, A. E. Reed, P. v. R. Schleyer and A. Streitwieser, *J. Am. Chem. Soc.*, **113**, 55 (1991).
284. R. S. McDowell and A. Streitwieser, *J. Am. Chem. Soc.*, **107**, 5849 (1985).
285. A. E. Reed and P. v. R. Schleyer, *Chem. Phys. Lett.*, **133**, 553 (1987).
286. H. Wasade and K. Hirao, *J. Am. Chem. Soc.*, **114**, 16 (1992).
287. J. Breidung, W. Schneider, W. Theil and H. F. Schaefer, *J. Mol. Spectrosc.*, **140**, 226 (1990).
288. J. Breidung, W. Thiel and A. Komornicki, *J. Phys. Chem.*, **92**, 5603 (1988).
289. C. S. Ewig and J. R. vanWazer, *J. Am. Chem. Soc.*, **111**, 1552 (1989).
290. J. A. Dieters, R. R. Holmes and J. M. Holmes, *J. Am. Chem. Soc.*, **110**, 7672 (1988).
291. H. H. Michels and J. A. Montgomery, *J. Chem. Phys.*, **93**, 1805 (1990).
292. M. O'Keefe, *J. Am. Chem. Soc.*, **108**, 4341 (1986).
293. W. Kutzelnigg and J. Wasilewski, *J. Am. Chem. Soc.*, **104**, 953 (1982).
294. R. Hoffmann, J. M. Howell and E. L. Muetterties, *J. Am. Chem. Soc.*, **94**, 3047 (1972).
295. A. Rauk, L. C. Allen and K. Mislow, *J. Am. Chem. Soc.*, **94**, 3035 (1972).
296. F. A. Westheimer in *Rearrangements in Ground and Excited States, Part 2* (Ed. P. de Mayo), Academic Press, 1980, pp. 229–271; *Acc. Chem. Res.*, **1**, 70 (1968).
297. R. Glaser and A. Streitwieser, *J. Comput. Chem.*, **11**, 249 (1990).
298. H.-J. Bestmann, J. Chandrasekhar, W. C. Downey and P. v. R. Schleyer, *J. Chem. Soc., Chem. Commun.*, 978 (1980).
299. P. Wang, D. K. Agrafiotis, A. Streitwieser and P. v. R. Schleyer, *J. Chem. Soc., Chem. Commun.*, 201 (1990).
300. A. Streitwieser and R. S. McDowell, *J. Mol. Struct. THEOCHEM*, **138**, 89 (1986).

301. F. Volatron and O. Eisenstein, *J. Am. Chem. Soc.*, **109**, 1 (1987).
302. F. Volatron and O. Eisenstein, *J. Am. Chem. Soc.*, **106**, 6117 (1984).
303. H. Rzepa, *J. Chem. Soc., Perkin Trans. 2*, 2115 (1989).
304. F. Mari and P. M. Lahti and W. E. McEwen, *J. Am. Chem. Soc.*, **114**, 813 (1992).
305. F. Mari and P. M. Lahti and W. E. McEwen, *Heteroatom Chem.*, **1**, 255 (1990).
306. H. Yamataka, T. Hanafusa, S. Nagase and T. Kurakake, *Heteroatom Chem.*, **2**, 465 (1991).
307. R. A. Aitken, G. Ferguson and S. V. Raut, *J. Chem. Soc., Chem. Commun.*, 812 (1991); *Tetrahedron*, **48**, 8023 (1992).
308. A. D. Abell, J. Trent and W. T. Robinson, *J. Chem. Soc., Chem. Commun.*, 362 (1991).

CHAPTER **2**

Preparation, properties and reactions of phosphonium salts

H. J. CRISTAU and F. PLÉNAT

Laboratoire de Chimie Organique, URA 458, Ecole Nationale Supérieure de Chimie, Université de Montpellier, 8 rue de l'Ecole Normale, 34053-Montpellier Cedex 1, France

The chemistry of organophosphorus compounds, Vol. 3
Edited by F. R. Hartley

I. PROPERTIES OF PHOSPHONIUM SALTS

A. Introduction

Phosphonium salts discussed in this chapter are represented by the general formula $R_4P^+X^-$, derived from the tetrahedral phosphonium ion PH_4^+, where:
– R may be a radical linked to phosphorus by a carbon or another atom and can be simple or functional, cyclic or acyclic substituent and even a part of a polymeric structure;
– X may be an anionic structure, simple or complex, inorganic or organic.

Since 1972, methods for the synthesis of phosphonium salts have been reviewed exhaustively in various treatises, particularly by Beck[1] and later by Jödden[2a]. Various reviews covering restricted aspects of the synthesis and/or reactivity of phosphonium salts have also appeared[3–8]. In addition, material since 1969 has been reviewed on a yearly basis in the Specialist Periodical Reports of the Royal Society of Chemistry entitled Organophosphorus Chemistry, successively by Trippett (Vol. 1)[9], then by Smith (Vols 2–7)[10] and lastly by Allen (Vols. 8–22)[11].

This chapter is organized in three parts.

The first part covers general, theoretical and structural aspects: analysis, purification, spectroscopy, thermo-, photo- and electro-chemistry and stereochemical and biochemical aspects.

In the second part, the synthetic routes to phosphonium salts are organized according to the nature of the substituents which are introduced on the phosphorus atom: carbon substituents (alkyl, aryl, unsaturated, functional or polyvalent groups), heteroatomic groups, followed by synthesis in which the phosphorus is introduced into a cyclic structure. Finally, synthesis in which the phosphonium cation is not altered but the associated anion is exchanged or modified are covered.

In the third and last section, the main reactivities described are the acidity, the behaviour toward organometallic reagents, the alkaline hydrolysis, the reductions and lastly, for the functional phosphonium salts, the reactivities induced by the positively charged phosphorus atom and those simply compatible with a phosphonio group. The last paragraphs in this section concern the reactivities brought on by the associated anion, either in phase-transfer catalysis or in homogeneous solution.

In this chapter, specific paragraphs are not devoted to the zwitterionic structures, such as betaines, or to the applications of phosphonium salts in organic synthesis, but both topics are mentioned incidentally.

B. Analysis and Purification

Phosphonium salts, which are generally stable in the form of iodide and tetraphenylborate crystalline solids, are frequently purified by recrystallization. Where this is not possible, thick-layer chromatography (0.2–1 mm Kieselguhr, silica) may afford a satisfactory method of purification[12,13]. In this manner, analysis using secondary ion mass spectrometry (SIMS) has a detection limit 10–50 ng per spot[14].

Liquid column chromatography on silica gel has been successfully used to purify a long-chain alkyltributylphosphonium methanesulphonate[15] where the vinylic phosphonium salts reacted with methanol during column chromatography[16]. The addition of HCOOH to eluents can make the chromatography of phosphonium salts on silica feasible on both an analytical and a preparative scale[17]. A quantitative HPLC method for the determination of mixtures of *cis*- and *trans*-β-ionylidenethyltriphenylphosphonium chloride has been studied: the cations were separated on reversed-phase columns using a cationic detergent mobile phase[18]. Ion-pair chromatography (HPLC) on an ODS column, filled with Inertsil, is recommended for optimum separation of mono- or bis-phosphonium salts, with either UV or conductivity detection; the eluent composition ($HCl-CH_3CN$) has a fairly dramatic effect on retention[19]. When pyrolysis–gas chromatography was used for the determination of quaternary phosphonium compounds, marked differences in behaviour appeared, depending on the nature of the anionic part (OH^-, Cl^-, I^-) of the salts[20].

A pH-controlled process for obtaining high-purity quaternary phosphonium chlorides has been patented[21]. An electrolytic process may also be used with the same aim in the case of phosphonium hydroxides[22]. Titration of water-soluble tetrakis hydroxymethylphosphonium chloride (**1**) with iodine in the presence of weak bases can be used for the analysis of this compound[23].

$$\underset{(\mathbf{1})}{(HOCH_2)_4P^+\,Cl^-} + I_2 + NaHCO_3 \longrightarrow (HOCH_2)_3P{=}O + CH_2O + 3\,CO_2 + NaCl + 2\,INa + 2\,H_2O \qquad (1)$$

C. Spectrometry

1. NMR spectrometry

a. ^{31}P NMR spectra. ^{31}P NMR spectra of phosphonium salts are most often obtained using FT-NMR spectrometers (T_1 value $R_4P^+X^-$ 2–9 s, $Ph_3P^+Me\,I^-$ 10.8 ± 0.2 s at 25 °C[24]); broad-band H decoupling (^{31}P–[1 H]) is usual, except in very special cases where narrow-band decoupling was used[25]. The solvents commonly used are $CHCl_3$, CH_3CN, dmso, etc., and their deuterated analogues. The ^{31}P signals seem unaffected by H–D exchange[26], but a dependence of ^{31}P chemical shifts on concentration, solvents and the presence of other compounds may be encountered, variations most often being within 1 ppm. It has been proved that 25-oleum (25 g SO_3 and 75 g H_2SO_4) is an extremely useful solvent for recording ^{31}P NMR spectra of a variety of halogenated and organic compounds

TABLE 1. ^{31}P NMR data for some phosphonium ions with aliphatic substituents in 25-oleum solution

Compounds	Ions	$\delta\,^{31}P$ (ppm)	$\delta\,^{31}P$ in solid (ppm)
$MePCl_4$	$MePCl_3^+$	120	119 ± 2
$MePCl_3^+ AlCl_4^-$	$MePCl_3^+$	120	117 ± 1
Me_2PCl_3	$Me_2PCl_2^+$	123	124 ± 5
$Me_2PCl_2^+ BCl_4^-$	$Me_2PCl_2^+$	123	119.5 ± 2
$Me_2PCl_2^+ SbCl_6^-$	$Me_2PCl_2^+$	123	123 ± 4
$Me_3PCl^+ BCl_4^-$	Me_3PCl^+	90	87 ± 1
$EtPCl_3^+ AlCl_4^-$	$EtPCl_3^+$	129	124.4 ± 1
$Et_2PCl_2^+ BCl_4^-$	$Et_2PCl_2^+$	138	137.4 ± 2
$Et_3PCl^+ BCl_4^-$	Et_3PCl^+	108	105 ± 2

containing phosphonium ions; both the solid-state and solution results[27] are in good aggreement (cf. Table 1).

Chemicals shifts are generally reported relative to the signal for 85% phosphoric acid which is used as an external reference and the sign convention, established in the mid-1970s, attributes a positive value downfield of the standard. Some authors refer to other standards such as trimethyl phosphite[28,29] at 141 ppm or $N_3P_3Cl_6$ at 19.9 ppm[30]; usually, the ^{31}P chemical shifts obtained were then converted to 80% H_3PO_4 at 0.00 ppm.

Empirical relationships allowed predictions of ^{31}P chemical shifts with a fair degree of accuracy[31]. The basic features, compilation and interpretation of experimental ^{31}P NMR chemical shifts in phosphonium salts were reviewed in 1967[32] and were recently significantly extended[33]. Their uses in stereochemical analysis were described in 1987[34]. According to the data, chemical shifts are usually found between -100 and $+150$ ppm (Figure 1). Considering the most stable $(C)_4P^+$ species, they are found between -25 and $+125$ ppm. Negatives values are less frequently encountered; they correspond to polyalkynyl compounds.

On the whole, compared with P^{III} compounds, quaternization of phosphorus produces a large ^{31}P downfield chemical shift which parallels the bond angle changes, including those due to the steric effects coming from phosphorus substituents (shielding γ-effect) (cf. Table 2).

The greater electronegativity of C_{ipso} in an aryl group as compared with an alkyl group (or other aryl groups) gives an upfield chemical shift, qualitatively in accordance with the possibility of the formation of dπ–pπ bonds[38]. The phosphorus resonances of

TABLE 2. ^{31}P NMR chemical shifts (ppm) of phosphonium and aminophosphonium halides[35,37] (also steric compression was shown to give upfield ^{31}P chemical shifts in cycloalkyltriphenylphosphonium salts)

Ph_3P	-4.8	Me_3P	-62
Ph_3P^+Me	21.1	Me_3P^+Me	25.3
Ph_3P^+Et	25.5	$Me_3P^+NMe_2$	62.5
$Ph_3P^+Pr^n$	24.1	$Et_3P^+NMe_2$	72.0
Ph_4P^+	23.2		
$Ph_3P^+NH_2$	36.1	$Pr_3P^+NMe_2$	66.0
Ph_3P^+NHMe	39.3		
$Ph_3P^+NMe_2$	47.5		

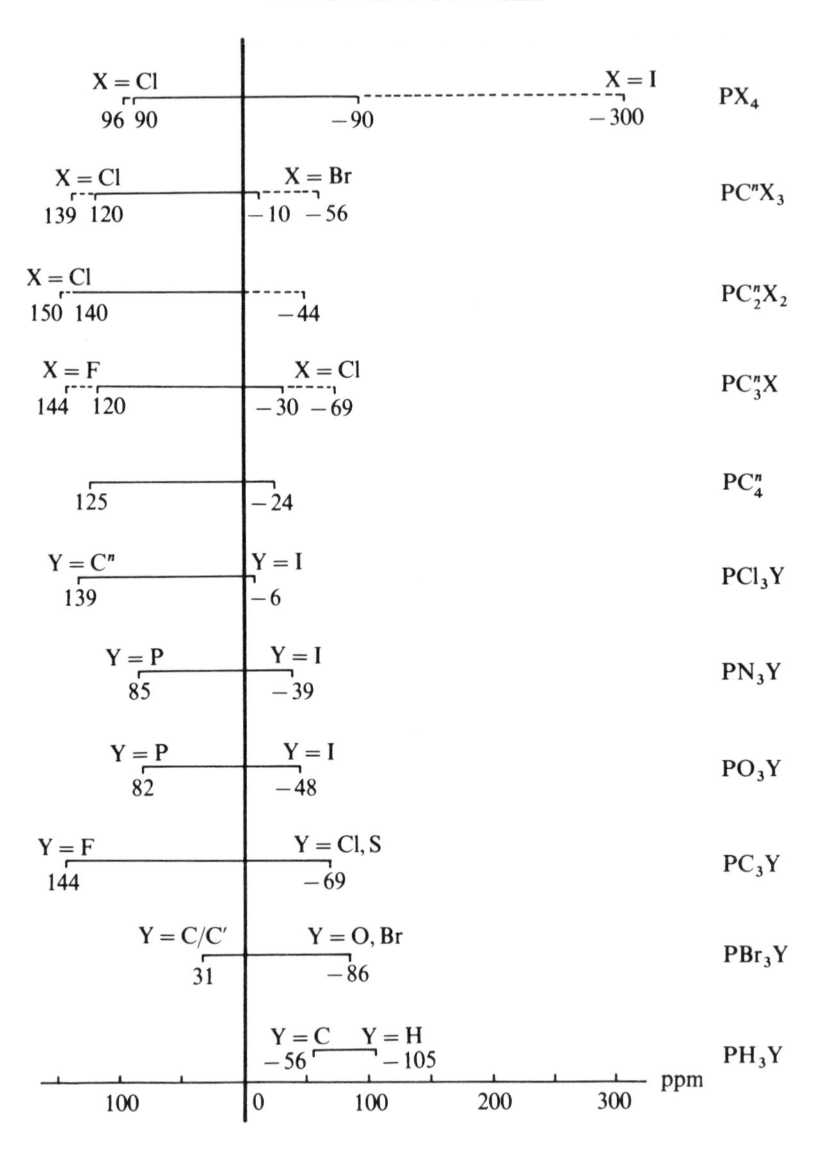

P-bound atoms are given above each bar line.
X = atoms other than carbon
Y = any atom
C'' = carbon of all coordination states

FIGURE 1. ^{31}P NMR chemical shifts of four-coordinate ($\lambda^5\sigma^4$) phosphonium salts (and betaines)

TABLE 3. ^{31}P chemical shifts (ppm) of unsaturated phosphonium salts

$Ph_3P^+CH_2CH_2CH_3Br^-$	24.1
$Ph_3P^+c\text{-}C_3H_5\,Br^-$	28.5
(E)-$Ph_3P^+CH{=}CHCH_3\,Br^-$	18.7
$Ph_3P^+C{\equiv}CCH_3\,Br^-$	5.3
$Ph_3P^+CH_2CH{=}CH_2\,Br^-$	20.8
$Bu^n_3P^+CH_2CH{=}CH_2\,Br^-$	31.6
$Ph_3P^+CH_2C{\equiv}CH\,Br^-$	21.3

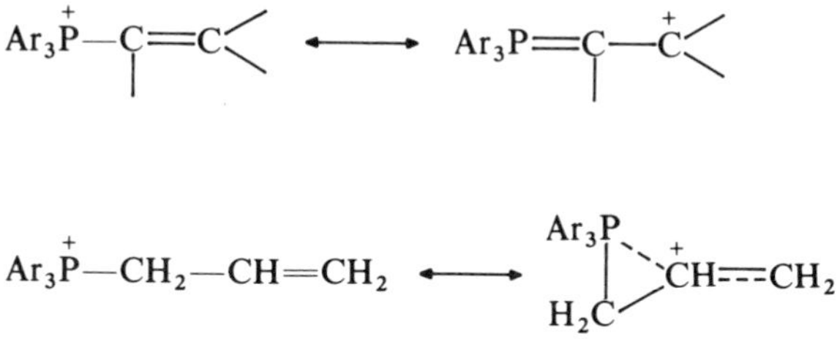

$$Ar_3\overset{+}{P}-\underset{|}{C}{=}C\!\!< \longleftrightarrow Ar_3P{=}\underset{|}{C}-\overset{+}{C}\!\!<$$

$$Ar_3\overset{+}{P}-CH_2-CH{=}CH_2 \longleftrightarrow \text{(cyclic: } Ar_3P\cdots\overset{+}{C}H{=}{=}{=}CH_2,\ H_2C\text{)}$$

FIGURE 2. Resonance structures in α,β- and β,γ unsaturated phosphonium salts

α,β-unsaturated phosphonium salts are also consistently shielded with respect to the saturated ones. This is due to the increased electron density on phosphorus via $d\pi$–$p\pi$ bonding; when the unsaturation is β to phosphorus, the observed shielding was attributed to overlap of the $p\pi$–$d\pi$ type through resonance structures[39] (cf. Table 3 and Figure 2).

α-, β- and γ-effects (no substituent effects beyond the γ-carbon atom) are operative in ^{31}P chemical shifts of alkyltriphenyl-[36] and tetraalkylphosphonium salts.[29,35] Hyperconjugative π bonding, involving the α-C—H bonds, and direct β-atom interaction with the 3d orbitals of phosphorus are both invoked[40].

In the case where tautomeric forms might exist (Figure 3), ^{31}P-[^1H]NMR spectra are reasonably able to confirm or deny the presence and the nature of any tautomer in the solutions under consideration[41]. In contrast, the proximity of γ-, δ-,...ω-functional groups on the alkyl chain relative to phosphorus seemed without any strong influence on ^{31}P chemical shifts[42].

$$Ph_3P^+(CH_2)_nCOOH\ X^- \quad (n = 2, 3, 5, 10){:}\quad \delta\ ^{31}P\ (CDCl_3) = 23.9\text{–}24.5\ \text{ppm}$$

The ^{31}P values for a series of o-dimethylamino-substituted benzyltriarylphosphonium halides (increasing upfield shift of P resonances with increasing electron density at

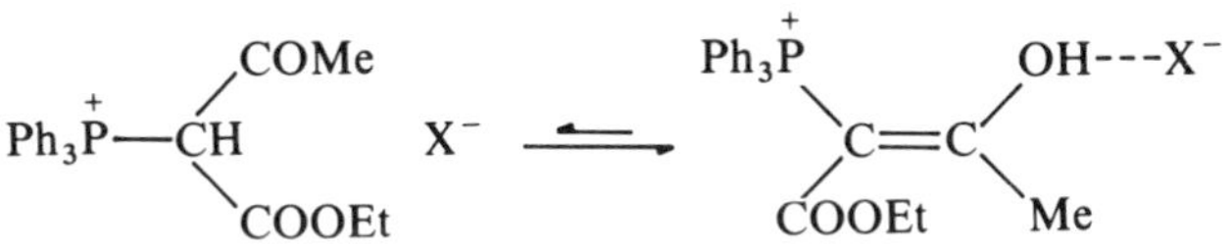

$$Ph_3\overset{+}{P}-CH(COMe)(COOEt)\ X^- \rightleftharpoons Ph_3\overset{+}{P}(COOEt)C{=}C(Me)OH\text{---}X^-$$

(**2**) (**3**)

$X = Cl, Br, CF_3COO$ $\delta\ ^{31}P(CHCl_3)$: 16.7–17.2 ppm

FIGURE 3. Hydrogen bonds in enol forms of α-(triphenylphosphonio)-substituted acetoacetic ester

phosphorus) have shown that the phosphorus atom is more shielded than in the corresponding *p*-dimethylamino compounds; this was explained on the basis of a through-space N_{2p}–P_{IV} interaction for the former[43] (field effect). For *para*-substituted compounds there would be an inductive effect superimposed on the resonance interaction between the substituent and the ring to which it is attached. Concerning this *para* effect, the ^{31}P chemical shifts of *N*-methyl-*N*-arylaminotriphenylphosphonium salts were shown to be unaffected by changing the substituents in the *para*-position on the *N*-aryl ring, which reflects no through conjugation[44] (Table 4).

Heterophosphonium species resonate downfield, although a few have marked upfield shifts (cf. Figure 1). The range of recorded chemical shifts for oxyphosphonium species varies from −26 ppm [$(PhO)_4P^+$ Br^- to +108 ppm [Et_3P^+OMe $SbCl_6^-$]. There is no simple correlation between substituent effects and chemical shifts, although the values for all cations of the same type generally lie within a narrow range. The disproportionation equilibria between oxyphosphonium salts through phosphoranes can be measured by ^{31}P NMR[45]. However, some care should be taken as the observed chemical shifts have been sometimes misleading[46] compared with the usual rule which gives positive values for phosphonium species (or betaines) and negative values for phosphoranes. Aminophosphonium salts resonate between 28 ppm [$(NH_2)_4P^+Cl^-$] and 95 ppm ($cHex_3P^+NMe_2$ Cl^-). In halophosphonium ions, replacement of Cl by Br caused a marked upfield shift which decreased gradually as organic groups become bound to phosphorus; the following values (ppm) were quoted[33] in the solid NMR spectra of halophosphonium tetrachloroborates:

Br_4P^+	Br_3P^+Cl	$Br_2P^+Cl_2$	BrP^+Cl_3	Cl_4P^+
−66	−29.5	5.8±2	49	85.5

In such a class of compounds, the technique has enabled the acceptor properties of halophosphonium salts toward Lewis bases[47,48] (Cl^-, pyridines) to be investigated and to offer convincing evidence that species did exist either as ionic (free ions or solvent-separated ion pairs), contact ion pairs or molecular forms, which may depend on solvent and temperature[27,49–52]. Fluorophosphonium salts may be highly deshielded ($Me_2PF_2^+$ PF_6^- 155 ppm)[53]; $^1J_{P-F}$ values are about 1000 Hz. Normally, $^2J_{P-F}$ in fluorinated phosphonium salts is much smaller and seems to be correlated with electronegativities of X, Y substituents as in R_3P^+CFXY Z^-[54,55].

TABLE 4. ^{31}P chemical shifts and Hammett constants of *N*-methyl-*N*-arylaminotriphenylphosphonium salts in $CHCl_3$[44]

$Ph_3P^+NMe-C_6H_4-X$

X	σ_p	$\delta\ ^{31}P$ (ppm)
OMe	−0.27	45.7
CH_3	−0.17	45.2
H	0.00	45.4
Cl	+0.23	45.7
Br	+0.23	45.5
I	+0.28	45.5
CO_2Et	+0.45	45.2
CN	+0.66	45.5
COMe	+0.50	45.1
NO_2	+0.78	45.5

Of course, phosphorus–phosphorus couplings do appear in unsymmetrical polyphosphonium salts ($^1J_{P-P} = 219$ Hz[56]; $^3J_{P-P} = 36$ Hz[57]; $^5J_{P-P} = 5.5$ Hz[58]), phosphinophosphonium salts ($^1J_{P-P} = 250$–350 Hz[56,59–63], but possibly lowered by ring strain[54,64,65]; $^2J_{P-P} = 30$–80 Hz[66–68]), and phosphonio-phosphonium ylides ($^1J_{P-P} = 440$–570 Hz[69,70]; $^2J_{P-P} = 10$–20 Hz[62,71]). $^4J_{P^+-P}$ values of *ca.* 4 Hz were found[72] in complexes such as $F_5C_6M(R_3P)_2(S_2CPR'_3)\ ClO_4^-$.

b. ^1H NMR spectra. Most of the investigations concerning ^1H NMR spectra of phosphonium salts refer to couplings with phosphorus. Quaternization of phosphorus usually produces the expected downfield shilt in δ_H for all the protons of groups attached to phosphorus, as exemplified in Figure 4[37,50,73,74]. As expected, bridge protons of symmetrical methylenebisphosphonium salts underwent substantial upfield shifts (2.5–4.8 ppm) after semi-ylidation[26]. Temperature dependence was of course important in deciding between tautomers and isomers[75–77]. Lastly, ion-pair interaction rather than hydrogen bonding has been invoked to explain the temperature and concentration dependence of δ_{NHR} in secondary aminophosphonium compounds[37].

$^1J_{P-H}$, $^2J_{P-H}$ and $^3J_{P-H}$ values are listed in ref.[33]. $^1J_{P-H}$ are usually higher than 430 Hz but $^1J_{P-D}$ was found to be 77 Hz[78]. ^1H NMR studies on the protonation of tertiary phosphines[79] or phosphonium halides[80] showed, in connection with the presence or not of a P–H doublet ($^1J_{P-H} = 480$–540 Hz), that a fast proton exchange between the competing Lewis bases R_3P and Br^- exists.

From a review on geminal $^2J_{PC-H}$ couplings (usually 12–17 Hz in $\equiv P^+CH_3$[3]), substitution at the α-carbon by an electronegative group should diminish the magnitude of the coupling[81]. In silicon-containing phosphonium salts, the larger values (ca 18 Hz) compared with those for neopentyl analogues (ca 13 HZ) are attributed to hyperconjugation of the $p\pi$–$d\pi$ type including the H_α and the 3d orbitals of silicon and phosphorus[40,82]. ^1H NMR of alkylene- and arylene-bisphosphonium salts, with regard both to the

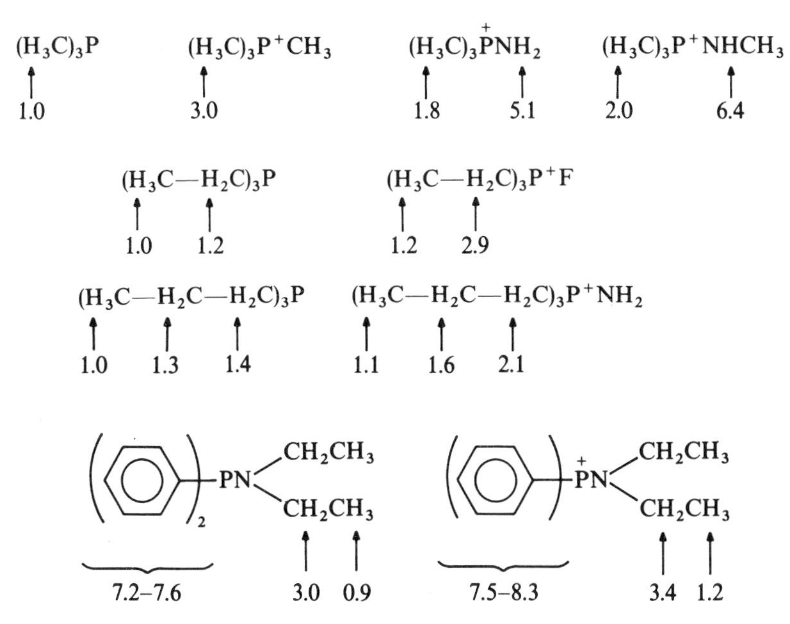

FIGURE 4. δ^1H of phosphonium salts as compared with phosphines

substituent protons (virtual coupling) and bridge protons, may give very complex or deceptively simple splitting[83,84]. In α,β-ethylenic phosphonium salts, high values have been found in the case of the diheterophosphonium salt **4**[85] (as compared with others[86,87]).

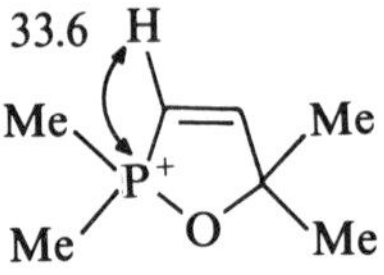

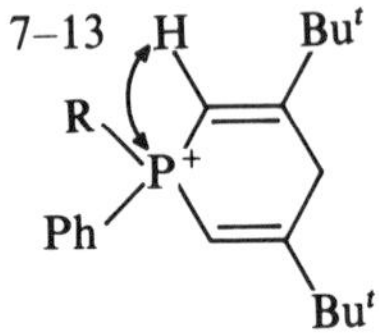

33.6 H Me Me P+ O Me Me 7–13 H Bu^t R P+ Ph Bu^t

(4)

Oversimplified spectra were found in the case of symmetrical vinylidene bisphosphonium salts[87,88]. Vicinal $^3J_{P-X-C-H}$ values are sometimes higher than geminal, in saturated species. In unsaturated phosphonium salts, their magnitude[89] was used as strong evidence in the assignment of the E(30–50 Hz) or Z(10–25 Hz) relationship[16,87,90,91] (as low as 23 Hz was found, however, for a *trans*-P—H relationship in heterocyclic structure[92]).

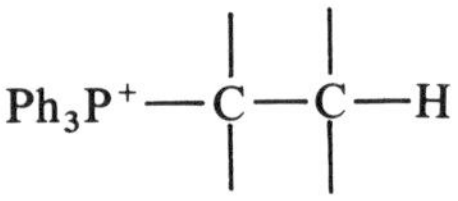

Ph_3P^+—C—C—H Ph_3P^+—N—C—H Ph_3P^+—S—C—H

$^3J_{P-H} = 14–19$ Hz 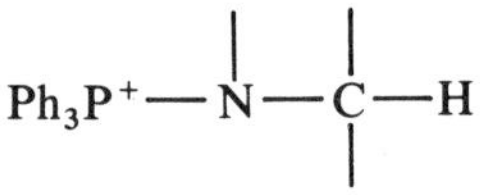 ~10 Hz 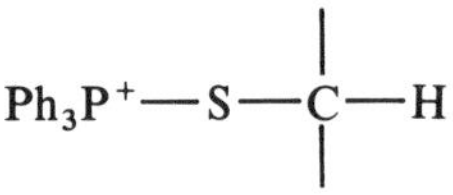 9–15 Hz

Ph_3P^+ 31.5 OPh 14.5 C=C H H 18 Ph_3P^+ H 18 C=C H SPh

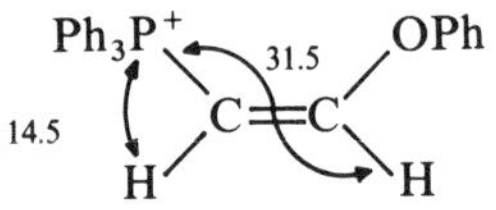

(5) 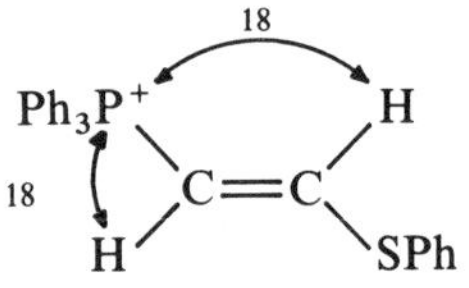 **(6)**

^{1}H NMR of the anionic counterpart has been used to evaluate the extent of the halogen exchange in $R_3P^+R'\ HX_2^-$ (reaction 2).[80]

$$R_3P^+R'\ \ HX_2^- \underset{X=Cl,Br}{\overset{X=I}{\rightleftharpoons}} R_3P^+R'\ \ X^- + HX \qquad (2)$$

c. *^{13}C NMR spectra.* Investigation of ^{13}C resonances for phosphonium salts started in the 1970s[35,36,39,93]. A collection of data appeared recently[33].

Bearing in mind that chemical shifts are usually not dependent on the solvent or concentration, but may be partly dependent on the anion[37], quaternization of alkylphosphines causes an upfield shift (1–7 ppm) for the near carbons of the alkyl chain, the effect becoming only slightly detectable on $C_{(\gamma)}$ carbons[37,94]. Where triarylphosphines are quaternized, the chemical shift of the phenyl *ipso* carbon C_i moves upfield whereas that of the C_{para} carbon moves downfield, as expected from the polarization of the π-electron density in the phenyl ring[36,37]. For the aromatic carbon atoms, the increments

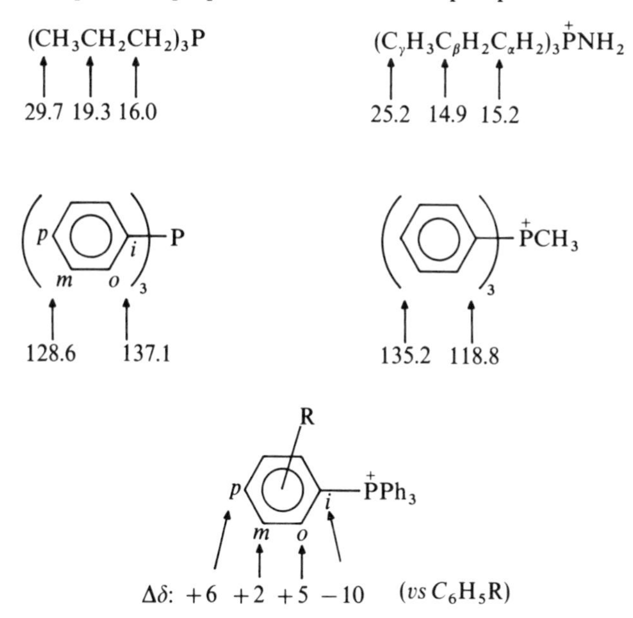

FIGURE 5. ^{13}C NMR of phosphines and phosphonium salts

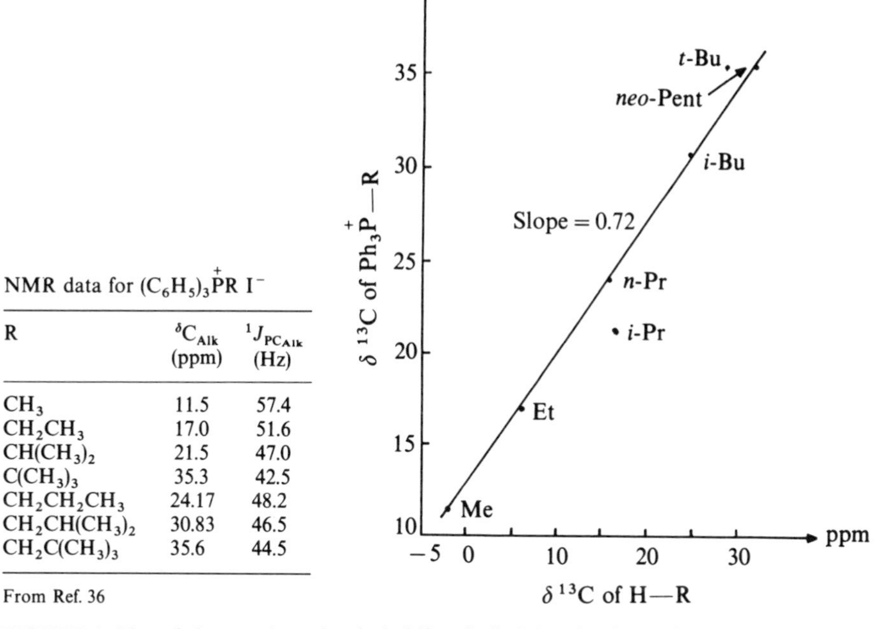

NMR data for $(C_6H_5)_3\overset{+}{P}R\ I^-$

R	$^{\delta}C_{Alk}$ (ppm)	$^{1}J_{PC_{Alk}}$ (Hz)
CH_3	11.5	57.4
CH_2CH_3	17.0	51.6
$CH(CH_3)_2$	21.5	47.0
$C(CH_3)_3$	35.3	42.5
$CH_2CH_2CH_3$	24.17	48.2
$CH_2CH(CH_3)_2$	30.83	46.5
$CH_2C(CH_3)_3$	35.6	44.5

From Ref. 36

FIGURE 6. Plot of the α-carbon chemical shifts of alkylphosphonium salts vs those in H—R alkanes[40]

due to the triphenylphosphonio group are approximately those given in Figure 5 as compared with benzene itself.

When compared with H–R alkanes (or cycloalkanes), a triphenylphosphonio group gives a deshielding constant effect on the α-carbon atom[36,39]. However a plot of the α-carbon chemical shifts versus those of carbon directly bonded in H–R alkanes, including the branched ones, gave a correlation with slope < 1[40] (Figure 6). This means that the deshielding effect of the quaternary phosphorus on the α-carbon decreases with increasing branching of alkyl groups. This has been attributed to either steric hindrance[36,39] or to an increase in hyperconjugative pπ–dπ bonding with more substituents on α- or β-carbons[40].

It shall be noted that in the case of tri- or tetra-alkylphosphonium salts, 'anomalous' shifts appear when three or four substituents change simultaneously (Figure 7): contributions from intra- and inter-atomic induced currents are suggested in such a case[29,37]. In the very particular case of a 1-triphenylphosphoniomethylene- (or alkylidene)-triphenylphosphorane, such as **7**, the $C_{(\alpha)}$ atom is strongly shielded compared with the symmetrical bisphosphonium salt **8**, according to the charge stabilization.[26,95]

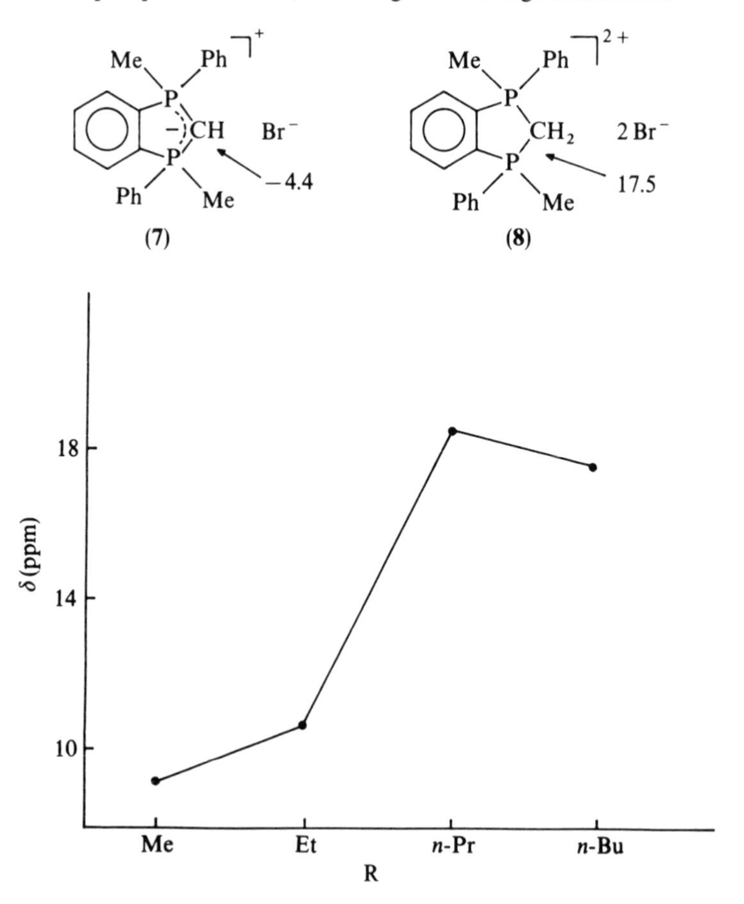

FIGURE 7. Variation of ^{13}C chemical shift of $C_{(\alpha)}$ in P^+R_4 with alkyl chain length

A notable deshielding β-effect (10 ppm) has been observed in an aminophosphonium salt on the P^+NMe ^{13}C chemical shift when the NHMe group is further alkylated to NMe_2[37]. Steric compression has been used, with some caution, to explain the relatively shielded β-carbon atoms in triphenylphosphonio cycloalkyl salts when they are compared with their methylcycloalkyl analogues[36].

In vinylic phosphonium salts, the β-carbon is found to be substantially deshielded ($\Delta\delta = 22.4$ ppm for the vinyltriphenylphosphonium bromide as compared with ethylene), although only a small inductive effect was expected. The extent of this was interpreted to be a consequence of pπ-dπ bonding between phosphorus and carbon[39]. This argument was used to identify structures when ^1H or ^{31}P NMR spectra failed[96].

$$\underset{}{Ph_3P^+}-\underset{119.2}{CH}=\underset{145.2}{CH_2} \quad Ph_3P^+-\underset{110.1}{CH}=\underset{159.5}{CH}-\underset{21.7}{CH_3} \quad Ph_3P^+-\underset{28.6}{CH_2}-\underset{126.3}{CH}=\underset{123.1}{CH_2}$$

The same situation is encountered with α,β-alkynic phosphonium salts in which $C_{(\alpha)}$ carbons are strongly shielded vs $C_{(\beta)}$ carbons: 47.8–73.0 vs 119–127 ppm[36,97]. The linear correlation of the $C_{(\alpha)}$ and $C_{(\beta)}$ chemical shifts with a *para* Hammet constant was illustrative of the polarization of the π-electron system in the C≡C triple bond (Figure 8). In the same sense, the observed $C_{(\alpha)}$ and $C_{(\beta)}$ chemical shifts were used to argue for the existence of either resonance hybrid structures as in the case of enaminophosphonium salts[98,99] or incipient ylide contribution[40].

Coupling constants. Although it seems that there is no correlation of directly bonded C–H couplings with corresponding P–C couplings, there are several systematic trends as deduced from the examination of alkyltriphenylphosphonium salts Ph_3P^+R (Table 5)[36]. In the case of saturated alkyl groups R, the unusual $^1J_{P-C} > {}^3J_{P-C} > {}^2J_{P-C}$ order is found. $^1J_{P-C}$ values decrease regularly on going from methyl to *tert*-butyl; this has been connected with the decreasing percentage of s character in the hybrid orbital comprising the P—C bond. Both electronic and steric factors were considered to have some influence in that respect (substitution of a more electronegative element than carbon for R increased $^1J_{P-C_{ipso}}$ further).

This is also evident from the large to very large $^1J_{P-C}$ in cyclopropylphosphonium salts[36,100] in which P—C bonds are expected to be approximately sp^2, and in alkenyl- and alkynyl-triphenylphosphonium salts (the change in the magnitude of $^1J_{P-C_{ipso}}$ was more important there). The increase in $^1J_{P-C}$ with more electron-donating *para*-substituents, in Ar_3P^+R, has been also explained by the change in the hybridization of phosphorus[40]. In the particular case of heterocyclic phosphirenium salts, $^1J_{P-C_{(\alpha)}}$ were

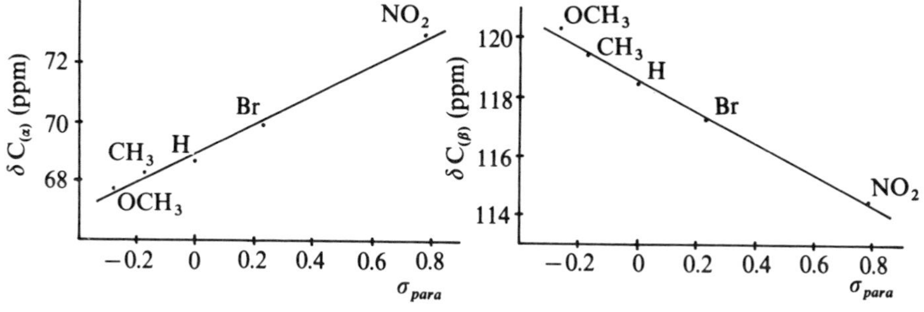

FIGURE 8. Correlation of the $C_{(\alpha)}$ and $C_{(\beta)}$ chemical shifts with σ_{para} in $(C_6H_5)_3P^+C_{(\alpha)}\equiv C_{(\beta)}p\text{-}C_6H_4R$[97]

TABLE 5. ^{31}P–^{13}C coupling constants for C_6H_5—P^+—C_α—C_β—C_γ[36,39]

Compound	^{31}P–^{13}C coupling constant (Hz)							
	α	β	γ	δ	*i*	*o*	*m*	*p*
$Ph_3P^+Me\,I^-$	57.1				88.6	10.7	12.9	3.0
$Ph_3P^+Et\,Br^-$	51.6	5.1			86.2	9.9	12.4	2.9
$Ph_3P^+Pr^i\,Br^-$	47.0	2.0			83.1	9.2	12.1	3.0
$Ph_3P^+{}_tBu^t\,I^-$	42.5	0.6			80.1	8.8	12.0	3.0
Ph_3P^+c-Pr Br^-	86.9	4.3			89.6	9.8	12.6	2.8
Ph_3P^+c-Bu Br^-	45.0	2.8	16.8		85.0	9.6	12.3	2.9
Ph_3P^+c-Pen Br^-	49.2	—[a]	8.9		84.5	9.5	12.1	2.7
Ph_3P^+c-Hex I^-	45.4	1.7	14.3	—[a]	83.0	9.5	11.5	—[a]
Ph_3P^+c-Hep Br^-	42.6	—[a]	15.8	—[a]	82.8	9.4	12.0	—[a]
$Ph_3P^+CH{=}CHMe\,Br^-$	86.1	2.6	19.9		90.7	10.5	12.9	2.9
$Ph_3P^+CH_2CH{=}CH_2\,Br^-$	49.7	13.4	9.8		85.9	9.8	12.4	2.6
$Ph_3P^+C{\equiv}CMe\,Br^-$	191.7	33.0	3.8		100.5	12.4	14.2	3.0
$Ph_3P^+NHPh\,Br^-$					102.5			

[a]Unresolved coupling.

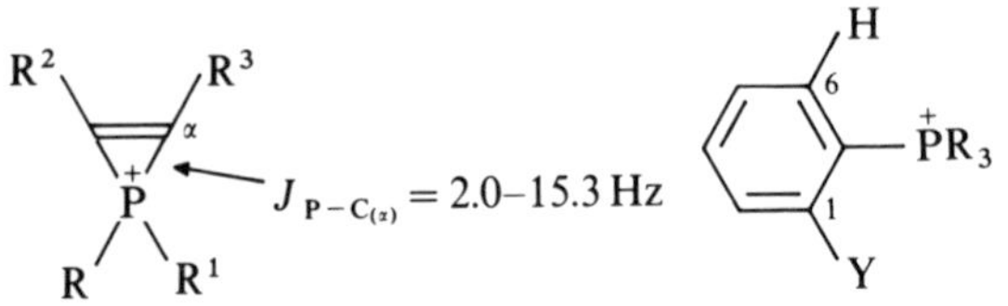

$J_{P-C(\alpha)} = 2.0$–15.3 Hz

R = Cl, Et, Ph
R^1, R^2, R^3 = H, Me, Et, Ph

Y = N=NAr, —CH=NAr
$^2J_{P-C_1} = 2$–5, $^2J_{P-C_6} = 7$–14 Hz

FIGURE 9. J_{P-C} in phosphirenium salts and *ortho*-donor substituted arylphosphonium salts

found to be of the low magnitude expected from a strained bond with high p character and predicted by calculations[28,30] (Figure 9). Finally, the high value of $^1J_{P-C}$ was an indication of an ylidic character of the α-carbon atom[26,95]. Unsaturated substituents *ortho* to phosphorus in the phenyl ring of arylphosphonium salts have been observed to have a marked effect on the value of $^2J_{P-C}$ involved[101,102] (Figure 9).

2. *Infrared spectra*

Bond stretching infrared correlations of P^+—Z(Z = C, N, O, X) bonds can extend to a wide range of values which very often overlap with the peaks of many other infrared spectra[4,103]. Among those correlated with Ph—P^+[1440(m to s), 1100(vs), 1000(w to m, sharp) and 730 (s to vs) cm^{-1}], the key bond around 1100 cm^{-1} may means of controlling the quaternization or not of phosphino groups in polymeric materials[104] (Figure 10).

Some tetralkylphosphonium salts can also absorb in the characteristic ranges very close to those quoted above. P^+—N bonds are visible near 760 and 700 cm^{-1} (m to s)[73].

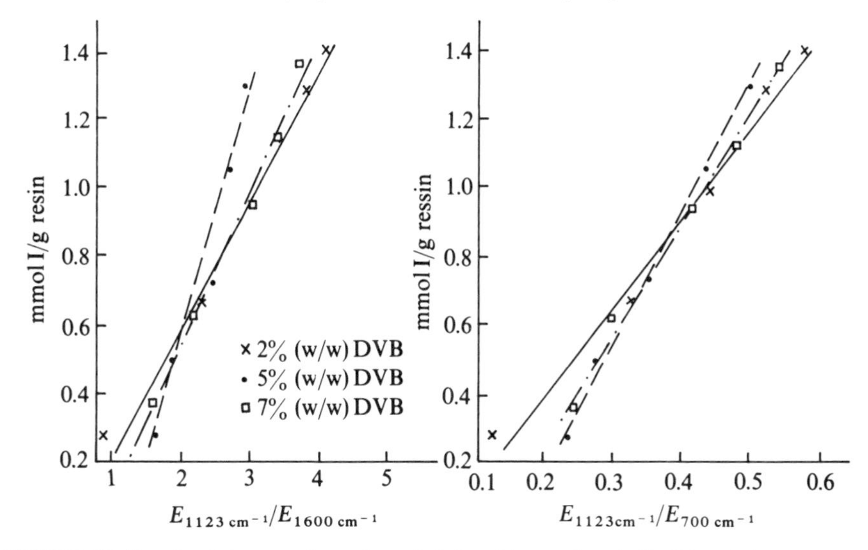

FIGURE 10. Reaction between polymer-bounded phosphines and methyl iodide: relative extinction of 1123 cm^{-1} phosphonium bond in comparison with the iodine concentration in copolymers (styrene–divinylbenzene)[104]

Intracyclic P^+—O—C bonds were found[85] at about 970 cm^{-1}. Infrared spectrometry has been used in the analysis of compounds of general formula R_4PF, which may exist in both ionic (phosphonium fluorides) and molecular (tetraorganofluorophosphoranes) forms, depending on their physical state[105]. Apart from the characterization of the phosphonio group itself, infrared spectrometry is used usually to identify other functional groups present in the cationic part, as well as in anionic counterparts[80,106,107].

3. *UV–visible spectra*

In addition to the absorption corresponding to the anionic part (tetraphenylborate, iodide, etc.), the electronic absorption spectra of phosphonium salts are typical of those of the independent chromophoric group present in the cationic part[108]. However, as the phosphorus atom is positively charged, its 3d orbitals become able to overlap with any nearby π orbital; this d–π interactions leads to bathochromic and hyperchromic

TABLE 6. Characteristics of the UV band λ_3 of p-$Me_2NC_6H_4Z$ (from[51])

	Dioxane		CH_3CN		$C_2H_4Cl_2$	
Z	λ(nm)	$\varepsilon \times 10^{-3}$ $(l\,mol^{-1}\,cm^{-1})^a$	λ(nm)	$\varepsilon \times 10^{-3}$ $(l\,mol^{-1}\,cm^{-1})^a$	λ(nm)	$\varepsilon \times 10^{-3}$ $(l\,mol^{-1}\,cm^{-1})^a$
$\overset{+}{P}(Me)(NEt_2)_2\,Cl^-$	292	19.2	295	23.7	303	22.0
$\overset{+}{P}(NH_2)(NEt_2)_2Cl^-$	283	28.0	288	26.6	287	27.9
$\overset{+}{P}(Cl)(NEt_2)_2\,Cl^-$	280	14.0	316	23.3	318	23.7

[a]Molar absorptivity.

effects in the UV absorption of the isolated chromophore[58,108–111]. There is no indication of a through-conjugation across phosphorus. As shown for [*p*-(dimethylamino)phenyl]-phosphonium salts, the π-acceptor effect of the phosphonium $^{+}P(Y)(NEt_2)_2$ group changes appreciably with Y[66]. The intramolecular transfer of charge from donor to acceptor on excitation is indicated by solvatochromic shifts[51,108]. Depending on both Y and the solvent, free ions and contact ion pairs were detected[51] (Table 6). UV spectrometry (phsophonium salt versus phosphonium ylide) may be used for the determination of the pK_a values of the phosphonium species[112].

4. *Mass spectrometry*

Field-desorption mass spectrometry (FDMS), where no evaporation prior to ionization is required, has been successfully used in the analysis of involatile phosphonium salts[113], although a direct thermal process gave similar spectra[114]. In the case where the FD spectra are complex, a chemical ionization technique may give wider applicability[115]. The cation is the base peak for monophosphonium salts when the $[2M + \text{anion}]^+$ cationic species is the one for bisphosphonium compounds.

Ionization may also be attained by the fast atom bombardment (FAB) method, which is useful in both negative- and positive-ion modes, and appears to be simpler and more sensitive than FD. Laser desorption with FT-MS detection[116] and positive-ion FAB desorption[116,117] are equally useful in determining the molecular weights of phosphonium ions; the relative abundances of fragment ions are higher in the former case. As consequence of xenon atom bombardment, reduction of bisphosphonium salts is a common process in FAB mass spectra. In both cases, high-energy collisional activation dissociation (CAD) is needed to complete the structural information. However, the mechanism of fragmentation may be deduced from low-energy CAD[118]. Finally, secondary ion mass spectrometry (SIMS) has been used for direct analysis from a sorbitol matrix of phosphonium salts separated by TLC[14].

5. *X-ray spectra*

In tetraalkylphosphoniums salts, the $P^+—C_{sp^3}$ bond length may become longer than standard dimensions, (ca 1.80 Å), reflecting steric crowding in the cation ($P^+—Pr^i \approx$ 1.82–1.83[119]; $P^+—Bu^t \approx 1.92$ Å[120]). Again for steric reasons, the tetrahedral geometry around the phosphorus atom may be distorted; the PCC angles may become greater (ca 113°) than the tetrahedral angle whereas the CCC angles may be compressed up to 106°. Twist between such bulky substituents as Bu^t attains 14°, leading to lower symmetry. In cases of α-functionalized alkyl chains, the $P^+—C(Z)$ distance significantly exceeds the usual $P^+—C_{sp^3}$ bond length (ca 1.9 versus 1.8 Å[75,121]), which may reflect the incipient donor–acceptor character of the bond, i.e., an $n \rightarrow \sigma^*$ interaction of the unshared electron pair in Z with the antibonding orbital of the $P^+—C$ bond (if $Z = NR_2$, there is no direct interaction between N and P^+).

$P—C_{sp^3}$ distances in phosphorus heterocycles are affected by strain and steric effects; endocyclic bonds may be shorter than exocyclic bonds (1.78 versus 1.80 Å[66]). In phosphetanium halides, the phosphetane ring is puckered with dihedral angles of 34°[122,123]. In five-membered phospholanium halides, the heterocycle conformation varies from twist to envelope [124,125], which probably indicates the small difference in energy between their conformations. In analogous compounds containing an endocyclic β–γ double bond[126] or a cyclopropane moiety[127], puckering is retained; if there is α–β unsaturation the ring may be considerably flattened[128]. In six-membered phosphorinanium salts, the chair forms are only slightly distorted from that observed for cyclohexane[129,130]; the familiar half-chair conformation is found in the case of endocyclic double bonds, apart from the disphosphoniacyclohexadiene salt where the cyclohexadiene moiety adopts a pronounced chair centrosymmetric conformation which is induced by the presence of bulky substituents located on the phosphorus atom[131]. A chair conformation which is substantially flattened at the phosphorus end is found in a phosphoniabicyclo[3.2.1]octane halide[132].

The geometry around the phosphorus atom in tetraaryl- or triarylalkyl-phosphonium salts is approximately tetrahedral with bond angles in the range 107–114°[133–136] and the propeller-like geometry is normal[137], except when there are bulky substituents[138]. The $P^+—C_{sp^2}$ length (1.78–1.83 Å) is similar to those reported for Ar_3P, while the unique $P^+–C_{sp^3}$ bonds are 1.84–1.88 Å long.

Some bisphosphonium disalts have been investigated, with either no[139], one[130,140], two[66], three[141] or four[142] carbons or heteroatoms, or with a phenyl group[131] inserted between the two phosphorus atoms; the most striking feature concerns a *meso*-1,1′-(ethane-1,2-diyl)bistetrahydrophospholinium disalt (9) where the potential symmetry is destroyed[66]: the relative configurations of substituents at $P_{(1)}$ and $P_{(1°)}$ are completely different (Figure 11).

In aminophosphonium salts, the angles at phosphorus are close to tetrahedral (from 107° to 113°), apart from when the structure is cyclic[56]. The $P^+—N$ bond length is 1.61–1.64 Å[126,143,144] whereas it is 1.71 Å for uncharged species. Contraction favours the idea of a strong π character for this bond, which is supported by the quasi-planar arrangement of bonds around nitrogen; this reflects a trend to sp^2 hybridization arising from involvement of the nitrogen lone pair in π bonding with phosphorus. The multiple bond character of the $P^+—N$ bond is especially evident in highly strained structures (1.58 Å)[56]. The $P—N—P^+$ cation may be bent or linear, with $P^+—N$ bond lengths of 1.57–1.58 Å[145,146].

Considering the anionic counterpart, their geometries, contact distances (including the possibility of hydrogen bonding with the cations), stacking and crystal packing have

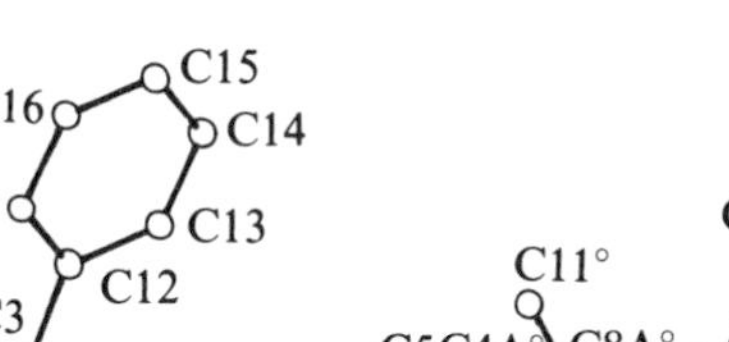

(9) $X = ClO_4$

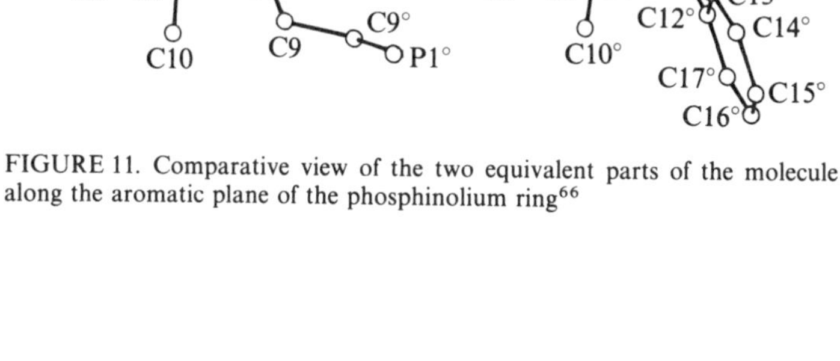

FIGURE 11. Comparative view of the two equivalent parts of the molecule along the aromatic plane of the phosphinolium ring[66]

been determined for various anions such as $X^- = Cl_2H^-$[147], BPh_4^-[119], Br_3^-[143], ArS^- and $ArSe^-$[136], $[MnO_4]^-$[137], $[Sb_3I_{10}]^-$[48] and $TCNQ^{-\bullet}$[149] for monophosphonium cations, $X^{2-} = [CdX_4]^{2-}$, $[CdCl_6]^{2-}$[135], $[Ag_2X_4]^{2-}$[150], $[WCl_6]^{2-}$[144], Tc_4^{2-}[151] and $TCNQ^{2-\bullet}$[131,145] for bisphosphoniums cations and $[Sb_8I_{28}]^{4-}$ for $(Ph_4P^+)_4$[150]. A bulky cation may be responsible for reduction in the symmetry in the anion[119].

6. Others

X-ray photoelectron spectroscopy has been used to determine the phosphorus 2p binding energies for quaternary phosphonium salts $RPh_3P^+\ Y^-$; a linear relationship has been established between the P(2p) binding energies (on solids) and the ^{31}P NMR spectra chemical shifts (in solution); lattice and sample charging effects were shown to be small or non-existent in ESCA studies of the quaternary phosphonium salts[152]. ESCA data have also been reported for tetrakis(hydroxymethyl)phosphonium polymeric urea derivatives[153].

Raman data, in connection with infrared spectra, were used to argue for cationic[26] or anionic[154] structures. Whereas the IR spectra did not yield significant structural information, the Raman spectra of $Ph_3P^+(CH_2)_2CO_2SnPh_3]^+\ X^-$ gave information about the true salt nature of these compounds[155].

D. Thermochemistry

The enthalpies of solution and apparent molar volumes in various solvents have been determined for methyltriphenylphosphonium iodide (**10**); rate constants, activation enthalpies, activation volumes and reaction enthalpies were also determined for its synthesis from triphenylphosphine and methyl iodide[156]. (Equation 3).

$$Ph_3P + MeI \xrightarrow[30\,^\circ C]{MeCN} \underset{\mathbf{(10)}}{Ph_3P^+Me\quad I^-} \qquad (3)$$

$$10^4 k = 76.41\ dm^3\,mol^{-1}\,s^{-1}$$
$$\Delta H^{\neq} = 47.4\ kJ\,mol^{-1}$$
$$\Delta H^\circ = -137.2\ kJ\,mol^{-1}$$

The thermal stability of phosphonium salts (except quasi-phosphonium species) indicates following stability order for those containing an alkenyl substituent[157];

$$Ar_3P^+CH_2CH{=}CH_2 < Ar_3P^+CH{=}CHMe$$
$$Ar_3P^+CH_2CH{=}CHR > Ar_3P^+CH{=}CHCH_2R$$

When endocyclic, a double bond α–β to the phosphonio group may isomerize in the corresponding exocyclic isomer[158] (reaction 4). Because of the weak basicity of triphenylphosphine ($pK_a = 2.73$), triphenylphosphonium bromide, $Ph_3P^+H\ Br^-$, is unstable towards thermolysis; in refluxing toluene, it gives a quantitative yield of hydrogen bromide[159].

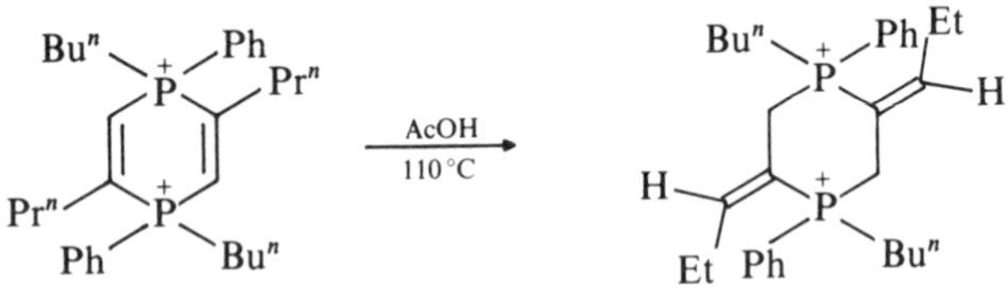

(4)

(11) (12)

Carbalkoxyalkyltriphenylphosphonium halides, either as free esters or as lactones, are able to release CO_2 when heated at, respectively, 100 °C in DMF or 190 °C without any solvent[160,161]. On the other hand, at higher temperatures, cleavage of the P—C bond is observed, leading to the appropriate unsaturated ester, the formation of which can be explained by a Hoffmann transylidation reaction **13**→**14**[162] (reaction 5). Heating methyl tris(*p*-methoxyphenyl)phosphonium iodide at 210 °C leads to methyl iodide owing to the stability of the phosphonium inner salt Ph_2MeP^+*p*-$C_6H_4O^-$[163].

250 °C/10 mm Hg; $-Ph_3P$ and or $-HX$

X, CHCOOMe, $^+PPh_3$ (**13**) → X, C, COOMe (**14**) (5)

Higher temperatures (>300 °C) induce decomposition of quaternary phosphonium salts with the loss of one of the organic groups bound to the phosphorus atom. With no polar substituents in the cation, volatilization is complete as deduced from TG/DTG thermal curves; increasing substitution by polar cyanoethyl groups leads to a decrease in decomposition temperature and an increase in the amount of non-volatile residue[164]. In the case of unsymmetrical methylphenylphosphonium salts, a methyl group is lost preferentially to a phenyl group[165]. The bisphosphonium structure seems thermally less stable than the monophosphonium structure[164]. At temperatures greater than 500 °C, radicals are probably formed[165]. Depending on the nature of the anionic part (Cl^-, Br^-, I^-), changes in the kinetics of thermal decomposition ($Cl^- > Br^-, I^-$) and in the pattern (I^-) are observed [164]. When the anion is a hydroxide ion, the first products of pyrolysis (including phosphine oxides) were detected at relatively low temperatures[165], as in **15**. The driving force for the formation of phosphine oxide would be the formation of the highly stable P—O bond.

$Me_3P^+Ph \quad X^-$ (**15**) $\quad X = I, Cl \quad 375°C$
$\quad X = HO \quad 125°C$

It should be mentioned that phosphonium salts with anions such as perchlorate, picrate, permanganate and other oxidizing anions may be very sensitive to decomposition (dangerous hazards)[137].

E. Photochemistry

As good yields of arylphosphonium salts were obtained in photochemical reactions from arylphosphines and aryl iodides[166], they do not appear to be very sensitive to UV radiation. However, it is known that tetraphenylphosphonium halides efficiently promote the radical photopolymerization of styrene[167]. Similarly, the fluorescence of a range of polycyclic aromatic hydrocarbons is found to be quenched in the presence of alkyltriphenylphosphonium salts (cf. Figure 12), via electron transfer from the hydrocarbon to the phosphonium ion with the formation of phosphoranyl radicals[168]. In the same way, the marked fluorescence observed for 9,10-anthracenylbis(methylenetriphenylphosphonium) dichloride at very high high dilutions is quenched at higher concentrations[169]. Laser-flash photolysis of naphthylmethylphosphonium chlorides in various solvents demonstrated their use as precursors for transient carbocations and/or radicals; added lithium perchlorate prolongates the cation yields and lifetimes[170].

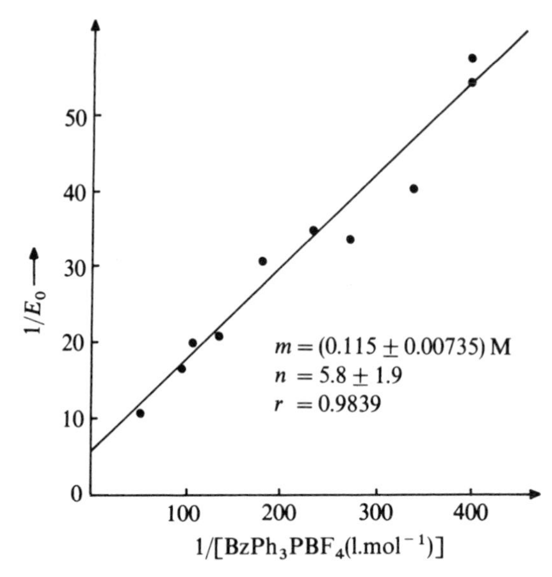

FIGURE 12. Extinction of the anthracene cation-radical by benzyltriphenylphosphonium tetrafluoroborate (in MeCN; $\lambda_{exe} = 340$ nm, $\lambda_{mes} = 730$ nm; [anthracene] = 5.10^{-5} mol.l^{-1})

It has been shown that triphenyl(*p*-cyanobenzyl)phosphonium tetrafluoroborate (**16**), which exhibits a σ^* LUMO level localized predominantly on the heteroatom and benzylic carbon, gives products derived from out-of-solvent cage chemical reactions on direct irradiation (reaction 6). This behaviour is connected with the nuclear hyperfine coupling constant of the heteroatom in triphenylphosphine radical-cation[171].

$$(C_6H_5)_3\overset{+}{P}CH_2C_6H_4CN \; BF_4^- \; \textbf{(16)} \xrightarrow[\text{MeCN}]{h\nu,\ \text{quartz cell}} \underset{25\%}{Ph_3P{=}O} + \underset{74\%}{Ph_3P{=}O \rightarrow BF_3} + \underset{49\%}{(NC{-}C_6H_4{-}CH_2)_2} \qquad (6)$$

F. Electrochemistry

Phosphonium (and quasi-phosphonium) salts generate phosphines under electrolytic reduction (Table 7). Mercury, lead, platinum, tin, copper and aluminium[172] cathodes were tried, mercury being the most often used[173]. Product dependence on cathode material, current, density and solvent has been observed[174]. In the case of unsymmetrical

TABLE 7. $E_{1/2}$ (V vs SCE) of phosphonium salts[168] (inert atmosphere, Bu_4N^+ BF_4^- as supporting electrolyte, in MeCN)

	Ph_4P^+	$BzPh_3P^+$	$EtPh_3P^+$	$MePh_3P^+$	$NCCH_2Ph_3P^+$
$E_{1/2}(PR_4^+/PR_4^\cdot)$	−2.00	−1.85	−2.04	−2.09	−1.55

TABLE 8. Cathodic cleavage of dimethyldiphenylphosphonium bromide (Hg cathode, 20 °C) selectivity versus potential[177].

Potential (V vs SCE)	Relative yield %	
	$PhPMe_2$	Ph_2PMe
−2.0	40	60
−2.1	54	46
−2.2	72	28
−2.3	63	37
−2.4	68	32
−2.5	65	35

phosphonium, salts, selectivity in the electrochemical phosphorus–carbon cleavage essentially depends on the ability of the leaving group to stabilize the first-formed radical, leading to the following order:[175,176]

$$PhCH_2 \approx RCH{=}CHCH_2 \approx CNCH_2CH_2 > Ph \gg R.$$

$$R^1R^2R^3R^4P^+ \longrightarrow R^1R^2R^3R^4P^\cdot \longrightarrow \begin{cases} R^1R^2R^3P \\ (R^4)^\cdot \longrightarrow \begin{cases} R^4-R^4 \\ \xrightarrow{e^-} R^{4\,-} \xrightarrow[H^+]{} R^4H \end{cases} \end{cases}$$

In the reduction of $R^1{}_2R^2{}_2P^+$, the influence of potential on the product ratio $R_2{}^1R^2P/R_1R^2{}_2P$ is significant (cf. Table 8). This selectivity is valuable in the synthesis of functional[178] and/or asymmetric tertiary phosphines, which moreover may be obtained stereospecifically as optically active species[179]. A competitive reduction may arise through ylide (as shown by the variation of the peak potential depending on the structure of the ylides)[176]; the carbanion species is in fact protonated by a molecular equivalent of the phosphonium salt which is thereby converted into the ylide (equation 7).

$$(R^4)^- \xrightarrow{R^1R^2R^3R^4P^+} R^4H + [R^1R^2R^3R(-H)^4]^-P^+ \qquad (7)$$

Such electrochemical experiments have been used to generate and study the reactivity of anionic species $[(R^4)^- = NCCH_2{}^-]$[112]. In another application, electrogenerated extra radical anions or dianions were used in the determination of pK_a values for common phosphonium ions, via double potential step chronoamperometry at a platinum cathode[180].

The electrosorption parameters of unsaturated cations, at the electrode–solvent interface were determined by phase selective a.c. polarography[181] on dropping and hanging mercury dropelectrodes. Unsaturated bisphosphonium salts show lower reduction potentials ($E_{1/2} = -1.2$ V) than the corresponding saturated structures. This can be related to the dπ–pπ conjugative interaction, in the first-formed radical, between phosphorus and the unsaturation; the possibility of delocalization of the electron gives to the radical an unstable ylidic character which is responsible for exclusive heterolytic P—(C=C) bond breaking, even in presence of other good leaving groups[182]. Then, depending on the media, mono- or new bis-phosphonium salts are obtained.

$$Ar_2-\overset{R}{\overset{|}{P^+}}-(CH{=}CH)_n-\overset{R}{\overset{|}{P^+}}-Ar_2 \xrightarrow{e^-}$$

$$Ar_2\overset{R}{\overset{|}{P}}{=}(\ldots)_{n-1}\overset{R}{\overset{|}{P^+}}Ar_2 \longleftrightarrow Ar_2\overset{R}{\overset{|}{P^\bullet}}(\ldots)_{n-1}\overset{R}{\overset{|}{P^+}}Ar_2 \longleftrightarrow Ar_2\overset{R}{\overset{|}{P}}{=}\ldots^\bullet(\ldots)_{n-1}\overset{R}{\overset{|}{P^+}}Ar_2$$

$$\xrightarrow[(-Ar_2PR)]{E_1cb} Ar_2\overset{R}{\overset{|}{P^+}}(\ldots)_{n-1}{}^\bullet \xrightarrow{e^-} \begin{cases} \xrightarrow{Ar_2Pr/H^+} Ar_2\overset{R}{\overset{|}{P^+}}(\ldots)_{n-1}\overset{R}{\overset{|}{{}^+PAr_2}} \\ \xrightarrow[(-HO^-)]{H_2O} Ar_2\overset{R}{\overset{|}{P^+}}(\ldots)_{n-1} \end{cases}$$

Electrolysis of phosphonium hydroxides in aqueous solution yields high-purity substrates[22]. Electrolysis has also been used as a deprotective method of γ-thioacetalated phosphonium nitrates[183]. Finally, 'TNCQ salts' have a semiconductor character[184,185]; solid tetraphenylphosphonium 2,4,6-trimethylbenzene selenolate and thiolate exist as 1:1 electrolytes in acetonitrile, electrochemical oxidation of which has been related to the degree of substitution of the phenyl ring of the anion[136]. Specific electrical conductivities were measured for some γ-keto α,β-unsaturated phosphonium salts[186].

G. Stereochemistry

Most of the problems related to the phosphonium salts stereochemistry are discussed under synthesis (Section II), reactivity (Section III) and spectrometric characteristics (Section I.C.). Because of the tetrahedral geometry of phosphorus, chirality arises from the presence of four different substituents. Enantiomeric forms may be separated by conventional procedures of fractional crystallization using optically active anions such as camphosulphonate and dibenzoyl hydrogentartrate[187–189]. Better yields and optical purities resulted from resolution via the corresponding phosphonium ylides[190] (equation 8). The salts are configurationally stable in solution. Base-catalysed equilibrations of the diastereoisomeric pairs of 2-phosphoniodithianes have been carried out: conformational preference of the 2-phosphonium substituent depends on normal anomeric and steric effects (1.3-*syn* diaxial interaction) operating in opposite directions[191]. The conformation of phosphonium salts where P is included in a ring has often been determined by X-ray diffraction (see Section I.C.5) in an attempt to predict their stereo chemical behaviour.

$$\mathrm{Me{-}\overset{Ph}{\underset{R^1}{P^+}}{-}CH_2Ph\ \ X^-}\ \xrightarrow{\mathrm{NaNH_2}}\ \mathrm{Me{-}\overset{Ph}{\underset{R^1}{P^+}}{-}\bar{C}H{-}Ph}$$

[(*R*, *S*)-**17**]

$$\xrightarrow{\mathrm{RCOOH}} \qquad (8)$$

$$\mathrm{MeO\underset{\underset{O}{\|}}{C}R + Me{-}\overset{Ph}{\underset{R^1}{P^+}}{-}CH_2Ph\ \ Br^-}\ \xleftarrow{\mathrm{MeBr}}\ \mathrm{Me{-}\overset{Ph}{\underset{R^1}{P^+}}{-}CH_2Ph\ \ {}^-OCOR}$$

[(*R*) or (*S*)-**17**] recryst.

R^1 = *n*-Pr, *i*-Pr, α-naphthyl
R = —CH(OCOPh)CH(OCOPh)COOH·H_2O [(*R*) or (*S*)]

E/*Z* stereochemistry of α,β-ethylenic phosphonium derivatives is essentially governed by steric factors apart from when there is extra stabilization, as is the case with betaines[87]. $Z \rightarrow E$ isomerization may be performed using sterically crowded amines (so that any Michael-type addition would be prevented)[192,193]. Spectrometry allows the progress of the reaction to be observed (Section I.C).

H. Biochemistry

Toxicological investigations of tetrakishydroxymethylphosphonium salts [$(HOCH_2)_4P^+$ X^-; X = Cl, THPC; X = $(SO_4)^{0.5}$, THPS] and their derivatives, used for the flame-retardant treatment of cotton, have been carried out (cf. Table 9). It was concluded that the salts and their urea condensation products are harmful to animals on oral or dermal application. Moreover, at sublethal doses, the salts gave no evidence of carcinogenesis in rodents when applied by the dermal or gavage routes. The polymers produced from them did not exhibit any adverse reactions[194].

[(Acylamino)methyl]phosphonium salts showed relatively high anticholinesterase activity[195]. Some poly(methylene)bis(triphenylphosphonium) and [3-(alkylamino)propyl]triphenylphosphonium salts have strong anticholinergic, effects, as observed on *Schistosoma mansoni*. Hexyltriphenylphosphonium bromide possibly exerts a cholinomimetic action.[196] Neuromuscular blocking activity of ketophosphonium salts containing a morpholino group seems to be related to the P···Oxygen atom distance in the morpholine group, in accordance with the known value for cholinergic receptors[197].

TABLE 9. Acute toxicity studies on THPC, THPS and related compounds[194]

Compound	THP (%)[a]	pH	LD_{50} ($mg\,kg^{-1}$)[b]	Route[c]
THPC	65	1–2	161–840	Oral
			1500–3200	Dermal
			280	Oral[d]
THPS	57	5	161–350	Oral
			2000	Dermal
Proban 210[e]	39	1–2	282–990	Oral
			4000–6000	Dermal
Proban CC[e]	51	5	>200	Oral
Proban NX[f]	51	5	390	Oral
Proban 210 Polymer[g]	116.5[i]		3200–5000	Oral
$THPOH-NH_3$ polymer[h]	139[i]		>10000	Oral
			>460	Dermal

[a]Expressed as tetrakis(hydroxymethyl)phosphonium ion.
[b]Fiducial limits.
[c]Rats were used for the oral tests and rabbits for the dermal route except where specified.
[d]Value for mice.
[e]Proban 210 and Proban CC are 2:1 (w/w) THPC–urea condensates.
[f]Proban NX is a 2:1 (w/w)THPS–urea condensate.
[g]The proban 210 polymer was prepared on a viscose acetate fabric and the fabric was removed by continuous extraction with acetone–methanol.
[h]The $THPOH-NH_3$ polymer was prepared in aqueous suspension.
[i]These values are calculated from the theoretical phosphorus contents of the polymers.

Phosphonium derivatives of aza crown ethers were effective as neuromuscular blocking agents in both *in vitro* and *in vivo* experiments[198] Analgesic properties were also found[199].

3-Oxopropyl quaternary phosphonium salts, $RCOCHR^1CHR^2P^+R_3X^-$, have been patented as microbiocides[200], and also phosphonium compounds containing a 1,4-benzodioxane system[201]. Antimicrobial properties of phosphonium sulphamides were evaluated[202]. Phosphonium compounds are useful anti-leprosy agents. Carcinostatic activity is also possible[203]. Patents protect the use of phosphonium salts in the agrochemical field where they are employed as herbicides[204], fungicides[205–207], algicides, microbicides[208] and bactericides[206,209]. 'Phosphon' ($2,4\text{-}Cl_2C_6H_3CH_2P^+Bu_3X^-$) is used as a plant growth regulator.

I. Others

Tetrakishydroxymethylphosphonium salts (THPC, THPS, etc.) are of major importance as raw material monomers in the flame-retardant treatment of cotton (industrial protective clothing, furnishing fabrics, etc.). Proban chemicals are produced from THPC or THPS with urea[194]. These compounds have been often patented as additives of natural and artificial fibres[210,211], coatings [212,213] and in fire-proofing solution for wood[214].

In polymer chemistry, phosphonium salts are useful in the manufacture of laminates[215], sealants[216] and coated cationic-dyeable polymers[217], as curing accelerators or catalysts[218–221], binders for ceramic powders[222] and acid scavengers[223]. They can also be used to modify the composition of cyclosiloxanes[224].

Anion-exchange extraction by triphenylnonylphosphonium salts has been studied for dye selection[225] and lead halo complex extraction[226]. Ethylene bis(trioctylphosphonium) cation extracted $M^{II}X^{2-}$ ($M = Zn, Cd, Cu, Pb$) better than $MeNR_3^+$; Co^{II}, Ni^{II} and Mn^{II} were not extracted[227]. Solvent extraction of dansylamino acid anions by dicationic and monocationic phosphonium salts has been investigated and their ability was shown to be strongly dependent on their structure[228]. In contrast, alkyl- and aryl-phosphonium nitrate extractants have little potential as decontaminating agents to remove technetium from concentrated uranyl nitrate solution[229].

Electrolytes for capacitors may include some phosphonium compounds with diverse anions[230–232]. When TCNQ is the anionic counterpart, semiconductor properties became evident[131,145] [as an example. $\sigma_{rt} \approx 10^{-4}\,S\,cm^{-1}$ with an energy of activation 170 meV, for $Ph_3P{=}N{=}PPh_3{}^+(TCNQ)_3(MeCN)_2$]. Electrostatic charging of powder coatings[233] and antistatic properties[99] are improved by adding $Bu_4P^+X^-$.

Finally, the addition of phosphonium salts can lead to positive effects concerning nitrate efflorescence, prevention or treatment of concrete and masonry[234], improved stability of IR-sensitive photographic materials[235], improvement of lubricating greases[236], effective disinfection of surgical goods, surfaces and sensitive components[237] (critical micelle concentrations in peculiar case[186]). Micellar structures are also obtained with various long-chain alkyl- and perfluoroalkyl-phosphonium salts in formamide solution[238].

II. PREPARATION OF PHOSPHONIUM SALTS

A. Synthesis of Phosphonium Salts by Alkylation of Phosphines

Most phosphonium salts are prepared by quaternization of phosphines with the appropriate alkyl halide. Generally, with normal S_N2 reactions, the rate of reaction increases from chloride to iodide[2c]. However, the alkylation with alkyl chlorides, simple or functional, can be improved by using not the tertiary phosphine itself but its bromohydrate[239] (reaction 9). In this case it is likely that the alkyl chloride is first converted *in situ* into alkyl bromide, which then yields the phosphonium bromide. Primary alkyl halides are normally used; with lower halides, the reaction occurs easily but becomes slower with increasing size of the alkyl group. Forcing conditions are needed for quaternization with secondary halides and these may require sealed tubes at high temperatures.

$$Ph_3\overset{+}{P}H\ Br^- + RCH_2Cl \xrightarrow[\text{of reflux, PhMe}]{70\,^{\circ}C,\text{dmf}} Ph_3\overset{+}{P}CH_2R\ Br^- + HCl \tag{9}$$

The nature of the phosphine can also determine the course of the alkylation reaction. Thus, in the kinetic study of benzylation of various methoxyphenyl- or diaminophenyl-phosphines[43], it has been shown that MeO and more significantly the Me_2N group, in the *ortho* or *para* position on the phenyl ring, is favourable to the alkylation reaction, increasing the rate of the reaction by up to a factor of 10 (reaction 10). For *ortho* substitutions, the rate is generally increased by a through- space N_{2p}–$P^{(IV)}$ interaction and a less negative value of $\Delta S^{\neq}$. Further, phosphines bearing several strongly electron-withdrawing perfluorinated substituents cannot be alkylated without losing one of these substituents[240] (reaction 11). In common, polar solvents such as formic acid, dmf and acetonitrile facilitate the alkylation reaction, which proceeds with the formation of ionic species.

A kinetic study of the influence of solvent on the quaternization of triphenylphosphine by benzyl chloride[241] or by methyl iodide[156] has been carried out (reaction 12). The relative order of solvent activities is $MeOH > AcOH > CH_2Cl_2 > Me_2CO > Et_2O > PhH > PhMe$.

$$Ar_nPPh_{3-n} + PhCH_2Br \xrightarrow[(CHCl_3)]{k} Ar_n\overset{+}{P}Ph_{3-n}CH_2Ph\ Br^- \tag{10}$$

$$Ar = o\text{- or } p\text{-}Me_2NC_6H_4 \text{ and } o\text{- or } p\text{-}MeOC_6H_4$$

$$R_{F_n}PMe_{3-n} \xrightarrow{MeI} (R_{F_n}\overset{+}{P}Me_{4-n}\ I^-)$$
$$\downarrow$$
$$(R_{F_{n-1}}PMe_{4-n}) \xrightarrow{MeI} R_{F_{n-1}}\overset{+}{P}Me_{5-n}\ I^- \quad (11)$$
$$n = 2\ (R_F = n\text{-}C_6F_{13}) \text{ or } n = 3\ (R_F = C_6F_5, CF_3)$$

$$Ph_3P + RCH_2X \underset{\text{(Solvent)}}{\xrightarrow{k}} Ph_3\overset{+}{P}CH_2R\ X^- \quad (12)$$

$$R = Ph, X = Cl \text{ (ref. 241)}; R = H, X = I \text{ (Ref. 156)}$$

It must be pointed out that the negative activation volume ($\Delta V^{\neq} = < 0$)[156] involved in the alkylation of tertiary phosphines allows the application[242] of high-pressure technology in order to achieve the alkylation reaction at room temerature under 15 kbar, even when the same reagents are ineffective under normal pressure (reaction 13). It is noteworthy that complications may arise if the alkyl halide contains, at the α-position, electron-withdrawing groups which favour nucleophilic displacement at halogen; however, the product may not differ from that obtained by direct displacement at carbon.

$$Ph_3P + RX \underset{15\,\text{kbar}}{\xrightarrow{3\,\text{days}, 20\text{–}40\,^\circ\text{C}}} Ph_3\overset{+}{P}R\ X^- \quad (13)$$

$$R = n\text{-Alk}, sec\text{-Alk}; X = Cl, Br, OMs, OTs$$

Tertiary phosphines, as bromohydrates, can also be alkylated by primary alcohols[243] (reaction 14). The esters from primary alcohols (acetates[243], lactones[243], carbonates[244], methanesulphonates[15]) can be also used to alkylate phosphines. In the preparation of some β,γ-unsaturated phosphoniums salts, various methods have been described using allylammoniums[245] or olefinic complexes of iron carbonyl[246] (reaction 15) and further, allylsilanes[247], cyclic or not, through an anodic oxidation procedure (reaction 16).

$$Ph_3\overset{+}{P}H\ Br^- + RCH_2OH \underset{-H_2O}{\xrightarrow{160\,^\circ\text{C}}} Ph_3\overset{+}{P}CH_2R\ Br^- \quad (14)$$
$$85\text{–}100\%$$

$$\text{(diene)}Fe(CO)_3 \xrightarrow[HBF_4/CO]{CF_3CO_2H} BF_4^-\ {}^+Fe(CO)_4\text{(allyl)} \xrightarrow[-30\,^\circ\text{C}]{Ph_3P} \text{allyl-}\overset{+}{P}Ph_3\ BF_4^- \quad 24\text{–}53\% \quad (15)$$

$$(CH_2)_n\text{(cyclic allylsilane, }SiMe_3) + Ph_3P \xrightarrow[\text{anodic oxidation } (Q^+\ BF_4^-)]{-e} (CH_2)_n\text{(exo-methylene, }\overset{+}{P}Ph_3)\ BF_4^- \quad (16)$$
$$n = 3\text{–}6$$

The alkylation of tertiary phosphines is, in general, compatible with elaborate structures bearing various functions or chiralities, as illustrated by the preparation of a phosphonium salt (**18**), intermediate in the synthesis of pseudomonic acid[248] (reaction 17). For the preparation of dialkylphosphonium salts, diphenylphosphine can be directly alkylated, but it is more advantageous to use triphenylphosphine as the starting material,

$$\mathrm{Me{-}CH(OH){-}CH(Me){-}CH_2CH_2I} + \mathrm{Ph_3P} \xrightarrow[\text{2 days}]{\text{reflux, PhMe}} \mathrm{Me{-}CH(OH){-}CH(Me){-}CH_2CH_2\overset{+}{P}Ph_3\ I^-} \quad (17)$$

(18)

generating *in situ* Ph_2PLi and performing the two successive alkylations in one pot[249] (reaction 18). When a diphosphine is used with a monohalogenated reagent[250], or a monophosphine with a dihalogenated derivative[251], the formation of either mono- or bis-phosphonium salt depends on the operating conditions, but the reaction is performed in common in order to obtain the corresponding bisphosphonium salts (reaction 19). In the case of a diphosphine and dihalogenated compound, the reaction can even be controlled to yield selectively a macrocyclic tetrakisphosphonium salt[252].

$$\mathrm{Ph_3P} \xrightarrow[\text{(ii) Bu}^t\text{Cl}]{\text{(i) 2Li/thf}} \mathrm{Ph_2PLi} \xrightarrow[\text{reflux, thf}]{\text{2RX}} \mathrm{Ph_2\overset{+}{P}R_2\ X^-} \quad (18)$$

$$\mathrm{RX + R'_2PQPR'_2} \longrightarrow \left(\mathrm{R'_2\overset{R}{P}{}^+QPR'_2\ X^-}\right) \xrightarrow{\text{RX}} \mathrm{R'_2\overset{R}{P}{}^+Q^+\overset{R}{P}R'_2\ \ 2X^-} \quad (19)$$

$$\mathrm{XQX + R'_2PR} \longrightarrow \left(\mathrm{R'_2\overset{R}{P}{}^+QX\ X^-}\right) \xrightarrow{\text{R}'_2\text{PR}} \mathrm{R'_2\overset{R}{P}{}^+Q^+\overset{R}{P}R'_2\ \ 2X^-}$$

Similarly, polyphosphonium salts linked to polymeric supports can be prepared either by the action of phosphines on polybromoalkylstyrenes[253,254] or by simple alkylation of polyphosphinoalkylstyrenes[104].

B. Synthesis of Phosphonium Salts by Arylation of Tertiary Phosphines

Aryl halides normally do not react with phosphines. However, under forcing temperature conditions, most frequently at 170–250 °C in the absençe of solvents, it is possible to achieve direct arylation of tertiary phosphines by several halogenated aromatic or heteroaromatic compounds, especially when they are activated[2d] (reactions 20 and 21). New examples of such arylations have been described for nitrogen heterocycles[255,256],

$$\mathrm{ArX + R_3P} \xrightarrow{\Delta} \mathrm{Ar\overset{+}{P}R_3\ X^-} \quad (20)$$

$$\mathrm{HetX + R_3P} \xrightarrow{\Delta} \mathrm{Het\overset{+}{P}R_3\ X^-} \quad (21)$$

$$\mathrm{X = Cl, Br, I}$$

chloroanthraquinones[257] and *p*-bromobutyrophenone[258]. However, this procedure is not well suited for use with compounds such as 1-bromonaphthalene (yield 10%)[259].

Another arylation method, in the case of nitrogen heterocycles, does not need a halogenated derivative but a heterocycle activated by triflic anhydride[260,261] (reaction 22). Simple aryl halides usually do not react with phosphines and special methods therefore have to be used for their arylation. The most widely used is the 'complex salt method', in which an aryl halide is heated with a phosphine in the presence of a transition metal such as nickel (II)[2e] (reaction 23). The catalytic cycle probably takes place by means of a reduced nickel(I) complex, generated *in situ* from the starting nickel(II) salt; this nickel(I) species could undergo an oxidative addition of the aryl halide to yield a transient nickel(III) adduct, which after the reductive elimination of the aryphosphonium affords the recovery of the first active-nickel(I) complex (reaction 24).

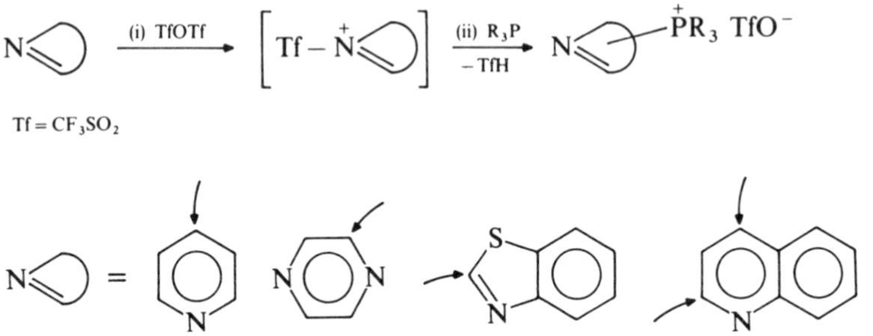

(i) TfOTf; (ii) R_3P, −TfH; $\overset{+}{P}R_3$ TfO^-

$Tf = CF_3SO_2$

(22)

$$R_3P + ArX \xrightarrow[150\text{–}200\,^\circ C]{cat., NiBr_2} R_3\overset{+}{P}Ar\ X^- \qquad (23)$$

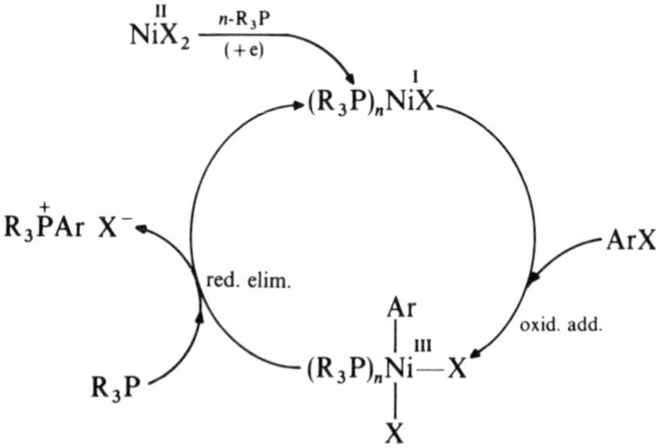

$\overset{II}{Ni}X_2$ (n-R_3P, (+e)) → $(R_3P)_n\overset{I}{Ni}X$; ArX, oxid. add. → $(R_3P)_n\overset{III}{Ni}(Ar)X_2$; red. elim. ($R_3P$) → $R_3\overset{+}{P}Ar\ X^-$

(24)

With normal aryl halides and $NiBr_2$ the reaction generally takes place at 150–200 °C without solvent, but with nitrogen derivatives of *o*-bromobenzaldehyde[102,262,263] or structural analogues[101,133] the reaction can easily occur at ethanol reflux temperature. Indeed, such structures determine a template effect in the metallic complexes which favour the phosphonium salt formation by promoting the substitution reaction (reaction 25). With some such *ortho*-substituted structures, the catalyst can be not only $NiBr_2$, but also $Cu(OAc)_2$ or Ni^{II} or Cu^{I} complexes. Moreover, it must be pointed out that with

$$\text{o-XC}_6\text{H}_4\text{-Z=N-R}' + R_3P \xrightarrow[\text{reflux, EtOH}]{10\,\text{mol\% NiBr}_2} \text{o-}(\overset{+}{P}R_3\ X^-)\text{C}_6\text{H}_4\text{-Z=N-R}' \quad \text{via } [\text{o-}((PR_3)_n\text{Ni-X})\text{C}_6\text{H}_4\text{-Z=}\ddot{N}\text{-R}'] \qquad (25)$$

$X = Cl, Br, I$; $Z = N, CH$; $R = n\text{-Bu}, Ph$;

$R' =$ non-sterically crowed alkyl or aryl

simple aryl bromides, and especially iodides, the phosphonium salt formation can be made easier by using Pd^{II} or $Pd^{(0)}$[264]. In this case, the reaction takes place at 120 °C in an aromatic hydrocarbon (reaction 26).

$$Ar'\ X + Ar_3P \xrightarrow[25\,h,\,120\,^\circ C]{1\,\text{mol\%}\,[Pd]} Ar'\overset{+}{P}Ar_3\ X^- \qquad (26)$$

$$[Pd] = Pd[PPh_3]_4 \text{ or } Pd(OAc)_2$$

C. Synthesis of α,β-Unsaturated Phosphonium Salts

α-Alkenylphosphonium salts, exhibiting a C=C double bond in the α,β-position to phosphorus, are obtained in three main synthetic ways (reaction 27), already well documented in previous reviews[2a]:

(a) by direct alkenylation of tertiary phosphines;
(b) by creation or modification of an unsaturation on a phosphonium salt;
(c) by quaternization of an α,β-unsaturated phosphine.

$$R_3P \xrightarrow{(a)};\quad R_3\overset{+}{P}\text{—}\overset{|}{\underset{|}{C}}\text{—}\overset{|}{\underset{|}{C}}\text{—}\ X^- \xrightarrow{(b)} R_3\overset{+}{P}\text{—}\underset{|}{C}\text{=}C\langle\quad X^-;\quad R_2P\text{—}\underset{|}{C}\text{=}C\langle \xrightarrow{(c)} \qquad (27)$$

These three synthetic methods can also be applied, generally, to α-alkynylphosphonium salts, with a C≡C triple bond α to the phosphorus. New results have recently completed and enlarged the scope of each pathway.

1. Introduction of an alk-1-enyl or alk-1-ynyl chain on a tertiary phosphine

a. Addition of phosphine to activated acetylenic compounds. α-Alkoxy-α-alkenyl-phosphonium salts have been obtained by the action of tertiary phosphines[265] or halophosphines[266] on acetylenic ethers, and then trapping of the adduct by various electrophilic reagents (reaction 28). γ-Functional α-alkenylphosphonium salts have been prepared in high yields by the Michael-type addition of phosphine on α,β-acetylenic carbonyl compounds in presence of acid. The reaction is favoured by using a microemulsion[16] (reaction 29).

$$R_2PY + HC{\equiv}COR' \longrightarrow \left| \begin{array}{c} R_2\overset{+}{P}{-}C{=}\bar{C}H \\ \quad | \quad\;\; | \\ \quad Y \quad OR' \end{array} \right| \xrightarrow{EX} \begin{array}{c} \qquad\quad X^- \\ R_2\overset{+}{P}{-}C{=}CHE \\ \quad | \quad\;\; | \\ \quad Y \quad OR' \end{array} \qquad (28)$$

$R = n\text{-Bu, Et; } Y = R, Br, Cl; EX = MeI, PhCH_2Br, AcBr, Me_3SiCl$, etc.

$$Ph_3P + RC{\equiv}C\overset{O}{\overset{\|}{C}}R' \xrightarrow[\text{microemulsion}\ (H_2O/HCl/hexane)]{H_2O\text{-}HCl(1\ equiv.)} Ph_3\overset{+}{P}\underset{R}{\underset{|}{C}}{=}CH\overset{O}{\overset{\|}{C}}R'\ Cl^- \qquad (29)$$

$R = H, Alk, Ph; \quad R' = H, Me, OMe$

b. Substitution by the phosphine of an activated vinylic halogen. The S_N 2t substitution reaction is strongly favoured by the presence on the vinylic carbon, β to the halogen, of one or several electron-withdrawing groups (reaction 30, $G_a = SO_2Ar$[87], CN[267], COR[186,268,269], $C{=}N^+Me_2$[109]). Commonly, the reaction conditions are mild and can even be applied to special structures such as derivatives of squaric acid[269].

$$R_3P + X{-}\underset{|}{C}{=}\underset{|}{C}\text{-}G_a \longrightarrow R_3\overset{+}{P}{-}\underset{|}{C}{=}\underset{|}{C}{-}G_a\ X^- \qquad (30)$$

$R = n\text{-Bu, Ph}; \qquad X = Br, Cl$

When the stereochemistry of the reaction was investigated[87], it was established that the reaction is fully stereospecific: the *Z* or *E* stereochemisty of the starting β-bromovinyl sulphone remains preserved in the resulting phosphonium salt (reaction 31).

$$ArSO_2CH{=}CHBr + P(NEt_2)_3 \xrightarrow[-30^\circ C]{\text{solvent}} ArSO_2CH{=}CH\overset{+}{P}(NEt_2)_3\ Br^- \qquad (31)$$

Solvent = MeCN, CH_2Cl_2, PhH or PhMe

c. Alkenylation of phosphines complexed on rhodium(I) by α,β-unsaturated acid chlorides. In presence of equimolecular amounts of Wilkinson's complex, the α,β-unsaturated carboxylic acid chlorides (except acryloyl chloride) lose 1 mol of carbon monoxide and yield the corresponding alk-1-enylphosphonium salts[270] (reaction 32). Starting from acid chlorides with *E* stereochemistry, the reaction affords stereoselectively the (*E*)-alk-1-enylphosphonium salts.

$$\begin{array}{c} R \\ H \end{array}\!\!>C{=}C\!\!<\!\!\begin{array}{c} H \\ COCl \end{array} + [ClRhH(PPh_3)_3] \xrightarrow[ClCH_2CH_2Cl]{85^\circ C} \begin{array}{c} R \\ H \end{array}\!\!>C{=}C\!\!<\!\!\begin{array}{c} H \\ \overset{+}{P}Ph_3 \end{array} Cl^- \quad 77\text{–}87\% \\ + [ClRh(CO)(PPh_3)_2] \qquad (32)$$

$R \neq H; \qquad R = Me, Bu, Ph$

d. Electrochemical alkenylation of phosphines by cycloalkenes. Anodic oxidation of triphenylphosphine has been performed by constant-current electrolysis with graphite electrodes in presence of cycloalkenes[271]. The process gives directly, in moderate yields, the corresponding cycloalk-1-enylphosphonium salts (reaction 33).

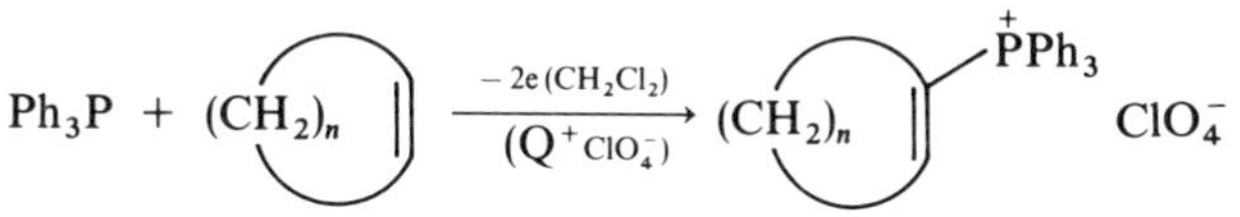

$$Ph_3P + (CH_2)_n \text{(cycloalkene)} \xrightarrow[(Q^+ClO_4^-)]{-2e\,(CH_2Cl_2)} (CH_2)_n \text{(cycloalk-1-enyl-}\overset{+}{P}Ph_3) \; ClO_4^- \qquad (33)$$

$n = 3$–6 50–60%

e. Alkynylation of phosphines by alk-1-ynyliodonium salts. Nucleophilic substitutions of haloalkynes with triphenylphosphine usually require 2–3 days at room temperature[272–274]. Using highly reactive alk-1-ynylphenyliodonium tetrafluoroborates, the reaction with Ph_3P proceeds smoothly even at $-78\,°C$ and affords quantitatively the corresponding alk-1-ynylphosphonium compound (reaction 34). The reaction, initiated by sunlight, proceeds through radical species.

$$Ph_3P + RC{\equiv}C\overset{+}{I}Ph\ BF_4^- \xrightarrow[78\,°C(-PhI)]{h\nu(\text{sunlight})} RC{\equiv}C\overset{+}{P}Ph_3\ BF_4^- \qquad (34)$$

2. Creation or modification of an unsaturation on a phosphonium group

a. Isomerization of alk-2-enyl- into alk-1-enyl-phosphonium salts. The prop-1-enylphosphonium salt (**20**) is prepared by base-catalysed migration of the double bond in the allylphosphonium salt **19**[275–277] (reaction 35).

$$Ph_3\overset{+}{P}CH_2CH{=}CH_2\ X^- \xrightarrow{\text{Cat., Cl}^- \text{ or } R_3N} Ph_3\overset{+}{P}CH{=}CHCH_3\ X^- \qquad (35)$$

(**19**) (**20**)

In the same way, the propargylphosphonium salt (**21**) can be rearranged to allenylphosphonium salt (**22**)[278,279] (reaction 36).

$$Ph_3\overset{+}{P}CH_2C{\equiv}CH\ X^- \xrightarrow{\text{Cat.}} Ph_3\overset{+}{P}CH{=}C{=}CH_2\ X^- \qquad (36)$$

(**21**) (**22**)

Nevertheless this kind of isomerization cannot be considered as a general method for the synthesis of all the alk-1-enylphosphoniums, because it often affords thermodynamic mixtures of α,β- and β,γ-unsaturated phosphonium salts[157,280] (equation 37).

$$Ph_3\overset{+}{P}CH_2CH{=}CHR\ X^- \rightleftharpoons Ph_3\overset{+}{P}CH{=}CHCH_2R\ X^- \qquad (37)$$

(**23**) (**24**)

$R = n\text{-}C_3H_7$[280], $n\text{-}C_5H_{11}$[157] (**23**/**24** $\approx$ 70:30)

Moreover, these results demonstrate that the stabilizing interaction of a positively charged phosphorus atom with the π-electron density of the unsaturation is not the main driving force for the isomerization of the double bond into the α,β-position. Therefore, the isomerization ratio generally depends on the actual structure of the alkenyl chain (reactions 38 and 39).

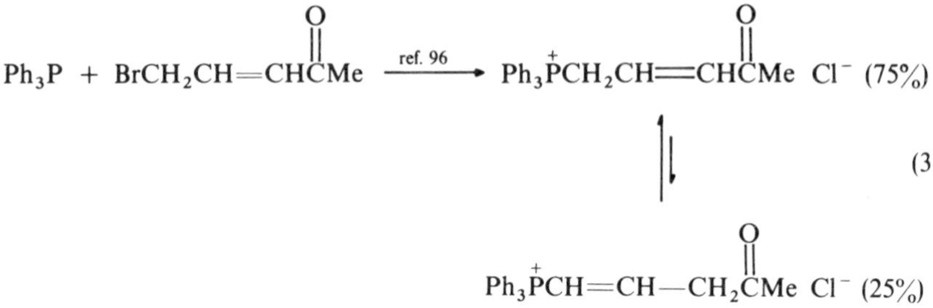

$$Ph_3P + BrCH_2CH{=}CHC(O)Me \xrightarrow{\text{ref. 96}} Ph_3\overset{+}{P}CH_2CH{=}CHC(O)Me\ Cl^-\ (75\%) \rightleftharpoons Ph_3\overset{+}{P}CH{=}CH{-}CH_2C(O)Me\ Cl^-\ (25\%) \quad (38)$$

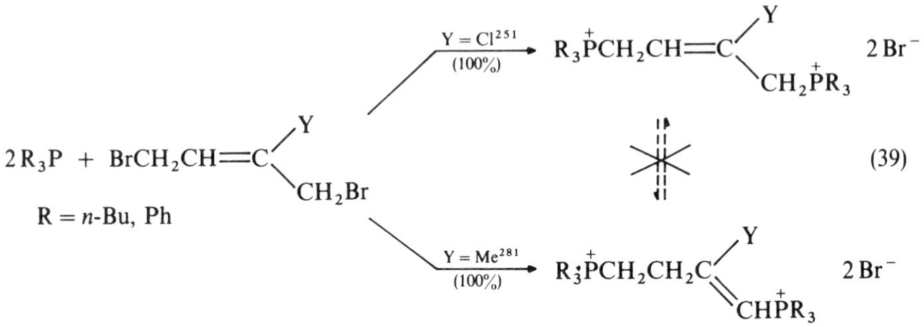

$$2R_3P + BrCH_2CH{=}C(Y)CH_2Br \xrightarrow[(100\%)]{Y = Cl^{251}} R_3\overset{+}{P}CH_2CH{=}C(Y)CH_2\overset{+}{P}R_3\ 2Br^-$$

$$2R_3P + BrCH_2CH{=}C(Y)CH_2Br \xrightarrow[(100\%)]{Y = Me^{281}} R_3\overset{+}{P}CH_2CH_2C(Y){=}CH\overset{+}{P}R_3\ 2Br^- \quad (39)$$

$R = n$-Bu, Ph

It depends also, of course, on the reaction conditions[282] (reactions 40 and 41). However this kind of isomerization has been used to obtain specifically some dienic phosphonium[283] or bisphosphonium[193] salts, because, in this case, the stability of the dienic system induced fully selective isomerization (reactions 42 and 43).

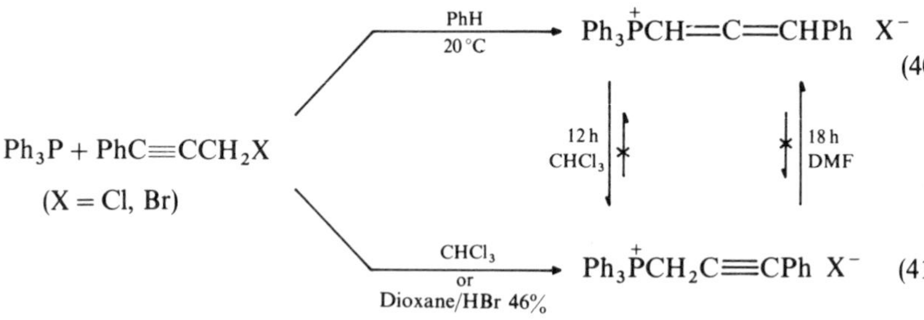

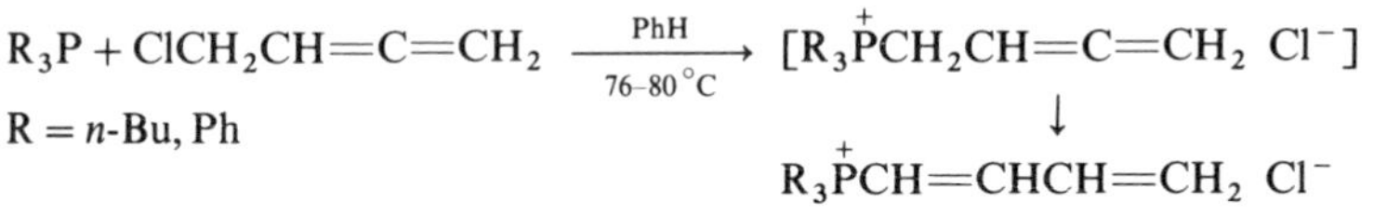

$$Ph_3P + PhC{\equiv}CCH_2X \xrightarrow[20\,°C]{PhH} Ph_3\overset{+}{P}CH{=}C{=}CHPh\ X^- \quad (40)$$

$$Ph_3P + PhC{\equiv}CCH_2X \xrightarrow[\text{Dioxane/HBr } 46\%]{CHCl_3 \text{ or}} Ph_3\overset{+}{P}CH_2C{\equiv}CPh\ X^- \quad (41)$$

$(X = Cl, Br)$

12 h $CHCl_3$; 18 h DMF

$$R_3P + ClCH_2CH{=}C{=}CH_2 \xrightarrow[76\text{–}80\,°C]{PhH} [R_3\overset{+}{P}CH_2CH{=}C{=}CH_2\ Cl^-] \rightarrow R_3\overset{+}{P}CH{=}CHCH{=}CH_2\ Cl^- \quad (42)$$

$R = n$-Bu, Ph

In the same way, in the case of the propargylphosphonium salt, two successive isomerizations of the unsaturation from the β,γ- to the α,β-position, combined with a Michael-type addition of nucleophiles on the intermediate allenyl phosphonium, afford various β-heterosubstituted α,β-unsaturated phosphonium salts, which have found interesting applications in organic synthesis (reaction 44).

$$Ph_3\overset{+}{P}CH_2C{\equiv}CCH_2\overset{+}{P}Ph_3\ 2X^- \xrightarrow{\text{cat., } R_3N} [Ph_3\overset{+}{P}CH{=}C{=}CHCH_2\overset{+}{P}Ph_3\ 2X^-]$$

$$\downarrow$$

$$Ph_3\overset{+}{P}CH{=}CHCH{=}CH\overset{+}{P}Ph_3\ 2X^- \longleftarrow [Ph_3\overset{+}{P}CH{=}CHCH{=}CH\overset{+}{P}Ph_3\ 2X^-] \qquad (43)$$

$$Ph_3\overset{+}{P}CH_2C{\equiv}CH\ X^- + \ddot{Y}H \longrightarrow [Ph_3\overset{+}{P}CH{=}C{=}CH_2\ X^-] \qquad (44)$$

$$\downarrow YH$$

$$Ph_3\overset{+}{P}CH{=}C(Y)CH_3\ X^- \longleftarrow [Ph_3\overset{+}{P}CH_2C(Y){=}CH_2\ X^-]$$

$YH = RCO_2H$[284], R_2NH[285], $R'RC{=}NNH_2$[286]

b. Synthesis by addition to alk-1-ynylphosphonium salts. Owing to the strong electron-withdrawing effect of phosphonio group alk-1-ynylphosphonium salts have an electrophilic triple bond which can add nucleophiles to afford β-heterosubstituted α-alkenylphosphonium salts. The ethynylphosphonium salt **26**, generated from the 1,2-vinylenebisphosphonium salt **25**, allows the *E*-stereoselective synthesis in high yields of a wide range of β-substituted vinylphosphonium salts[91,287] (reaction 45). The same kind of addition reaction, but in an intramolecular way, affords heterocyclic unsaturated phosphonium salts[86] (reaction 46).

$$\underset{(\mathbf{25})}{Ph_3\overset{+}{P}(H)C{=}C(H)\overset{+}{P}Ph_3\ 2Br^-} \xrightarrow[-Ph_3P,\ -Et_3NH^+Br^-]{Et_3N} \underset{(\mathbf{26})}{[Ph_3\overset{+}{P}C{\equiv}CH\ Br^-]} \xrightarrow{ZH} Ph_3\overset{+}{P}(H)C{=}C(H)Z\ Br^- \qquad (45)$$

$ZH = ROH, ArOH, RSH, ArSH, RNH_2, R_2PH$

c. Synthesis by elimination from an alkylphosphonium salt. Unsaturation can be created through base-catalysed elimination of a leaving group, Y, from the β-position on an alkylphosphonium salt[157] or from the δ-position on an allylphosphonium salt[288,289] (reactions 47 and 48). Elimination can also result from α-seleniation of an alkylphosphonium followed by oxidation to the corresponding selenoxide. This method has been used to obtain cyclobut-1-enylphosphonium[290], but also various salts with a juxtacyclic double bond[291] (reaction 49). In the same way, through elimination reactions, alk-1-ynylphosphonium salts can be prepared from unsaturated structures, arising generally from β-ketoylides (reactions 50[97] and 51[292]).

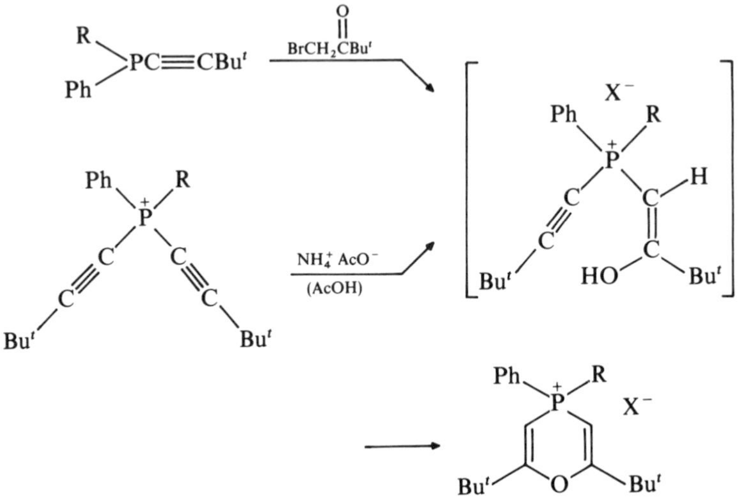

$$\begin{array}{l} \mathrm{R(Ph)PC{\equiv}CBu^t} \xrightarrow{\mathrm{BrCH_2COBu^t}} \\ \mathrm{Ph(R)\overset{+}{P}(C{\equiv}CBu^t)_2} \xrightarrow[\mathrm{(AcOH)}]{\mathrm{NH_4^+\,AcO^-}} \end{array} \left[\mathrm{Ph(R)\overset{+}{P}(C{\equiv}CBu^t)CH{=}C(OH)Bu^t}\ X^- \right] \longrightarrow \mathrm{Ph(R)\overset{+}{P}\text{-}[CH{=}C(Bu^t)\text{-}O\text{-}C(Bu^t){=}CH]}\ X^- \quad (46)$$

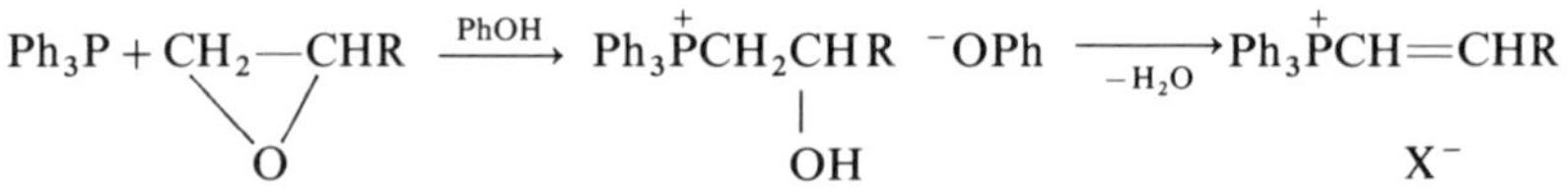

$$\mathrm{Ph_3P + CH_2\text{—}CHR\ (epoxide,\ O)} \xrightarrow{\mathrm{PhOH}} \mathrm{Ph_3\overset{+}{P}CH_2CH(OH)R\ {}^-OPh} \xrightarrow[-\mathrm{H_2O}]{} \mathrm{Ph_3\overset{+}{P}CH{=}CHR}\ X^- \quad (47)$$

$$\mathrm{Ph_3\overset{+}{P}CH_2CH{=}C(R)CH_2Y}\ X^- \xrightarrow{\mathrm{Et_3N}} \mathrm{Ph_3\overset{+}{P}CH{=}CHC(R){=}CH_2}\ X^- \quad (48)$$

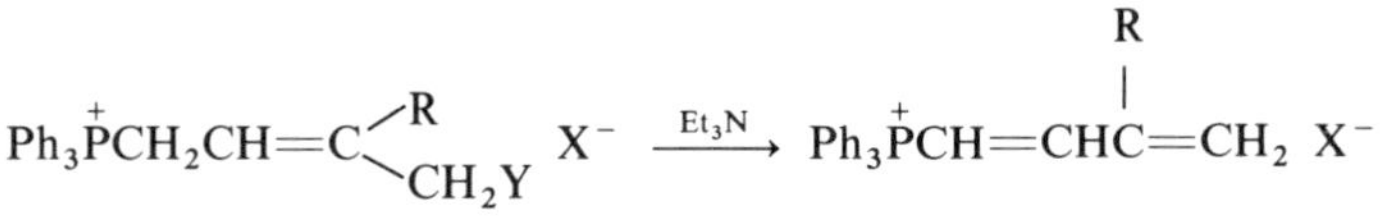

R = Cl, X = Y = Br[288]; R = H, X = Br, Y = $\overset{+}{P}Ar_3$[289]

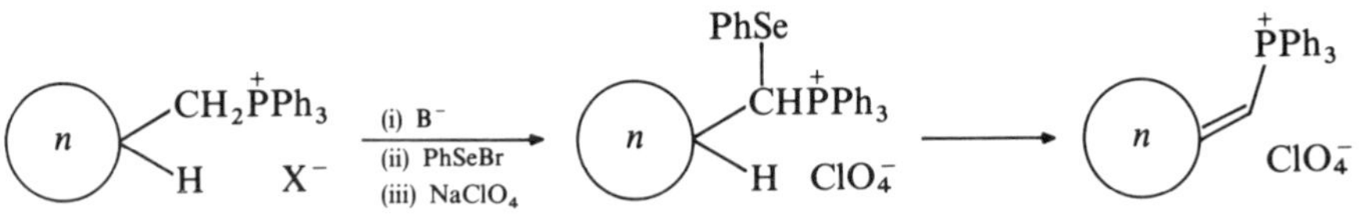

$$(\mathrm{ring}\ n)\mathrm{C(H)CH_2\overset{+}{P}Ph_3}\ X^- \xrightarrow[\mathrm{(iii)\ NaClO_4}]{\mathrm{(i)\ B^-;\ (ii)\ PhSeBr}} (\mathrm{ring}\ n)\mathrm{C(H)CH(SePh)\overset{+}{P}Ph_3}\ \mathrm{ClO_4^-} \longrightarrow (\mathrm{ring}\ n)\mathrm{C{=}CH\overset{+}{P}Ph_3}\ \mathrm{ClO_4^-} \quad (49)$$

$n = 3$–6

$$\mathrm{Ph_3\overset{+}{P}CH{=}C(\overset{-}{O})R} \xrightarrow{\mathrm{Ph_3PBr_2}} \left| \mathrm{Ph_3\overset{+}{P}CH{=}C(R)O\overset{+}{P}Ph_3}\ 2\mathrm{Br^-} \right| \xrightarrow[-\mathrm{Ph_3PO}]{\mathrm{Et_3N}} \mathrm{Ph_3\overset{+}{P}C{\equiv}CR\ Br^-} \quad (50)$$

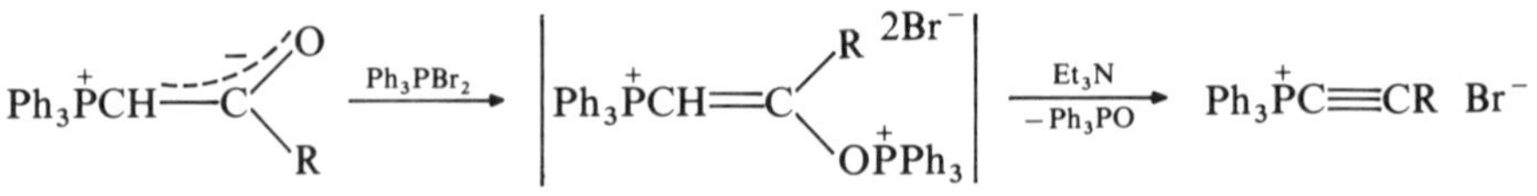

R = Ar, NR$'_2$

$$\mathrm{Me_2C{=}C(SMe)\bar{C}(\overset{+}{P}Ph_3)_2} \xrightarrow{HBF_4-H_2O} \left[\mathrm{Me_2CH{-}C(=O)\bar{C}(\overset{+}{P}Ph_3)_2}\ \ BF_4^- \right] \xrightarrow{-Ph_3PO} \mathrm{Me_2CHC{\equiv}C\overset{+}{P}Ph_3}\ BF_4^- \qquad (51)$$

d. Synthesis by modification of phosphonium ylides. As in the preceding syntheses, the formation of alk-1-enylphosphonium salts can be achieved by reaction of simple ylides with electrophilic reagents capable of later β-elimination. However, now the electrophilic reagent is used to introduce also a part of the alkenyl group (equation 52). The reaction of carbonyl compounds on α-heterosubstituted ylides can also give the α,β-unsaturated phosphonium salts (equation 53). Finally, in a classical way, α,β-keto ylides can be *O*-alkylated leading to the corresponding β-alkoxyvinylphosphonium salt[296] (reaction 54).

3. *Alkylation of α,β-unsaturated phosphines*

This method is used essentially for unstable unsaturated structures. The synthetic difficulty lies mainly in the preparation of the starting unsaturated phosphines, since alkylation takes place generally with high reactive agents (reactions 55 and 56).

D. Functional Phosphonium Salts

1. *α-Functionalized phosphonium salts*

a. Alcohols and derivatives. α-Hydroxyalkylphosphonium salts are obtained by a classical addition of tertiary phosphines to aldehydes or ketones, in the presence of acids (reaction 57). Anhydrous hydrobromic acid used can be generated *in situ*[121] by the action of bromine on acetone in the trialkylphosphines case. α-Alkoxyalkylphosphonium salts can also be obtained by addition of tertiary phosphines to enol ethers in the presence of acid (reaction 58). The synthesis could be extended to the precursors of enol ethers or alkoxycarbocations[299,300,303–305] and also alkoxyacetylene[266]. In the case where the enol ether is a polyfunctional structure, eventually chiral, the α-alkoxyalkphosphonium salt can be an excellent tool for the synthesis of natural products[302,303].

Finally, α-acyloxyalkylphosphonium salts can be prepared in a classical way either by the esterification of an α-hydroxyalkylphosphonium salt[306] or by alkylation of a tertiary phosphine by an α-chloroester[306]; however, the easiest way is by acylation of the adduct obtained in the reaction of a tertiary phosphine with an aldehyde (reaction 59)[306,307]. Further, this method of trapping allows α-silyloxyalkylphosphonium salts to be obtained[121].

b. Sulphurated or seleniated compounds. α-Functional salts can be obtained by sulphuration or seleniation of the corresponding phosphonium ylides. This preparation has been illustrated again in a cyclic series[308,309] (reaction 60). They can also be obtained by substitution of an α-bromoalkylphosphonium salt[310] (reaction 61). A less conventional

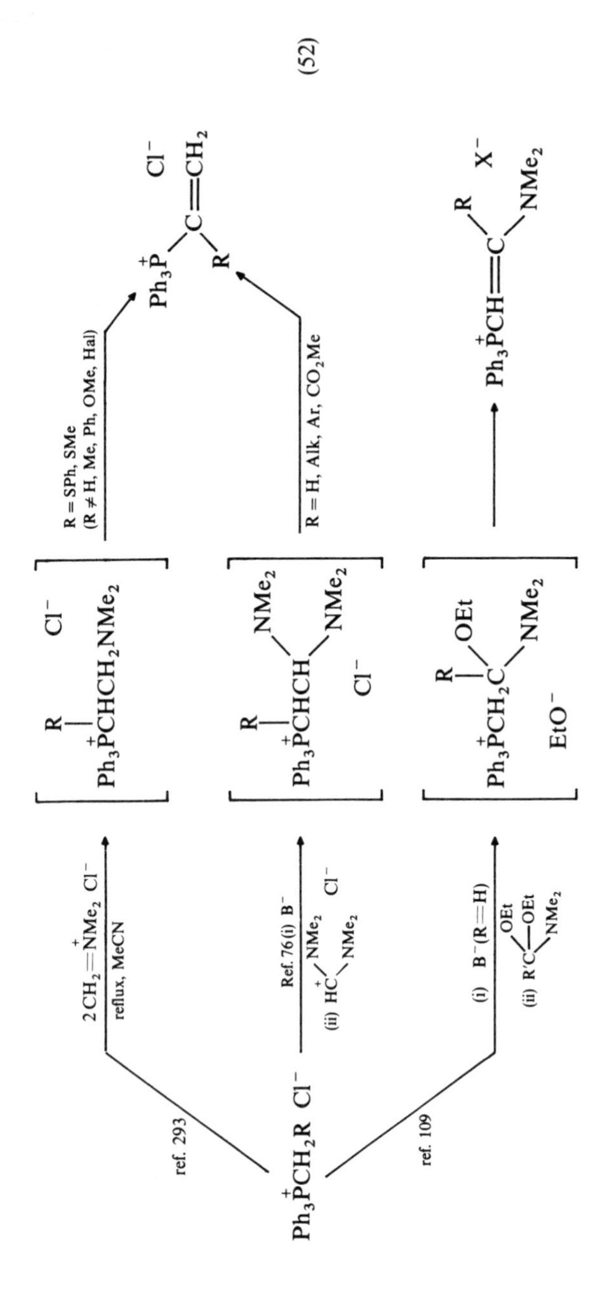
(52)
ref. 293
ref. 109
2 CH2=NMe2 Cl−
reflux, MeCN
Ref. 76 (i) B−
(ii) HC NMe2 NMe2 Cl−
(i) B− (R=H)
(ii) R′C OEt OEt NMe2
Ph3PCH2R Cl−
Ph3PCHCH2NMe2 R Cl−
Ph3PCHCH R NMe2 NMe2 Cl−
Ph3PCH2C R OEt NMe2 EtO−
R = SPh, SMe
(R ≠ H, Me, Ph, OMe, Hal)
R = H, Alk, Ar, CO2Me
Ph3P C=CH2 R Cl−
Ph3PCH=C R NMe2 X−

$$R_3\overset{+}{P}\overset{-}{C}(Y)(Z) + R'R''C{=}O \longrightarrow \left| R_3\overset{+}{P}{-}C(Z)(Y){-}C(R')(R''){-}O^- \right| \xrightarrow{-YO^-} R_3\overset{+}{P}(Z)C{=}CR'R''\ X^- \quad (53)$$

ref. 294: R = Ph, Y = $SiMe_3$, Z = H, R′ = alk-1-enyl, R″ = Alk, Ar

ref. 295: R = n-Bu, Y = $\overset{+}{P}Bu_3$, Z = F, R′ = F, R″ = perfluoroalkyl

$$2\ Ph_3\overset{+}{P}\overset{-}{C}H{-}Ar \xrightarrow[-\,Ph_3\overset{+}{P}CH_2Ar\ Cl^-]{ClCO_2Me} \left| Ph_3\overset{+}{P}C(Ar){-}C(=O)OMe \right| \xrightarrow{MeOSO_2F} Ph_3\overset{+}{P}(Ar)C{=}C(OMe)_2\ FSO_3^- \quad (54)$$

$$PhP(C{\equiv}CBu^t)_2 \xrightarrow[(R = H,\ X = I)]{RCH_2X} RCH_2(Ph)\overset{+}{P}(C{\equiv}CBu^t)_2\ X^- \quad (55)^{297}$$

$$(Ph_2P)_2C{=}CH_2 \xrightarrow[(X = I,\ SO_3F)]{2\,MeX} (Ph_2\overset{+}{P}Me)_2C{=}CH_2\ 2X^- \quad (56)^{140}$$

$$R_3P + R'CHO \xrightarrow{HX,\,anhyd} R_3\overset{+}{P}{-}CH(OH){-}R'\ X^- \quad (57)$$

$$\text{cyclic enol ether (O–CH=CH in ring)} + Ph_3P \xrightarrow{HX,\,anhyd.} \text{2-}(\overset{+}{P}Ph_3)\text{-oxacycle}\ X^- \quad (58)$$

X = Cl[298–300], Br[301,302], OTs[301] (85–99%)

$$RCHO + ClY + R'_3P \longrightarrow R'_3\overset{+}{P}{-}CH(OY)(R)\ Cl^- \quad (59)$$

R = Alk, Ar, R′ = n-Bu, Ph, Y = COAr, $SiMe_3$

method involves a radical addition of a 2-pyridylthio group and a vinylphosphonium salt[311] (reaction 62).

$$(CH_2)_n\ CH\overset{+}{P}Ph_3\ X^- \xrightarrow[\text{(ii) RYX}]{\text{(i) }B^-} (CH_2)_n\ C(\overset{+}{P}Ph_3)(Y)\ X^- \quad (60)$$

$n = 2^{308}$ (– 30 °C), RYX = MeSSMe, Pr^iSSPr^i, PhS—N(succinimide)

$n = 3^{309}$ (–75 °C), RYX = PhSeBr

$$ArC(=O)S^-K^+ + BrCH_2\overset{+}{P}Ph_3\ Br^- \xrightarrow[\text{CHCl}_3]{\text{4h, reflux}} ArC(=O)SCH_2\overset{+}{P}Ph_3\ Br^- \quad (61)$$

Ar = Ph, *p*-Tol, *p*-Anis   64–68%

$$\text{N-(acyloxy)pyridine-2-thione (N–OCOR)} + CH_2{=}CH\overset{+}{P}Ph_3\ Br^- \xrightarrow[\text{(-CO}_2)]{\Delta \text{ or } h\nu} RCH_2CH(S\text{-2-pyridyl})\overset{+}{P}Ph_3\ Br^- \quad (62)$$

(R = Adamantyl, C-Hex, Ph_2CHCH_2)   71–88%

c. Nitrogenous function. Phosphonium salts with an amino function, in the α-phosphorus position, either (i) protonated[312] or protected as (ii) an imine [313,314] or (iii) a lactam[315] can be obtained by direct alkylation of a tertiary phosphine with suitable nitrogen reagents (reaction 63). With some α-chloroimines, the alkylation reaction can

$$Ph_3P \xrightarrow{\text{(i)}^{312}\ PhCH=NPh/HCl,\ anhyd.} Ph_3\overset{+}{P}CH(Ph)\overset{+}{N}H_2Ph\ \ 2Cl^-$$

$$Ph_3P \xrightarrow{\text{(ii)}^{313}\ ClCH(Ph)N{=}C(Ph)Cl} Ph_3\overset{+}{P}CH(Ph)N{=}C(Ph)Cl\ \ Cl^- \quad (63)$$

$$Ph_3P \xrightarrow{\text{(iii)}^{315}\ ClCH_2N(\text{pyrrolidin-2-one})} Ph_3\overset{+}{P}CH_2\text{—N(pyrrolidin-2-one)}\ \ Cl^-$$

lead to a rearranged structure[314] (reaction 64). When α-aminoalkylphosphonium salts are also α,β-unsaturated, they can be obtained by simple substitution of the corresponding α-halogenated equivalent enamine[141,172] (reaction 65).

$$CF_3CH(Cl)N{=}CHPh \xrightarrow{Ph_3P} \left[CF_3CH{=}NCH(Ph)\overset{+}{P}Ph_3 \; Cl^- \right] \xrightarrow{20^{\circ}C} CF_3CH_2N{=}C(Ph)\overset{+}{P}Ph_3 \; Cl^- \qquad (64)$$

$$R_3P + (Me_2N)(X)C{=}C(R')(R'') \longrightarrow (Me_2N)(R_3\overset{+}{P})C{=}C(R')(R'') \; X^- \qquad (65)$$

$R = n$-Bu, Ph, $X = I$, Cl

d. α-Halogenated salts. α-Haloalkylphosphonium salts can be obtained by halogenation of a phosphonium ylide (reaction 66). α-Fluoroalkylphosphonium salts are not prepared in this way but can be obtained by fluorination with SF_4 or better with Et_2NSF_3[317] (reaction 67). α-Fluorohalomethylphosphonium salts can be obtained by the alkylation of a phosphine with a polyhalomethane (reaction 68): the facility and the course of the reaction (which can also produce a bisphonium ylide)[295] are highly dependent on the fluorohalomethane used[55,318] and also on the phosphine[319].

$$Ph_3\overset{+}{P}\overset{-}{C}HR + XY \longrightarrow Ph_3\overset{+}{P}CH(X)R \; Y^- \qquad (66)$$

$R \neq H$, $X = Cl$, Br, Y, $XY = CuCl_2$[84], $X(CF_2)_2X$[316] ($X = Cl$, Br, I)

$$Ph_3\overset{+}{P}CH_2OH \; BF_4^- + Et_2NSF_3 \xrightarrow[CH_2Cl_2]{0^{\circ}C} Ph_3\overset{+}{P}CH_2F \; BF_4^- \quad 88\% \qquad (67)$$

$$R_3P + CFXY_2 \longrightarrow R_3\overset{+}{P}{-}CFY_2 \; X^- \qquad (68)$$

$R = Ph$, n-Bu, Me_2N, Et_2N, $X = Br$, Cl, $Y = F$, Br, Cl

e. Carbonyl functions and derivatives. Corresponding phosphonium salts are not usually very stable and can be obtained by the action of appropriate acylating agents (reaction 69). When the acylation takes place with a chloroformate, the intermediate adduct is unstable, even at room temperature, and is spontaneously decarboxylated[321] reaction 70).

$$Ph_3P \xrightarrow[\text{ref. 110}]{ArC(=O)-C(Br)=N-NHAr} Ph_3\overset{+}{P}C(=NNHAr)C(=O)Ar \quad Br^-$$

$$Ph_3P \xrightarrow[\text{ref. 320};\ (-Ph_3PS)]{0.5\ R_2\overset{+}{N}=C\langle S_4 \rangle C=\overset{+}{N}R_2\ \ ClO_4^-} Ph_3\overset{+}{P}-C(=S)NR_2 \quad ClO_4^- \tag{69}$$

$$R_3P + ClC(=O)OR' \xrightarrow{20\,^\circ C} \left[R_3\overset{+}{P}C(=O)OR' \quad Cl^- \xrightarrow{-R'Cl} R_3\overset{+}{P}C(=O)O^- \right] \xrightarrow{-CO_2} R_3P \tag{70}$$

Thioacetalized formylphosphonium salts, which are excellent precursors for tetrathiafulvalene and analogous derivatives, are obtained in high yields directly from tertiary phosphines (reaction 71).

$$\text{(1,3-dithiolane)}C(Y)H \xrightarrow[\text{or (ii) } R_3P]{\text{(i) } Ph_3P/HBF_4} \text{(1,3-dithiolane)}C(\overset{+}{P}Ph_3)H \quad X^- \tag{71}$$

(i) Y = OEt[322], SMe[323]
(ii) Y = Cl[191]

2. β-Functionalized phosphonium salts

a. Phosphine alkylation by a β-functionalized alkyl halide. Usually β-functionalized phosphonium salts can be easily obtained by the direct alkylation of phosphine with the appropriate functional reagent (reaction 72).

$$R_3P + X-\overset{|}{\underset{|}{C}}-\overset{|}{\underset{|}{C}}-Y \longrightarrow R_3\overset{+}{P}-\overset{|}{\underset{|}{C}}-\overset{|}{\underset{|}{C}}-Y \quad X^- \tag{72}$$

$$Y = OH, OR, SR, NR_2, COR, CO_2H, \text{Hal, etc.}$$

The function thereby introduced can be an alcohol[324], an ether[325,326], a sulphide[289], an amine[327]a ketone[186,328–330], an acid[42] or a halide[327] Further, the Y function initially introduced can be modified later[324,327,331,332].

b. Addition of YH compounds, with mobile hydrogen, to α,β-unsaturated phosphonium salts. This method takes advantage of the Michael-olefinic character of the vinyl and ethynylphosphonium salts, allowing the nucleophilic addition of mobile hydrogen compounds to C=C or C≡C multiple bonds activated by the phosphonio group (reaction 73).

$$\mathrm{Ph_3\overset{+}{P}C{=}C\!\!<\ X^- + YH \longrightarrow Ph_3\overset{+}{P}CH{-}C{-}Y\ X^-} \qquad (73)$$

$\mathrm{YH = ROH}$[91,287,333], RSH[91,287], $\mathrm{RNH_2}$[91,178,287,334], $\mathrm{RCO_2H}$[284]

The same kind of addition can be carried out on the ethynylphosphonium salt **26**, generated *in situ* from the 1,2-vinylenebisphosphonium salt **25**, and, depending on the experimental conditions, the vinylphosphonium salt produced can eventually undergo a second addition of YH to produce the β-acetalized or-thioacetalized salt. The same type of β-heterosubstituted vinylphosphonium salt can also undergo an acidic hydrolysis that leads to the ketonic salt itself[335] (reaction 74).

$$\underset{(\mathbf{26})}{\mathrm{Ph_3\overset{+}{P}C{\equiv}CH\ Br^-}} \xrightarrow{\mathrm{YH}} \mathrm{Ph_3\overset{+}{P}CH{=}CHY\ Br^-} \left(\xrightarrow{\mathrm{YH}} \mathrm{Ph_3\overset{+}{P}CH_2CH{<}^{Y}_{Y}\ X^-} \right) \qquad (74)$$

$$\uparrow \mathrm{YH,\ Et_3N\ (-Ph_3P)}$$

$$\underset{(\mathbf{25})}{\mathrm{Ph_3\overset{+}{P}CH{=}CH\overset{+}{P}Ph_3\ 2\,Br^-}}$$

β-carbonyl derivatives can also be obtained by direct prototropy of the intermediate enol form, when the ethynylphosphonium salt reacts with water[97].

Buta-1,4-dienylenebisphosphonium salts behave like bisvinylphosphonium salts and can undergo a double addition in the 2,3-position (reaction 75)[335].

$$\mathrm{Ph_3\overset{+}{P}CH{=}CHCH{=}CH\overset{+}{P}Ph_3\ 2X^-} + \mathrm{YH{\frown}YH} \longrightarrow \text{(cyclic 2,3-bis(}\mathrm{CH_2\overset{+}{P}Ph_3}\text{) Y,Y-heterocycle)}\ \mathrm{2X^-} \qquad (75)$$

$\mathrm{Y = O,\ NH}$

c. Functional chain creation from a phosphonium ylide. This method is particularly suitable for the formation and separation[336,337] of the β-hydroxyalkylphosphonium diastereoisomers, obtained by acidic cleavage, at low temperature, of the oxaphosphetanes formed by reacting an ylide with an aldehyde, under conditions favourable either to the *erythro* isomer or to the *threo* isomer (reaction 76). Usually the action of an acylating agent (acid chloride[338], carboxylic acid anyhydride[338,339], chloroformate[296]) on a phosphonium ylide forms a β-ketophosphonium salt (after transylidation and acid treatment when there is a hydrogen in the α-position to phosphorus) (reaction 77).

$$\mathrm{Ph_3\overset{+}{P}CH_2Pr^n} \xrightarrow[\text{(b) PhCHO, } -78\,^\circ\mathrm{C}]{\text{(a) LiHMDS}} \underset{cis/trans\,=\,72:28}{\mathrm{Ph_3P{-}CHPr^n,\ O{-}CHPh}} \xrightarrow{\text{(i) or (ii)}} \mathrm{Ph_3\overset{+}{P}CH(Pr^n)CH(OH)Ph\ Br^-} \qquad (76)$$

(i) $\mathrm{HBr}(-78\,^\circ\mathrm{C})$: *erythro*/*threo* = 72:28

(ii) (a) $\mathrm{Bu^nLi}$ ($-78\,^\circ\mathrm{C}$); (b) HBr ($-78\,^\circ\mathrm{C}$): *erythro*/*threo* = 2:98

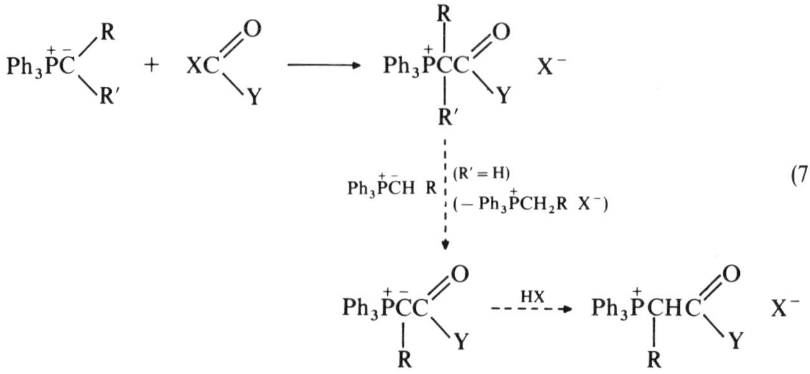

$$\mathrm{Ph_3\overset{+}{P}\overset{-}{C}RR'} + \mathrm{XC(=O)Y} \longrightarrow \mathrm{Ph_3\overset{+}{P}CRR'C(=O)Y\ X^-} \xrightarrow[(-\,\mathrm{Ph_3\overset{+}{P}CH_2R\ X^-})]{\mathrm{Ph_3\overset{+}{P}\overset{-}{C}H\,R}\ (R'=H)} \mathrm{Ph_3\overset{+}{P}\overset{-}{C}(R)C(=O)Y} \xrightarrow{\mathrm{HX}} \mathrm{Ph_3\overset{+}{P}CH(R)C(=O)Y\ X^-} \quad (77)$$

The same acylating agents, but also other less powerful reagents (esters, carbonates, carbamates, carbodiimides, isocyanates), can react with phosphonium diylides[340] to give β-oxoalkylphosphonium salts and derivatives without wasting half of the initial diylide thanks to the intramolecular prototropy (reaction 78). The addition of N-methylnitrilium to a stabilized ylide forms, in moderate yields, enaminophosphonium salts by isomerization of the intermediate imine[98] (reaction 79).

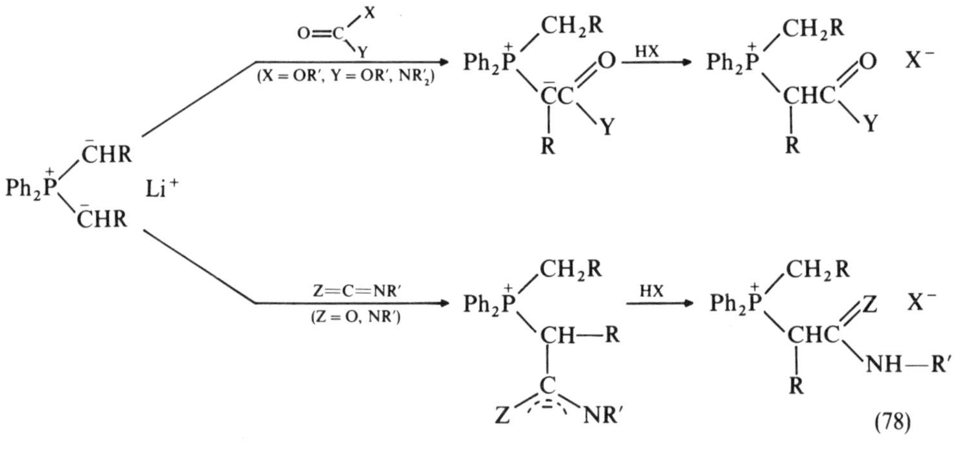

$$\mathrm{Ph_2\overset{+}{P}(\overset{-}{C}HR)_2\ Li^+} \xrightarrow[(X = OR',\ Y = OR',\ NR'_2)]{\mathrm{O{=}C(X)Y}} \mathrm{Ph_2\overset{+}{P}(CH_2R)\overset{-}{C}(R)C(=O)Y} \xrightarrow{\mathrm{HX}} \mathrm{Ph_2\overset{+}{P}(CH_2R)CH(R)C(=O)Y\ X^-}$$

$$\mathrm{Ph_2\overset{+}{P}(\overset{-}{C}HR)_2\ Li^+} \xrightarrow[(Z = O,\ NR')]{\mathrm{Z{=}C{=}NR'}} \mathrm{Ph_2\overset{+}{P}(CH_2R)CH(R)\text{—}C(\text{⋯}Z)(\text{⋯}NR')^-} \xrightarrow{\mathrm{HX}} \mathrm{Ph_2\overset{+}{P}(CH_2R)CH(R)C(=Z)NH\text{—}R'\ X^-} \quad (78)$$

$$\mathrm{Ph_3\overset{+}{P}\overset{-}{C}HCO_2Me} + \mathrm{RC{\equiv}\overset{+}{N}Me}\ \mathrm{TfO^-} \longrightarrow \left[\mathrm{(MeO_2C)(Ph_3\overset{+}{P})CHC(=NMe)R}\right]\ \mathrm{TfO^-} \longrightarrow \mathrm{(MeO_2C)(Ph_3\overset{+}{P})C{=}C(NHMe)R}\ \mathrm{TfO^-} \quad (79)$$

$R = Pr^i$, Ph, Tol, Me

31–64%

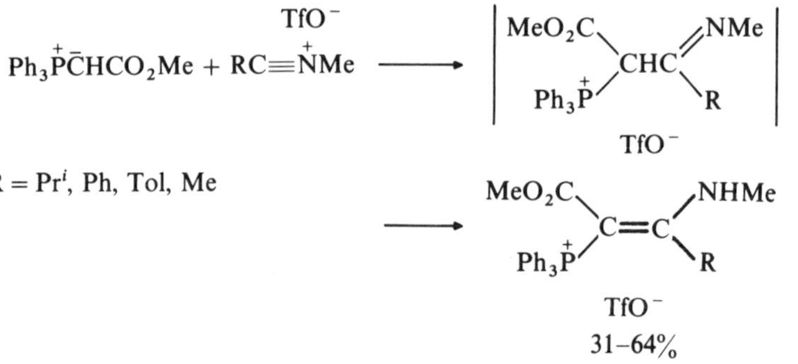

3. n-Functionalized phosphonium salts (n ⩾ γ)

By analogy with one of the syntheses used for the β-hydroxylalkylphosphonium salts, the γ-hydroxylalkyphosphonium salts can be obtained by the opening of an oxirane ring with a phosphonium ylide[341] (reaction 80). However, most of the methods achieve phosphine alkylation by using functional alkylating reagents (reactions 81–84). This approach can also be used for benzylating agents with substituents (—NHR[343], —N=C=S[77], —N$^+$≡C$^-$ [344]) on the aromatic ring.

$$Ph_3\overset{+}{P}\overset{-}{C}H_2 + CH_2\underset{O}{—}C(R')(R) \xrightarrow[\text{(ii) HX}]{\text{(i) LiBr or LiI}} Ph_3\overset{+}{P}CH_2CH_2CR(R')(OH) \quad X^- \qquad (80)$$

R = H, Me, R′ = H, Me, CH_2=CH— 32–61%

$$Ph_3P \xrightarrow[\text{ref. 342}]{(CH_2)_nO\ (n = 3, 4),\ CF_3CO_2H \text{ or } CF_3SO_3H} Ph_3\overset{+}{P}(CH_2)_nOH\ X^- \qquad (81)$$

$$Ph_3P \xrightarrow[\text{ref. 327}]{HO(CH_2)_nNR_2,\ HBr\ (n = 2 \text{ or } 4)} Ph_3\overset{+}{P}(CH_2)_nNR_2\ X^-(R = H, Alk) \qquad (82)$$

$$Ph_3P \xrightarrow[\text{refs 196, 327}]{Br(CH_2)_nNR_2\ (n = 2\text{–}5)} Ph_3\overset{+}{P}(CH_2)_nNR_2\ X^-$$

$$Ph_3P \xrightarrow{Br(CH_2)_nBr\ (n = 2\text{–}4)} Ph_3\overset{+}{P}(CH_2)_nBr\ Br^- \xrightarrow[\text{refs 196, 327}]{R_2NH} Ph_3\overset{+}{P}(CH_2)_nNR_2\ X^-$$

$$Ph_3P \xrightarrow[\text{ref. 82}]{I(CH_2)_nSiR_3\ (n = 3)} Ph_3\overset{+}{P}(CH_2)_nSiR_3\ I^- \qquad (83)$$

$$Ph_3P \xrightarrow[\text{ref. 42}]{X(CH_2)_nCO_2H;\ X = Cl\ (n = 2, 3);\ X = Br\ (n = 5, 10, 11)} Ph_3\overset{+}{P}(CH_2)_nCO_2H\ X^- \qquad (84)$$

Two standard routes to γ-functional phosphonium salts with electron-withdrawing groups consist in using the corresponding Michael alkenes to achieve either halide substitution in the β-position to the functional group, in order to obtain γ-functional vinylphosphonium salts, or Michael addition in the presence of acid to produce the γ-functional alkylphosphonium salt (reaction 85).

The Michael type of addition, in the presence of acids, has been extended successfully to cyclopropyl ketones[347] (reaction 86). Of course, the functional group can also undergo subsequently a transformation[202,346,347,349], compatible with the phosphonium structure (acetalization, thioacetalization, imination, etc.).

$$R_3P \xrightarrow[\substack{(X = Cl, Br;\\ G_a = COR^{186,268,269,345}, CN^{267})}]{XC{=}CG_a} R_3\overset{+}{P}C{=}CG_a\ X^-$$

$$R_3P \xrightarrow[(G_a = COR^{346-348}, CHO^{349})]{>C{=}C<^{G_a} / HX} R_3\overset{+}{P}CCHG_a\ X^- \quad (85)$$

$$Ph_3P + \triangleright\!-C(=O)R \xrightarrow{HBr} Ph_3\overset{+}{P}(CH_2)_3\overset{O}{\overset{\|}{C}}R\quad Br^- \quad (86)$$

E. Synthesis of Polyphosphonium Salts

The case of molecular structures with several phosphonium units, the 'multiphosphoniums', can be distinguished from the macromolecular structures involving organic polymers with many phosphonium units, called the 'polymeric phosphoniums'.

1. *Multiphosphonium synthesis*

Bis- (or poly-) phosphonium salts can be prepared in a classical way by the action of tertiary phosphines on a di- (or poly-) halogenated hydrocarbon backbone[350–352] or fluorinated hydrocarbon[318], or heteroatomic chain[353] or aliphatic derivative, but more easily on di- (or poly-) halogenated allylic[281,289], propargylic[193] or benzylic[169,354] compounds (reaction 87). Polyhalogenated derivatives of the benzylic type were used to

$$X\frown Q\frown X + 2R_3P \longrightarrow R_3\overset{+}{P}\frown Q\frown\overset{+}{P}R_3\ 2\ X^- \quad (87)$$
$$(X = Cl, Br, I)$$

$$P\left(-C_6H_4-CH_2OMe\right)_3 \xrightarrow{RX} R\overset{+}{P}Ar_3\ X^-$$

'Ar_3P' RX = MeI or $PhCH_2Br$

$$\xrightarrow[(ii)\ 3\,Ar_3P]{(i)\ 3\,Me_3SiI} R\overset{+}{P}\left(-C_6H_4-CH_2\overset{+}{P}Ar_3\right)_3\ 4\ I^-$$

$$\downarrow (i)\ 9\,MeSiI;\ (ii)\ 9\,Ar_3P \quad (88)$$

$$R\overset{+}{P}\left[-C_6H_4-CH_2\overset{+}{P}\left(-C_6H_4-CH_2\overset{+}{P}Ar_3\right)_3\right]_3\ 13\ I^-$$

work out an attractive recurrent synthesis of the so-called 'phosphonium cascade molecules'[355,356] (reaction 88). Unsaturated structures can occasionally produce secondary reactions during the formation of bisphosphonium salts: isomerization of the double bond[281] or elimination–addition[289] for allylic systems, addition reactions on the triple bond[193,335] for propargylic structures. Finally, for the 1,1-bis(bromomethyl)cyclopropane,[357] the reaction proceeds to the monophosphonium salt after ring opening (reaction 89).

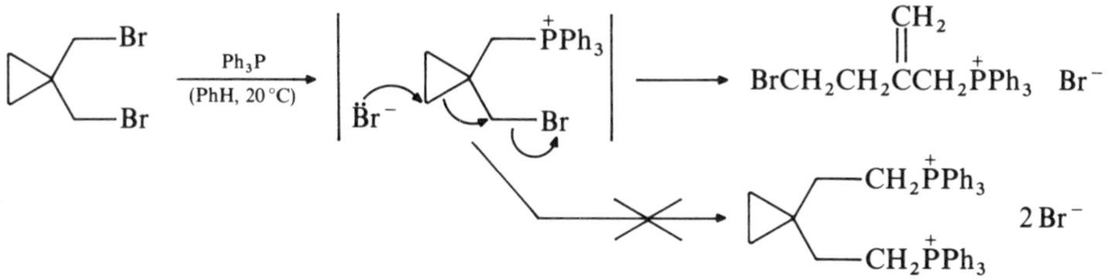

$$BrCH_2CH_2C(=CH_2)CH_2\overset{+}{P}Ph_3\ Br^- \qquad \text{(not: } 1,1\text{-}(CH_2\overset{+}{P}Ph_3)_2\text{cyclopropane}\ 2Br^-) \tag{89}$$

Ph$_3$P (PhH, 20 °C)

For α-halo ketones, bisphosphonium salt formation can also take place by vinylation or arylation of the phosphine (reaction 90). Bisphosphonium salts can also be obtained, in moderate yield, by the introduction of two phosphines on the same aromatic or heterocyclic ring, at high temperature[358] (reaction 91). For pyridine, activation by triflic anhydride allows successive additions of two phosphonio groups in the 2,4-positions[247] (reaction 92).

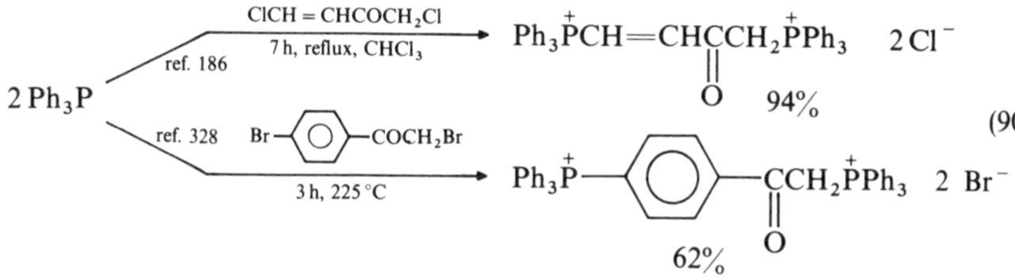

$$2\,Ph_3P \xrightarrow[\text{7 h, reflux, CHCl}_3]{\text{ClCH=CHCOCH}_2\text{Cl (ref. 186)}} Ph_3\overset{+}{P}CH{=}CHC(O)CH_2\overset{+}{P}Ph_3\ \ 2\,Cl^-\quad 94\%$$

$$2\,Ph_3P \xrightarrow[\text{3 h, 225 °C}]{\text{Br-C}_6\text{H}_4\text{-COCH}_2\text{Br (ref. 328)}} Ph_3\overset{+}{P}\text{-C}_6\text{H}_4\text{-}C(O)CH_2\overset{+}{P}Ph_3\ \ 2\,Br^-\quad 62\% \tag{90}$$

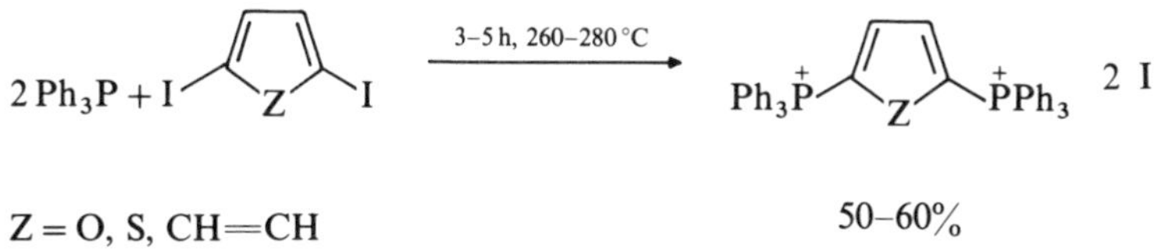

$$2\,Ph_3P + I\text{-(C}_4\text{H}_2\text{Z)-}I \xrightarrow{\text{3–5 h, 260–280 °C}} Ph_3\overset{+}{P}\text{-(C}_4\text{H}_2\text{Z)-}\overset{+}{P}Ph_3\ \ 2\,I^- \tag{91}$$

Z = O, S, CH=CH 50–60%

$$\text{pyridine} \xrightarrow[\text{(ii) Ph}_3\text{P (MeCN)}]{\text{(i) Tf}_2\text{O}} \text{4-}\overset{+}{P}Ph_3\text{-pyridine TfO}^-\ (91\%) \xrightarrow[\text{(ii) R}_3\text{P}]{\text{(i) Tf}_2\text{O}} \text{2-}R_3P^+\text{-4-}\overset{+}{P}Ph_3\text{-pyridine}\ \ 2\,TfO^- \tag{92}$$

R = *n*-Bu: 26%
R = Ph: 88%

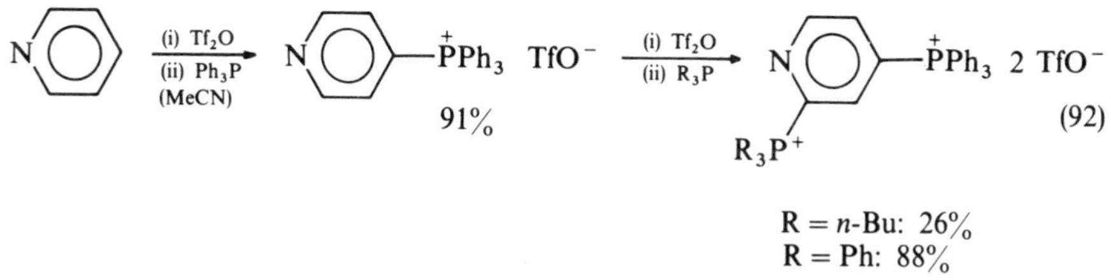

1,2-Vinylenebisphosphonium salts were obtained, in very good yields, by the reaction of a tertiary phosphine with either acetyl bromide[359–362] or β-bromovinyl sulphone[363] (reaction 93). The second method was extended to aminophosphines $(R_2N)_3P$[87], whereas the first was extrapolated to but-1,4-dienylenebisphosphonium salts[58,362] (reaction 94). Another classical route to bis- or poly-phosphonium salts consists in the use of a di- or poly-phosphine with a suitable alkylating agent[66,189,331,364,365] (reaction 95). This method can present some difficulties for methylene[250,366] and 1,1-vinylene[130] bisphosphines, for which, probably because of steric hindrance, the second alkylation is often more difficult than the first.

$$2R_3P \xrightarrow{MeCOBr} \left[R_3\overset{+}{P}C(=O)Me\ Br^- \xrightarrow[-HBr]{AcBr} R_3\overset{+}{P}C(OAc)=CH_2\ Br^- \xrightarrow{R_3P,\ HBr} R_3\overset{+}{P}CH(OAc)CH_2\overset{+}{P}R_3\ \ 2Br^-\right] \xrightarrow{-AcOH}$$

$$2R_3P \xrightarrow{ArSO_2CH=CHBr} \left[ArSO_2CH=CH\overset{+}{P}R_3\ \ Br^-\right] \xrightarrow{R_3P} R_3\overset{+}{P}(H)C=C(H)\overset{+}{P}R_3\ \ 2X^- \tag{93}$$

$$2R_3P + CH_2=CHCH_2C(=O)Br \longrightarrow R_3\overset{+}{P}CH=CHCH=CH\overset{+}{P}R_3\ \ 2\,Br^- \tag{94}$$

$$R_2P\frown Q\frown PR_2 \ + \ 2R'X \longrightarrow R_2\overset{+}{P}(R')\frown Q\frown\overset{+}{P}R_2(R')\ \ 2X^- \tag{95}$$

ref. 250: R = Ph, $Q = (CH_2)_n$ $(n = 1-3)$, R′X = 9-bromofluorene (H, Br)

ref. 66,189: R = Ph, $Q = (CH_2)_n$ $(n = 1-6)$, R′X = $Me_2C=CHCH_2X$

When a di- or a poly-phosphine is reacted with a di or poly-halogenated compound, the reaction can take place with the formation of cyclic polyphosphonium salts when the structure[26,365,367,368] or the experimental conditions[352,369,370] favour cyclization in regard to the linear polycondensation (in some cases[366], the reaction takes place to give simultaneously the two cyclic and open adducts) (reaction 96). It must be pointed out that when the phosphorus atoms of the phosphonio groups are chiral centres, the bisphosphonium salts, such as **8** and **9**, are diastereoisomeric structures which can be separated[26,66] or even the enantiomeric forms[66,189].

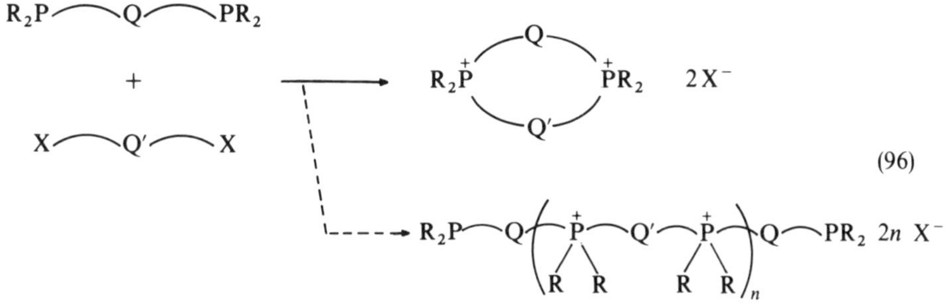

$$R_2P\frown Q\frown PR_2 + X\frown Q'\frown X \longrightarrow R_2\overset{+}{P}(-Q-)(-Q'-)\overset{+}{P}R_2 \; 2X^- \quad \text{or} \dashrightarrow R_2P\frown Q\left(\overset{+}{P}R_2\frown Q'\frown\overset{+}{P}R_2\right)_n Q\frown PR_2 \; 2n\,X^- \qquad (96)$$

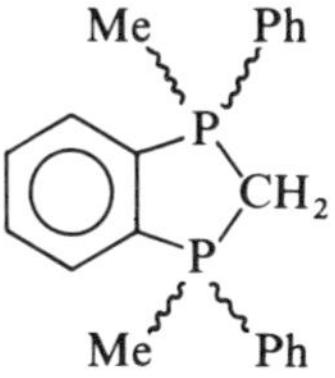

(8)[26] $2X^-$

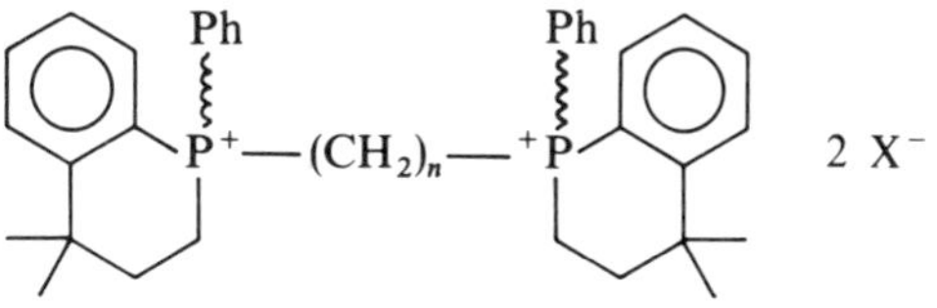

(9)[66,189] $Ph\overset{+}{P}-(CH_2)_n-\overset{+}{P}Ph$ $2X^-$

Another synthetic route to 1,2-ethylenebisphosphonium salts was developed by oxidative coupling of two phosphonium ylide units[84,144], with the creation of ethylene bridges by C—C bond formation (reaction 97). This C—C coupling takes place, in moderate yields, by monoelectron transfer with stoechiometric quantities of the metallic salt and, in some cases (particularly for α-disubstituted ylides[84]), the reaction affords only the ylide chlorination. For a monosubstituted ylide (R = Ph, R′ = Me), the reaction gives a diastereoisomeric mixture with some diastereoselection (ca 3:1)[84].

$$2\ R_3\overset{+}{P}\overset{-}{C}HR \xrightarrow{MCl_n} R_3\overset{+}{P}CH(R')CH(R')\overset{+}{P}R_3\ 2\ X^- \quad (22\text{–}67\%) \qquad (97)$$

Ref. 84: $R = Me, Et, Ph$, $R' = H, Me$, $MCl_n = CuCl_2$

Ref. 144: $R = NEt_2$, $R' = H$, $MCl_n = NbCl_5, WCl_6$

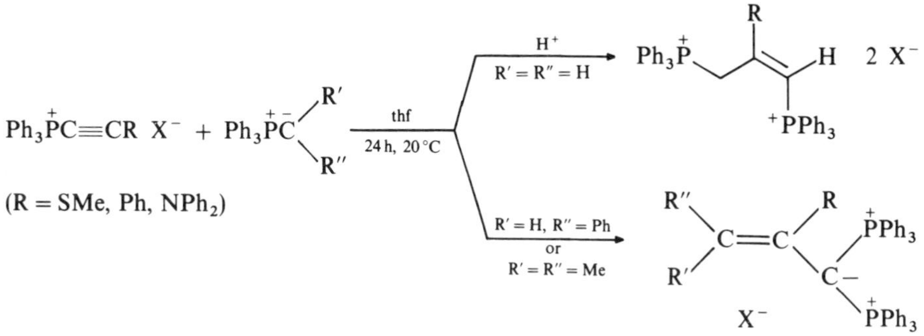

$$Ph_3\overset{+}{P}C{\equiv}CR\ X^- + Ph_3\overset{+}{P}\overset{-}{C}R'R'' \xrightarrow[24\,h,\ 20\,^\circ C]{thf}$$

(R = SMe, Ph, NPh₂)

$\xrightarrow[R' = R'' = H]{H^+}$ $Ph_3\overset{+}{P}CH_2C(R){=}CH\overset{+}{P}Ph_3$ $2\ X^-$

$\xrightarrow{R' = H,\ R'' = Ph \text{ or } R' = R'' = Me}$ $R''R'C{=}C(R)\overset{-}{C}(\overset{+}{P}Ph_3)_2$ X^-

(98)

Phosphonium ylides can also be used to prepare from substituted ethynylphosphonium[371], 1,3-propenylidenebisphosphonium salts or, more curiously, 1,1-propenylidenebisphosphonium ylides (reaction 98).

2. Synthesis of polymeric phosphonium salts

The most straightforward way to obtain polymeric phosphonium salts involves introducing the phosphonio groups on to a suitable polymeric structure, for example by reacting tertiary phosphines with a poly(chloromethylstyrene) (reaction 99). The polymeric phosphonium salts obtained in this way are mostly used as polymer-supported phase-transfer catalysts for nucleophilic substitutions reactions under triphase conditions.

$$\text{(polystyrene)}_n\text{-}C_6H_4CH_2Cl + n\,R_3P \longrightarrow \text{(polystyrene)}_n\text{-}C_6H_4CH_2\overset{+}{P}R_3Cl^- \qquad (99)$$

$R = n$-Bu[372–375], Ph[376]

Another way to prepare polymeric phosphonium salts involves the radical-initiated polymerization of allylphosphonium salts[377] or their copolymerization with methacrylates[378].

Finally, polyphosphonium salts with the phosphorus atoms as a part of the polymeric chain can be obtained starting from oxaphospholanes[379] (reaction 100).

$$\text{PhP(oxaphospholane)} \xrightarrow{\text{cat. E}^+} \left[\overset{O}{\overset{\|}{P}}(Ph)CH_2CH_2CH_2 \right]_n \xrightarrow[\text{(ii) } Bu^i_2AlH \text{ (iii) RI}]{\text{(i) } COCl_2} \left[\overset{R}{\underset{Ph}{P^+}}CH_2CH_2CH_2 \; I^- \right]_n \qquad (100)$$

F. Synthesis of Cyclic Phosphonium Salts

Two kinds of synthesis can be used to prepare phosphonium salts in which the phosphorus atom forms part of a ring: either syntheses with cyclization on the phosphorus itself, or syntheses with cyclization taking place between two chains already linked to the phosphorus.

1. Synthesis with cyclization on the phosphorus

The most direct synthesis is performed by reaction of a dihalo compound with a tricoordinated phosphorus compound including a reactive P—Y bond (reaction 101). This synthesis, first used with secondary phosphines, the corresponding phosphides or diphosphines ($Y = H$, Na, K, PPh_2)[2b,380], has subsequently been developed also with silylphosphines ($Y = SiMe_3$),[132,381–383] which are easier to handle (reaction 102). The same kind of cyclization by intramolecular alkylation can be achieved using a functional phosphine prepared in a multi-step synthesis[380,384,385] (reaction 103).

$$R_2PY + X\text{–}Q\text{–}X \xrightarrow{-YX} [R_2P\text{–}Q\text{–}X] \longrightarrow R_2\overset{+}{P}\frown Q \; X^- \quad (101)$$

$Y = H, Na, K, PPh_2, SiMe_3$, etc

$$Ph_2PSiMe_3 + Q(CH_2X)_2 \xrightarrow[20\,^\circ C \text{ or reflux } (-Me_3SiX)]{PhMe} Ph_2\overset{+}{P}(CH_2)_2Q \; X^- \quad (102)$$

CH_2Br / CH_2Br (cyclopentane) 67%[132]; CH_2Br / CH_2Br (cyclopentene) 73%[382]; CH_2Br / CH_2Br (naphthalene) 86%[383]; CH_2Cl / CH_2Cl (benzene) 91%[383]

Et, Ph—P, OMe, R, R′ $\xrightarrow[\text{(ii) } Na_2CO_3 \text{ (iii) } CHCl_3, \Delta]{\text{(i) HBr (g) HBr–AcOH}}$ Et, Ph—P^+, R, R′ Br^- (103)

R, R′ = H: 50%[384]
R, R′ = $(CH_2)_3$: 43%[385]

$Q(PR_2)_2 + Q'X_2 \longrightarrow$ cyclic $[Q(PRR')_2Q']^{2+}$ $2X^-$ and/or cyclic tetraphosphonium $4X^-$ (104)

Q =	CH=CH	CH_2	CH_2OCH_2	$(CH_2)_n$	$(CH_2)_3O(CH_2)_3$	CH_2CH_2
Q′ =	CH_2CH_2	CH_2OCH_2	CH_2CH_2	$(CH_2)_m$	$(CH_2)_3O(CH_2)_3$	CH_2-*p*-$C_6X_4CH_2$
Ref.:	386	387	387	388	389, 390	391

The reaction of a diphosphine with a dihalo compound, resulting in the formation of a di- or tetra-phosphonium salt[2c], is a special case because if fits well with a cyclization on phosphorus, but also at the same time with a cyclization between two chains already linked to the second phosphorus (reaction 102). Another kind of ring closure on the phosphorus results from the biphilic character of halophosphines toward dienic[392] or acetylenic[28,393] systems (reaction 105).

R_2PCl —(ref. 392; R'–CH=CH–CH=CH–R')→ 2,5-di-R'-phospholenium $\overset{+}{P}R_2$ Cl^-

R_2PCl —(ref. 28, 393; $R'C{\equiv}CR'$ ($AlCl_3$))→ $R_2\overset{+}{P}$ (phosphirenium, R', R') Cl^- (105)

2. Synthesis by cyclization between two chains linked to phosphorus

A first possibility for this kind of cyclization is an intramolecular *C*-alkylation of the anionic position in a phosphonium ylide[394] (reaction 106). Actually, as shown for the preparation of a spirophosphonium salt[395], this synthesis affords mixtures and leads to purification difficulties as soon as there are several, non-equivalent, possibilities for the anionic position α to the phosphorus (reaction 107). In the same way, a similar synthesis using the dialkylation of phosphonium diylide with a α,ω-dihaloalkane presents transylidation and elimination problems[395] (reaction 108). A second kind of ring closure between two phosphorus substituents is achieved in a phenyl- or benzyl-phosphonium salt by Friedel–Crafts alkylation or acylation[396–400] (reactions 109–111). Advantage has been taken of this method with bisphosphonium salts to obtain the corresponding diastereoisomers and even to separate the enantiomers of the 'like' diastereoisomer[66,189] (reaction 112).

o-($BrCH_2CH_2$)$C_6H_4\overset{+}{P}(CH_2Ph)_3$ Br^- —(Bu^tOK, dmf)→ [o-($BrCH_2CH_2$)$C_6H_4\overset{+}{P}(CH_2Ph)_2$–$\overset{-}{C}HPh$] → 1,1-bis(benzyl)-2-phenyl-phosphinolinium ClO_4^- (PhCH$_2$, CH$_2$Ph, Ph) 80% (106)

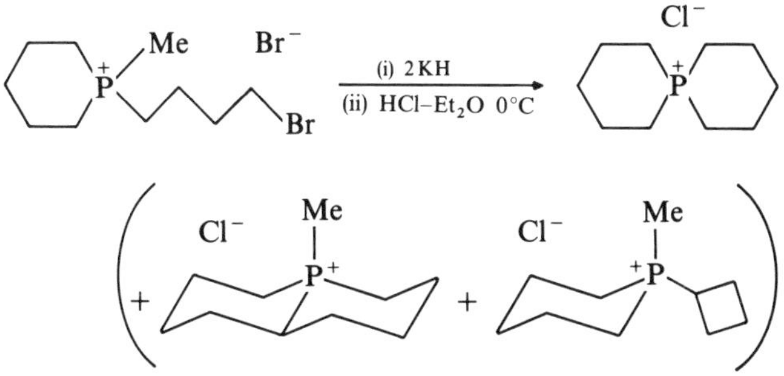

Me Br⁻ P⁺ Br —(i) 2 KH; (ii) HCl–Et_2O 0 °C→ Cl⁻ P⁺

+ (Cl⁻ Me P⁺ + Cl⁻ Me ⁺P) (107)

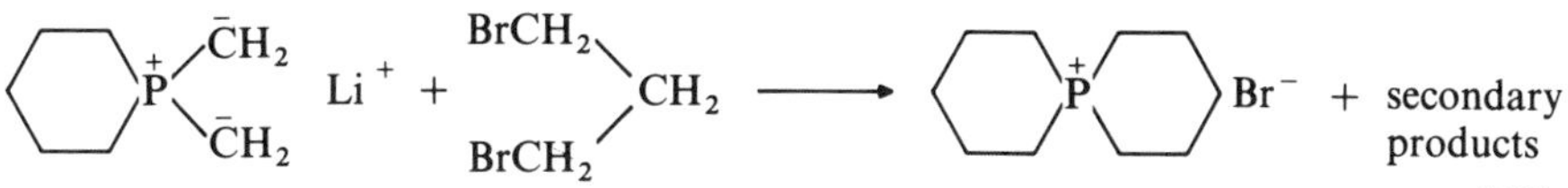

P⁺ $\bar{C}H_2$ $\bar{C}H_2$ Li^+ + $BrCH_2$ CH_2 $BrCH_2$ ⟶ P⁺ Br^- + secondary products (108)

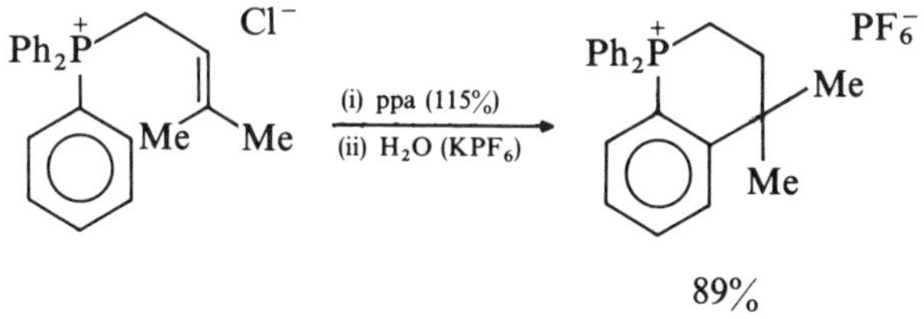

$Ph_2\overset{+}{P}$ Cl⁻ Me Me —(i) ppa (115%); (ii) H_2O (KPF_6)→ $Ph_2\overset{+}{P}$ PF_6^- Me Me

89% (109)

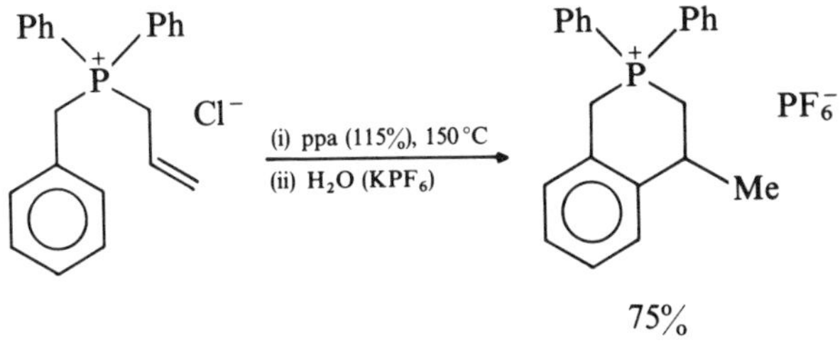

Ph Ph P⁺ Cl⁻ —(i) ppa (115%), 150 °C; (ii) H_2O (KPF_6)→ Ph Ph P⁺ PF_6^- Me

75% (110)

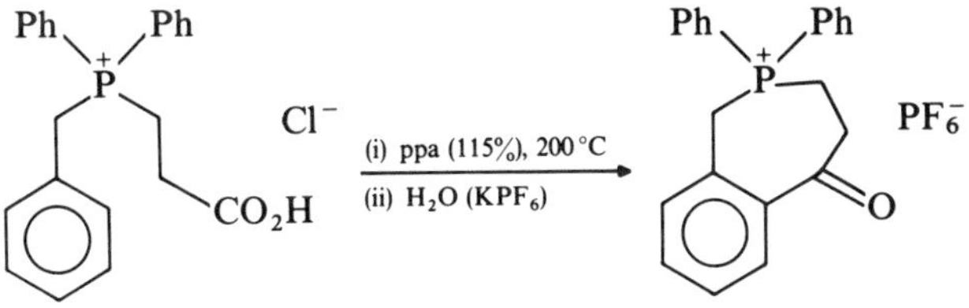

Ph Ph P⁺ Cl⁻ CO_2H —(i) ppa (115%), 200 °C; (ii) H_2O (KPF_6)→ Ph Ph P⁺ PF_6^- O (111)

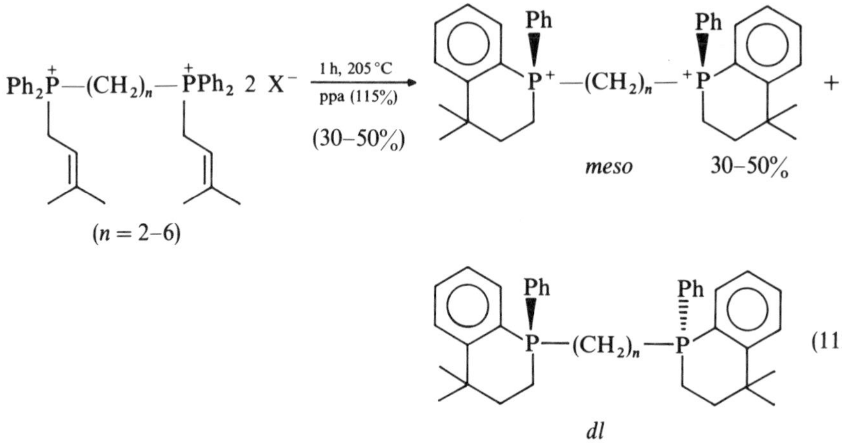

$$Ph_2\overset{+}{P}-(CH_2)_n-\overset{+}{P}Ph_2\ 2\,X^- \xrightarrow[ppa\ (115\%)]{1\ h,\ 205\,°C} \text{(30–50%)}$$

$(n = 2–6)$

meso 30–50% + dl (112)

G. Heterophosphonium Salts

Heterophosphonium salts, $R_3P^+Z\ X^-$, sometimes called quasi-phosphonium[1,2,6] or pseudophosphonium salts, are phosphonium salts in which the central phosphorus atom is not bonded to four carbon atoms but has at least one bond to an heteroatom Z (O, N, S, halide, etc.). They are frequently formed and then used *in situ* as halogenating or dehydrating agents; some reviews are devoted to their applications in organic synthesis[401–404]. Because of the polarity and the fragility of these P—Z bonds, heterophosphonium salts are generally more reactive than phosphonium salts with only P—C bonds. They can favour equilibria with pentacoordinated structures and ligand exchanges on the phosphorus atom, which give them some sensitivity to the solvolysis (reaction 113).

$$R_3\overset{+}{P}Z\ X^- \rightleftharpoons R_3P\begin{smallmatrix}Z\\X\end{smallmatrix} \rightleftharpoons R_3\overset{+}{P}X\ Z^- \tag{113}$$

Alkoxyphosphonium salts are a special case, in that they show thermal instability, due to the thermodynamic stability of the P=O bond formed (reaction 114). This enables them to be isolated only in favourable cases (reaction 114).

$$R_3\overset{+}{P}OR'\ X^- \longrightarrow R_3P{=}O + R'X \tag{114}$$

Various structural factors are generally favourable for the stability of these salts, as follows:

An important steric hindrance of the carbon in the α-position to the oxygen (or a bicyclic structure) reduces or even inhibits formation of the phosphoryl group by substitution reaction on this carbon. It must be pointed out that phosphoryl formation occurs, generally, in Arbuzov-type reactions, by a substitution process, but can take place also by a β-elimination which is favoured by major steric hindrance.

A weak nucleophilicity of the X^- anion (SbF_6^-, PF_6^-, BF_4^-, ClO_4^-, $CF_3SO_3^-$) causes the same favourable stabilization effect.

Electron-donating R groups, which decrease the positive charge on phosphorus, inhibit also phosphoryl formation, but also have another stabilizing effect, because they decrease the solvolysis or ligand-exchange reactions occurring by nucleophilic attack on the phosphorus.

Bulky R and R′ groups have for steric reasons the same stabilizing effect for nucleophilic attack on the phosphorus atom.

Accordingly, heterophosphonium salts are generally more difficult to prepare, to handle and to characterize than phosphonium salts themselves. The salt stability increases generally in the order $R_3\overset{+}{P}Hal < R_3\overset{+}{P}OR' < R_3\overset{+}{P}SR' < R_3\overset{+}{P}NR'_2$.

Four methods are used for the heterophosphonium salt preparation (reaction 115): (a) quaternization of a compound with a tricoordinated phosphorus by a heteroatomic oxidizing reagent: formation of a phosphorus–heteroatom bond, P–Z; (b) heteroatomic groups exchange on the phosphorus of a heterophosphonium salt; (c) alkylation or arylation of a tricoordinated phosphorus compound including a heteroatomic group: formation of a phosphorus–carbon bond; and (d) reaction of phosphine oxide and their heteroatomic analogues $R_3P{=}Y$, with electrophilic reagents.

$$\left.\begin{array}{l}\text{(a) } R_3P + ZX \\ \text{(b) } R_3\overset{+}{P}X\ Z^- \\ \text{(c) } R_2PZ + R{-}X \\ \text{(d) } R_3P{=}Y + EX \xrightarrow{(Z = YE)}\end{array}\right\} \longrightarrow R_3\overset{+}{P}Z\ X^- \qquad (115)$$

(N.B.: the remanent group R initially linked to the phosphorus can be itself an heteroatomic group)

1. Reaction of tricoordinated phosphorus compounds with heteroatomic oxidizing agents

For more than a century, the direct reaction of tricoordinated phosphorus compounds, R_3P (R = Alk, Aryl, OR, OAr, SR, NR_2, Cl, Br etc.), with the halogens (or with halogenating agents) has been known[405c,406,407] to give, initially, generally at low temperature, either a halophosphonium (P^{IV}) or a dihalophosphorane or their mixture in equilibrium[408] (equation 116). The result of this addition is strictly dependent on the exact nature of the compound R_3P, on the halogen and on the solvent. A halophosphonium structure is generally promoted by polar solvents and by the halogens in the order I > Br > Cl ≫ F[409]. In the case of fluorophosphoranes R_nPF_{5-n}, which are almost completely in the covalent form (whereas other halophosphoranes are much more inclined to be in the ionic form), however, the fluorophosphonium form can be promoted by additon of a strong enough Lewis acid (reaction 117). Moreover, these adducts are often only intermediates (generally identified by ^{31}P NMR) and the reaction can proceed either by phosphoryl P=O formation or by ligand exchange on the phosphorus. For example, with trialkyl phosphites, alkoxyphosphonium salts are stable and identifiable below −85 °C but decompose rapidly above this temperature (between −85 and −50 °C)[413–415], without any detectable formation of dihalophosphorane (reaction 118). In the case of triphenyl phosphite, at −80 °C, halogen addition gives only the halophosphonium salt[416] but for X ≠ I this undergoes ligand exchange at 0 °C[417] (reaction 119).

$$R_3P + XX \longrightarrow R_3\overset{+}{P}X\ X^- \rightleftharpoons R_3P\begin{smallmatrix}X\\X\end{smallmatrix} \qquad (116)$$

$$R_nPF_{5-n} + MX_m \longrightarrow [R_n\overset{+}{P}F_{4-n}][MX_mF]^-$$

$R = $ Alk, Ar, R'_2N, R'O

$MX_m = BF_3$ [410,411], AsF_5 [410], $Et_3O^+BF_4^-$ [411], $R_3\overset{+}{P}Br\ Br^-$ [50], AsF_3 [412]

$$(RO)_3P \xrightarrow[-85\,^{\circ}C]{X_2} (RO)_3\overset{+}{P}X\ X^- \xrightarrow[-RX]{-50\,^{\circ}C} (RO)_2\overset{O}{\overset{\|}{P}}X \quad (118)$$

$$(PhO)_3P \xrightarrow[-80\,^{\circ}C]{X_2} (PhO)_3\overset{+}{P}X\ X^- \xrightarrow[-PX_3]{0\,^{\circ}C} (PhO)_4\overset{+}{P}\ X^- \quad (119)$$

A detailed review has been published[407] of halogenation reactions for phosphorous acid derivatives including other types of phosphorus–heteroatom bonds (P—SR, P—NR_2, P—Hal). In fact, as shown in the next section (II.G.2), the halogenation of tricoordinated phosphorus structures is, most often, used to generate halophosphonium salts as precursors for other heterophosphonium salts by consecutive exchange of ligands, even if some halophosphonium salts are of interest as transient isolated reagents[418].

Alkoxyphosphonium salts can be obtained and isolated by direct alkoxylation of a trialkyl phosphite with alkyl hypochlorite when the structure of the alkoxy group is such as to prevent transformation to phosphonate[419,420] or when it is possible to trap the resulting chloride ion by $SbCl_5$ [421,422] (reaction 120). This chloride ion trapping is not necessary in the similar reaction of tertiary phosphines with chloramine-T for the preparation of aminophosphonium salts widely developed by Sisler and coworkers[423–430] and other groups[431,432] and recently extended further[37] (reaction 121).

$$(MeO)_3P \xrightarrow{MeOCl} (MeO)_3\overset{+}{P}OMe\ Cl^- \xrightarrow{SbCl_5} \underset{38\%}{(MeO)_4\overset{+}{P}\ SbCl_6^-} \quad (120)$$

$$R_3P + \begin{matrix} R' \\ R'' \end{matrix}\!\!>NCl \xrightarrow[-76\,^{\circ}C]{(Et_2O)} \underset{21\text{–}96\%}{R_3\overset{+}{P}N\!\!<\!\!\begin{matrix} R' \\ R'' \end{matrix}\ Cl^-} \quad (121)$$

R = Me, Et, *n*-Pr, Ph; R′, R″ = H, Me

In the direct synthesis of thiophosphonium salts, the oxidizing agent can be a disulphide using either constant-current electrolysis (c.c.e) conditions or by simple addition with benzoic acid[433] (reaction 122).

$$R_3P + R'SSR' \xrightarrow[\text{20 °C(MeCN), PhCO}_2\text{H + LiX (88–94\%)}]{\text{c.c.e(HClO}_4\text{—MeCN) (53–77\%)}} R_3\overset{+}{P}SR'\ X^- \quad (122)$$

R = *n*-Bu, Ph; R′ = Me, Et, Pr, $PhCH_2$, Ph; $X^- = ClO_4^-$, BF_4^-

By constant-current electrolysis of triphenylphosphine with alcohol and benzoic or succinic acid the alkoxyphosphonium perchlorates can be obtained[434] (reaction 123).

$$Ph_3P \xrightarrow[NaClO_4 + ROH + R'CO_2H]{c.c.e.(MeCN, 20^\circ C)} Ph_3\overset{+}{P}OR\ ClO_4^- \quad (123)$$

$$R = Me, Et, Pr \qquad 45\text{–}50\%$$

2. Heteroatomic ligand exchange in heterophosphonium salts

This synthetic route is based on heterophosphonium salt formation, as an intermediate reagent in which the labile heteroatomic Y group is easily exchanged *in situ* by an other heteroatomic Z group (reaction 124).

$$R_3P \longrightarrow (R_3\overset{+}{P}Y\ X^-) \xrightarrow[-YH]{ZH} R_3\overset{+}{P}Z\ X^- \quad (124)$$

$$R = Alk, Ar, OR', SR', NR_2', \quad etc.$$

The reactive intermediate is often a halophosphonium salt formed by direct halogenation (most often bromination) of the tricoordinated phosphorus compound. Thus, a series of relatively stable alkoxyphosphonium salts were prepared either when the alkoxy structure had sufficient steric hindrance to inhibit the evolution[435–439], or when the associated anion is not nucleophilic[412] (reaction 125). In the same way, but without particular conditions for the stabilization of aminophosphonium salts thus prepared, the reaction can be performed, generally in presence of a tertiary base, with primary amines[440–442], secondary amines[443], hydrazines[440,444,445], ammonia[446] or directly with an amide[447] or an azide[448]. The aminophosphonium salts are therefore obtained in high yields (reaction 126). For the formation of the intermediate halophosphonium the halogen can be replaced by CCl_4, which gives access to the equivalent synthesis of alkoxy- and amino-phosphonium salts, often performed at low temperature in THF (reaction 127).

$$Ph_3P \xrightarrow[(ii)\ ROH(-HX)]{(i)\ X_2} Ph_3\overset{+}{P}OR\ X^- \quad (125)$$

$$R_3P \xrightarrow[(ii)\ R'R''NH,\ NEt_3]{(i)\ Br_2} R_3\overset{+}{P}N\begin{matrix} R' \\ R'' \end{matrix}\ Br^- \quad (126)$$

$$R = Ph, NR_2;\ R', R'' = H, Alk, Ar, NH_2, N{=}CAr_2$$

$$R_3P + CCl_4 + ZH \xrightarrow[(-CHCl_3)^*]{} R_3\overset{+}{P}Z\ X^- \quad (127)$$

$R_3P =$	Ph_3P	Ph_3P	$(Me_2N)_3P$	$(R_2N)_2PN\triangleleft$
$ZH =$	$PhOH$	$NH_3, RNH_2, HN\triangleleft$	ROH	NH_3
Ref.:	449	450	451–453	454

(In addition to $CHCl_3$, the chloromethylphosphonium salt can be a by-product of the reaction[450].)

The halophosphonium salt can be also generated by action of $POCl_3$ on a phosphoramide. Its trapping by hydroxybenzotriazole therefore gives a good synthetic route to the corresponding benzotriazolyloxyphosphonium salts, used as peptidic coupling agents (reaction 128). It must be pointed out that, when a compound ZH, with a labile hydrogen, is not used as a trapping reagent for the intermediate halophosphonium salt, but an aldehyde or a ketone is present, the heterophosphonium salt formed can have a more complex structure resulting from secondary addition reactions on the carbonyl (reactions 129 and 130).

$$(R_2N)_3PO \xrightarrow[\text{(ii) HCl}]{\text{(i) POCl}_3} (R_2N)_3\overset{+}{P}Cl\ Cl^- \xrightarrow[\text{(ii) K[PF}_6]]{\text{(i) 1-hydroxybenzotriazole, NEt}_3} (R_2N)_3\overset{+}{P}O{-}N(\text{benzotriazolyl})\ PF_6^-$$

$$R_2N = Me_2N^{455}, \text{ piperidino}^{456} \qquad (128)$$

$$R_2PCl + 2\,PhCHO \longrightarrow \text{2,2-}R_2\text{-4,5-}Ph_2\text{-1,3,2-dioxaphospholanium}\ Cl^- \qquad (129)^{457}$$

$$Ar\overset{O}{\overset{\|}{C}}Y + P(NR_2)_3 + CF_3X \xrightarrow[-70\,^\circ C]{CH_2Cl_2} Ar{-}C(CF_3)(Y){-}O\overset{+}{P}(NR_2)_3\ X^- \qquad (130)^{458}$$

$$Y = H, R, R_F(=C_nF_{2n+1});\ X = Br, Cl$$

Halophosphonium salt formation, immediately followed by its transformation to enoxyphosphonium, takes place in the reaction of tertiary phosphines with the positive halogen of some secondary α-haloketones. Indeed, the reaction, which can also lead to a 2-oxoalkylphosphonium salt by a classical S_N2 reaction (route a), is oriented to the enoxyphosphonium salt (route b) by different factors[459–461] (reactions 131): an increasing steric hindrance on the phosphine and/or the carbon linked to the halogen, or an increasing stability of the enolate and an increasing electrophilicity of the halogen.

$$R_3P: + R'CH(X)C(O)R'' \xrightarrow{(a)} R_3\overset{+}{P}CH(R')C(O)R''\ X^-$$

$$R_3P: + R'CH(X)C(O)R'' \xrightarrow{(b)} \left[R_3\overset{+}{P}X\ \ {}^-O{=}C(R''){=}CHR'\right] \longrightarrow R_3\overset{+}{P}OC(R''){=}CHR'\ X^- \qquad (131)$$

The enoxyphosphonium salts are relatively stable, even if they are very sensitive to solvolysis; they have been considered as the intermediates in Perkow's reaction[462]. A very similar reaction leads to enaminophosphonium salts (or to iminophosphonium salts when there is a subsequent isomerization possibility of the double bond by prototropy)[463] (reaction 132).

Halophosphonium salts are the useful precursors for the preparation of heterophosphonium by ligand exchange, although some relatively stable alkoxyphosphonium salts

$$\mathrm{Ar_3P} \xrightarrow[\text{(PhH, reflux)}]{\mathrm{MeNHC(CCl_3){=}NMe}} \mathrm{MeN{=}C(CHCl_2)N(Me)\overset{+}{P}Ar_3\ Cl^-}$$

$$\mathrm{Ar_3P} \xrightarrow{\mathrm{RR'NC(CCl_3){=}NH}} \mathrm{RR'NC(CHCl_2){=}N\overset{+}{P}Ar_3\ Cl^-} \quad 62\text{–}98\% \tag{132}$$

Ar = Ph, *p*-Tol
R = H, Me, Et, Ph
R′ = Me, Et, Pr, *t*-Bu, $PhCH_2$, Ph

have been obtained by reaction of an alcohol either with the adduct of triphenylphosphine with DEAD (Mitsunobu's reagent)[464,465] (reaction 133) or with phosphonium anhydride[466] [obtained (cf. section II.G.5) by the action of triflic anhydride on triphenylphosphosphine oxide][467] (reaction 134).

$$\mathrm{Ph_3P + EtO_2CN{=}NCO_2Et} \longrightarrow \left|\mathrm{Ph_3\overset{+}{P}N{<}^{\bar{N}CO_2Et}_{CO_2Et}}\right| \xrightarrow[\mathrm{-DEADH_2}]{\mathrm{ROH, HY}} \mathrm{Ph_3\overset{+}{P}OR\ Y^-} \tag{133}$$

$$\mathrm{2Ph_3P{=}O} \xrightarrow{\mathrm{Tf_2O}} \mathrm{Ph_3\overset{+}{P}O\overset{+}{P}Ph_3\ 2\ TfO^-} \xrightarrow[(-\mathrm{Et_3\overset{+}{N}HTfO^-})]{\mathrm{ROH, NEt_3,\ CH_2Cl_2, -20\,^\circ C}} \mathrm{Ph_3\overset{+}{P}OR\ TfO^-} \tag{134}$$

3. *Alkylation or arylation of heterosubstituted tricoordinated phosphorus compounds*

Alkylation or arylation of heterosubstituted tricoordinated phosphorus esters corresponds to the first stage of the Michaelis–Arbuzov reaction (reaction 135). Generally, the intermediate alkoxyphosphonium salt is not isolable and is rapidly transformed into the corresponding phosphoryl derivative. Nevertheless, some favourable factors can allow the alkoxyphosphonium salts to be isolated, for example when the alkylation brings in a very weakly nucleophilic X^- anion (BF_4^-)[468,469], $(CF_3SO_3^-)$[470] (reactions 136 and 137). When the alkoxy groups are very hindered[471,472] or when overcrowded amino groups are fixed on the phosphorus[473], the alkylation can be achieved with methyl iodide at 0 °C. The alkoxyphosphonium salts prepared in this way allowed a better understanding of the mechanism of the Arbuzov reaction owing to kinetic and thermodynamic studies[474,475] on the true intermediate of this reaction.

$$\mathrm{R_nP(OR')_{3-n} + R''X \longrightarrow R_n\overset{+}{P}(R'')(OR')_{3-n}\ X^- \dashrightarrow R'X + R_nP(O)(R'')(OR')_{2-n}} \tag{135}$$

$$\mathrm{(RO)_3P + Me_3O^+\ BF_4^-} \xrightarrow{\mathrm{CH_2Cl_2\,(20\,^\circ C)}} \mathrm{(RO)_3\overset{+}{P}Me\ BF_4^-} \tag{136}$$ [469]

(R = Me, Et, *i*-Pr, *i*-Oct)

$$\mathrm{Me_nP(OMe)_{3-n} + CF_3SO_2OMe} \xrightarrow[\mathrm{1\,h, 0\,^\circ C}]{\mathrm{Et_2O}} \mathrm{Me_{n+1}\overset{+}{P}(OMe)_{3-n}\ CF_3SO_3^-} \tag{137}$$ [470]

In the case of the aromatic esters of tricoordinated phosphorus acids, the alkylation, performed with methyl iodide[476] or methyl triflate[477], yields the corresponding aryloxyphosphonium salts (reaction 138), thermally stable even though it is sensitive to solvolysis. The aminophosphines can also be alkylated at phosphorus, and the resulting aminophosphonium salts are generally stable[478–481], if they do not include alkoxy groups bonded to the phosphorus[481,482] (reaction 139). Moreover, the thermal stability of the aminophosphonium salts allows them to be obtained by catalytic arylation of aminophosphine[73] (reaction 140)

$$(PhO)_3P + MeX \xrightarrow[(X=I,OTf)]{} Me\overset{+}{P}(OPh)_3\ X^- \tag{138}$$

$$R_2PN\begin{matrix}R' \\ R''\end{matrix} + R'''X \longrightarrow R_2\underset{R'''}{\overset{+}{P}}N\begin{matrix}R' \\ R''\end{matrix}\ X^- \tag{139}$$

$R'''X = ArCH_2Cl, (ClCH_2CH_2)_2O, R_2NCH_2Cl, Ar\underset{\underset{O}{\|}}{C}NHCH_2Cl$, etc.

$$Ph_2PNR_2 + PhBr \xrightarrow[\text{cat., }NiBr_2]{24\,h,\ 200\,°C} Ph_3\overset{+}{P}NR_2\ Br^- \tag{140}$$

68–81%

$R_2N = Et_2N, Bu^n_2N,$ (piperidino) N

The arylation reaction can be extended to bis- and tris-(dialkylamino)phosphines[73], but in the case of *N*-ethylanilinophosphine the reaction takes place with a rearrangement of the phosphasemidine type[483]. A structural transposition also occurs in the alkylation of some *N*-silylaminophosphines[484–487] (reaction 141). Finally, it is noteworthy that, as far as the tris(dimethylamino)phosphine is concerned, many polyhalomethyltris-(dimethylamino)phosphonium salts have been prepared by the action of the appropriate polyhalomethanes (reaction 142).

$$(Me_3Si)_2NPMe_2 \xrightarrow{MeI} (Me_3Si)_2N\overset{+}{P}Me_3\ I^- \tag{141}$$

$$(Me_3Si)_2NPMe_2 \xrightarrow[(CH_2Cl_2,\ 0\,°C)]{BrCH_2CO_2Et} Me_3SiNH\overset{+}{P}(Me)(Me)CH(SiMe_3)(CO_2Et)\ Br^-$$

$$(Me_2N)_3P + YCX_3 \longrightarrow (Me_2N)_3\overset{+}{P}CYX_2\ X^- \tag{142}$$

$YCX_3 = CHCl_3$[488], CCl_4[488], CF_2Br_2[318], $CFCl_3$[318]

In the formation of halophosphonium salts, the chlorophosphine alkylation occurs easily with diverse alkyl halides[405a–c]. Generally, the reaction proceeds without any solvent or in CH_2Cl_2, with a Lewis acid such as $AlCl_3$, especially in the case of PCl_3 or $RPCl_2$ (reaction 143). Finally in heterophosphonium synthesis by the alkylation of heterosubstituted phosphorus compounds, mention should be made of the specific synthesis for the phospholenium systems by direct 1,3-diene addition to halophosphines[405c,489,490] or phosphenium salts[126] (reaction 144).

$$R_nPCl_{3-n} + R'X + AlCl_3 \longrightarrow R_nP^+(X)(R')Cl_{2-n}\ AlCl_4^- \qquad (143)$$

$n = 0, 1, 2;\quad X = Cl, Br, I$

$$ClP(NPr^i_2)_2 \xrightarrow{AlCl_3} (Pr^i_2N)_2P^+\ AlCl_4^- \xrightarrow[12\text{–}24\,h,\ 0\text{–}25\,°C]{RCH=CR'-CR''=CHR'''} \text{1,1-bis(diisopropylamino)phospholenium (R, R', R'', R''')}\ AlCl_4^- \quad 55\text{–}95\% \qquad (144)$$

R, R′, R″, R‴ = H, Me

4. Reaction of $R_3P{=}Y$ compounds with electrophilic reagents

In the case of phosphine oxides and acids derivatives of tetracoordinated phosphorus acids, the reaction of the phosphoryl group with an electrophile resulting in an oxyphosphonium salts (reaction 145) corresponds formally to the opposite reaction of the Michaelis–Arbuzov second step. If we consider the high energy of the phosphoryl bond, this reaction can only take place with hard electrophiles and especially the more powerful: trialkyloxonium salts[491–493] (reaction 146), the intramolecular carbocationic systems generated from allenylphosphine oxides[494,495] (reaction 147) or phosphorylating agents[496] (reaction 148).

$$R_3P{=}O + EX \longrightarrow R_3\overset{+}{P}OE\ X^- \qquad (145)$$

$$R_3P{=}O + R'_3O^+\ X^- \xrightarrow{-(-R'_2O)} R_3\overset{+}{P}OR'\ X^- \qquad (146)$$

$R' = Me, Et;\quad X^- = BF_4^-, SbCl_6^-$

$$Me_2P(O)CH{=}C{=}CMe_2 \xrightarrow{EX} \text{2,2-dimethyl-5,5-dimethyl-4-E-1,2-oxaphospholenium}\ X^- \quad 85\text{–}90\% \qquad (147)$$

$EX = HCl$[494] or Br_2, I_2, RSCl, RSeCl[495]

$$R_3P{=}O + X_3P{=}O \xrightarrow{(PhH)} [R_3\overset{+}{P}OP(O)X_2]\ X^- \qquad (148)$$

$R = Alk, Ar;\quad X = Cl, Br$

This kind of reaction has also been used for, through a ligand exchange reaction, the *in situ* preparation of the chlorophosphonium salt with $R = Me_2N$ from hmpt[455,497]. The reaction of silylating reagents with tertiary phosphine oxides[496] or alkyl phosphates[498] yields silyloxyphosphonium salts (reaction 149) (in the latter case, the reaction proceeds with silanolysis of the alkoxy groups and yields the tetrakis(silyloxy)-phosphonium salts (reaction 150)). Acylating reagents react with hmpt to yield the acyloxyphosphonium salts, which are isolable in some cases with a non-nucleophilic anion[499] (reaction 151).

$$Ph_3PO + Me_3SiCl \longrightarrow Ph_3\overset{+}{P}OSiMe_3\ Cl^- \tag{149}$$

$$(MeO)_3PO + 4\,Me_3SiI \xrightarrow[-3MeI]{} (Me_3SiO)_4\overset{+}{P}\ I^- \tag{150}$$

$$(Me_2N)_3P{=}O + RCOCl \rightleftharpoons (Me_2N)_3\overset{+}{P}OC(=O)R\ Cl^- \xrightarrow[\text{exchange}]{\text{anion}} (Me_2N)_3\overset{+}{P}OC(=O)R\ X^- \tag{151}$$

$$X^- = SbCl_6^-,\ BPh_4^-$$

Sulphonylation reagents such as triflic anhydride (Tf_2O) are particularly effective towards tertiary phosphine oxides (reaction 152). The resulting 'phosphonium anhydride triflates' can be isolated, but mostly they are used *in situ* as dehydrating reagents which are specially interesting for organic syntheses[466,467,474,475,502,503,504]. With regard to the other heterophosphonium salts, the preparative reaction generally needs less powerful electrophiles because the P=Y bond has a lower energy than the phosphoryl bond. The alkylthiophosphonium salts, for instance, can be obtained by alkylation with a trioxonium salt and an alkyl triflate or tosylate[505–512], but also with an alkyl iodide or even bromide[513] (reaction 153).

$$R_3P{=}O + (CF_3SO_2)_2O \xrightarrow[15\,min,\,0\,^\circ C]{CHCl_3} R_3\overset{+}{P}O\overset{+}{P}R_3\ 2\,CF_3SO_3^- \tag{152}$$

$R = Ph$[467,500,501], n-Bu[502]

$$R_3P{=}S + R'X \longrightarrow R_3\overset{+}{P}SR'\ X^- \tag{153}$$

$R = Alk, Ar, R''O$

The seleno[514–516] and telluro-phosphonium[517] salts can also be obtained by simple alkylation, using methyl iodide (reaction 154). The phosphinimines are alkylated at the nitrogen by methyl or ethyl iodide to give the corresponding aminophosphonium salts[441,518,519] (reaction 155). The alkylation with more diverse alkyl iodides can be performed on *N*-metalated phosphinimines of the diaminophosphonium diylide type[520–522] (reaction 156) or of the aminophosphonium yldiide type[446,521,522] (reaction 157). These compounds can also be acylated[523]. Finally, it is noteworthy that the *N*-arylation of some phosphinimines can be achieved with benzenediazonium salts[524] (reaction 158). This method has been successfully applied to prepare cationic polymers.

$$R_3P{=}Z + MeI \longrightarrow R_3\overset{+}{P}ZMe\ I^- \tag{154}$$

$R = NMe_2$; $Z = Se$[515,516]

$R = Me$, i-Pr, n-Bu, t-Bu, NMe_2; $Z = Te$[517]

$$Ph_3P{=}NR + R'I \xrightarrow{PhH,\,3\,h,\,80\,^\circ C} Ph_3\overset{+}{P}N(R)(R')\ I^- \tag{155}$$

$R = n$-Alk, i-Alk, NR_2; $R' = Me, Et$

$$Ph_2\overset{+}{P}(\bar{N}R)_2\ Li^+ + 2R'X \longrightarrow Ph_2\overset{+}{P}(NRR')_2\ X^- \quad (156)$$

$$Ph_3P{=}NLi \xrightarrow{RX} Ph_3P{=}NR \begin{cases} \xrightarrow{HX} Ph_3\overset{+}{P}NHR\ X^- \\ \xrightarrow{R'X} Ph_3\overset{+}{P}NRR'\ X^- \end{cases} \quad (157)$$

$$Ph_3P{=}NPh + PhN_2^+\ BF_4^- \xrightarrow[\text{MeCN or PhH}]{24h, 60^\circ C} Ph_3\overset{+}{P}N(Ph)_2\ BF_4^-\ 84\% \quad (158)$$

H. Synthesis by Anion Exchange or Modification

In syntheses using anion exchange the phosphonium ion, already present, is not modified, but only the associated anion is exchanged or modified[2b,525,526] (reaction 159). Depending on the synthesis, the goal of this transformation can be different. First, most often, a higher stability of the crystal network for the new ion pair, $R_4P^+\ Y^-$, is sought in order to allow a better isolation and purification in the solid state ($Y^-{=}I^-$, Ph_4B^-, BF_4^-, etc.). In some cases, the interest in this stabilization is less related to the phosphonium ion than to an uncommon inorganic anion ($Y^- = [Te_4]^{2-}$, $[Sb_3I_{10}]^{3-}$, $[ZrCl_6]^{2-}$, etc.). Second, for the new anion Y^-, a lack of nucleophilic reactivity toward the phosphonium ion is sought in order to avoid its decomposition by the X^- anions (e.g., for the alkoxyphosphonium ions, $Y^- = [MX_n]^-$: BF_4^-, SbF_6^-, PF_6^-, etc.). Third, for a chiral phosphonium ion, there is a possibility of enantiomeric separation by introducing an optically active $^*Y^-$ anion (such as dibenzoyl tartrate or camphorsulphonate), and, then separating the two diastereomeric ion pairs (reaction 160).

$$R_4P^+\ X^- \longrightarrow R_4P^+\ Y^- \quad (159)$$

$$abcdP^+\ X^- \longrightarrow \begin{cases} (abcdP^+)_R\ \ ^*Y^- \\ (abcdP^+)_S\ \ ^*Y^- \end{cases} \quad (160)$$

Fourth, the introduction of anion Y^- ($Y^- = CN^-$, $[MnO_4]^-$, $[CR_2O_7]^{2-}$, X_3^-, etc.), reactive towards organic substrates but not towards the phosphonium ion, is sought in order to produce a new reagent with particular properties for organic syntheses. Finally, the association with an anion which can introduce interesting physical properties [$Y^- = (TCNQ)_x^-$] is sought.

1. Anion exchange (Table 10)

The most widely used method for anion exchange is based on the redistribution of ion pairs in solution with another ionic species, most often inorganic, so that sufficient solubility differences between the ions pairs induce a shift of this exchange towards the formation of the wanted salt owing to a higher insolubility either for the phosphonium salt, $R_4P^+\ Y^-$, or for the inorganic salt, $M^+\ X^-$ (equation 161). When one of the salts is particularly insoluble (e.g. AgCl, $BaSO_4$, $R_4P^+\ BPh_4^-$), the exchange takes place by mixing two concentrated homogeneous solutions (in H_2O, ROH, etc.); filtration of the

TABLE 10. Examples of anion exchanges $R_4P^+ X^- \xrightarrow[-MX]{+MY} R_4P^+ Y^-$

R_4P^+	X^-	M^+	Y^-	Solvent	Ref.
Ph_4P^+	Cl^-	Na^+	$(RO)_2PS_2^-$	ROH (reflux)	529
Ph_4P^+	Br^-	$K_4S_n^{2+}$	Te_4^{2-}	MeOH	151
Ph_4P^+	Br^-	K^+	NO_2^-	dmso	528
Ph_4P^+	Cl^-	Na^+	NO_2^-	H_2O	528
Ph_4P^+	Cl^-	Na^+	RS^- or RSe^-	MeCN	136
Ph_3P^+Bu	Cl^- or Br^-	Na^+	PhO^-	$MeOH-H_2O$ (NaOH)	530
Bu_3P^+R	$MeSO_3^-$ or Cl^-	Na^+	N_3^-	$PhH-H_2O$	531
Bu_3P^+R	N_3^-	H^+	SO_4^{2-}	H_2O	531
Ph_3P^+R	Br^-	Ag^+	NO_3^-	$H_2O-CHCl_3$	183
Ph_3P^+R	I^-	Na^+	ClO_4^-	H_2O	255
Ph_3P^+R	Cl^- or Br^-	K^+	MnO_4^-	H_2O or CH_2Cl_2	137
$(HOCH_2)_4P^+$	Cl^-	Na^+	I^-	EtOH	532
$(HOCH_2)_4P^+$	SO_4^{2-}	Ba^{2+}	X^-	H_2O	533
$(R_3PN{=}PR_3)^+$	$Br^- + F^-$	Na^+	I^-	MeCN	534
$(R_3PN{=}PR_3)^+$	$Br^- + F^-$	NH_4^+	PF_6^-	H_2O	269, 535
$(R_3PN{=}PR_3)^+$	$Br^- + F^-$	K^+	ClO_4^-	H_2O	534
$(R_3PN{=}PR_3)^+$	Cl^-	Na^+	RS^- or RSe^-	MeCN	136
$(R_3PN{=}PR_3)^+$	Cl^-	K^+	CN^-	H_2O	536
$(R_2N)_4P^+$	Cl^-	Ag^+	F^-	MeCN	537
$Ph_3\overset{+}{P}CH_2\overset{+}{P}Ph_3$	Br^-	K^+	$Cr_2O_7^{2-}$	H_2O	538
$Ph_3\overset{+}{P}H$	$Cl^- + Br^-$	H^+	Br^-	dmf	239
Ph_4P^+	Br^- or I^-	Amberlite IRA-401 or -410	HF_2^-	MeCN	105, 539
$(HOCH_2)_4P^+$	Cl^-	Dowex 1	SO_4^{2-}		540
R_4P^+	Br^-	Amberlite IRA-401 or -900	HO^-		541, 542
Polym.–C_6H_4–$CH_2\overset{+}{P}Bu_3$	Cl^- or Br^-	Na^+	CN^-		373, 374
	Cl^-	K^+	$PtCl_6^-$		543
Polym.–C_6H_4–$CH_2\overset{+}{P}Ph_3$	Cl^-	Na^+	ArO^-, RCO_2^-, $PhSO_2^-$		376, 544
	Cl^-	H^+	$Cr_2O_7^{2-}$		107
Polym.–C_6H_4–$(CH_2)_n\overset{+}{P}Bu_3$	Br^-	Na^+	CN^-		545

precipitated salt allows the isolation of the phosphonium salt either in the precipitate or in the solution. On the other hand, when there is only a moderate solubility difference between the various ions pairs, anion exchange often takes place in biphasic conditions, between two liquid phases (CH_2Cl_2–H_2O, etc.), by washing repeatedly a phosphonium salt $R_4P^+ X^-$ solution with concentrated solutions of the exchanging salt MX (e.g. in exchanges such as Cl^-/Br^- or Br^-/I^-).

$$R_4P^+ X^- + M^+ Y^- \rightleftharpoons R_4P^+ Y^- + M^+ X^- \tag{161}$$

In some cases, the anion exchange can also take place in solid–liquid biphasic conditions either for monomeric or dimeric phosphonium salts, in suspension or by using an anion-exchange resin, or for polymeric phosphonium salts, by washing with solutions of salt MX. Finally, it must be pointed out that the new anion Y^- can be generated *in situ* by a classical preparation of such anions (e.g. addition of cyanide anion to CS_2 for $NCCS_2^-$[527] or aromatic substitution with fluoride anion on substituted nitrobenzene derivatives for NO_2^-[528]). Several recent examples of the various anions exchanges are shown in Table 10.

2. *Modification of anion*

The simplest modification occurs on addition of nucleophilic anions X^- to Lewis acids MX_n, resulting in the creation of an ate-complex as anionic adduct (Table 11) (reaction 162). The formation of complex anions can also be the result of halogen addition, with the formation of mixed or symmetrical trihalides $[YX_2]^-$, or else of hydracid addition, with formation of anions $[XHY]^-$, which are not very stable and readily dissociated in solution but are characterizable in the solid state (reaction 163).

$$R_4P^+ X^- + MX_n \longrightarrow R_4P^+ [MX_{n+1}]^- \tag{162}$$

$$R_4\overset{+}{P}\ Y^- \xrightarrow[\text{refs. 154, 546, 547}]{X_2} R_4\overset{+}{P}\ [YX_2]^-$$

$$R_4\overset{+}{P}\ Y^- \xrightarrow[\text{ref. 80}]{HX} R_4\overset{+}{P}\ [YHX]^- \tag{163}$$

X, Y = Cl, Br, I.

Another kind of anion transformation results from chemical reaction, with a reagent introduced in the medium, corresponding to a neutralization of the hydroxide anions

TABLE 11. Examples of anion modification with formation of ate-complexes

R_4P^+	X^-	Reagent	Complex anion	Ref
Ph_4P^+	Br^-	$SbBr_3$	$[Sb_2Br_8]^{2-}$	549
Ph_4P^+	I^-	SBI_3	$[SbI_4]^-$	550
Ph_4P^+	I^-	SbI_3	$[Sb_3I_{10}]^{3-}$ or $[Sb_8I_{28}]^{4-}$	148
Ph_3P^+R	Cl^-	$ZrCl_4$	$[ZrCl_6]^-$	551
Ph_3P^+R	Cl^-	$PhTeCl_3$	$[PhTeCl_4]^-$	552
Ph_4P^+	Br^-	Te_2Br_8	$[Te_2Br_{10}]^{2-}$	554
$[R_3PN{=}PR_3]^+$	CN^-	Me_3Sn	$[MeSnClCN]^-$	536
$[Ph_6P_3N_4H_2]^+$	Cl^-	WCl_6, $NbCl_5$	$[WCl_6]^-$, $[NbCl_6]^-$	553
Polym. $CH_2\overset{+}{P}Bu^n_3$	Cl^-	MCl_n	$[MCl_{n+1}]^-$ (Fe^{II}, Ni^{II}, Cu^{II}, Zn, Rh, Pd, Sn, Sb, In)	543

TABLE 12. Examples of anions modifications with charge transfer

R_4P^+	X^-	Reagent	Complex anion	Ref.
Me_4P^+	I_3^-	LiTCNQ	$(TCNQ_2I)^-$	555
$Ph_2\overset{+}{P}(Me)-C_6H_4-\overset{+}{P}Ph_2(Me)$	I^-	TCNQ	$(TCNQ)_2^-$	131, 556
$Ph_2\overset{+}{P}$ (cyclic) $\overset{+}{P}Ph_2$	I^-	TCNQ	$(TCNQ)_2^-$	131, 556
$[Ph_3PN=PPh_3]^+$	I^-	TCNQ	$(TCNQ)_3^{2-}$	185
$Ph_2\overset{+}{P}(Me)-C_6H_4-NMe_2$	I^-	TCNQ	$(TCNQ)^-$ or $(TCNQ)_5^-$	149

by an acid[244,542] (reaction 164). In this way, it is possible to introduce many different anions Y^-, but the method is often limited by the high instability of the starting phosphonium hydroxides. Alternatively, it corresponds to a substitution reaction on aliphatic substrates[548] (reactions 165), or a redox reaction for the formation of anionic charge-transfer complexes (Table 12) (reaction 166).

$$R_4P^+\ HO^- + HY \xrightarrow[-H_2O]{} R_4P^+\ Y^- \qquad (164)$$

$$R_4P^+\ Cl^- + MeI \xrightarrow[-MeCl]{} R_4P^+\ I^- \qquad (165)$$

$$R_4P^+\ X^- + x\text{TCNQ} \longrightarrow R_4P^+(\text{TNCQ})_x^- \qquad (166)$$

III. REACTIVITY OF PHOSPHONIUM SALTS

A. Acidity of Phosphonium Salts

Phosphonium salts with a hydrogen on the carbon α to phosphorus are the conjugated acids of the corresponding phosphonium ylides (equation 167). It is noteworthy that the formation of phosphonium ylides by a transylidation process is based on the superposition of two such acid–base equilibria (reaction 168). Transylidation has been used in order to establish a qualitative scale of relative acidities in the case of the most acidic phosphonium salts[557,558] and is still very useful for obtaining various ylides[26,67,162,366,383]. The most usual way for the preparation of phosphonium ylides is the 'salt method'[559,560], in which the phosphonium salt is reacted with a base strong enough to shift the acid–base equilibrium fully to the side of the phosphonium ylide (equation 169). Therefore, the acidities of phosphonium salts are important in determining the ease of formation of phosphonium ylides in this way. They are also important as an indication of the stability of the conjugate base, allowing three classes of phosphonium ylides to be distinguished: 'stabilized, semi- and non-stabilized' or 'stable, moderate and reactive' (corresponding respectively to phosphonium salts with high, medium and weak acidity). The various classes of phosphonium ylides can exhibit very different reactivities, particularly in the Wittig reaction[561].

$$R_3\overset{+}{P}CH\begin{matrix}R'\\R''\end{matrix}\ X^- \underset{+HX}{\overset{-HX}{\rightleftharpoons}} R_3\overset{+}{P}\overset{-}{C}\begin{matrix}R'\\R''\end{matrix} \qquad (167)$$

$$R_3\overset{+}{P}CH_2R'\ X^- + R''_3\overset{+}{P}\overset{-}{C}HR''' \rightleftharpoons R_3\overset{+}{P}\overset{-}{C}HR' + R''_3\overset{+}{P}CH_2R'''\ X^- \quad (168)$$

$$R_3\overset{+}{P}CH(R')(R'')\ X^- + M^+\ B^- \longrightarrow R_3\overset{+}{P}\overset{-}{C}(R')(R'') + BH + M^+\ X^- \quad (169)$$

TABLE 13. pK_a values for various β-ketophosphonium salts

$$Ph_3\overset{+}{P}CH(R)C(=O)R'\quad X^-$$

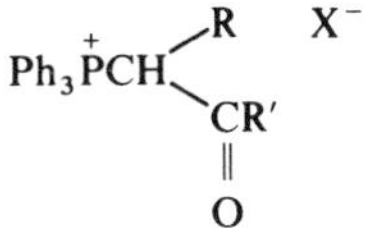

R	R′	pK_a	Solvent	Ref.
H	Me	6.65	EtOH–H_2O (80:20)	562
H	Me	7.5	H_2O	566
H	Ph	6.0	MeOH	567
H	Ph	6.01	EtOH–H_2O (80:20)	562
H	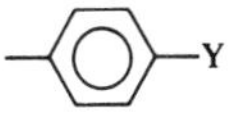 $-C_6H_4-Y$ (p)	6.01–2.35 σ	EtOH–H_2O (80:20)	562
H	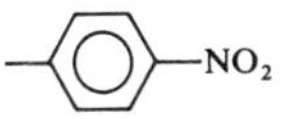 $-C_6H_4-NO_2$ (p)	4.2	EtOH–H_2O (80:20)	568
H	OEt	8.85	EtOH–H_2O (80:20)	562
H	OEt	8.95	EtOH–H_2O (80:20)	568
H	NH_2	11	EtOH–H_2O (80:20)	568
COPh	COPh	2.69	EtOH–H_2O (80:20)	562
COPh	N=NPh	4.06	EtOH–H_2O (80:20)	110
COMe	COPh	2.76	EtOH–H_2O (80:20)	562
COMe	CO_2Et	2.76	EtOH–H_2O (80:20)	562
CO_2Et	CO_2Et	2.76	EtOH–H_2O (80:20)	562
$-(CH_2)_3C(=O)-$		7.4	H_2O	566
$-(CH_2)_4C(=O)-$		7.4	H_2O	566

Most of the pK_a values for the phosphonium salts have been determined by potentiometric measurements in the case of β-ketophosphonium salts[562] (Table 13). The acidity of these phosphonium salts depends not only on the R and R′ groups of the carbonyl chain, but also on the other substituents linked to the phosphorus[562–564]. Particularly for the cyclic phosphonium salts, the pK_a value can vary significantly depending on the different steric and electronic (even transannular) effects on the phosphorus atom[564]:

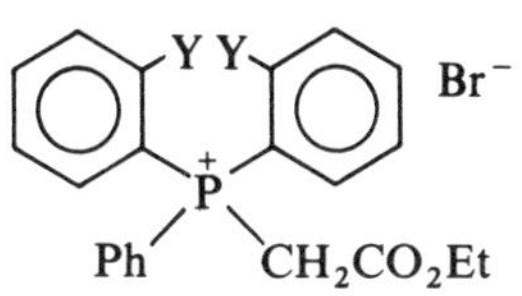

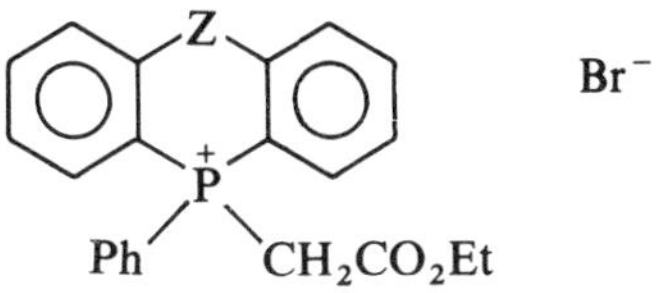

[Structures: 2,2′-(Y)(Y)-biphenyl-bridged $\overset{+}{P}(Ph)(CH_2CO_2Et)$ Br^-; Z-bridged dibenzophosphoninium $\overset{+}{P}(Ph)(CH_2CO_2Et)$ Br^-]

Y =	H	Me
pK_a(EtOH) =	8.5	8.4

Z =	—	$-C(=O)-$	$-CH{=}CH-$	CH_2SCH_2	CH_2OCH_2
pK_a(EtOH) =	3.5	6.5	8.5	8.6	9.9

In addition, the acidities of phosphonium salts vary in relation to the solvent in a typical way for onium salts: in the case of various water–ethanol mixtures, the acidity decreases as the proportion of ethanol increases and then reincreases in pure ethanol[562]. Further, there are great differences between the pK_a values measured in two different solvents[562]:

$$pK_a(MeNO_2) = 10.26 + 1.03\,pK_a[EtOH\text{–}H_2O\ (80{:}20)]$$

The kinetic acidity of phosphonium salts was first studied by NMR measurements of the hydrogen–deuterium exchange rate for a series of phosphonium salts[565]:

$$R_3\overset{+}{P}Me\ X^- \qquad Bu_3^t\overset{+}{P}CH_2\text{–}C_6H_4\text{–}Y\ X^-$$

	R = Me	Bu^t		Y = H	Me	Cl
$k_{(50\,^\circ C, CD_3OD)} =$	8.8×10^{-3}	5.15×10^{-4}	$k_{(20\,^\circ C, D_2O)} =$	0.066	0.021	0.426

The results indicate that the kinetic acidity is markedly lower for phosphonium salts than for the corresponding nitroalkanes, but cannot be correlated with pK_a values. In contrast, the measurement of kinetic acidity using double potential step chronoamperometry[180] allowed the determination of pK_a values for a series of phosphonium salts corresponding to semi-stabilized or non-stabilized ylides:

$$Ph_3P^+CH_2R\ X^-$$

R =	*n*-Pr	CH=CH$_2$	CH=CHMe	C≡CH	Ph
pK_a(25 °C; DMSO) =	15.4	14.3	13.6	12.7	12.6

B. Reactivity Towards Organolithium compounds

When the phosphonium salt has a hydrogen on the carbon α to the phosphorus atom, organolithium compounds, particularly phenyllithium, act chiefly as bases to generate the corresponding phosphonium ylide[560b] (reaction 170). However, towards the tetra-arylphosphonium salts, they react as nucleophiles to afford the phosphoranes[569a] by creating a fifth P—C bond, and with some dilithium compounds such as 2,2′-dilithiumbiphenyl the reaction can even go further to form the phosphorate anion with six P—C bonds[569b] (reaction 171). Actually, even in the case of phosphonium salts with a hydrogen α to the phosphorus atom, the formation of the phosphorane competes with the formation of the phosphonium ylide, as shown by ligand exchange on phosphorus[560b,570–573] resulting from the decomposition of the intermediate alkylphosphorane (reaction 172). This ligand exchange, in fact secondary, can nevertheless produce an ylide mixture, so that phenyllithium is generally used rather than *n*-butyllithium for the preparation of phosphonium ylides in the Wittig reaction.

$$Ph_3\overset{+}{P}CH(R)(R')\ X^- + R''Li \xrightarrow[-LiX]{} Ph_3\overset{+}{P}\overset{-}{C}(R')(R') + R''H \qquad (170)$$

$$Ar_4P^+\ X^- + Ar'Li \xrightarrow[-LiX]{} Ar_4PAr'(\overset{Ar'Li}{\dashrightarrow} [Ar_4PAr'_2]^-\ Li^+) \qquad (171)$$

It is noteworthy that in some cases the alkylphosphorane obtained by the addition of an organolithium compound to a phosphonium salt is sufficiently stabilized, owing to

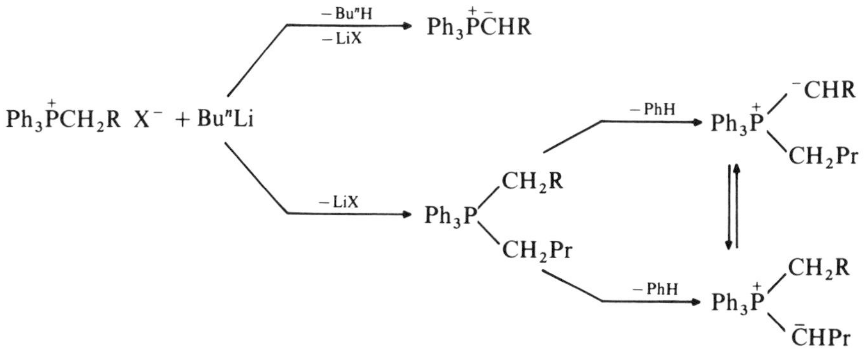

$$Ph_3\overset{+}{P}CH_2R\ X^- + Bu^nLi \xrightarrow[-LiX]{-Bu^nH} Ph_3\overset{+}{P}\overset{-}{C}HR$$

$$Ph_3\overset{+}{P}CH_2R\ X^- + Bu^nLi \xrightarrow{-LiX} Ph_3P(CH_2R)(CH_2Pr)$$

$$Ph_3P(CH_2R)(CH_2Pr) \xrightarrow{-PhH} Ph_3\overset{+}{P}(\overset{-}{C}HR)(CH_2Pr) \rightleftharpoons Ph_3\overset{+}{P}(CH_2R)(\overset{-}{C}HPr) \xleftarrow{-PhH} Ph_3P(CH_2R)(CH_2Pr) \quad (172)$$

the presence of a ring (such as spirophosphafluorene or homocubane)[569a], to be isolated and handled fairly easily:

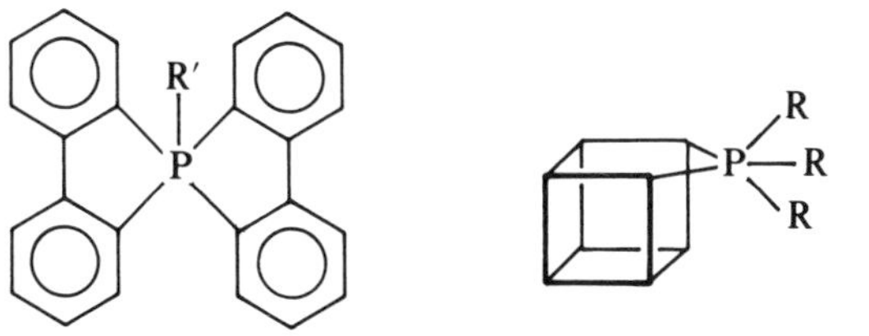

A third kind of reactivity of organolithium compounds towards phosphonium salts does not relate to the reaction with the hydrogen α to the phosphorus or with the phosphorus atom itself, but to the addition reaction on the β- or δ-carbon atom of an unsaturated system activated by the phosphonio group[560c]. With vinylphosphonium salts, for example, the organolithium compounds and the enolic carbanions give a Michael addition reaction resulting in the formation of phosphonium ylides generally used *in situ* in a Wittig reaction (reaction 173). Related to this reactivity, the addition reaction of Bu^tLi at the *para* position of a benzene ring activated by a phosphonio substituent should also be mentioned[574] (reaction 174).

$$Ph_3\overset{+}{P}CH{=}CHR\ X^- + R'Li \xrightarrow[-LiX]{} Ph_3\overset{+}{P}{-}\overset{-}{C}HCH(R)(R') \quad (173)$$

$$Ph_3\overset{+}{P}R\ X^- + Bu^tLi \xrightarrow[-LiX]{-65\,^\circ C\ (Et_2O)} Bu^t{-}C_6H_5^{-}{-}\overset{+}{P}Ph_2(R) \quad (174)$$

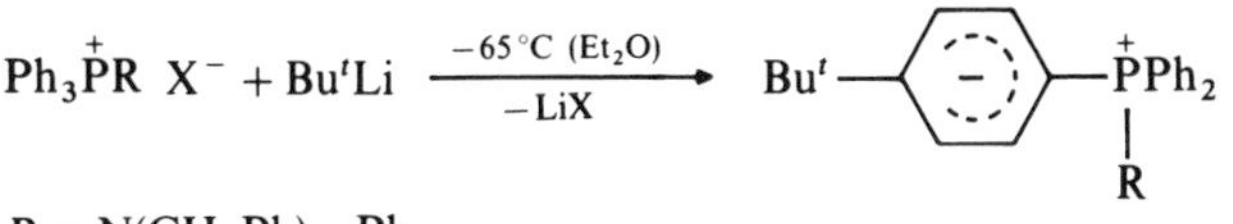

$R = N(CH_2Ph)_2$, Ph

C. Alkaline Hydrolysis of Phosphonium Salts

Alkaline hydrolysis of phosphonium salts is one of the oldest[575] and of the best investigated reactions in organophosphorus chemistry and several excellent reviews are devoted to it[3,5,576–578]. Generally the reaction affords the formation of phosphine oxide and hydrocarbon (equation 175). The reaction is of great practical interest for the preparation of phosphine oxides[579], but also from the theoretical point of view as a typical example of nucleophilic substitutions at tetracoordinated phosphorus: $S_{N(P)}$. Apart from the main reaction shown in equation 175, several secondary reactions are specifically related to various R groups and have been generally less investigated than the main reaction.

$$R_4P^+X^- + M^+HO^- \longrightarrow R_3P{=}O + RH \qquad (175)$$

The different evolution possibilities for the alkaline hydrolysis of phosphonium salts are shown together in the general Scheme 1. Two major parts can be distinguished: on the one hand, $S_{N(P)}$, $S_{N(P)mig}$ and E_p reactions, which result from the initial attack on the phosphorus atom by hydroxide anion acting as a nucleophile; and on the other, $E_{H\alpha}$ and $E_{H\beta}$ reactions (and also bearing in mind the formation of phosphonium ylides), which result from the initial attack of the hydroxide anion on the hydrogen in the α- or β-position to the phosphorus.

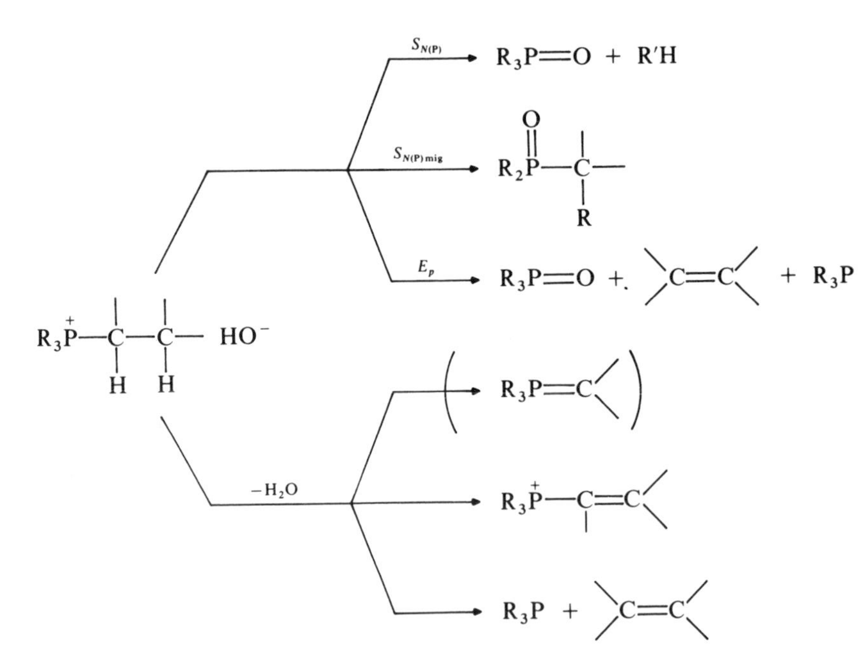

SCHEME 1. Different mechanisms in the alkaline hydrolysis of phosphonium salts

A few examples of uncommon hydrolyses, corresponding formally to an $S_{N(C\alpha)}$ reaction should also be mentioned (reaction 176). Indeed with particularly stable carbenium ions such as diphenylcyclopropenium[580] or tropylium[581,582], spontaneous dissociation of the phosphonium ion into the phosphine and the carbenium ion occurs before reaction with the hydroxide anion can take place.

$$R_3\overset{+}{P}R' + HO^- \longrightarrow R_3P + R'OH \qquad (176)$$

1. Main reaction in the hydrolysis of phosphonium salts: $S_{N(P)}$ *mechanism*

The mechanism of the hydrolysis of phosphonium salts was formulated by McEwen and coworkers[583,584], after an initial proposal as early as 1929 by Fenton and Ingold[585] (Scheme 2). This $S_{N(P)}$ mechanism accounts for most results, even though in some cases[584] the results can also be explained by variations of the classical mechanism.

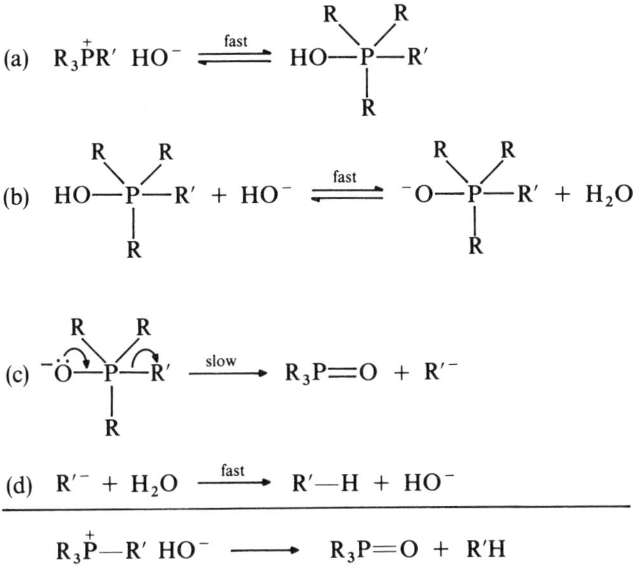

$$\text{(a)}\quad R_3\overset{+}{P}R'\ HO^- \underset{}{\overset{\text{fast}}{\rightleftharpoons}} HO{-}PR_3{-}R'$$

$$\text{(b)}\quad HO{-}PR_3{-}R' + HO^- \overset{\text{fast}}{\rightleftharpoons} {}^-O{-}PR_3{-}R' + H_2O$$

$$\text{(c)}\quad {}^-\ddot{O}{-}PR_3{-}R' \xrightarrow{\text{slow}} R_3P{=}O + R'^-$$

$$\text{(d)}\quad R'^- + H_2O \xrightarrow{\text{fast}} R'{-}H + HO^-$$

$$R_3\overset{+}{P}{-}R'\ HO^- \longrightarrow R_3P{=}O + R'H$$

SCHEME 2. Mechanism of $S_{N(P)}$ reaction for the alkaline hydrolysis of phosphonium salts

a. Nature of the leaving group. According to the mechanism, the leaving group cleaved to afford the hydrocarbon is generally the group corresponding to the most stable carbanionic species. This has been established by two different kinds of investigation: either by the determination of the products ratio in the alkaline hydrolysis of phosphonium salts including different groups linked to the phosphorus (reaction 177), or by the kinetic data determined in the alkaline hydrolysis of unsymmetrical phosphonium ions in which only the leaving R′ group is changed (reaction 178). The investigations have led to several series of leaving groups (Table 14), in which the facility of cleavage parallels exactly the stability of the corresponding anion. The series mainly follow the order allyl and benzyl > aryl > alkyl. However, as shown in some series, the three classes can overlap depending on the exact nature of the groups; further, it must be pointed out that two groups of the same class are cleaved competitively.

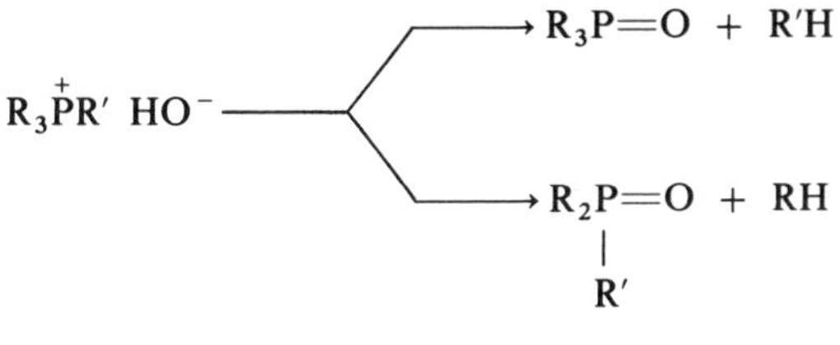

$$R_3\overset{+}{P}R'\ HO^- \longrightarrow \begin{cases} R_3P{=}O + R'H \\ R_2R'P{=}O + RH \end{cases} \qquad (177)$$

TABLE 14. Decreasing facility order[a] for cleavage of the leaving group

Order	Ref.
$CH_2{=}CHCH_2$ and $PhCH_2 > Ph > Me > Et$, n-Pr, $PhCH_2CH_2$	588
Ph > Alk, Me	589
$PhCH_2 >$ Ph, Alk	590
Ph > $(CH_2)_n$ (n = 2–6)	271, 290
Ph > $(CH_2)_n$ O (n = 4, 5)	300
$PhCH_2 \gtrsim$ [naphthyl] $\gtrsim$ Ph	591
$Me \gtrsim Et \gtrsim n\text{-}C_{12}H_{25}$	592
O_2N-C6H4– > Cl-C6H4– > C6H5-C6H4– > [naphthyl] > Ph > H_2N-C6H4– > MeO-C6H4–	593
$Et \gtrsim n\text{-}Pr \gtrsim n\text{-}Bu$	594
Cymantrenyl $\gtrsim$ ferrocenyl $\gtrsim PhCH_2$	595, 596
[2-thienyl] $\gtrsim$ [2-furyl] $> PhCH_2 >$ Me	597–599
[2-pyridyl] $> PhCH_2$	600
[2-furyl]CH_2– $\gtrsim$ [2-thienyl]CH_2– $\gtrsim PhCH_2$– $\gtrsim$ [3-thienyl]CH_2^- [3-furyl]CH_2^-	598, 601
OHC-[thienyl]– > OHC-[furyl]– > Ph > OHC-C6H4–	602
R(R')N– (R' = Alk, Ar, NR''_2) > Ph	154, 483, 520
MeS– > RO– > Ph, Alk	603–605

[a] $R \gtrsim R'$ means that the R groups is cleaved more easily than the R' group, but in a competitive way.

$$R_3\overset{+}{P}R'\ HO^- \longrightarrow R_3P{=}O + R'H \tag{178}$$

The kinetic results obtained for the alkaline hydrolysis of phosphonium salts **27**[586] and **28**[587] establish a direct correlation between the Hammett constant values for the Z substituents and the ease of cleavage of the substituted groups: electron-withdrawing Z substituents facilitate the cleavage, whereas electron-donating Z substituents are unfavourable. It has been pointed out that heteroatomic substituents in the arylic *ortho* position of arylbenzylphosphonium (Me_2N[43,606], OMe[607]) can modify the normal cleavage order of the groups linked to the phosphorus as a consequence of a through-space $N_{2p} \rightarrow P^{IV}$ (or $O_{2p} \rightarrow P^{IV}$) orbital overlap (reaction 179).

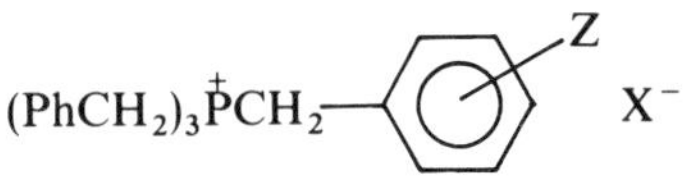

$(PhCH_2)_3\overset{+}{P}CH_2-C_6H_4-Z\ X^-$

(27)

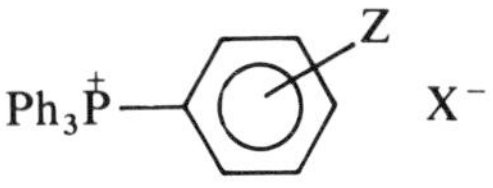

$Ph_3\overset{+}{P}-C_6H_4-Z\ X^-$

(28)

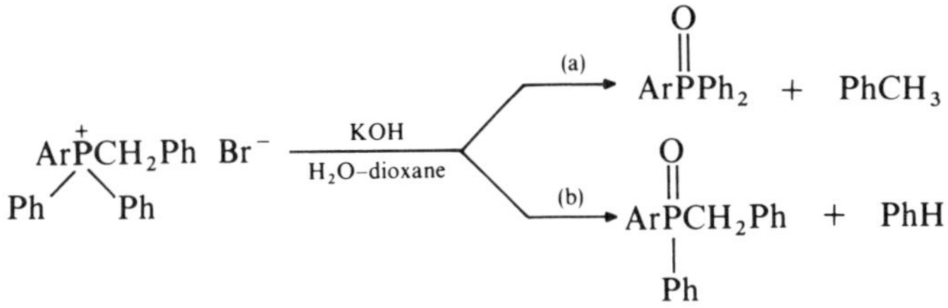

$$Ar\overset{+}{P}(Ph)_2CH_2Ph\ Br^- \xrightarrow[H_2O\text{-dioxane}]{KOH} \begin{cases} \text{(a)} & ArP(O)Ph_2 + PhCH_3 \\ \text{(b)} & ArP(O)(Ph)CH_2Ph + PhH \end{cases} \quad (179)$$

	(a)	(b)	k_{rel}
Ar = Ph	100	—	1
Ar = $Me_2N-C_6H_4-$ (para) 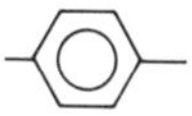	100	—	4.76
Ar 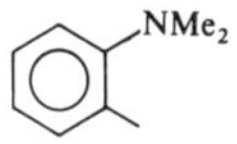 = C_6H_4 with ortho NMe_2	3.5	96.5	4.3×10^{-3}

b. Kinetic characteristics of the reaction. The first fundamental kinetic investigations, by McEwen and coworkers[583,586] and Hoffmann[587] showed that most alkaline hydrolyses obey third-order kinetics, being first order in phosphonium salts and second order in hydroxide ion:

$$v = k_{obs}[R_4P^+][HO^-]^2$$

This result has been corroborated by numerous other examples[43,593,597,601,608–614] and fits well with the classical mechanism described in Scheme 2. Only a few examples of second-order reactions are known; they concern in particular the alkaline hydrolysis of phosphonium salts **29**[586,615] and **30**[616]. In both cases, the second-order kinetics are very likely the result of a 'direct substitution' induced by the very high stability of the carbanion resulting from the P—C bond cleavage.

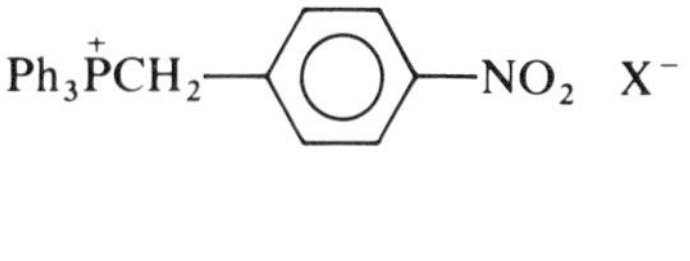

$Ph_3\overset{+}{P}CH_2-C_6H_4-NO_2\ X^-$

(29)

1-Ph-1-Me-2,5-Ph-3,4-R-phospholium X^-

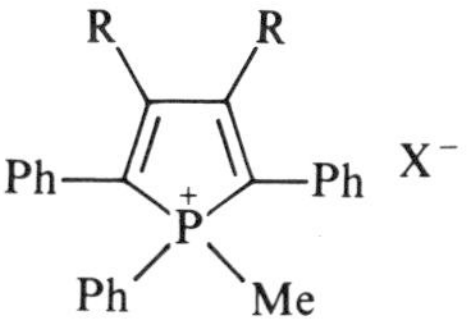

(30)
R = H, Ph

Investigation of the influence of the solvent on the kinetics of the reaction corroborate, in the same way as the order of the reaction, the formation of hydroxyphosphorane. Kinetic investigations have been performed with various solvent mixtures: EtOH–H_2O[601,608,616–620], MeOH–H_2O[610,611], $MeOCH_2CH_2OMe$–H_2O[583,586,595,621], dioxane–H_2O[587,609], thf–H_2O[610,611,622], dmso–H_2O[612,613], dmso–MeOH[623]; it appears that the rate of the reaction increases with decrease in the polarity of the solvent[587,610,611]. This conclusion is corroborated by the variation of hydrolysis rate with the water content in the solvent mixture: the rate can be increased by 10^5–10^7 when the water content is decreased in EtOH–H_2O[615], dioxane–H_2O[609], thf–H_2O[622] or dmso–H_2O[612,613]. It was also shown[617] that the hydroxide ion is the only effective nucleophile in the reaction, even for very low water contents in EtOH mixtures; therefore, only one mechanism takes place in this case.

The influence of the groups linked to phosphorus on the rate differs depending on the nature of the group: either, the leaving group is cleaved to give a hydrocarbon, or a residual group which remains linked to phosphorus in the resulting phosphine oxide. The residual groups exert their influence on the hydrolysis rate as a consequence of their influence on the positive charge of the various phosphorus species occurring in the course of the reaction; for the same leaving group, the rate increases with increasing electron withdrawing ability of the residual groups[586,597,608,610,611,617,620] (reaction 180). Of course, the leaving groups exert the same effects as the residual groups in steps (a) and (b), but this effect is secondary[586] in regard to the direct influence on step (c), which will be all the easier as the carbanion formed is stable. It is noteworthy that heteroatomic groups (NMe_2[43,606] OMe[607]) in the *ortho* position, which can exert a through-space electron-donating effect to the phosphorus reduce the rate of alkaline hydrolysis (equations 179) and in modifying the nature of the leaving group.

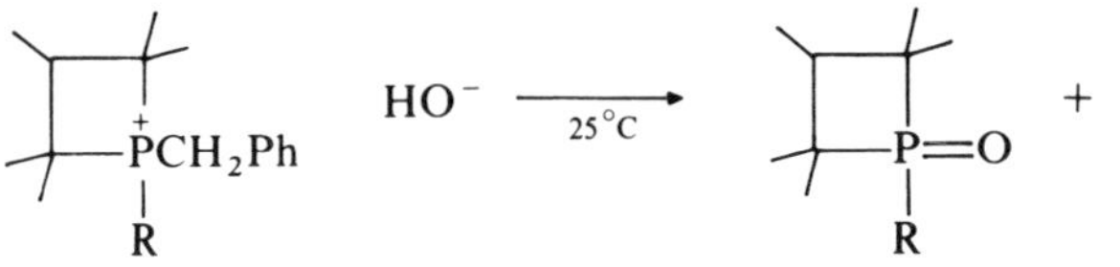

$\overset{+}{P}CH_2Ph$ (R) $\quad$ HO^- $\xrightarrow{25\,^\circ C}$ $P{=}O$ (R) + PhMe (180)[620]

	R	k_{rel}
(a)	Ph	81
(b)	$PhCH_2$	4
(c)	Me	1

The electronic factors, although important, are not the only factors to exert influence on the hydrolysis rate: steric factors also play an important part. Steric hindrance of the groups linked to phosphorus is of minor importance, even though in some cases[624,625] it induces a significant slowing of the rate. In contrast, the ring strain is crucial for the four- or five-membered cyclic phosphonium salts[608,620] when the ring is not opened during the reaction: the rate is higher when the ring is small; from six-membered rings upwards the ring strain is not significant and the hydrolysis rate is almost the same as for the acyclic compounds:

ref. 620:

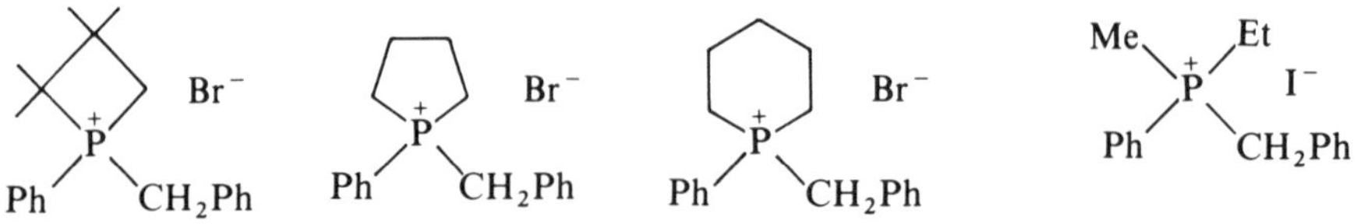

Four-membered ring $\overset{+}{P}$(Ph)(CH_2Ph) Br^-	Five-membered ring $\overset{+}{P}$(Ph)(CH_2Ph) Br^-	Six-membered ring $\overset{+}{P}$(Ph)(CH_2Ph) Br^-	Me, Et, Ph, CH_2Ph $\overset{+}{P}$ I^-
$k_{rel} = 6.9 \times 10^6$	520	1.7	1

ref: 608: 1-methyl-1-phenylphospholanium ($k_{rel} = 1100$), 1-methyl-1-phenylphosphinanium (0.73), Me_3P^+Ph (1)

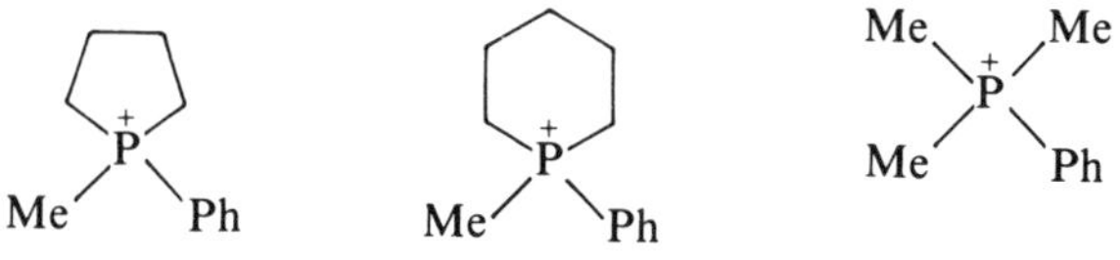

The kinetic acceleration noted in the case of small rings is generally ascribed to the release of the ring strain occurring during the formation of the intermediate hydroxyphosphorane, in which the ring occupies an apical–equatorial position in the trigonal bipyramid (TBP) corresponding to a geometrical $\widehat{CPC}$ angle of 90°, which is less constraining than the tetrahedral angle of 109.3° theoretically necessary for the initial phosphonium structure (sp^3).

The activation energies determined from kinetic investigations are in good agreement with the results already described. The mean values of E_a for the most common leaving groups in the alkaline hydrolysis of phosphonium salts are ca 15 kcal mol^{-1} for $R = CH_2Ph$[597,620] and ca 35 kcal mol^{-1} for $R = Ph$[597,608,617]. Of course, the activation energies vary according to the residual groups[617,620], the solvent[609,612,617,623] and the cyclic or acyclic structure of the phosphonum salt[608,620].

Volumes of activation have been measured for the alkaline decomposition of benzylphosphonium salts[614]. The rates of reaction are very strongly retarded by pressure; the volumes of activation are 32 cm^3 mol^{-1} while the volume of reaction is 16 cm^3 mol^{-1}. This is in good agreement with a two- or three-step mechanism, involving desolvation and fragmentation, each of which contributes to increasing the volume.

c. Stereochemistry of the reaction. If the number 1 is assigned arbitrarily to the entering hydroxyl group, 2 to the leaving group, and 3, 4 and 5 to the distinct residual groups, the alkaline hydrolysis of an enantiomer can take place, in the case of a chiral phosphonium salt, either with configuration inversion or retention (equation 181). The stereochemical outcome of the reaction depends essentially on the formation path, the lifetime, the isomerization possibilities and the decomposition path of the pentacoordinate intermediate hydroxyphosphorane[3,5,576–578].

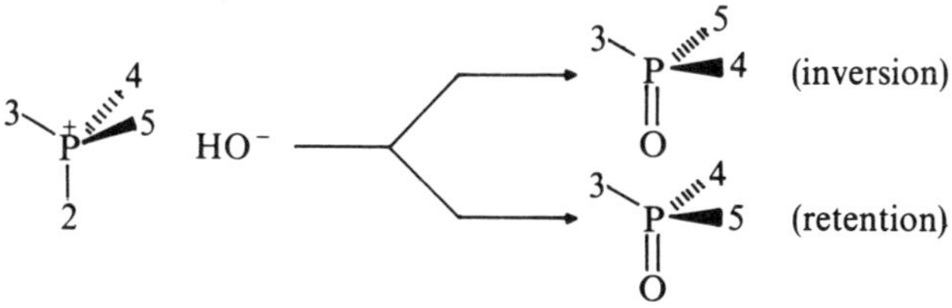

$$P^+(2)(3)(4)(5) \xrightarrow{HO^-} O{=}P(3)(5)(4) \text{ (inversion)} \quad / \quad O{=}P(3)(4)(5) \text{ (retention)} \qquad (181)$$

The initial trigonal bipyramid (TBP) can be formed in two ways from the tetrahedral phosphorus atom: either by nucleophilic attack of the hydroxide ion at any of the six edges of the tetrahedron, leaving the nucleophile 1 in an equatorial position (equatorial attack), or by attack at any of the four faces of the tetrahedron, which would set the nucleophile in an apical position (apical attack). Because apical bonds are longer, and therefore weaker, than equatorial bonds, apical attack is favoured since the formation energy of the initial TBP is then lower. In the same way, since apical bonds are weaker they should be more easily broken than their equatorial counterparts, hence the apical departure of the leaving group 2 is favoured.

Between the formation of the initial TBP (1 apical) and the decomposition of the final TBP (2 apical) into the corresponding phosphine oxide, if its lifetime is sufficient, the initial phosphoranes can undergo ligand reorganization [following a Berry pseudorotation[626] (BPR) or less likely a turnstile rotation[627]] to give, if all five groups attached to phosphorus are different, ten enantiomeric pairs of phosphoranes, each of which with substituent 2 in an apical position can undergo evolution to the products. The most useful model to discuss all the possibilities for the stereochemical outcome of the alkaline hydrolysis of phosphonium salts is the Desargues–Levi graph[628–630] using, for the enantiomers of the TBP, Mislow's notation[628] (Scheme 3).

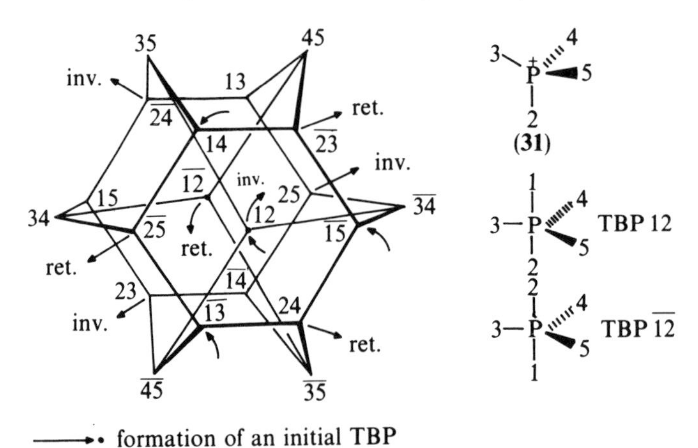

→• formation of an initial TBP

←• decomposition of a final TBP (with inversion or retention of configuration).

SCHEME 3. Formation, isomerization and decomposition of the hydroxyphosphoranes from the chiral salt **31**

i. Acyclic phosphonium salts. Without any particular constraint, the stereochemical course of the reaction depends mainly on the cleavage facility for the leaving group. Indeed, according to the electronegativity of the leaving group, the TBP preferentially formed initially is 12. This is favourable to evolution with inversion of configuration corresponding to the direct cleavage of the leaving group 2 from the apical position in which it is already located. When the leaving groups is easily cleaved (e.g. $PhCH_2$, allyl), the initial TBP, 12, undergoes no pseudorotation and alkaline hydrolysis affords almost full inversion of configuration (Table 15). When the leaving group is less easily cleaved (e.g. aryl), pseudorotation takes place and decomposition can occur through the ten TBP isomers exhibiting the leaving group 2 in apical positions. The result is then almost a full racemization (Table 16).

In the presence of geometrical constraints or electronegativity effects, the stereochemical outcome of the reaction depends on the relative importance of each factor, which will determine on one hand the relative ratio of the initial TBPs 12 and $\overline{13}$ (an overcrowded residual group, arbitrarily indicated here as 3, favours TBP $\overline{13}$ at the expense of TBP 12), and on the other hand the ratio of pseudorotation to decomposition of TBPs (a very good leaving group not only favours TBP 12 but is also unfavourable for the pseudorotation). This balance of factors accounts for the following results: for the same leaving group 2, the inversion ratio is decreasing with increasing steric hindrance of

TABLE 15. Stereochemistry of alkaline hydrolysis of phosphonium salts with a good leaving group

	2				
	$PhCH_2$				$CH_2{=}CHCH_2$
R	Et	*i*-Pr	cyclohexyl	cyclopentyl	Et
Inversion (%)	100	100	100	100	92
Ref.	582, 631–634	635–637	635–637	635–637	638

TABLE 16. Stereochemistry of alkaline hydrolysis of phosphonium salts with a moderate leaving group

	2			
	Ph	α-Naphth	β-Naphth	Ph–C_6H_4–
R	cyclohexyl	Ph	Ph	Ph
Inversion/retention (%)	23 (inv.)	41 (ret.)	10 (ret.)	12 (ret.)
Ref.	636	636	636	636

TABLE 17. Stereochemistry of alkaline hydrolysis of phosphonium salts with overcrowded residual groups

	2	
	$PhCH_2$	$PhCH_2$
3	α-Naphth	*t*-Bu
Inversion (%)	28	19
Ref.	635	637

the remanent group 3(Table 17); and for the same overcrowded remanent group 3, the inversion ratio is decreasing with increasing facility of cleavage for the leaving group 2 (Table 18).

It must be pointed out than, in the case of phosphonium salts in which the phosphorus atom is a prochiral centre, attempted alkaline hydrolyses with asymmetric induction

TABLE 18. Stereochemistry of alkaline hydrolysis of phosphonium salts with overcrowded residual group, depending on the facility of cleavage for the leaving group.

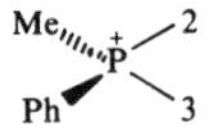

	2			
	OEt	CH_2Ph	$CH_2CH{=}CH_2$	α-Naphth
3	*t*-Bu	*t*-Bu	*t*-Bu	*t*-Bu
Inversion/retention (%)	100 (inv.)	19 (inv.)	49 (ret.)	63 (ret.)
Ref.	639	637	637	637

either by a menthyl group[589] (linked to phosphorus or generated through a quick *in situ* epimerization of a neomenthyl group) or by a chiral ammonium salt[590] used as a phase transfer catalyst have failed[589] or have been very poor (ee < 8%)[590], expect in the case of tetramethylenebisphosphonium salts (de 33%)[589] (reaction 182). In these cases also it is very likely that the stereoselectivity of the reactions depends not only on the asymmetric inducing group but also on the nature of the residual groups at the phosphorus.

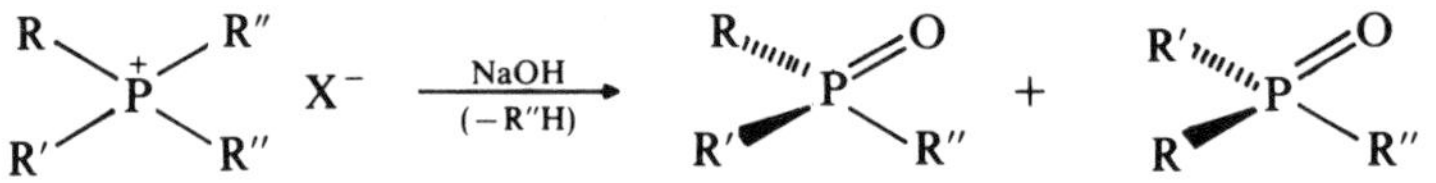

$$RR'P^+R''_2\ X^- \xrightarrow[(-R''H)]{NaOH} RR'P(O)R''\ (x) + R'RP(O)R''\ (1-x) \quad (182)$$

ref. 589: R = Me, Et, *i*-Pr, $-(CH_2)_4-$; R′ = menthyl; R″ = Ph
conditions: 18 h reflux, $MeOH-H_2O$

ref. 590: R = Me, Et, Pr; R′ = Ph; R″ = CH_2Ph, $-CH_2-C_6H_4$(MeO)
conditions: 22 °C, $H_2O-CHCl_3$ (phase transfer catalysis: quininium, etc.)

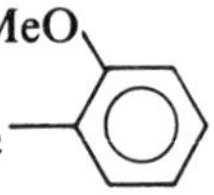

ii. Cyclic phosphonium salts. The stereochemistry of alkaline hydrolysis has been mainly investigated for cyclic phosphonium salts in which the ring is not opened during the reaction. Since the leaving group is not a cyclic bond, the two cyclic P—C bonds can be considered as 3 and 4 groups. In the corresponding X-ray structural analysis, the cyclic C—P—C angles have been determined:

82–85 ° for the phosphetanium salts[640–642] (four-membered rings);
95 ° (± 1 °) for the phospholanium salts[125,643] (five-membered rings);
104 ° (± 2 °) for the phosphiranium salts[129,132,382] (six-membered rings).

Therefore, the first two classes of salts exhibit important ring strains. Accordingly, in the corresponding hydroxyphosphoranes, the ring will occupy very preferentially an apical–equatorial position within the TBP structures. In entries 1–8 in Table 19 this factor primarily determines the stereochemical course of the reaction. Indeed, this geometrical constraint results in excluding in Scheme 3 TBPs 12, $\overline{12}$, 15, $\overline{15}$, 25 and $\overline{25}$ (in which the two groups 3 and 4 would be together in the equatorial position), and also

TABLE 19. Stereochemistry of the alkaline hydrolysis of cyclic phosphoniums salts

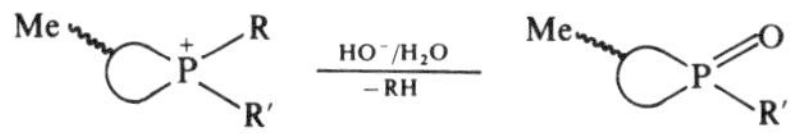

$$\xrightarrow[-\mathrm{RH}]{\mathrm{HO^-/H_2O}} \qquad (183)$$

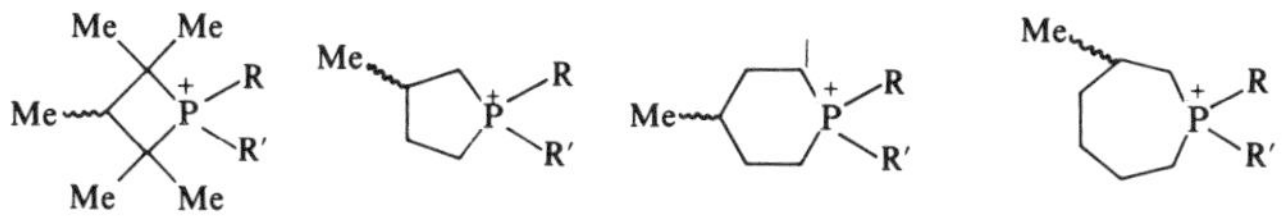

(32) phosphetanium (33) phospholanium (34) phosphorinanium (35) phosphepanium

Compound	Entry	R	R′	Structure Me/R′	Retention (%)	Inversion (%)	Ref.
32a	1	CH_2Ph	Ph	*cis*	10	90	643
32a	2	CH_2Ph	Ph	*trans*	90	10	643
32b	3	OEt	Ph	*cis*	100	—	644, 645
32b	4	OEt	Ph	*trans*	100	—	644, 645
33a	5	CH_2Ph	Me	*cis*	100	—	593, 646–648
33a	6	CH_2Ph	Me	*trans*	100	—	593, 646–648
33b	7	CH_2Ph	Ph	*cis*	100	—	593, 647, 649
33b	8	CH_2Ph	Ph	*trans*	100	—	593, 647, 649
33c	9	Ph	Me	*cis*	50	50	593, 647
33c	10	Ph	Me	*trans*	50	50	593, 647
33d	11	OMe	Ph	*cis*	51	49	650
33d	12	OMe	Ph	*trans*	42	58	650
34a	13	CH_2Ph	Ph	*cis*	48	52	651
34a	14	CH_2Ph	Ph	*trans*	78	22	651
34b	15	OMe	Ph	*cis*	—	100	650
34b	16	OMe	Ph	*trans*	—	100	650
35a	17	CH_2Ph	Ph	*cis*	—	100	652
35a	18	CH_2Ph	Ph	*trans*	—	100	652

TBPs 34 and $\overline{34}$ (in which they would be together in the apical position). Consequently, the only possibilities for the initial TBPs are $\overline{13}$ and 14, and decomposition in the phosphine oxide can only occur after one or two pseudorotations resulting in retention or inversion of configuration, respectively.

When the leaving group 2 is easily cleaved, then it has a high electronegativity and the two routes are not equivalent: owing to the favoured apical position of the group 2, TBP 24 (or $\overline{23}$) is favoured relative to TBP $\overline{45}$ (or 35), and the reaction takes place with retention of configuration. Accordingly, the reaction takes place with retention of configuration for the phosphetanium salts **32a** and **32b** and also for the phospholanium salts **33a** and **33b** (entries 1–8). However, in the case of a less good leaving group (entries 9 and 10), electronegativity factors are less important and the two competitive ways afford the same mixture independently from the starting *cis* or *trans* diastereoisomer. On the other hand, when the electronegativity (or apicophilicity) of the leaving group becomes very important (for the alkoxy group, for example), this factor can be competitive with the ring strain (not for the four-membered rings, but partly for the five-membered rings and fully for the six-membered rings). The result is then a new structural possibility for the initial hydroxyphosphorane with the ring in a diequatorial position, resulting directly in a partial (entries 11 and 12) or full (entries 15 and 16) inversion of configuration. The same competition between the geometrical and electronic factors, even with weaker effects, occurs with benzylic leaving groups and the six-membered (entries 13 and 14) or seven-membered rings (entries 17 and 18).

It is noteworthy that the different results obtained with the *cis* and *trans* isomers (entries 11–14) can be explained, as corroborated[125,129] by X-ray structural analysis and theoretical calculations, taking into account the steric interactions between the hydroxide ion and the ring methyl during the hydroxyphosphorane formation. Finally, with regard to cyclic phosphonium salts, it must be pointed out that stereoelectronic effects[653–655] can take place in the alkaline hydrolysis of rigid heterophosphoniums[512], in which the heteroatoms such as oxygen are part of the ring; the stereoelectronic effects occur in this case through $n_O \rightarrow \sigma^*_{P-X}$ interactions (X-leaving group), which contribute to the specific cleavage of the P—X bond:

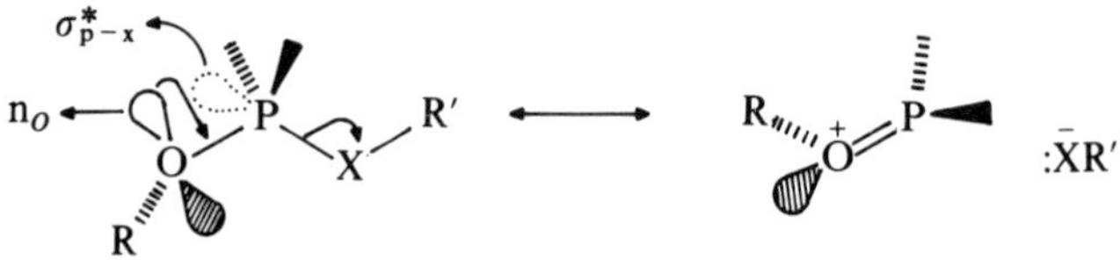

$$n_O \quad \sigma^*_{P-X} \qquad \longleftrightarrow \qquad :\bar{X}R'$$

The stereoelectronic effects can invert the normal order of cleavage for the various groups linked to the phosphorus. For example, in the case of a bicyclic phosphonium salt[512], the alkaline hydrolysis occurs with a P—O bond cleavage, even when in the case of the acyclic analogues only the thio group, the best leaving group, is cleaved (reaction 184).

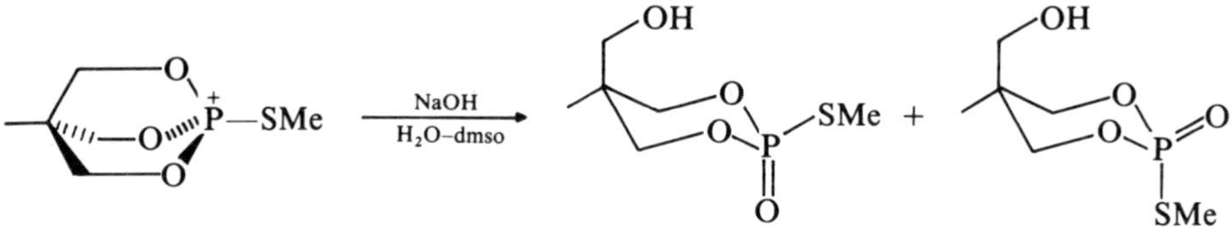

$$\overset{+}{P}\text{—SMe} \xrightarrow[\text{H}_2\text{O–dmso}]{\text{NaOH}} \text{(OH)}\ldots\text{P(=O)–SMe} + \text{(OH)}\ldots\text{P(=O)–SMe}$$

$$(MeO)_3\overset{+}{P}SEt \xrightarrow[\text{H}_2\text{O–dmso}]{\text{NaOH}} (MeO)_3P{=}O \qquad (184)$$

2. *Secondary reactions in the alkaline hydrolysis of phosphonium salts*

As already mentioned, apart from the main reaction shown in equation 175, several secondary reactions (Scheme 1) are specifically related to various R groups linked to the phosphorus.

a. Alkaline hydrolysis of bisphosphonium salts with fragmentation: E_p mechanism. Generally bisphosphonium salts react as monophosphonium salts[656,657] (equation 185). Only P^+—C—C—P^+ bisphosphonium salts react particularly with fragmentation of the bridge (equation 186). Depending on the particular phosphonium group, this reaction should be considered either as an $S_{N(P)}$ reaction, in which the resulting carbanion is fragmented by elimination of a leaving group (R_3P) in the β-position, or as an E_β reaction, in which the hydroxide ion is not attacking on the hydrogen but on the phosphorus atom in the β-position to the leaving group.

$$Ph_3\overset{+}{P}CH_2CH_2CH_2\overset{+}{P}Ph_3\ \ 2\,HO^- \xrightarrow[-2\,PhH_4]{} Ph_2\overset{O}{\overset{\|}{P}}CH_2CH_2CH_2\overset{O}{\overset{\|}{P}}Ph_2 \qquad (185)$$

$$R_3\overset{+}{P}-\overset{|}{\underset{|}{C}}-\overset{|}{\underset{|}{C}}-\overset{+}{P}R_3\ \ 2\,HO^- \longrightarrow R_3P + \;>C=C<\; + R_3P{=}O \qquad (186)$$

The duality of approach, on the one hand, and the combination of the resulting compounds, unique in the basic hydrolysis of phosphonium salts, on the other, lead to this reaction being said to occur by an 'E_p mechanism', described in a separate section.

i. Acyclic bisphosphonium salts. The decomposition of acyclic bisphosphonium salts **36** and **37** by alkaline hydrolysis can occur depending on the nature of the R group, either by a pure $S_{N(P)}$ mechanism, or by a pure E_p mechanism, or by a combination of the two (Scheme 4). As shown in Table 20, the E_p mechanism is the normal route for 1,2-ethylene-

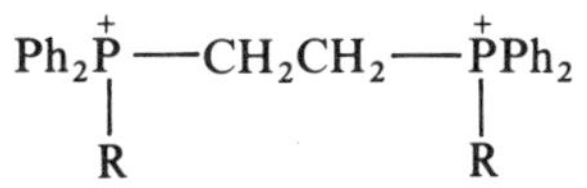

$$Ph_2\underset{R}{\overset{+}{P}}-CH_2CH_2-\underset{R}{\overset{+}{P}}Ph_2 \qquad (36)$$

$$Ph_2\underset{R}{\overset{+}{P}}-CH{=}CH-\underset{R}{\overset{+}{P}}Ph_2 \qquad (37)$$

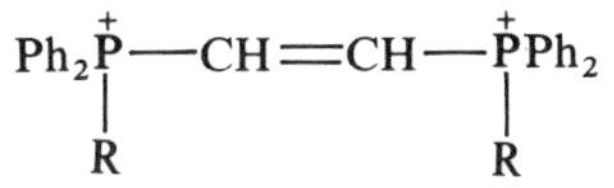

$$R_3\overset{+}{P}CH_2CH_2\overset{+}{P}R_3\ \ 2X^- \underset{}{\overset{HO^-}{\rightleftharpoons}} HOP(R)_3CH_2CH_2\overset{+}{P}R_3\ X^-$$

$$HOPR_3CH_2CH_2\overset{+}{P}R_3\ X^- \underset{H_2O}{\overset{HO^-}{\rightleftharpoons}} {}^-O-PR_3-CH_2-CH_2-\overset{+}{P}R_3\ X^- \longrightarrow R_3P=O + CH_2=CH_2 + PR_3 \qquad E_p$$

$$HOPR_3CH_2CH_2\overset{+}{P}R_3\ X^- \underset{H_2O}{\overset{HO^-}{\rightleftharpoons}} {}^-O-PR_3-CH_2CH_2\overset{+}{P}R_3\ X^- \longrightarrow R_2\overset{O}{\overset{\|}{P}}CH_2CH_2\overset{+}{P}R_3\ X^- \xrightarrow{HO^-} R_2\overset{O}{\overset{\|}{P}}CH_2CH_2\overset{O}{\overset{\|}{P}}R_2 \qquad S_{N(P)}$$

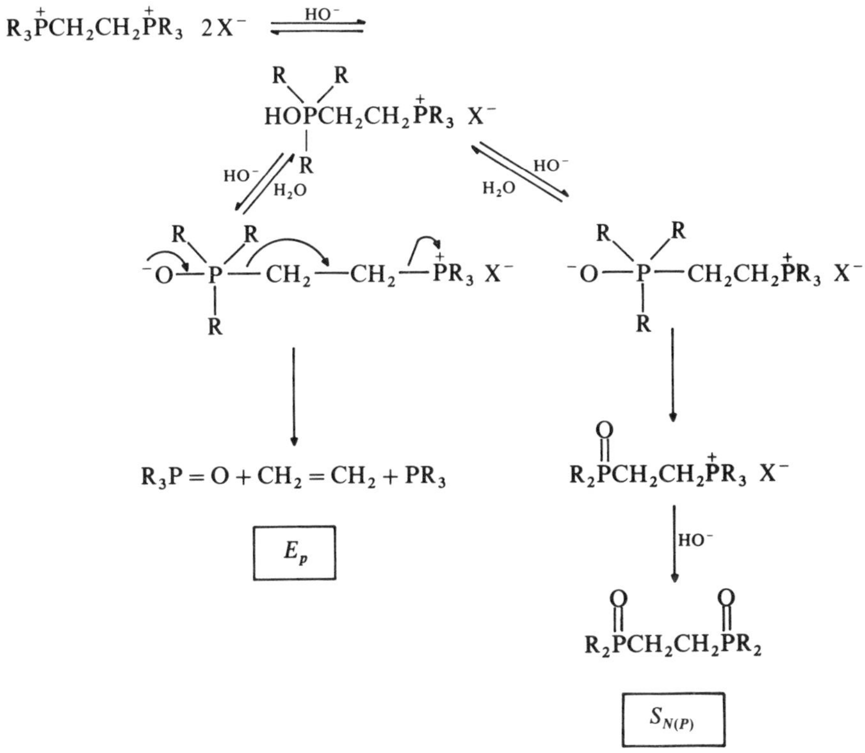

SCHEME 4

TABLE 20. Alkaline hydrolysis of 1,2-ethylene- or 1,2-vinylenebisphosphonium salts

	Compound							
	36a	**36b**	**36c**	**36d**	**36e**	**37a**	**37b**	**37c**
R	Ph	Me	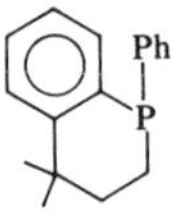 (cyclic P–Ph structure)	CH_2Ph	 CH_2–C_6H_4–NO_2	Ph	Me	CH_2Ph
E_p	100	100	100	30	—	100	100	100
$S_{N(P)}$	—	—	—	61	100	—	—	—
Ref.	657	657	66	657	657	359	657	386

or 1,2-vinylene-bisphosphonium salt decomposition; the $S_{N(P)}$ mechanism occurs only when the R group is a good leaving group (*p*-nitrobenzyl and to a lesser extent benzyl).

The ease of decomposition according to the E_p mechanism was explained by Brophy and Gallagher[657] on the basis of an intermediate conformation particularly in favour of the elimination reaction:

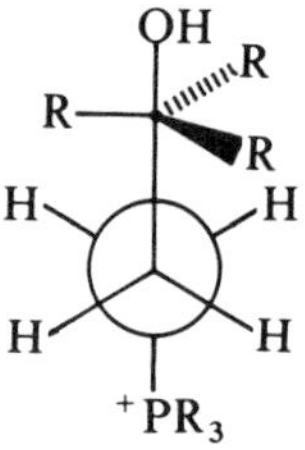

Indeed, on account of the important steric hindrance and electrostatic repulsion of the two phosphonio groups, the initial disalt conformation is essentially antiperiplanar. Immediately on hydroxyphosphorane formation, with introduction of the hydroxide ion in an axial position opposite to the P–CH_2 bond, there emerges a system in which all the groups interfering in the next fragmentation reaction are in 'transcoplanar' position. Such a structure is probably the most suitable for the concerted fragmentation reaction. The easy intervention of the E_p mechanism was also corroborated by kinetic studies; the reaction, of third order, is almost as fast as the $S_{N(P)}$ reaction for very reactive monosalts as the tetrabenzylphosphonium salts[657]. Other structures can also decompose by an E_p mechanism, e.g. the vinylogue salts[658] (reaction 187). We should emphasize that in the case of tetrakisphosphonium salts **39** which have both 1,2-ethylene and 1,4-but-2-enylidene groups on each phosphorus atom, the butenylidene bridge is cleaved selectively[252] (reaction 188). On the other hand, under the same conditions, the corresponding saturated cyclic salt, with tetramethylene bridges, gives only E_p cleavage of the 1,2-ethylene bridge[252].

$$Ph_3\overset{+}{P}CH_2CH{=}CHCH_2\overset{+}{P}Ph_3 \xrightarrow{3\text{ equiv. }^-OH} Ph_3P + CH_2{=}CH_2 + Ph_3P{=}O \quad (187)$$

Finally, it should be noted that other vinylogue structures such as **40** and **41** decompose

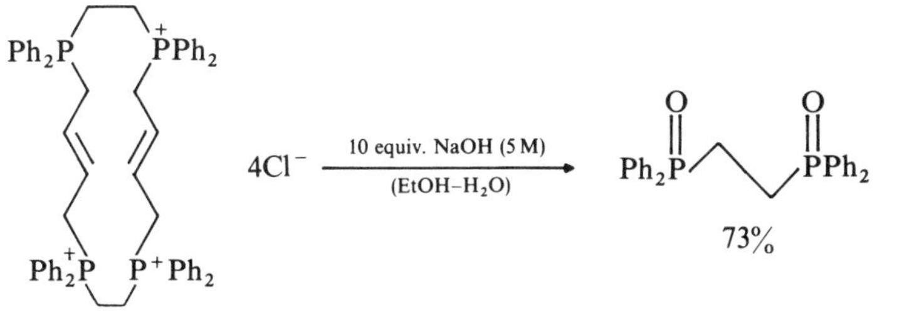

$$4Cl^- \xrightarrow[(EtOH-H_2O)]{10 \text{ equiv. NaOH (5 M)}} Ph_2P(O)CH_2CH_2CH_2P(O)Ph_2 \quad 73\% \qquad (188)$$

(39)

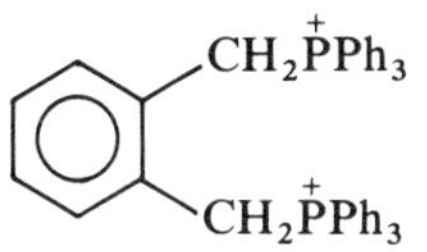

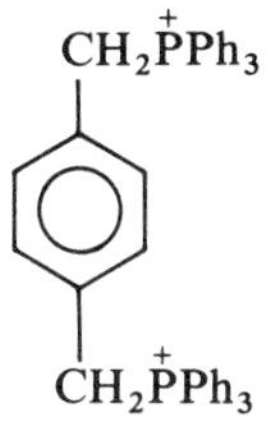

(40) (41)

only by an $S_{N(P)}$ mechanism, with xylene formation[657]. This result is expected, because the unfavourable E_p mechanism would destroy the benzene aromaticity.

Another kind of bridge between two phosphonio groups corresponding to the $P^+—C—C—O—C—C—P^+$ chain, is favourable to the same type of E_p fragmentation (reaction 189)[353]. Here the leaving group fragmentation is probably favoured by the interaction $n_O \rightarrow P^+$.

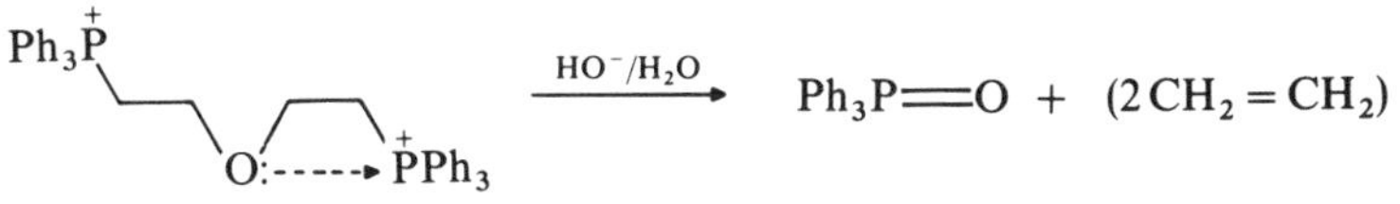

$$Ph_3\overset{+}{P}CH_2CH_2OCH_2CH_2\overset{+}{P}Ph_3 \xrightarrow{HO^-/H_2O} Ph_3P{=}O + (2\,CH_2{=}CH_2) \qquad (189)$$

ii. Cyclic bisphosphonium salts. Studies[386,657,659–662] on the alkaline hydrolysis of cyclic disalts **42** and **43** show the different behaviours of these disalts according to the base concentration used (Scheme 5). With an excess of base, they both react according to the E_p mechanism to give the same compound **46**. With a deficiency of base, the hydrolysis gives the monosalts **44** and **47** by the $S_{N(P)}$ mechanism (in the case of the salt **43**, the vinylphosphonium salt formed by the $S_{N(P)}$ mechanism undergoes a classical hydration reaction to give the monosalt **47** as the reaction product). These monosalts are not intermediates in the formation of **46**, since the alkaline hydrolysis of salt **44** also takes place by an S_{NP} mechanism and gives the bisphosphine dioxide **45**.

It is noteworthy that, for the cyclic disalts **42** and **43**, alkaline hydrolysis breaks down the cycle in all cases, whereas in the case of the cyclic monosalt **48** the best leaving group, phenyl, is eliminated normally with preservation of the cycle[608,663] (reaction 190). In the case of disalts **42** and **43** the steric hindrance between aromatic hydrogen atoms *ortho* to the phosphorus and the cyclic methylenic hydrogens could explain the

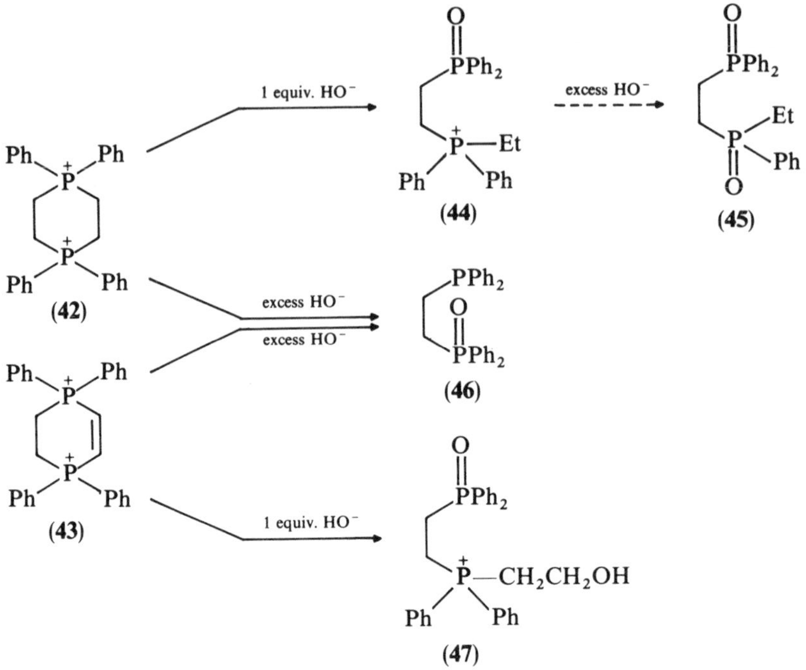

SCHEME 5

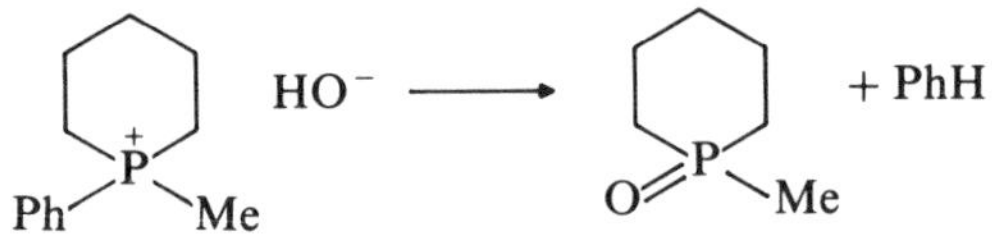

(190)

non-formation of the hydroxyphosphorane involving a ring in a diequatorial position[657]. This interaction should be strong enough to prevent the formation by pseudorotation of an hydroxyphosphorane with a phenyl group in the apical position as well as to prevent the cleavage of this group. This interpretation is corroborated by the alkaline hydrolysis of the disalt **49** according to the $S_{N(P)}$ mechanism[664]. The hindrance disappears and the better leaving group, benzyl, is eliminated with preservation of the ring (reaction 191).

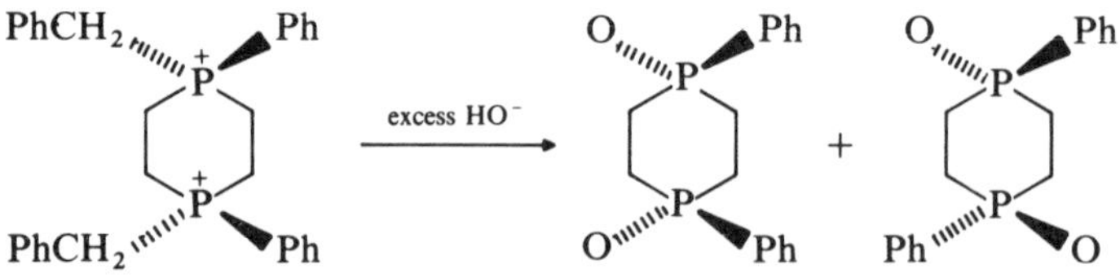

(191)

In the same way, decomposition of the disalt **50** follows a $S_{N(P)}$ mechanism with *o*-xylene formation[665]. In contrast, alkaline hydrolysis of the unsaturated homologue **51** occurs[665] by an E_p mechanism and shows the higher instability of the 1,2-vinylene bridge relative to the 1,2-ethylene bridge. This has been shown already for the decomposition of salt **43** with two types of chain. Finally, for the cyclic bisphosphonium salts **52**[666,667] and **53**[668], decomposition takes place by an $S_{N(P)}$ mechanism inducing the cleavage of the benzylic groups.

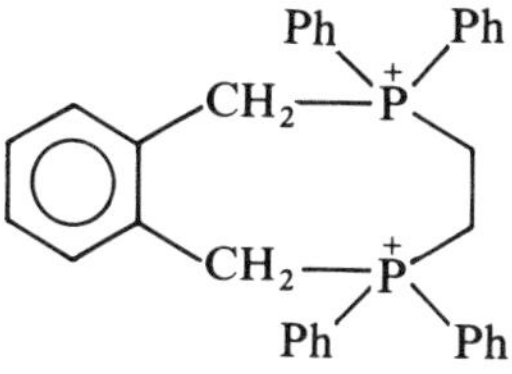

(50)

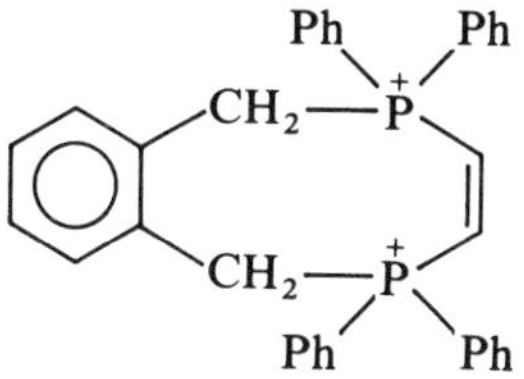

(51)

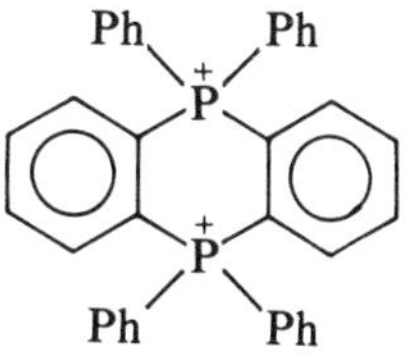

(52)

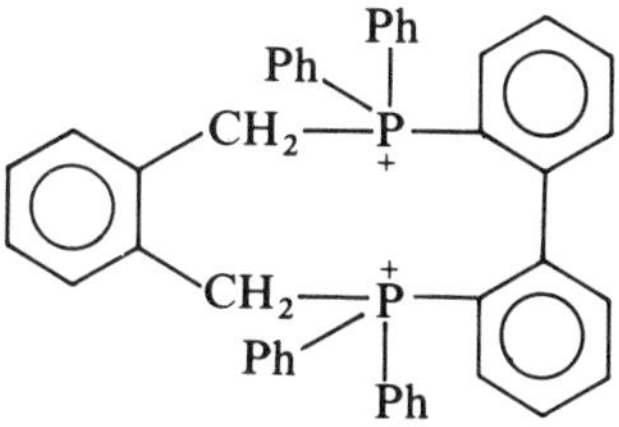

(53)

b. Alkaline hydrolysis with migration: $S_{N(P)mig}$*: mechanism.* The alkaline hydrolyses of some phosphonium salts take place with migration of one of the groups from the phosphorus atom to a carbon in the α-position. These salts are of two kinds: the α-halomethylphosphonium salts (reaction 192) and the vinylphosphonium salts (reaction 193).

$$R_3\overset{+}{P}CH_2X + HO^- \xrightarrow[-X^-]{} R_2\overset{O}{\overset{\|}{P}}CH_2R \tag{192}$$

$$R_3\overset{+}{P}CH{=}CHZ + HO^- \longrightarrow R_2\overset{O}{\overset{\|}{P}}\underset{R}{\underset{|}{C}}HCH_2Z \tag{193}$$

i. α-Halomethylphosphonium salts. The alkaline hydrolysis of α-halomethylphosphonium salts usually occurs with the formation of a mixture of products corresponding to the competition of three different mechanisms (Scheme 6). The $E_{H\beta}$ mechanism results from the decomposition of the α-hydroxymethylphosphonium salt formed by a substitution reaction on the methylenic carbon α to the phosphorus. The normal $S_{N(P)}$

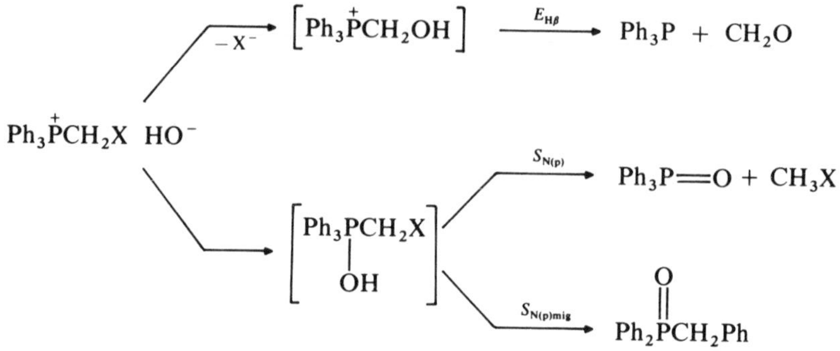

$$Ph_3\overset{+}{P}CH_2X \; HO^- \xrightarrow{-X^-} [Ph_3\overset{+}{P}CH_2OH] \xrightarrow{E_{H\beta}} Ph_3P + CH_2O$$

$$Ph_3\overset{+}{P}CH_2X \; HO^- \longrightarrow [Ph_3P(OH)CH_2X] \xrightarrow{S_{N(p)}} Ph_3P{=}O + CH_3X$$

$$[Ph_3P(OH)CH_2X] \xrightarrow{S_{N(p)mig}} Ph_2P(O)CH_2Ph$$

SCHEME 6

TABLE 21. Alkaline hydrolysis of α-halomethylphosphonium salts[658]

X	$E_{H\beta}$ (%)	$S_{N(P)}$ (%)	$S_{N(P)}$ mig (%)
Cl	15	74	11
Br	8	66	22
I	—	83	17

and $S_{N(P)mig.}$ mechanisms correspond to different decomposition routes for the hydroxyphosphorane intermediate. Of course, the relative importance of each mechanism depends on the nature of the halogen[658,669,670] (Table 21).

The proportion of the $E_{H\beta}$ mechanism decreases as the steric hindrance of the halogen increases, because the substitution reaction at the carbon is then disfavoured. The proportion of the $S_{N(P)mig.}$ mechanism increases with increasing capacity of the halide to be eliminated. However, in the case of iodine, its steric hindrance, favouring the apical position of the iodomethyl group in the initial phosphorane, also helps the cleavage of this group and consequently promotes the $S_{N(P)}$ mechanism, to the prejudice of the $S_{N(P)mig}$ mechanism. The rate of the $S_{N(P)mig}$ mechanism is still low or null, when salts **54**[591] and **55**[671] undergo alkaline hydrolysis. However, the iodomethyl group stands first in an equatorial position because an overcrowded group or a small ring is present;

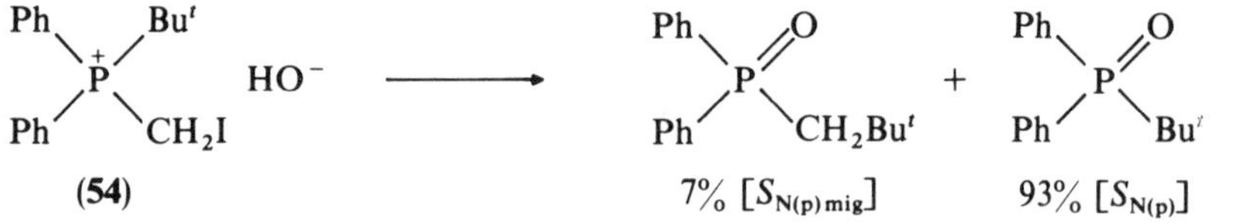

$$Ph_2\overset{+}{P}(Bu^t)CH_2I \; (\mathbf{54}) \quad HO^- \longrightarrow Ph_2P(O)CH_2Bu^t \; 7\% \, [S_{N(p)mig}] + Ph_2P(O)Bu^t \; 93\% \, [S_{N(p)}] \qquad (194)$$

$$(\mathbf{55}) \quad HO^- \longrightarrow \text{phenoxaphosphine oxide } P(O)R \qquad (195)$$

(55): 10-R-10-(iodomethyl)phenoxaphosphinium (ring P bearing R and CH_2I)

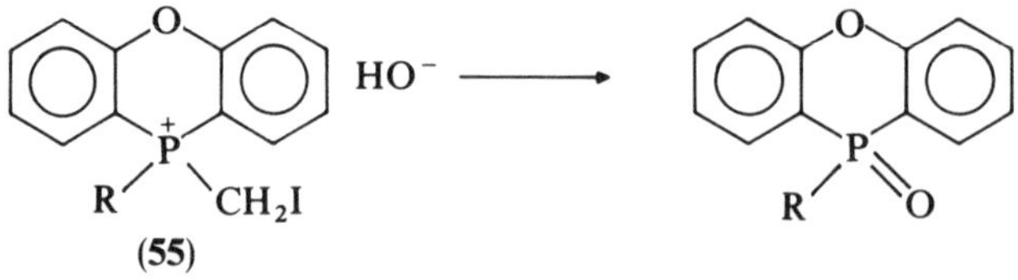

R = Ph or Me

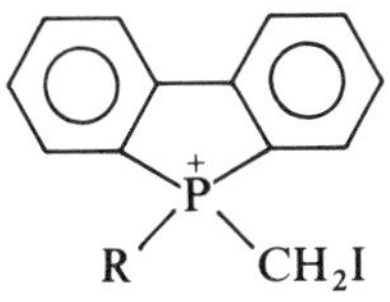

(56) (57) (58)

R = Ph, Me

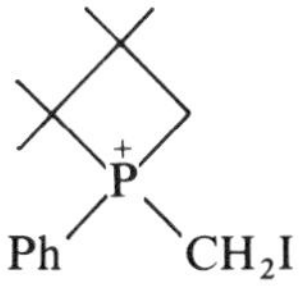

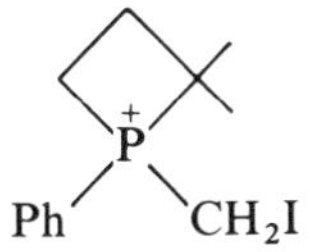

however, since a good leaving group is missing, the pseudorotation process is allowed and the iodomethyl group is cleaved from an apical position.

On the other hand, the $S_{N(P)mig}$ mechanism is the only mechanism taking place with a ring expansion in the hydrolysis of salts **56**[646,671,672], **57**[673] and **58**[674]. For all these salts, the iodomethyl group has an equatorial position in the initial phosphorane, because of the presence of a small ring. However, the migration reaction corresponds here to a ring expansion, and therefore to a release of the ring strain. The reaction is then fast; there is no permutational isomerization process, and so no cleavage of the iodomethyl group from an axial position. The absence of the permutational isomerization is supported by the stereochemical results[674] concerning the alkaline hydrolysis of the two enantiomers **59a** and **59b** (reaction 196).

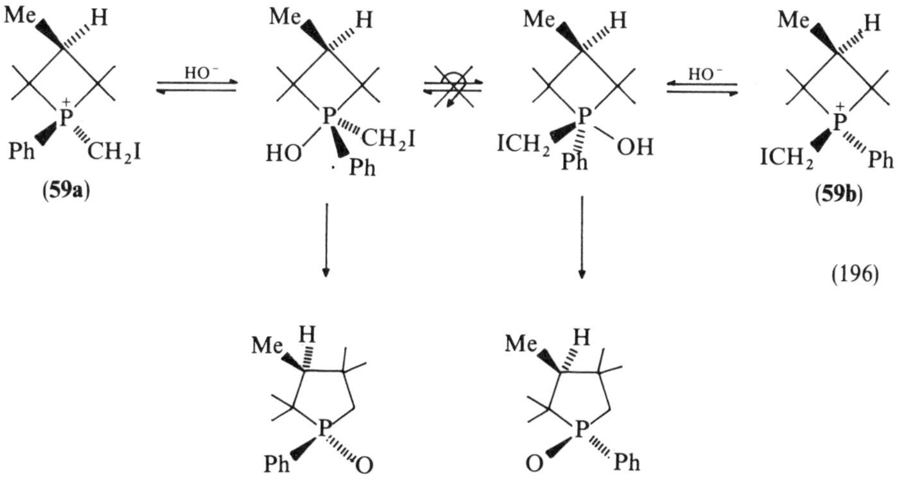

(59a) (59b)

(196)

As far as the actual mechanism of the migration is concerned, it follows from the preceding examples that the migrating group is eliminated from an axial position of the intermediate phosphorane. In fact, in acyclic structures with no restraint, the group forming the most stable carbanion species moves preferentially[669]. It must be pointed out that the $S_{N(P)mig.}$ mechanism ratio increases relative to that of the $E_{H\beta}$ or $S_{N(P)}$ mechanisms, with the facility of cleavage of the migrating group (reaction 197). Accordingly, the migration is anionotropic. The mechanism usually admitted for migration at the

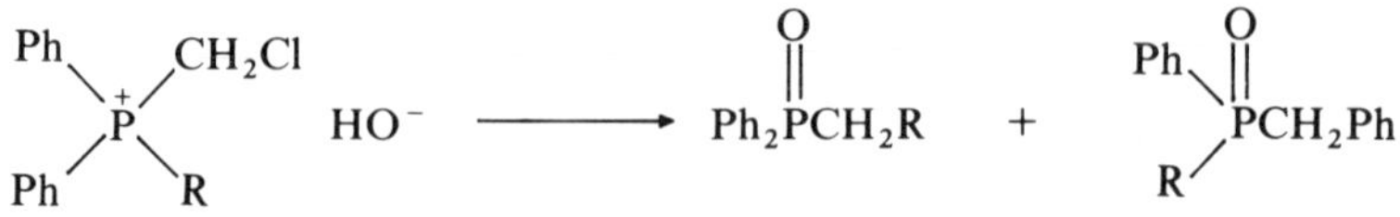

$$\underset{\mathbf{(60)}}{Ph_2\overset{+}{P}(CH_2Cl)R} \xrightarrow{HO^-} \underset{\mathbf{(61)}}{Ph_2P(O)CH_2R} + \underset{\mathbf{(62)}}{Ph(R)P(O)CH_2Ph} \qquad (197)$$

60a: R = Ph; $S_{N(P)mig} = 11\%$
60b: R = CH_2Cl; $S_{N(P)mig} = 55\%$ [**61**(45%) + **62**(10%)]

phosphorane **63** stage corresponds to the migration of the group with its bonding electron pair.

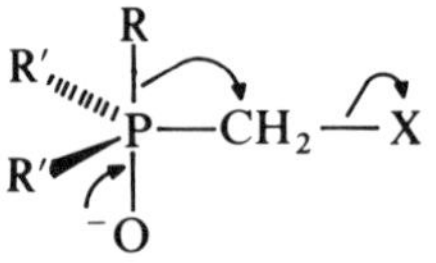

$$\underset{\mathbf{(63)}}{R'_2(R)(^-O)P—CH_2—X} \qquad \underset{\mathbf{(64)}}{R'_2(R)(^-O)P—\bar{C}H—X}$$

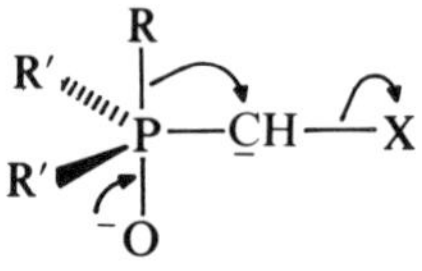

Schlosser[670] proposed an equivalent mechanism at the phosphorane **64** stage, formed from the corresponding ylide and not from the phosphonium salt; the driving force for the migration comes from the carbenoid nature of the carbon in the α-position to the phosphorus. This is consistent with the anionotropic nature of the migration. The mechanism is corroborated by a similar migration from the ylide **65**[670] (reaction 198).

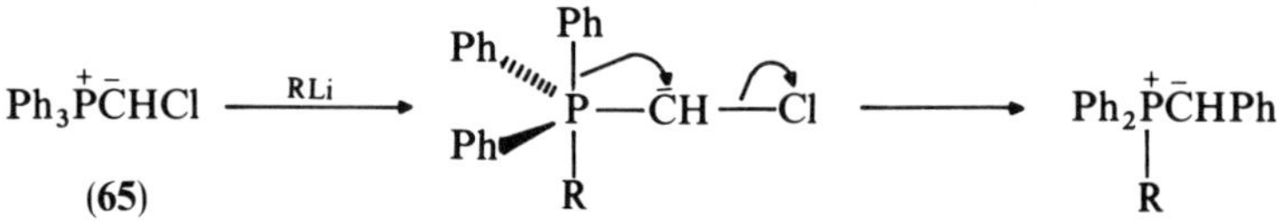

$$\underset{\mathbf{(65)}}{Ph_3\overset{+}{P}\overset{-}{C}HCl} \xrightarrow{RLi} Ph_3(R)P—\bar{C}H—Cl \longrightarrow Ph_2(R)\overset{+}{P}\overset{-}{C}HPh \qquad (198)$$

ii. Vinylphosphonium salts. The alkaline hydrolysis of the vinylphosphonium salts can yield three different phosphine oxides depending on the nature of the substituents

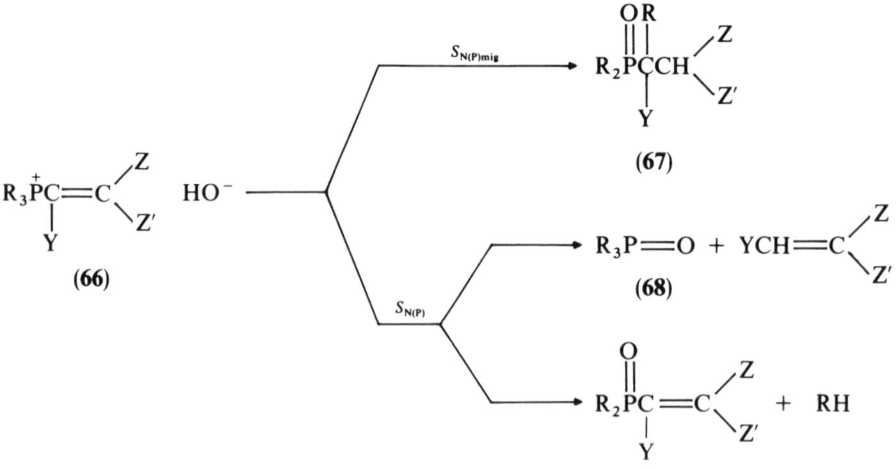

$$\underset{\mathbf{(66)}}{R_3\overset{+}{P}C(Y)=CZZ'} \xrightarrow{HO^-} \begin{cases} \xrightarrow{S_{N(P)mig}} \underset{\mathbf{(67)}}{R_2P(O)C(R)(Y)CHZZ'} \\ \xrightarrow{S_{N(P)}} \begin{cases} \underset{\mathbf{(68)}}{R_3P=O} + YCH=CZZ' \\ \underset{\mathbf{(69)}}{R_2P(O)C(Y)=CZZ'} + RH \end{cases} \end{cases}$$

SCHEME 7

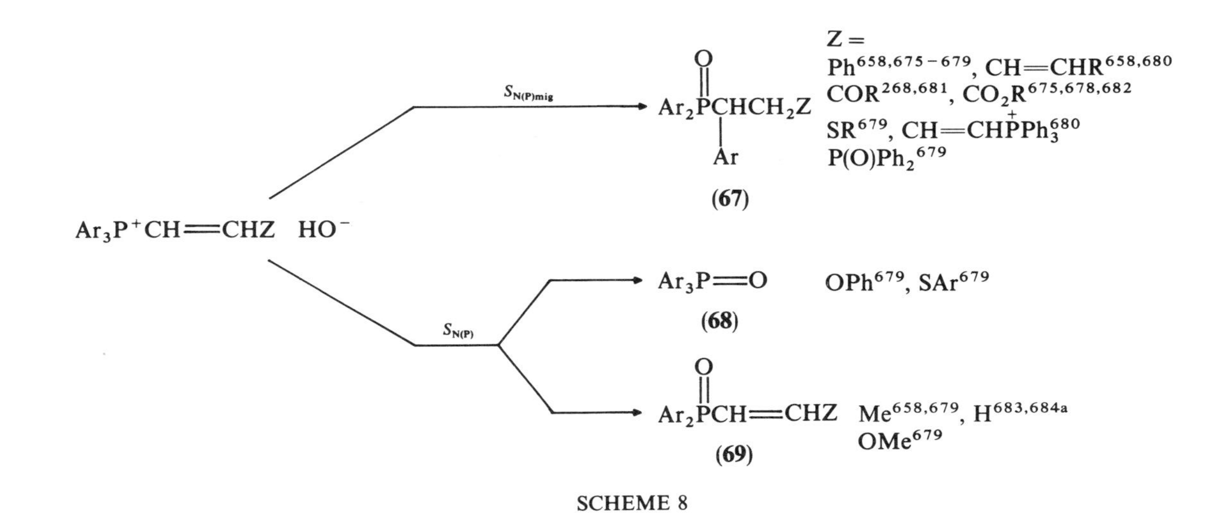

Ar3P+CH=CHZ HO−
SN(P)mig
SN(P)
Ar2P(O)CH(Ar)CH2Z
(67)
Z =
Ph658,675−679, CH=CHR658,680
COR268,681, CO2R675,678,682
SR679, CH=CHP+Ph3 680
P(O)Ph2 679
Ar3P=O
(68)
OPh679, SAr679
Ar2P(O)CH=CHZ
(69)
Me658,679, H683,684a
OMe679

SCHEME 8

on the vinylic group (Scheme 7). For the α-unsubstituted (Y = H) vinylphosphonium salts, migration occurs mostly for the Z substituents, conjugated with the vinylic group, exhibiting a more or less strong $-M$ effect. The migration is only a secondary reaction for the vinylphosphonium salt itself (Z = H). It does not seem to occur in the case of substituents with $+I$ or $+M$ effects (Scheme 8). On the other hand, migration does not occur, even for Z electron withdrawing groups, if the vinylphosphonium salts carry a substituent on the vinylic carbon α to the phosphorus (reaction 199). Other alkaline hydrolyses also occur without migration for various α-substituted vinylphosphonium salts[271,687–690].

$$Ph_3\overset{+}{P}(Ph)C{=}CHZ\ HO^- \longrightarrow Ph_3P{=}O + PhCH{=}CHZ \qquad (199)$$

Z = COR[685], CO_2H[685], CO_2R[685], H[686]

The mechanism for the migration of one of the groups from the phosphorus atom to the carbon in the α position apparently corresponds to an anionotropic migration; the best leaving group goes with its bonding doublet, either at the stage of oxyphosphorane B (route a)[691] or, most probably, at the stage of its conjugated base D (route b)[679] which presents a carbenoid character for the α-carbon. The second route (b), corroborated by isotopic exchange (H–D) experiments for the hydrogen on the carbon α to the phosphorus, further justifies the lack of migration for the α-substituted salts (Scheme 9).

c. Hofmann elimination reactions: $E_{H\beta}$ mechanism. The classical decomposition reaction of phosphonium salts by alkaline hydrolysis corresponds, as we have seen, to an $S_{N(P)}$ substitution reaction on phosphorus (reaction 200). This result is in contrast to the normal decomposition of a substituted ammonium hydroxide following the Hofmann mechanism which takes place by an elimination reaction (reaction 201). The difference in the behaviour between the two types of onium salts appears to be logical when one compares the respective ability of ammonium and phosphonium salts to decompose either by substitution or elimination mechanisms. Indeed, the ammonium salts cannot decompose by a substitution mechanism because the nitrogen atom has no vacant d orbitals of low enough energy and therefore cannot form a five-coordinate intermediate analogous to the hydroxyphosphoranes. In contrast, the mechanism of elimination is more difficult for phosphonium salts than for ammonium salts, because initial attack of the hydroxide ion on the hydrogen situated on the β-carbon to the heteroatom is unfavoured and, further, the R_3P group is not as good a leaving group as the R_3N group.

$$R_4P^+\ HO^- \longrightarrow R_3P{=}O + RH \qquad (200)$$

$$R_3\overset{+}{N}CH_2CH(R')(R'')\ HO^- \xrightarrow[-H_2O]{} R_3N + CH_2{=}C(R')(R'') \qquad (201)$$

In the case of phosphonium salts, some examples of decomposition by Hofmann elimination reactions are known. The phosphonium salts which normally undergo such a decomposition possess an activated hydrogen β to the phosphorus: the salts **70**, with a strong electron-withdrawing Z group on the carbon β to the phosphorus (reaction 202); and the α-hydroxyalkylphosphonium salts **71** (reaction 203)

$$\underset{(\mathbf{70})}{R_3\overset{+}{P}CH_2CH_2Z}\ HO^- \xrightarrow[-H_2O]{} R_3P + CH_2{=}CHZ \qquad (202)$$

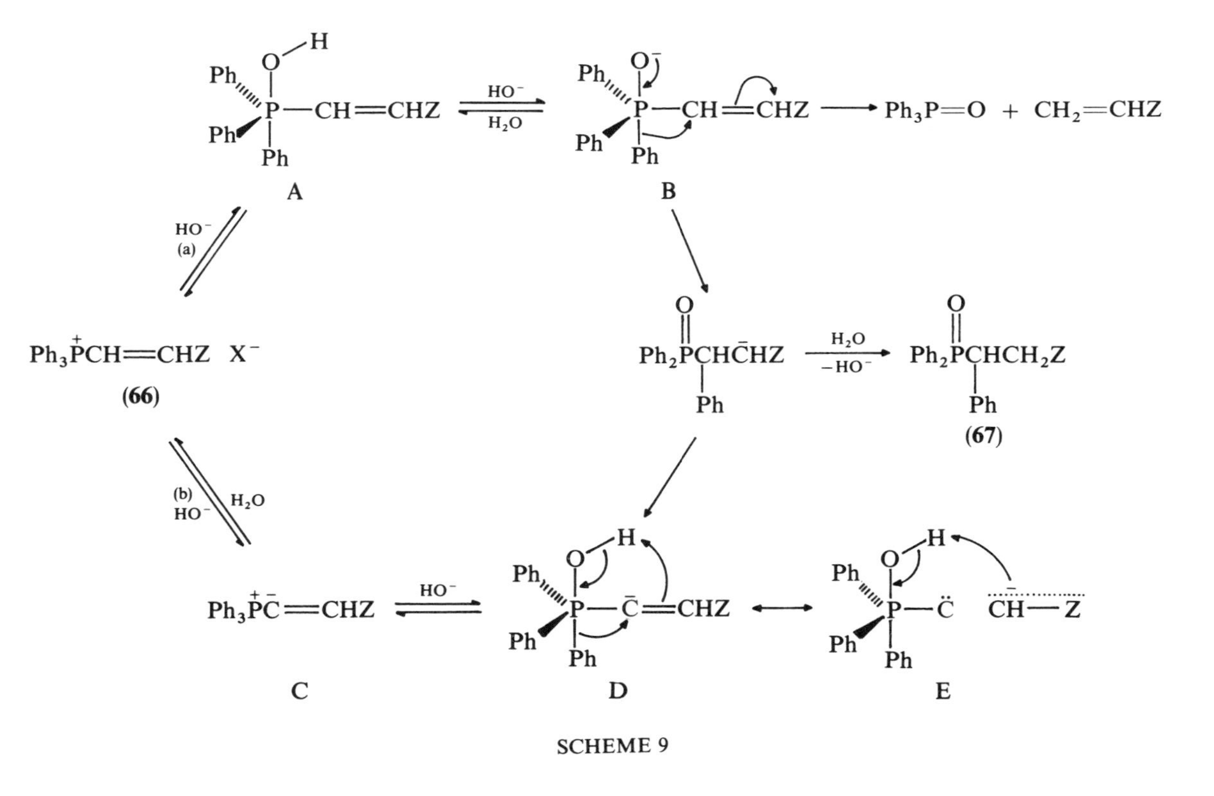

SCHEME 9

$$\underset{\underset{\textbf{(71)}}{\overset{|}{R'}}}{R_3\overset{+}{P}CHOH}\ HO^- \xrightarrow[-H_2O]{} R_3P + R'CH{=}O \qquad (203)$$

It is noteworthy that structures like such as **71** are generally more favourable to elimination reactions, because of the particularly high acidity of the hydroxylic hydrogen. Indeed, for the mixed salt **72**, the hydroxymethyl group is preferentially eliminated[692] (reaction 204).

$$(NCCH_2CH_2)_2\overset{+}{P}(CH_2OH)(CH_2CH_2CN)\ \textbf{(72)}\quad HO^- \xrightarrow[-H_2O]{} (NCCH_2CH_2)_3P + CH_2O \qquad (204)$$

i. α-Hydroxyalkylphosphonium salts. Although the elimination mechanism is particularly favoured for α-hydroxyalkylphosphonium salts by the hydroxylic hydrogen acidity, there is in fact most often competition between the elimination (a) and substitution (b) mechanisms[671,693] (equation 205). The phosphine oxide cannot proceed from a simple oxidation of the phosphine, subsequent to its formation in the elimination mechanism, because simultaneous formation of hydrogen takes place[694–697]. The formation of phosphine oxide together with hydrogen and aldehyde corresponds to a particular substitution reaction, similar to the E_p mechanism in which the carbanion arising from the hydroxyphosphorane [classical $S_{N(P)}$ mechanism] undergoes spontaneous fragmentation (reaction 206).

$$R_3\overset{+}{P}CH(O{-}H)(R')\ \textbf{(71)} + HO^- \begin{cases} \xrightarrow{(a)} R_3P + R'CHO + H_2O \\ \xrightarrow{(b)} R_3P{=}O + R'CHO + H_2 \end{cases} \qquad (205)$$

$$^-O{-}PR_3{-}CH_2{-}O{-}H \quad H{-}OH \xrightarrow[-HO^-]{} R_3P{=}O + CH_2O + H_2 \qquad (206)$$

Studies on the influence of molar ratios between the base and the phosphonium salt have shown[696,697] that the substitution reaction is the main reaction when an excess of base is used; however, in contrast, the elimination reaction is clearly favoured when an insufficient amount of base is used. These results are not surprising if it is assumed that the substitution mechanism is second order with respect to the hydroxide ion concentration[696,697], whereas the elimination reaction $E_{H\beta}$ is only first order.

This reaction was used for the resolution of racemic triarylphosphines[698], transformed temporarily as (+)-camphorsulphonates of the corresponding hydroxymethylphosphonium. The two diastereoisomeric salts were separated by recrystallization and their reaction with KOH or NaOH (or simply NEt_3) gave back each of the two enantiomers of the chiral phosphine.

ii. Phosphonium salts with electron-withdrawing group. Examples of the Hofmann elimination for the salts **70** are numerous in the literature. However, in most cases, the base used is not the hydroxide ion but an alcoholate[699–702], an organolithium[702,703], an ylide[699,704], an amine[700], an amide[700], a cyanide[700], etc.

In the alkaline hydrolysis of phosphonium salts, the action of the hydroxide ion can sometimes be paralleled[705] to each of the other nucleophiles, but it can also be different as is shown by the decomposition of the salt **73** (equation 207). We shall mention only

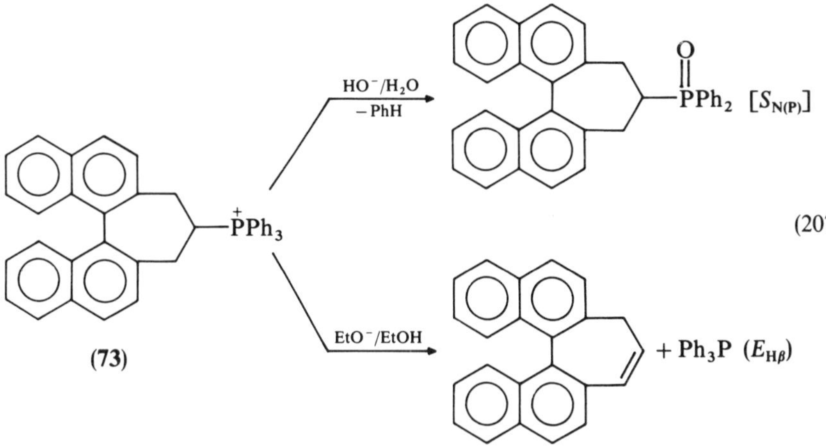

$\overset{+}{P}Ph_3$ **(73)** $\xrightarrow[-PhH]{HO^-/H_2O}$ $P(O)Ph_2$ $[S_{N(P)}]$

$\xrightarrow{EtO^-/EtOH}$ $+ Ph_3P$ $(E_{H\beta})$ (207)

cases where the hydroxide ion is the actual agent of decomposition. The following examples (equation 208) concern the decomposition of salts **70**. In the case of an aryl substituent, activation of the β-hydrogen is not sufficient to orientate the reaction mainly to the Hofmann elimination. It is necessary for such orientation that several groups are present or that the unsaturated system formed is stabilized by conjugation (Table 22).

$$R_3\overset{+}{P}CH_2CH_2Z\ HO^- \longrightarrow R_3P + CH_2{=}CHZ \qquad (208)$$

(70)

$Z = Ar$[706,707], CO_2R[708], CN[708], $CONH_2$[708], COR[705]

For the other electron-withdrawing groups[708], a high proportion of elimination is generally observed. The only limitation observed results from a secondary hydrolysis of the group Z which is transformed into a carboxylate group. The formation of this new group deactivates the β-hydrogen and orientates the reaction towards an $S_{N(P)}$ mechanism.

TABLE 22.

	$Et_3\overset{+}{P}CH_2CH_2Ph$	$Bu^n_3\overset{+}{P}CH_2CHPh_2$	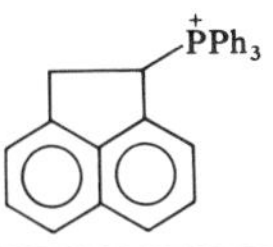 $\overset{+}{P}Ph_3$	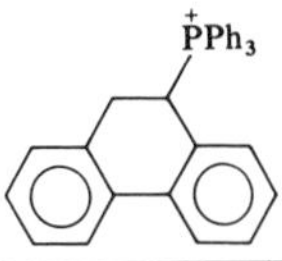 $\overset{+}{P}Ph_3$	**73**
$E_{H\beta}$	5	95	—	100	—
$S_{N(P)}$	95	5	100	—	100
Ref.	707	707	706	706	709

This is verified by the alkaline hydrolysis of salt **70** in which the group Z is CO_2H (reaction 209).

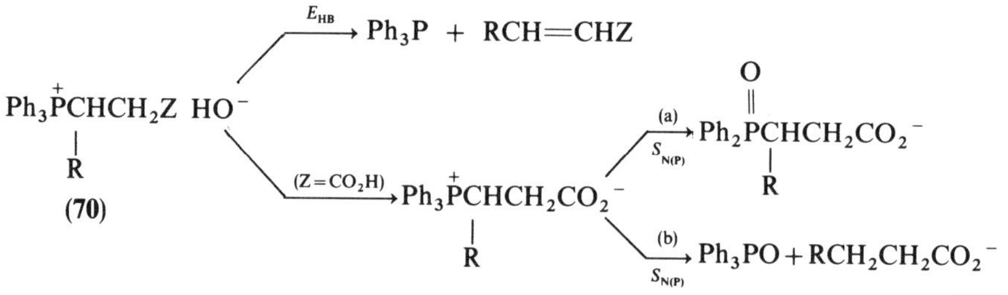

$$Ph_3\overset{+}{P}CH(R)CH_2Z\ HO^- \xrightarrow{E_{HB}} Ph_3P + RCH{=}CHZ$$

$$Ph_3\overset{+}{P}CH(R)CH_2Z\ HO^- \xrightarrow{(Z=CO_2H)} Ph_3\overset{+}{P}CH(R)CH_2CO_2^- \xrightarrow[S_{N(P)}]{(a)} Ph_2P(=O)CH(R)CH_2CO_2^-$$

$$Ph_3\overset{+}{P}CH(R)CH_2CO_2^- \xrightarrow[S_{N(P)}]{(b)} Ph_3PO + RCH_2CH_2CO_2^- \quad (209)$$

(70)

Z	R	$E_{H\beta}$	$S_{N(P)}$
CO_2Et	H	71	26(a)
CN	H	93	—
$CONH_2$	Ph	62	24(b)
CO_2H	H	—	95(a)
CO_2H	Ph	—	97(b)

iii. Elimination resulting from steric hindrance. The foregoing two types of Hofmann reactions are a consequence of the particularly high acidity of the hydrogen β to the phosphorus which favours the $E_{H\beta}$ mechanism. Another type of elimination occurs where the $S_{N(P)}$ mechanism is clearly unfavoured. This concerns the alkaline hydrolysis of phosphonium salts with particularly overcrowded groups (reaction 210). It must be

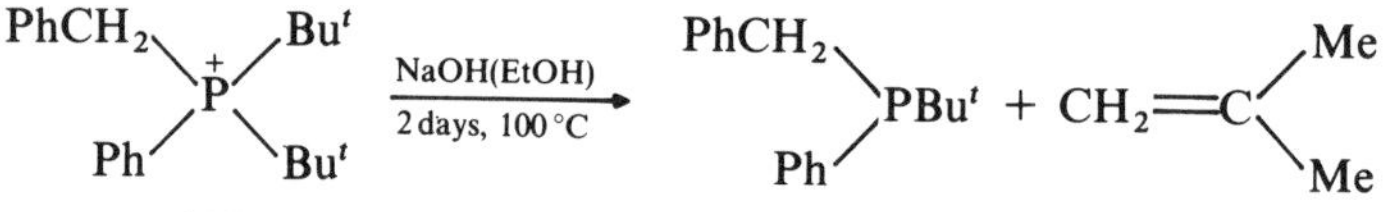

$$(PhCH_2)(Ph)\overset{+}{P}(Bu^t)_2 \xrightarrow[2\ days,\ 100\,^\circ C]{NaOH(EtOH)} (PhCH_2)(Ph)PBu^t + CH_2{=}C(Me)_2 \quad (210)$$

(74) 73%

pointed out that the reaction is incomplete since 27% of the initial salt is recovered unreacted[591]. This give a clear indication that the elimination reaction is difficult when the hydrogen β to the phosphorus is not activated. Elimination takes place here only because the $s_{N(P)}$ mechanism is extremely difficult. The steric hindrance of the two *tert*-butyl groups prevents the approach of the hydroxide ion and the formation of a hydroxyphosphorane. In the case of salt **75** this impossibility disappears because the only *tert*-butyl group present can take the axial position enabling the reaction to follow the $s_{N(P)}$ classical mechanism[591].

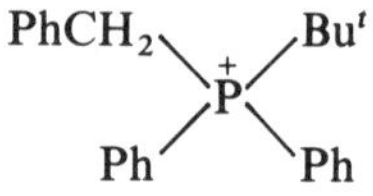

$$(PhCH_2)(Bu^t)\overset{+}{P}(Ph)_2$$

(75)

In another example[710], the phosphonium salt **76** decomposes competitively by the two mechanisms of substitution and elimination (reaction 211). This salt behaves in a manner intermediate between salts **74** and **75**; indeed, the $S_{N(P)}$ mechanism is not completely excluded as for the salt **74** but the $E_{H\beta}$ mechanism is still favoured with regard to the salt **75** since the alkene formed is stabilized.

$$\mathrm{Ph_3\overset{+}{P}C(Me)_2Ph\ HO^-}\ (\mathbf{76}) \longrightarrow \mathrm{Ph_3PO + PhCH(Me)_2}; \quad \xrightarrow[(50\%)]{E_{H\beta}} \mathrm{Ph_3P + Ph(Me)C{=}CH_2} \tag{211}$$

d. Elimination with vinylphosphonium formation: $E_{H\alpha}$ *mechanism.* An E_{Hz} mechanism can be observed in phosphonium salts cantaining a Y group which is removable from the carbon β to the phosphorus (reaction 212). Many similar elimination reactions have already been discussed in the chemistry of phosphonium salts, but generally they do not occur in aqueous phases and they are induced by nucleophiles other than the hydroxide ion, for example the formation of vinylphosphonium salts[711,712] and the decomposition of bisphosphonium salts by tertiary amines[713,714]. Some examples of $E_{H\alpha}$ elimination concern the alkaline hydrolysis of phosphonium salts[657,715–719]. The mechanism can be proved either directly[718], by isolation of the substituted vinylphosphonium salts formed as intermediates by interrupting their decomposition to the corresponding phosphine oxide (reaction 213), or indirectly[718], when the intermediate vinylphosphonium salt is not substituted in the position β to the phosphorus, so that it can trapped by the addition of a ZH compound with mobile hydrogen (reaction 214).

$$\mathrm{R_3\overset{+}{P}{-}CH{-}C{-}Y\ HO^-} \xrightarrow[-H_2O]{} \mathrm{R_3\overset{+}{P}{-}C{=}C\langle\ Y^-} \dashrightarrow \rightarrow \text{evolution} \tag{212}$$

$$\mathrm{Ph_3\overset{+}{P}CH_2CH(SEt)_2\ X^-} \xrightarrow[E_{H\alpha}]{HO^-(-H_2O)} \mathrm{Ph_3\overset{+}{P}CH{=}CHSEt} \left(\xrightarrow[S_{N(P)mig}]{HO^-} \mathrm{Ph_2P(O)CH(Ph)CH_2SEt} \right) \tag{213}$$

$$\mathrm{Ph_3\overset{+}{P}CH_2CH_2\ Y} \xrightleftharpoons{HO^-} \mathrm{Ph_3\overset{+}{P}CH{=}CH_2\ Y^-} \xrightleftharpoons{ZH} \mathrm{Ph_3\overset{+}{P}CH_2CH_2Z\ Y^-} \tag{214}$$

Y	OAc	OCOR	OH	PPh_3	OPh	Br
Z	CN	$H_2NCH(CO_2^-)R'$	OMe	OMe	OMe	OMe
Ref.	716, 720	717	718	719	720	720

The ability of the Y group to be eliminated results apparently from the superposition of two factors: on the one hand, the specific leaving ability of the Y group, and on the other, the possibility of intramolecular interactions between the Y group and the phosphorus atom. Further, the $E_{H\alpha}$ mechanism is favoured by low hydroxide ion concentration[680,719], since it is first order with respect to hydroxide ion, whereas the competitive mechanisms [$S_{N(P)}$ or E_p] are second order.

3. *Application of alkaline hydrolysis of phosphonium salts*

The alkaline hydrolysis of phosphonium salts has been particularly studied from a theoretical point of view as a model for nucleophilic substitution reactions on tetracoordinated phosphorus. However, it also has interesting applications, since it corresponds to the cleavage of the P—C bond with the formation of a hydrocarbon and a phosphine oxide (reaction 215). Recently this reaction has still been used to obtain various arenes and aromatic heterocycles by reduction of the corresponding halides derivatives[162,256,721,722] by regioselective electrophilic substitution on an arylphosphonium salt prior to alkaline hydrolysis[259] (reaction 216), or by building the heterocycle (starting with a functional phosphonium salt) before alkaline hydrolysis[723,724] (reaction 217). In contrast, an alkaline hydrolysis reaction has been used to obtain certain types of organophosphorus compounds, e.g. macrocyclic polyheteroatomic phosphorus compounds[353], optically active phosphine oxides[589] and tertiary polyphosphine monoxides[725].

$$R_3\overset{+}{P}R'\ X^- \xrightarrow[-X^-]{HO^-/H_2O} R_3P{=}O + R'H \qquad (215)$$

1-Methylnaphthalene —(i) CH_2O/HCl; (ii) Ph_3P; (iii) HO^-/H_2O ($-Ph_3PO$)→ 1,4-dimethylnaphthalene (216)

$$Ph_3\overset{+}{P}CH_2\overset{O}{\overset{\|}{C}}CH_2\overset{O}{\overset{\|}{C}}R \xrightarrow[\text{(ii) } HO^-/H_2O\ (-Ph_3PO)]{\text{(i) } NH_2OH,\ HCl}$$ 5-methyl-3-R-isoxazole (217)

Another important use of the susceptibility of phosphonium salts to undergo alkaline hydrolysis concerns their use as liquid–liquid phase–transfer catalysts. Phosphonium salts can be decomposed much more easily than ammonium salts, under alkaline conditions[726–729], and they have to be used under much milder conditions [i.e. low temperature ($< 25\,^\circ C$) and moderate aqueous base concentration ($< 15\%$)]; in all cases reaction conditions should be used which prevent the extraction of OH^- into the organic phase or minimize its reactivity[730].

D. Reduction of Phosphonium Salts

1. Reduction using hydrides

Mono- and bis-phosphonium disalts are known to be reduced with high yields to tertiary phosphines using metallic hydrides, such as alkali metal-aluminium hydrides, their alkyl/alkoxy derivatives or sodium hydride[731–733]; thf is the recommended solvent. The ease of the P^+—C cleavage parallels the ability of the leaving group to stabilize the incipient negative charge (reaction 218). This order conforms with a mechanism implying addition of the hydride or AlH_4^- ion to the positively charged phosphorus; the isolation of spirocyclic phosphoranes is in agreement with this assumption[734], as is the fact that racemic phosphines result from reduction of optically active phosphonium salts[584,731,735,736]. A mechanism involving ylide intermediates has also been invoked[737].

Exceptions happen in the cases of sterically crowded substrates ($R^1 = i$-Pr, c-C_3H_5)[732,738] in which the hydride attack is assumed to be at hydrogen[739].

$$R_3P^+{-}R^1 \xrightarrow[\text{thf}]{\text{LiAlH}_4} R_3P + R^{1-} \tag{218}$$

$$R^1: PhCH_2 > Ph > \text{alkyl}$$

Such hydride reductions have been used in the synthesis of phosphines[731,732,737,740], through sequential cleavage for unsymmetrical phosphines[731], diphosphines[732,741,742] and di- or polyphospha compounds[369,388,742]. However, in the case of $LiAlH_4$ reduction of ethylene bisphosphonium disalts[66,743], cleavage of the two-carbon bridge occurs preferentially to that of a phenyl (or alkyl) group (reaction 219) and, even more, to that of a benzyl group, using NaH[743]. In the same way, the unsaturated bridge of 1, 2-vinylene benzylbisphosphonium disalts is selectively cleaved with $LiAlH_4$, while the buta-1,4-dienylene analogue gave both saturated phosphines and bisphospines[362] (reaction 220). The formation of such saturated species has been considered as proof of a partly ylide mechanism through hydride attack at the vinylic β-carbon atom.

$$2\ I^-\ Ph_2\overset{+}{P}(R)CH_2CH_2\overset{+}{P}(R)Ph_2 \xrightarrow[(-C_2H_4)]{H^-} Ph_2PR \tag{219}$$

$$R = Me\ (AlLiH_4);\ CH_2Ph(NaH)$$

$$2\ Br^-\ R_2\overset{+}{P}(R^1){+}CH{=}CH{+}_n\overset{+}{P}(R^1)R_2 \xrightarrow{H^-} \begin{cases} \xrightarrow{n=1} R_2PR^1 + R_2PEt \\ \xrightarrow{n=2} R_2PR^1 + R_2P{+}CH_2{+}_4PR_2 \end{cases} \tag{220}$$

In the case of phosphonium salts substituted by an heteroaryl group, the furfuryl group is easily cleaved[744]; 2,4-pyridindiyl bis(phosphonium) salts have been used as source of 2-pyridylphosphonio compounds[745] (reaction 221). Clearly, the presence of

$$\text{2-}(R_3P^+)\text{-4-}(^+PPh_3)\text{pyridine}\ \ 2\,CF_3SO_3^- \xrightarrow[(-PPh_3)]{NaBH_4} \text{2-}(R_3P^+)\text{pyridine}\ \ CF_3SO_3^- \tag{221}$$

$R = n$-Bu, Ph   ~70%

oxygen in the alkyl substituents of phosphorus can be prejudicial; phosphonium salts containing $\overset{+}{P}(C)_nO(C)_nP^+$ chain ($n = 2$, 3) are also cleaved at the bridge[353] whereas $\overset{+}{P}(C)_nP^+$ ($n = 4, 6$) gave diphosphines[732]; ω-hydroxyalkylphosphonium salts can give complexes mixtures, using NaH reduction, depending on the alkyl chain length[746]. Interestingly, sodium hydride reduction of triphenyl ω-carboxyalkyl phosphonium salts in dmso, which could be used in the synthesis of ω-(diphenylphosphinyl)alkyl carboxylic acids, strongly competes with the formation of triphenylphosphine when $n = 3$[42] (reaction 222).

$$X^- \, Ph_3\overset{+}{P}(CH_2)_nCOOH \xrightarrow[(2)\,H_3O^+]{(1)\,2NaH\text{-}dmso} Ph_2\overset{O}{\overset{\|}{P}}(CH_2)_nCOOH \qquad (222)$$

n	
3	17% (20% in Ph_3P)
5	32%
10	75%
11	62%

Finally, heterophosphonium salts are reduced with cleavage of the phosphorus–heteroatom bond: optically active aminophosphonium compounds are usefully reduced to phosphines with retention of configuration[747]. Methyltrineopentoxyphosphonium trifluoromethanesulphonate gave the quantitative formation of dineopentylmethyl phosphonite, which is indicative of P—O bond cleavage[748] (reaction 223)

$$(Bu^tCH_2O)_3\overset{+}{P}Me \;\; F_3CSO_3^- \xrightarrow[(2)\,H_3O^+]{(1)\,LiHEt_3B\,thf(78\,^\circ C)} (Bu^tCH_2O)_2 \;\; PMe \qquad (223)$$

2. Reduction using metals

Information on the reduction of phosphonium salts with alkali metals is limited. In the reduction of the tribenzylmethylphosphonium bromide, sodium has been used either in liquid ammonia, ethanol or benzene, which gave the estimated best yield (77%) of dibenzylmethylphosphine; however, under the same conditions, no phosphine was detected from the dibenzyldimethyl analogues[731]. A phenyl group can be removed in preference to an alkyl or benzyl group by sodium in toluene or naphthalene. However, when the phosphonium salt was trialkylated, the alkyl group was preferentially split off, the order being t-Bu > Ph > $PhCH_2$ > Et > Me[749].

Sodium–naphthalene reduction of organotrineopentoxyphosphonium salts led to the instantaneous loss of phosphonium ion; phosphonates and phosphites were obtained[748] (reaction 224). Alkali metal amalgams are efficient reagents for the reductive cleavage of both achiral and optically active phosphonium salts; configuration is retained[750] (Table 23).

$$(Bu^tCH_2O)_3\overset{+}{P}CH_2R\,X^- \xrightarrow[\text{Na-naphth}]{} (Bu^tO)_2\underset{\underset{O}{\|}}{P}CH_2R + (Bu^tO)_3P \qquad (224)$$

R = H, alkyl, vinyl, phenyl

In the reductive cleavage of salts containing both benzyl and *tert*-butyl substituents, cleavage of the latter predominates (85% versus 15%). That solvents are influential in

TABLE 23. Reductive cleavage of benzylmethylphenyl-*n*-propylphosphonium bromide to methylphenyl-*n*-propylphosphine

Amalgam	Solvent	Time	Yield (%)	Unreacted (%)
Ba–Hg	MeOH	16 h	—	95
K–Hg	MeOH	30–40 min	81	—
Na–Hg	MeOH	20 min	83	—
Li–Hg	Pr^iOH	40 min	40	50

the case of the Li–Hg amalgam as compared with Na–Hg and K–Hg has been shown in the reduction of β-aminoalkylphosphonium salts in MeCN, MeOH and PiOH[178].

3. Others

Unsuccessful attempts have been made to use $SnCl_2$–AcOH and H_2–Pd[731]. Potassium diaryl phosphides have reduced tetraphenylphosphonium bromide to the corresponding aryl phosphines[751] (reaction 225). A range of β-carbonylalkyltriphenylphosphonium salts[752] have been reduced to triphenylphosphine by an excess of Grignard reagents[753] (nucleophilic attack on phosphorus) (reaction 226).

$$Br^- Ph_4P^+ \xrightarrow[\text{thf}]{(R\text{-}C_6H_4)_2P^-K^+} \begin{cases} R = H: & Ph_3P \ (21\% \ Ph \ transfer) \quad 70\% \\ R = Me: & Ph_3P + (Me\text{-}C_6H_4)_2PPh \quad 80\% \quad 55\% \end{cases} \tag{225}$$

$$X^- \ Ph_3P^+\text{—}C(R^1)(R^2)\text{—}C(=O)\text{—}Y + 2R^3MgX \xrightarrow[(2)\ H^+]{(1)\ -(R^3R^3)} Ph_3P + R^1R^2C{=}C(OMgX)Y \tag{226}$$

40–67%

Y = Ph, OEt; R^1, R^2 = Me, cycloalkyl, Ph; R^3 = alkyl, Ph

Electrolytic cleavage of phosphonium salts leads to tertiary phosphines[178,754–757] with, where phosphorus is asymmetric, conservation of configuration at phosphorus[188,755]. The main limitation lies in obtaining mixtures of tertiary phosphines, of which the composition depends upon the nature of the cathode (Hg, Pb, Pt, Zn, Cu, Al), the solvent, temperature, potential and chemicals added[177,741,757]. Quasi-phosphonium salts are also cleaved electrochemically (the order of precedence in the cleavage of ligands at phosphorus is not always forecastable) and optically active species may also be reduced with retention of configuration[758] (reactions 227 and 228). Products of coupling (bibenzyl starting from

$$Br^- \ \text{(cyclic } Bu^t(Ph)P^+ \text{ with X and two Me)} \xrightarrow[H_2O/\text{toluene}]{\text{HgCathode } (2e^-)} \text{(cyclic } PhP \text{ with X and two Me)} \tag{227}$$

X	Yield
O	82%
S	56%
MeN	84%

$$\mathrm{PhCH_2}\overset{+}{\mathrm{P}}(\mathrm{Ph})(\mathrm{Me})-\mathrm{NR^1R^2}\ \mathrm{Br^-} \xrightarrow[\mathrm{H^+}]{2e^-} \begin{cases} (\pm):\ R^1=\mathrm{Me};\ R^2=\mathrm{Ph} \longrightarrow \mathrm{Ph(Me)PCH_2Ph}\ \ 80\% \\ (+)\,(S):\ R^1=R^2=\mathrm{Et} \longrightarrow \mathrm{Ph(Me)PNEt_2}\ \ 70\% \end{cases} \tag{228}$$

benzyltriphenylphosponiums salts) are best formed on an aluminium cathode[759,760] (d.c. power supply), using high current densities, in dmf or hmpa solution. This suggested that reactions for polyene synthesis could be developed (reaction 229).

$$\mathrm{Ph_3\overset{+}{P}CH_2R\ NO_3^-} \xrightarrow[(-\,\mathrm{Ph_3P})]{\mathrm{Al\ cathode/0.7\ A\ cm^{-2}}} \mathrm{RCH_2CH_2R} \tag{229}$$

dmf: $R = Ph$ 56%
hmpa: $R = CH{=}CHPh$ 30% ($E + Z$)

Studies using a differential-pulse polarographic technique, which is recommended in place of classical direct current polarography, yielded the following results. For $Ar_3\overset{+}{P}CH_2Ph\,Br^-$, the observed order in the ease of reduction, in aqueous solution at pH 7.00, correlates with the electron-withdrawing ability of the aryl groups (Ar = 2-furyl > 2-thienyl > phenyl > 1-methylpyrrolyl); the first reduction wave would correspond mainly to the addition of an electron to the phosphorus of the phosphonium ion to form a phosphoranyl radical. In contrast, for $Ph_3\overset{+}{P}CH_2Ar\,Br^-$, the ease of reduction correlates best with the order of stability of the carbanion the formation of which seems to be involved in the reduction process[761].

In media of low proton availability (MeCN, dmf, hmpa)[762], ylide formation was found and products derived from the radical cleavage of the phosphonium ion have been observed. The initial cation would interfere in the reaction process as an acid. A competition can exist between the one-electron pathway (dimerization, disproportionation of $R^{\bullet}$) and the two-electron pathway (ylide formation, Hofmann degradation, phosphine oxide formation) (Table 24).

Unsaturated bisphosphonium salts have been also studied[182,763,764]. On mercury cathodes, in unbuffered or acidic aqueous–organic media, the electrochemical reduction was shown to be controlled by the ylide character of the first-formed radicals, which then exclusively induced a heterolytic P—C bond breaking on the unsaturated bridge[182] (Scheme 10).

E. Reactivity of Functional Chains

1. Reactivity induced by the phosphonio group

a. Reaction with formation of P═Element double bond (α-proton abstraction and basic hydrolysis excluded). Dealkylation of alkoxyphosphonium salts through *O*-alkyl fission corresponds to the final step in the Michaelis–Arbusov reaction, the high rate of which is due to the energetically favourable formation of a P═O double bond. When the anionic counterpart is only weakly nucleophilic, as in the case of tetraphenylborate, hexachloroantimonate, hexafluorophosphate and, to a lesser extent, tetrafluoroborates

TABLE 24. Various reaction paths in the electrochemical reduction of phosphonium cations in media of low proton availability[762]

Initial electrode electron transfer	$RP^+Ph_3 + 1e \rightleftharpoons RP^{\bullet}Ph_3$	
Competitive radical cleavage	$R\dot{P}Ph_3 \rightarrow R^{\bullet} + PPh_3$	$R\dot{P}Ph_3 \rightarrow Ph^{\bullet} + RPPh_2$
Dimerization	$R^{\bullet} \rightarrow 1/2RR$	$Ph^{\bullet} \rightarrow 1/2Ph\text{-}Ph$
H-atom transfer disproportionation	$R^{\bullet} \rightarrow 1/2RH + 1/2R(-H)$	
Second wave electrode electron transfer	$R^{\bullet} + 1e \rightleftharpoons R^-$	$Ph^{\bullet} + 1e \rightleftharpoons Ph^-$
Second wave solution electron transfer	$R^{\bullet} + R\dot{P}Ph_3 \rightleftharpoons R^- + R\overset{+}{P}Ph_3$	$Ph^{\bullet} + RP{\cdot}Ph_3 \rightleftharpoons Ph^- + RP^+Ph_3$
Hofmann degradation	$R^- + R\overset{+}{P}Ph_3 \rightarrow R(-H) + RH + PPh_3$	$Ph^- + RP^+Ph_3 \rightarrow R(-H) + PhH + PPh_3$
Ylide formation	$R^- + R\overset{+}{P}Ph_3 \rightarrow R(-H){=}PPh_3 + RH$	$Ph^- + RP^+Ph_3 \rightarrow R(-H){=}PPh_3 + PhH$
Phosphine oxide formation	$R^- + H_2O \rightarrow RH + OH^-$	$Ph^- + H_2O \rightarrow PhH + OH^-$
	$R(-H){=}PPh_3 + H_2O \rightleftharpoons R\overset{+}{P}Ph_3 + OH^-$	
	$R\overset{+}{P}Ph_3 + OH^- \rightarrow Ph_3PO + RH$	$R\overset{+}{P}Ph_3 + OH^- \rightarrow RPh_2PO + PhH$

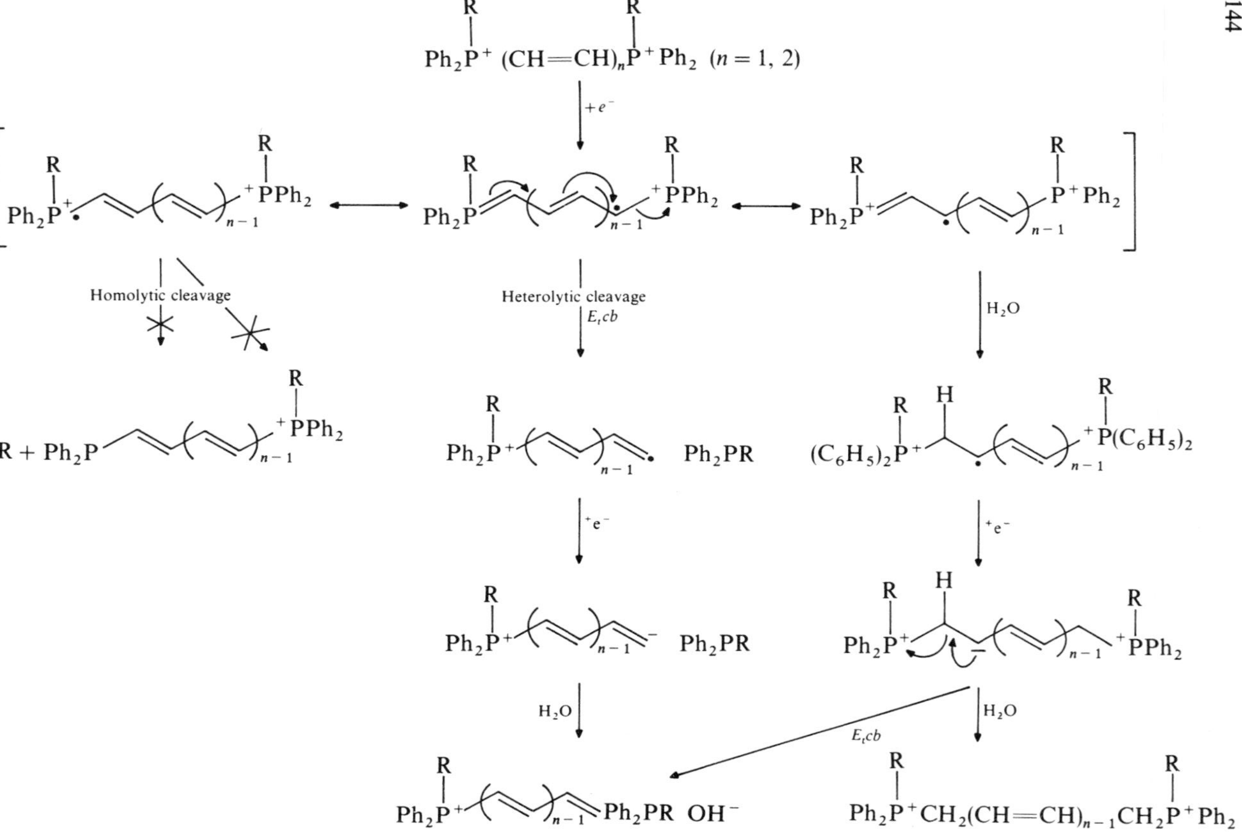

SCHEME 10. Reduction mechanism of 1,2-vinylene and buta-1,4-dienylene bisphosphonium salts (counterion omitted)[182].

or triflates, their stability is improved and such salts can be used as alkylating agents[6]. Thus, some methoxyphosphonium salts are very powerful methylating agents of nucleophiles[493], as evidenced and measured[474] by the rates of methyl transfer to the 2,4-dinitrophenoxide ion (reaction 230). Trialkoxyphosphonium tetrafluoroborates can be

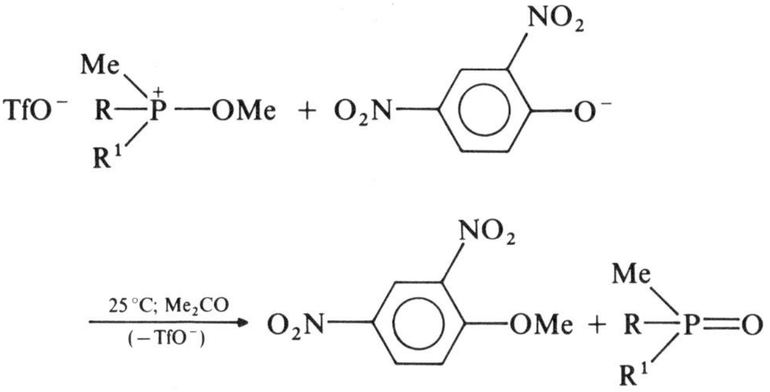

$$TfO^- \; R(Me)(R^1)\overset{+}{P}{-}OMe + O_2N{-}C_6H_3(NO_2){-}O^- \xrightarrow[(-TfO^-)]{25\,^\circ C;\; Me_2CO} O_2N{-}C_6H_3(NO_2){-}OMe + R(Me)(R^1)P{=}O \qquad (230)$$

$k = 0.2$–$40\,mol^{-1}\,s^{-1}$, depending on R and R^1 (alkyl, Ph, OAlk)

considered as excellent synthetic alternative for the formation of ethers, thioethers, esters, amines and alkyl halides, even when the alkoxy group is secondary[269] (reaction 231). Nucleophilic attacks occur at the α-carbon atom by an S_N2 and/or S_N1 mechanism. In fact, because the phosphonium salts are not volatile at room temperature, such methoxy reagents appear to be safer than compounds such as CH_2N_2, F_3CSO_3Me or FSO_3Me and the reaction is not very sensitive to the nature of the substituents or the medium.

Alkoxy(trisdimethylamino)phosphonium salts (PF_6^-, ClO_4^- or BF_4^- as anions)[765] are also used for the same purpose with good yields; of course, alkoxy groups can be chosen as optically active[451,766,767]. Moreover, when the alkoxy chain is functionalized in the appropriate position by oxyanions, intramolecular etherification is made possible; 3,6-anhydrohexosides were prepared in this way[537] (reaction 232).

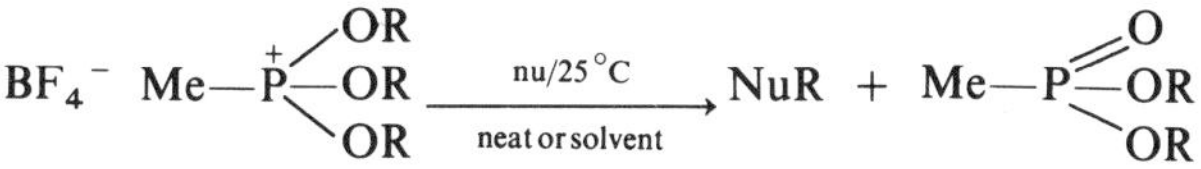

$$BF_4^- \; Me{-}\overset{+}{P}(OR)_3 \xrightarrow[\text{neat or solvent}]{nu/25\,^\circ C} NuR + Me{-}P(=O)(OR)_2 \qquad (231)$$

51–99%

Nu = alkyl—OH, alkyl—COOH, alkyl—COO-, aryl-COOH, aryl—SH, alkyl-NH_2, dialkyl—NH, halides
R = primary or secondary alkyl

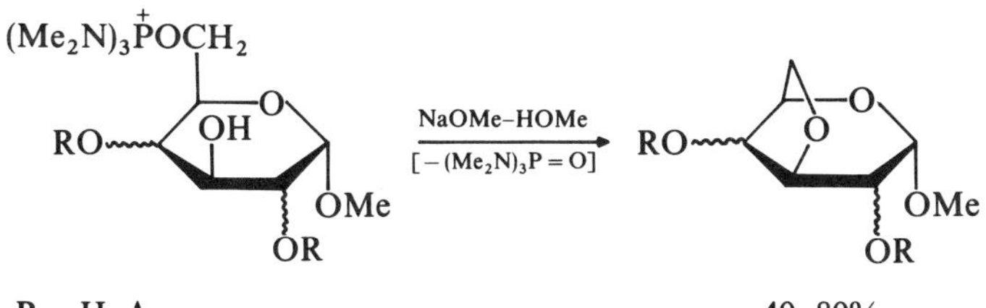

$$(Me_2N)_3\overset{+}{P}OCH_2\text{-hexoside (RO, OH, OMe, OR)} \xrightarrow[{[-(Me_2N)_3P=O]}]{NaOMe{-}HOMe} \text{3,6-anhydrohexoside (RO, OMe, OR)} \qquad (232)$$

R = H, Ac

40–80%

Compared with dealkylation, dearylation of aryloxyphosphonium salts occurs only at elevated temperatures (200–250 °C)[6,768,769], at which some disproportionation was also observed. Nevertheless, methyltriphenoxyphosphonium iodide has been shown to be a very useful and versatile reagent in the conversion of alcohols, epoxides or halides into alkyl halides[770,771] or alkenes[772–776] because of the stability of the phenoxy group, which is readily displaced in the primary step of the reaction, through phosphorus oxygen cleavage (reaction 233). In addition, in the case of non–nucleophilic anions, alcoholysis can be a method for the synthesis of alkoxyphosphonium salts,[772,777] the process being complete in an excess of alcohol. Connected with this, hydrolysis of aryloxyphosphonium salts is usually very fast whereas the behaviour of alkoxy analogues varies widely with changes in structure[769]. Aniline is also able to displace aryloxy groups[778].

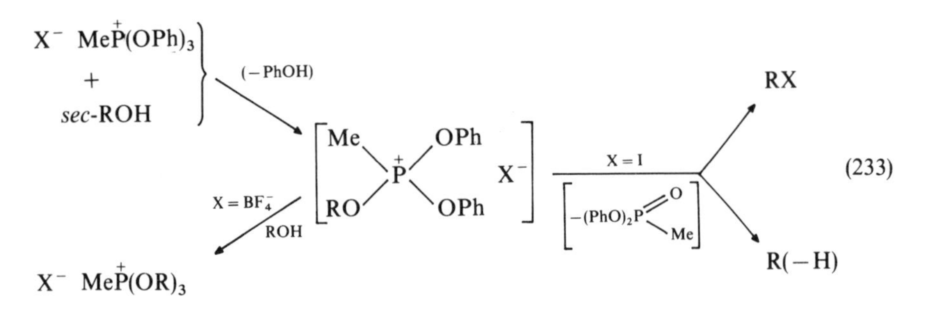

The high energy of the P=O bond also governs the reactivity of 'phosphonium anhydrides'. These compounds have been shown to be very efficient in the 'extraction' of oxygen from organic molecules by an equivalent of dehydration; OH groups in carboxylic acids, amides, enols, alcohols and other compounds are activated by their transformation into intermediates containing the good leaving group $OPPh_3^+$; triphenyl or tributylphosphonium anhydrides seem to be the most efficient[500–502,504,779]; the latter has led to improved stereoselectivity in the formation of the final products (reaction 234). Mono- and di- (dialkylamino) phosphonium salts have been found to be rather unstable in the presence of water and to lead to phosphone oxides[154,481]. Chlorotris(diisobutylamino)phosphonium halides are hydrolytically more stable than expected[780].

$$(R_3\overset{+}{P})_2O\ 2\ TfO^- \longrightarrow \begin{cases} \xrightarrow[(-2\ R_3P=O)]{R^1COOH/R^2OH} R^1COOR^2 \\ \xrightarrow[(-2\ R_3P=O)]{R^1COOH/R^2_2NH} R^1CONR^2_2 \\ \xrightarrow[(-2\ R_3P=O)]{R^1CONHR^2/Nu} R^1C(Nu)=NR^2 \\ \xrightarrow[(-2\ R_3P=O)]{R^1C(OH)=CHR^2} R^1C\equiv CR^2 \\ \xrightarrow[(-2\ R_3P=O)]{R^1OH/R^2OH} R^1OR^2 \end{cases} \qquad (234)$$

The formation of a P=O bond is the driving force in the reaction of thioalkoxyphosphonium salts, having non-nucleophilic counter anions, with carboxylates or alcoholates, thereby cirumventing the inconvenience of using thiols or thiolates in the synthesis of thiol esters or sulphides; the yields are fair to excellent[433,781] (reaction 235).

$$X^- \ R_3\overset{+}{P}SR^1 \xrightarrow[(-R_3P{=}O)]{20\,^\circ C/\text{base}} \begin{cases} \xrightarrow{R^2COO^-} R^2C({=}O)SR^1 \\ \xrightarrow{R^2O^-} R^2SR^1 \end{cases} \tag{235}$$

R = Ph, Bu; X = BF_4, ClO_4; R^1,R^2 = alkyl, aryl

Phosphonium salts having an oxygen atom α or β to phosphorus are also capable of being transformed to P=O bond-containing compounds. Thus, thermolysis of quaternary 1-alkoxy substrates can proceed through a phosphorus-oxygen coordination which finally gives tertiary phosphine oxides as the main products[782] (reaction 236).

$$Br^- \ Bu_3\overset{+}{P}CH(OEt)CH_2Br \xrightarrow{150\text{-}200\,^\circ C} \underset{70\%}{Bu_3P{=}O} + EtBr + \underset{X{=}H,\,Br}{H_2C{=}CHX} \tag{236}$$

On prolonged heating (in acidic medium), α-hydroxymethylphosphonium compounds rearrange to phosphine oxide[783]. Nucleophilic additions on activated and many non-enolizable β-ketophosphonium species have yielded olefins, via elimination of phosphine oxide[784,785] from betaine species; in the case of hydride anions, unsymmetrical precursors gave predominantly (Z)-alkenes in contrast to the products observed starting from α,α-disubstituted ylides and aldehydes. Intramolecular Wittig reactions commonly occur when carbanion is generated α' to the carbonyl group[339] (reaction 237).

$$R_fCOO^- \ Ph_3\overset{+}{P}CR^1R^2C({=}O)R' \xrightarrow[(-Ph_3P{=}O)]{Nu^-} R^1R^2C{=}C(Nu)R' \tag{237}$$

R^1, R^2 = alkyl, allyl, benzyl, alicyclic; $R' = R_F(=C_nF_{2n-1})$, Ph; Nu = Ph, H

Phosphirenium salts are instantaneously opened on hydrolysis, affording the corresponding phosphine oxides[28] (reaction 238).

$$BF_4^- \ \left[\text{cyclo-}R^2C{=}CR^2\overset{+}{P}R^1_2\right] \xrightarrow{H_2O} R^1_2P({=}O)C(R^2){=}CHR^2 \tag{238}$$

R^1, R^2 = alkyl, phenyl

A P=N bond formation can be observed in the case of substrates containing $\overset{+}{P}$—N bonds; aminohalophosphonium halides[780] and, in more severe conditions, tetrakisdialkylaminophosphonium bromides[786] are thus thermally dealkylated, analogously to the second stage of the Arbusov reaction, and the thermal stability is lowered by the introduction of an electronegative group on one of the nitrogen atoms. Phosphorimidic triamides are also obtained by reaction with sodium amide, through attack on the α-carbon atom (in addition to aminophosphines if there are alkyl substituents on the nitrogen)[786] (reaction 239). Owing, in part, to the electron-withdrawing ability of the phosphonio group, the halogen atom of α-polyhalophosphonium salts[787] is sufficiently electrophilic to be abstracted by phosphorus nucleophiles, in an equilibrium which can be displaced by the appropriate aldehyde or ketone[318,788]. Cadmium, zinc and mercury were also found to be suitable as dehalogenation reagents[318,789] (reaction 240). The halogen atom is even more reactive in the case of halofluoromethylene bisphosphonium species[318].

$$Br^- \ (R_2N)_4P^+ \xrightarrow[\text{or (b) NaNH}_2]{\text{(a)}\Delta} (R_2N)_3P{=}NR \ + \ RBr \qquad (239)$$

$NR_2 = NEt_2$	(a) 250–280 °C/0.02 mmHg	38%	
	(b) 180 °C/0.02 mmHg	52%	$[+34\%(R_2N)_3P]$
$NR_2 = N(Et)SO_2Ph$	(a) 80–100 °C/10 mmHg	74%	

$$Br^- \ Y_3\overset{+}{P}CFX_2 \underset{\text{or M/(}-\text{MBrX)}}{\overset{Y_3P/(-Y_3PBrX)}{\rightleftharpoons}} [Y_3\overset{+}{P}\overset{-}{C}FX] \rightarrow\rightarrow \ \text{evolution} \qquad (240)$$

$Y = Ph, Me_2N$; $X = Br, Cl, F$; $M = Zn, Cd, Hg$

b. Reactions with formation of tricoordinated phosphorus. Phosphines can be displaced from their phosphonium salts by their own counter anions as in the case of 1-iminobenzyl[314] or 1-alkoxycarbonyl[321] phosphonium substrates (reactions 241 and 242) or when a good electronwithdrawing group is β to the phosphorus[790]. The same kind of behaviour has been observed by heating at 120 °C 2-alkoxycarbonylalkyl analogues, neat or in solution, which gave carbonic anhydride and phosphonium ylides[791]. External nucleophiles have also been observed to give displacement of phosphine in 4-pyridinylphosphonium salts[745].

$$Cl^- \ Ph_3P^+(Ph)C{=}NCH_2CF_3 \xrightarrow[(-Ph_3P)]{20\,^\circ C} Cl(Ph)C{=}NCH_2CF_3 \qquad (241)$$

$$X^- \ R^1R^2_2\overset{+}{P}COOR^3 \xrightarrow[(-R^1R^2_2P)]{20\,^\circ C} R^3X \ + \ CO_2 \qquad (242)$$

Lastly, mono- or bis-phosphonium salts can be cleaved to phosphines when a negative charge is created at a suitable position either with nucleophiles[792,793] or using a base to abstract labile hydrogen atoms[540,701,794–798] (reaction 243). In the case of 1.2-ethylene,

$$X^- \ Ph_3\overset{+}{P}CH_2\text{-(4-}R^1\text{-2-}R^2\text{-azole, Y, N)} \xrightarrow[(-Ph_3P)]{Nu^-} NuCH_2\text{-(4-}R^1\text{-2-}R^2\text{-azole, Y, N)} \qquad (243)$$

45–80%

$Y = NH, S$; $R^1 = Me, Ph$; $R^2 = H, SMe, NH_2$; $Nu = MeO, CH(COOEt)_2, CN$

1,2-vinylene, 1,4-but-2-enylene or buta-1,4-dienylene bisphosphonium salts, the nature of the other substituents on the phosphorus strongly influences the reduction to phosphine species with or without cleavage of the bridge[364,700,713].

c. Michael type addition. In α,β-unsaturated phosphonium salts, carbon atoms β to phosphonio groups are activated with respect to nucleophilic additions. This has been widely used in the synthesis of β-functionalized phosphonium species (cf. Sections II.c.2. II.D.), or in the synthesis of Wittig reagents for external carbonyl compounds. We shall consider here the only case where the added nucleophiles contain a carbonyl group in the β- or γ-position relative to the nucleophilic centre; in such cases, further intramolecular Wittig reaction readily occurs, leading to cycloalkenes, annelated cycloalkenes or unsaturated heterocyclic compounds, depending on the nature of nucleophiles[4] (reaction 244; Table 25). Sequential Michael–Michael ring closure has allowed an extremely efficient method for one-pot, three-component, 2 + 2 + 2 construction of functionalized cyclohexenes, using various enolates as substrates[807] (reaction 245).

$$X^- \; Ph_3\overset{+}{P}C(R^1)=C(R^3)(R^2) + R^4C(=O)(A)Y \xrightarrow{-Ph_3P=O} \text{cyclo}[R^4C=C(R^1)-C(R^2)(R^3)-Y-A] \quad (244)$$

$$Br^- \; Ph_3P^+CH=CH_2 + RC(=O)CH_2^- + CH_2=CHCH_2\overset{+}{P}Ph_3Br^- \xrightarrow{(-Ph_3P=O)} \text{4-R-cyclohex-3-enyl-}\overset{+}{P}Ph_3 \; Br^- \quad (245)$$

TABLE 25. Synthesis of cyclic compounds from vinylic phosphonium salts and carbonyl-containing nucleophiles

R^1	R^2	R^3	R^4	A	Y	Ref.
H	H	H	Me	$(CH_2)_n$	$(EtO_2C)_2C^-$	799
$-CH_2CH_2-$		H	H, Me, Ph	CH_2	$(EtO_2C)_2C^-$	800
H	H	H	Ph	N(R)	$Ar(CN)C^-$	801
H	H	H	H, Me	2-pyrrolylyl		802
H	H	H	H	o-$C_6H_4O^-$		803
H	H	H	Me, Ph	—	R_2CO^-	804
H	H, Me	H, Me, Ph	Me, *i*-Pr, H	—	R_2CS^-	275, 276
H	H	H	H, Me, Ph	o-$C_6H_4NHR^5$		805
H	H	H	Ph	—	$R^5C=NX$ ($X=O^-$, NHR^6)	806[a]

[a]The formation of pyrroles or dihydropyridazines appears to be dependent on the *E* or *Z* geometry of the α-imino ketone.

During the formation of cyclic compounds, phosphonium salts α-[793] or β-substituted in the vinyl group do not generally react as quickly or in such high yield as do the unsubstituted compounds[808–810] and anomalous reactions can be observed[808]; this may be attributed to steric hindrance probably in the Wittig step. However, α-alkyl (or -aryl) thiovinylphosphonium salts react fairly well with carbanionic species[293,811,812]. In the case of β *O*-acyl substituents, such activated vinylic esters are very sensitive to water or other nucleophiles, forming β ketophosphonium salts[284] in place of the usual Michael-type adducts (reaction 246).

$$X^- \; Ph_3\overset{+}{P}CH{=}C(OCOR)Me \xrightarrow[(-NuCOR)]{NuH} Ph_3\overset{+}{P}CH_2C(=O)Me \; X^- \qquad (246)$$

$NuH = H_2O, AlkOH, AlkNH_2, ArNH_2$

The presence of a carbonyl group γ to the phosphorus in the initial vinylphosphonium substrates make them versatile starting materials for the synthesis of heterocyclic systems, because of the electron-withdrawing effect of the carbonyl which appeared able to compete with that of the positive phosphorus[813]; it is occasionally possible to influence the direction of addition by varying the solvent (reaction 247). When ethenylphosphonium salts are β,β-disubstituted by heteroatomic groups, anionic removal of one of them has been observed during nucleophilic addition of alcoholates or sulphides[814] (reaction 248). This is analogous to the behaviour of 2-ethoxyvinyltriphenylphosphonium bromide with thiophenol[277] or in an exchange reaction[333], in which a *Z–E* isomerization occurs.

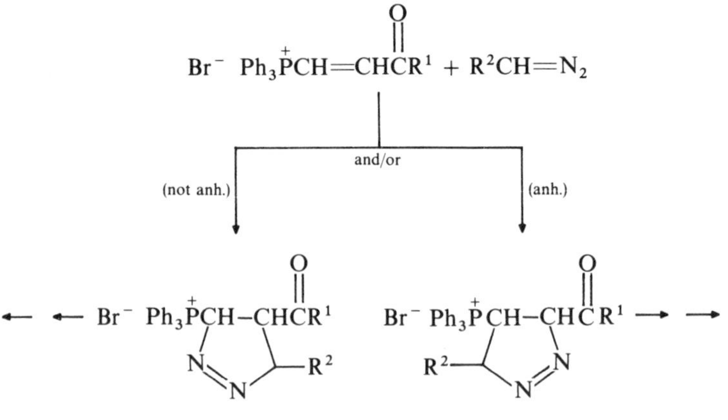

$$Br^- \; Ph_3\overset{+}{P}CH{=}CHC(=O)R^1 + R^2CH{=}N_2 \qquad (247)$$

and/or

(not anh.) ← ← $Br^- \; Ph_3\overset{+}{P}CH{-}CHC(=O)R^1$ (pyrazoline ring: N=N, R^2)

(anh.) $Br^- \; Ph_3\overset{+}{P}CH{-}CHC(=O)R^1$ (pyrazoline ring: N=N, R^2) → →

$X^- \; Ar_3P^+$, Z (azole ring: N, Y, R) $\xrightarrow[(-Z^-)]{R^1Y^-}$ [$X^- \; Ar_3P^+$, YR^1 (azole ring: N, Y, R)] → → (248)

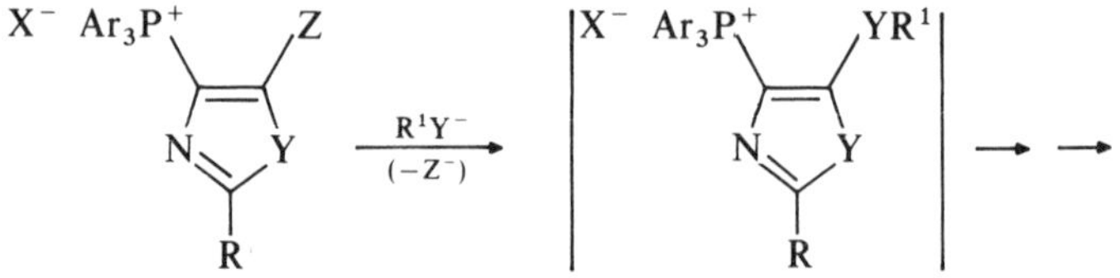

$Y = S, O; Z = SCH_2COPh$

As expected, buta-1,3-dienylphosphonium salts add ketonic nucleophiles in the δ-position[792]. A butadienylphosphonium salt is also the intermediate accounting for the products obtained in the reaction of but-2-enyl-1,4-bistriphenylphosphonium dichloride with carbanionic species in presence of an excess of base[289] (reaction 249).

$$Cl^- \; Ph_3\overset{+}{P}CH_2CH{=}C(SPh)CH_2\overset{+}{P}Ph_3 \; Cl^- \xrightarrow[(2)\ \text{2-methoxycarbonylcyclopentanone anion};\ (-Ph_3P{=}O)]{(1)\ K_2CO_3\ \text{excess};\ (-Ph_3PH^+Cl^-)} \text{SPh-substituted bicyclic diene, COOMe} \quad (249)$$

$$Br^- \; Ph_3\overset{+}{P}CH_2C{\equiv}CR^1 + \text{2-}R^2\text{-aniline} \xrightarrow[(-Ph_3P{=}O)]{(1)\ R^1 = H;\ R^2 = COPh;\ (2)\ \text{Base}} \text{4-}R^3\text{-2-methylquinoline}$$

$$\xrightarrow[R^1 = Ph;\ R^2 = SH,\ CONH_2]{(-Ph_3P{=}CH_2)} \text{2-phenyl benzo-fused heterocycle}, \; Y = S,\ NHCO \quad (250)$$

Allenyl- and propargyl-phosphonium salts have also been used as precursors of heterocyclic compounds in the presence of various functionalized nucleophiles[279], via the intermediacy of vinylphosphonio compounds; however, an ylide extrusion has been sometimes observed[815,816], depending on the nature of the functions in the nucleophile (reaction 250). Allenylphosphonium salts are able to add such weak CH acids as in ketones, even in absence of basic catalysts[817,818], without any modification of the keto group (reaction 251).

$$Br^- \; Ph_3\overset{+}{P}CH{=}C{=}CHPh \xrightarrow{PhCOMe} Br^- \; Ph_3\overset{+}{P}CH_2C({=}CHPh)CH_2C(O)PhPhCOMe \xleftarrow{PhCOMe} Br^- \; Ph_3\overset{+}{P}CH_2C{\equiv}CPh \quad (251)$$

Schweitzer's exploratory studies on the reaction of cyclopropylphosphonium bromide showed that this reagent is of a limited utility as an annelation reagent[819,820]. However,

$$PF_6^- \; \text{1-}(Ph_3P^+)\text{-1-}(EtO_2C)\text{cyclopropane} \xrightarrow[(-Ph_3P{=}O)]{RC(O)\overset{-}{C}(COOEt)R^1} \text{EtO}_2\text{C-, R-, R}^1\text{-, COOEt-substituted cyclopentene} \quad 88\text{–}95\% \quad (252)$$

$R = Me, Ph; R^1 = H, Me$

when the cyclopropyl ring bears one extra geminal electron-withdrawing group, the ring opening is easier, and this reagent has been used successfully for cycloalkenylation of compounds such as β-keto-esters, 1,3-diketones[821], simple esters[822] and imides[823,824] (reaction 252)

d. Others. The activation of the unsaturation in alkenyl and alkynylphosphonium salts is responsible for their behaviour towards 1,3-dipoles such as diazoalkanes[825,826], azides[827–829] nitrile ylides[830], nitrile oxides[826] and nitrones[826,831]. It usually undergoes regiospecific cycloaddition; strain energies account for the difference in the reactivities of 1-cycloalkenylphosphonium substrates[826]. Vinylic phosphonium salts also act as good dienophiles in Diels–Alder reactions with a variety of dienes[268,345,711,829,832,833], including anthracenes[13].

2. Reactivity consistent with the phosphonio group

Phosphonium salts, when functionalized (with e.g. keto, hydroxy, halo groups), are able to give characteristic reactions of such specific functions. This has been used in the synthesis of modified species; nevertheless, the absence of any strong basic agent is the required condition (see 3A–C).

a. Phosphonium salts containing keto groups. β-Oxoalkyltriphenylphosphonium salts undergo condensation with hydrazine derivatives; the formation and yield of hydrazones are said to be dependent on the solvent used and the reaction temperature[834]. Ketophosphonium salts in which the carbonyl group is separated from the phosphonio group by either aromatic or ethylenic substituents also react fairly readily with hydrazines (brief heating in chloroform solution for aldehydes; 3–5 h for ketones); nearly quantitative yields of hydrazones were often obtained[111,258,259,268,345,835]. In the case of a chloroacetyl group, a basic hydrazine such as phenylhydrazine or an excess of any hydrazine (or hydroxylamine) may lead to the fully substituted hydrazino hydrazono phosphonium species[345] (reaction 253). Subsequent modifications of the derivatives may also occur; a quantitative cyclization of a phosphonioalkylthiocarbazone to a thiazoylmethylphosphonium compound (as a disalt) has been observed[836]. Cyclization also results from the regiospecific reaction of (2,4-dioxoalkyl)triphenylphosphonium salts with hydroxylamine or hydrazines; isoxazole and pyrazole derivatives are obtained and the direction of cyclization is the opposite to the observed in the case of free 1,3-dicarbonyl compounds[723] (reaction 254). Thiazoylmethylphosphonium salts were obtained from 2-acylvinylphosphonio–thioamide adducts[837]. Formyl-containing phosphonium salts were able to give chalcones by condensation with ketones[111].

$$Cl^- \ Ph_3\overset{+}{P}CH{=}CH\overset{O}{\overset{\|}{C}}Cl \xrightarrow[\text{r.t. to } 60\,^\circ C]{ZNH_2} Cl^- \ Ph_3\overset{+}{P}CH{=}CH\overset{NZ}{\overset{\|}{C}}CH_2Cl \tag{253}$$

$$Cl^- \ Ph_3\overset{+}{P}CH{=}CH\overset{O}{\overset{\|}{C}}Cl \xrightarrow[ZNH_2 \text{ excess}]{60\text{–}80\,^\circ C} Cl^- \ Ph_3\overset{+}{P}CH{=}CH\underset{NZ}{\underset{\|}{C}}CH_2NHZ$$

$$X^- \; Ph_3\overset{+}{P}CH(R^1)C(=O)CH(R^2)C(=O)R^3 \xrightarrow[20\text{–}140\,^\circ C]{Z\overset{+}{N}H_3\,Cl} X^- \; \text{[5-membered ring: } Ph_3\overset{+}{P}CH(R^1)\text{–C=C(R}^2\text{)–C(R}^3\text{)=N–Z]} \quad 7\text{–}80\% \qquad (254)$$

$R^1, R^2, R^3 = H$, alkyl, aryl; $Z = OH, NR^4$

The reaction of formylaryl (or formylethenyl) triphenylphosphonium bromide with aniline to form the azomethine product required more severe conditions (heating for 10 h in chloroform solution) than the reaction of aniline with benzaldehyde; this has been attributed to the deactivating effect of the phosphonio group on elimination of water, which is the rate-determining step of the reaction[835]. Under the usual conditions, the azomethines obtained are stable. Such salts also react with quinaldinium salts to give styryl dyes[835]. Condensation of a [*p*-(bromoacetyl)benzyl]phosphonium bromide with either *o*-phenylenediamine or 2-amino derivatives of thiazole and pyridine goes as far as the formation of either quinoxaline- or imidazole-containing phosphonium salts[838] (reactions 255 and 256). The smooth oxidation of a terminal formyl group is accomplished by heating for a short time in aqueous potassium permanganate solution. However, elimination of the triphenylphosphonio group, as triphenylphosphine oxide, may occur; potassium hydroxide formed during the reaction is probably involved in this undesirable side-reaction[835].

$$Br^- \; Ph_3\overset{+}{P}ArCHO + Me\text{–[}\overset{+}{N}\text{(Ph) heterocycle] } Y^- \xrightarrow[100\,^\circ C]{Ac_2O} Br^- \; Ph_3\overset{+}{P}ArCH{=}CH\text{–[}\overset{+}{N}\text{(Ph) heterocycle] } Y^- \quad 56\text{–}81\%$$

$Ar = 1,4\text{-}C_6H_4,\ 2,5\text{-}C_4H_2O,\ 2,5\text{-}C_4H_2S$ (255)

$$Br^- \; Ph_3\overset{+}{P}CH_2C_6H_4COCH_2Br + \text{2-aminopyridine (}NH_2\text{)} \longrightarrow \text{imidazo[1,2-}a\text{]pyridinium (N–H) } Br^- \text{ with 2-}C_6H_4CH_2\overset{+}{P}Ph_3\,Br^- \quad 72\% \qquad (256)$$

The keto group of 3-oxoalkyltriphenylphosphonium salts could be protected as acetal (usual acidic deprotection[790]) or dithioacetal, using glycol or thioglycols, with excellent yields[346] (γ-thioacetalated phosphonium as nitrates are monooxidizable at the sulphur atom, using Ce^{IV}[839]). α-Ketobromination works well with dioxane dibromide as brominating agent. When PBr_5/irradiation is applied, the reaction proceeds differently[258] (reaction 257).

$$\mathrm{Br^- \; p\text{-}Ph_3\overset{+}{P}C_6H_4COPr} \begin{cases} \xrightarrow[\text{dmf, 50 °C}]{\text{Br}_2\text{–dioxane}} \mathrm{Br^- \; p\text{-}Ph_3\overset{+}{P}C_6H_4COCHBrEt} \quad 33\% \\ \xrightarrow[\text{PBr}_5/h\nu]{\text{CHCl}_3,\text{ r.t.}} \mathrm{Br^- \; p\text{-}Ph_3\overset{+}{P}C_6H_4CO(CH_2)_3Br} \quad 60\% \end{cases} \tag{257}$$

Lastly, α-diazotization of ketophosphonium salts with azidinium tetrafluoroborate as a diazo group transfer agent has been described as a general process[840] (reaction 258).

$$\mathrm{Y^- \; Ph_3\overset{+}{P}CH_2COR} + \mathrm{BF_4^- \; {}^+N_2N{=}\!\!\!\bigcirc\!\!(N{=}R)} \xrightarrow[19\text{–}100\%]{20\text{–}80\,°\text{C}} \mathrm{Y^- \; Ph_3\overset{+}{P}C(=N_2)COR} \tag{258}$$

$$\mathrm{Y^- = X^-, BF_4^-; \quad R = alkyl, CH_2\overset{+}{P}Ph_3\,X^-, OR^1}$$

b. Phosphonium salts containing hydroxy groups. The chemistry of hydroxymethylphosphonium salts [including the very common tetrakishydroxymethylphosphonium chloride (thpc) and the corresponding sulphate (thps)] and their derivatives has been extensively reviewed, with emphasis on reactions in which the quaternary structure of the phosphonium salts is preserved[540]: acylation, PCl_5 chlorination, reaction with epoxides, addition to electrophilic olefins and to isocyanates, condensation with urea (the PROBAN process), carbamates, etc. The chemistry of thpc in solution is dominated by its sensitivity to bases and oxidizing agents; in non-hydroxylic media, the quaternary phosphonium structure is usually preserved. Weak bases such as aniline react without loss of quaternary structure, unlike more basic amines, which induce a P—C bond cleavage (cf. Section III.E.1). A clean, high yield of fluorinated compound is obtained using diethylaminosulphur trifluoride (reaction 259); this approach is suitable for small-scale preparations[317]. *sec*-Hydroxyalkylphosphonium salts behave normally relative to phosgene leading to chloroformylalkyl derivatives[324]. [2-(alkoxycarbonyl)ethyl]phosphonium salts gave substitution reactions with alcoholates, thiolates and amines[2f]. [(2-Chlorocarbonyloxy)alkyl]triphenylphosphonium chlorides are used as protecting groups in peptide synthesis. The 2-(triphenylphosphonio)ethoxycarbonyl (Peoc) chain has also been used as a protecting group for alcohols, and cleavage is carried out using dimethylamine in methanol at 0 °C[841]. *o*-Acyloxybenzylphosphonium salts are readily formed from phenolic precursors[842]; by analogy, (2-mercatophenyl)methyltriphenylphosphonium bromide behaves in the same way when it is reacted with acyl chlorides or α-halo ketones[843]. (Polyhydroxyalkyl)phosphonio compounds react successfully with dimethoxypropane as a protecting group[844].

$$\mathrm{BF_4^- \; Ph_3\overset{+}{P}CH_2OH} + \mathrm{Et_2NSF_5} \xrightarrow[\mathrm{CH_2Cl_2}]{0\,°\mathrm{C}} \mathrm{BF_4^- \; Ph_3\overset{+}{P}CH_2F} \quad 88\% \tag{259}$$

c. Phosphonium salts containing halo groups. (Aroylthiomethyl)triphenylphosphonium bromides (which are equivalent to mercaptomethyltriphenylphosphonium species) are easily obtained by means of addition of potassium arenethiocarboxylate to the bromomethylphosphonio precursor[310] (reaction 260). ω-Aminoalkylphosphonium salts have been prepared from ω-bromoalkyl substrates and amines[327]. Triphenylphosphonium salts containing heterocycles such as oxazoles, imidazoles and thioimidazoles can be prepared from very specific mono- or poly-halogenated phosphonium salts using primary or secondary aliphatic amines[845,846], sodium hydrogensulphide[845] or alkali metal thiocyanates[847] (α-haloketoalkylphosphonium salts have already been mentioned as substrates; see above).

$$Br^- \; Ph_3\overset{+}{P}CH_2Br + ArC(=O)S^- \; K^+ \xrightarrow[CHCl_3]{reflux, 4h} Br^- \; Ph_3\overset{+}{P}CH_2SC(=O)Ar \quad \sim 65\% \qquad (260)$$

Dehydrobromation of halogenated phosphonium salts is currently used in the synthesis of unsaturated compounds, (cf. Section II.C) provided that a labile carbanion is created in a suitable position relative to the halogen atom (reaction 261). Such bases as Et_3N (for β- or δ-halogenated phosphonium salts)[288] or bromide anion[357] may induce this elimination. Fluorine atoms of the cationic part of 3-methyl-1,1-difluorocyclopent-2-ene-1-phosphonium hexafluorophosphate were substituted when the compound was treated with absolute methanol under mild conditions[412].

$$\text{cyclopropane-}C(CH_2\overset{+}{P}Ph_3)(CH_2Br) \; Br^- \xrightarrow{r.t.} BrCH_2CH_2C(=CH_2)CH_2\overset{+}{P}Ph_3 \; Br^- \quad 65\% \qquad (261)$$

d. Phosphonium salts containing ethers or thioethers. Aryl ethers included in phosphonium salts can be cleaved to the corresponding phenolic compounds on acidic treatment[400]. Benzylic ether functional groups are efficiently cleaved by iodotrimethylsilane, which is used as an activating approach to phosphonium dendrites with up to 40 cationic salts[356] (reaction 262).

$$R\overset{+}{P}\left(C_6H_4-CH_2OMe\right)_3 \xrightarrow{(i)} R\overset{+}{P}\left(C_6H_4-CH_2I\right)_3 \xrightarrow{(ii)} R\overset{+}{P}\left[C_6H_4-CH_2-\overset{+}{P}\left(C_6H_4-CH_2OMe\right)_3\right]_3 \qquad (262)$$

(i) $ISiMe_3$; (ii) $P\left(C_6H_4-CH_2OMe\right)_3$

The sulphur atom of alkyl(thioalkyl)phosphonium salts forms a new onium centre on triethyloxonium tetrafluoroborate alkylation in nitromethane[848,849] (thiocetals; see above). Phosphonium ketene acetals are potential alkylating agents for phosphorus dithioic acid anions in non-aqueous, aprotic and aqueous media and in phase-transfer catalysis conditions[296] (reaction 263). It is suggested that onium ketene acetals react by nucleophilic attack on the methyl group of the acetal.

$$\mathrm{FSO_3^-}\ \ \mathrm{Ph_3\overset{+}{P}(R)C{=}C(OMe)_2} + \mathrm{EtO(Me)P(S)S^-Y^+} \longrightarrow \mathrm{EtO(Me)P(S)SMe} \quad (263)$$

R = Ph; Y = Na; MeOH–MeCN; 20 °C/190 h	75%
R = Ph; Y = Na; H_2O–MeCN; 38 °C/600 h	84%
R = Ph; Y = Na; CH_2Cl_2–$C_{16}H_{33}\overset{+}{N}Me_3\ Br^-$; 20 °C/100 h	96%
R = *p*-$MeOC_6H_4$; Y = H; $CHCl_3$; 20 °C/48h	98%

e. Phosphonium salts containing amino groups. Hydrazinophosphonium salts react in high yield with aldehydes or ketones in methanol[850]. Deprotection of amino groups in phosphonium salts has been currently achieved[790]. Among the aminophosphonium salts, (1-aziridinyl)phosphonium halides are very reactive and often of low thermal stability. The more stable amino(1-aziridinyl)phosphonium chlorides have been shown to react very quickly, at 20 °C, with primary (including *t*-$BuNH_2$) and secondary amines; opening of the aziridinyl ring has been attributed to the direct attack of amine (and not chloride) on a carbon atom of the aziridine ring. Both nitrogen atoms of piperazine participate in the formation of a bisphosphonium compound[454] (reaction 264).

$$\mathrm{Cl^-}\ (\mathrm{Et_2N})_2\overset{+}{P}(\mathrm{NH_2})(\text{N-aziridinyl}) + \text{HN(piperazine)NH} \xrightarrow[85\%]{20\,^\circ C} \left(\mathrm{Cl^-}\ (\mathrm{Et_2N})_2\overset{+}{P}(\mathrm{NH_2})\mathrm{NH{-}(CH_2)_2{-}N}\right)_2 \quad (264)$$

f. Others. Phosphonium salts containing a carboxyl group are converted into the corresponding acid chloride by treatment with oxalyl chloride[851]. Phosphonium salts containing both an aryl group and a carboxylalkyl chain at the phosphorus atom are able to form heterocylic compounds in the presence of polyphosphoric acid[852]. Aminolysis and hydrazinolysis of triphenylphosphonio acetic acid esters have been carried out at high temperature[331,332]; amido and hydrazidoalkylphosphonium salts were obtained with better yields (up to 96%) with bromide than chloride counter ions. A first step towards phosphonio peptidic compounds was the condensation of phenoxycarbonylmethyl-triphenylphosphonium chloride with *N,O*-bissilylated amino acids[853]. A thiono group included in phosphonium salts underwent as much as 60% oxidative desulphurization with H_2O_2[846].

It should be noted that, because of the strong negative induction effect of the triphenylphosphonio group, salts such as $Ph_3\overset{+}{P}CH{=}CHC(O)CH_2Cl\ Cl^-$ scarcely react by electrophilic addition of halogens at the ethylenic bond, even under fairly severe conditions[345]. Electrophilic substitution on the aliphatic or aromatic carbon atom of phosphonium salts has been carried out; as usual, *N*-bromosuccinimide α-brominates aralkylphosphonium salts[394,854]. Mono- and bis-nitrosomethylene phosphonium salts are normally obtained using isopropyl nitrite in the presence of hydrogen chloride[328,201]. The influence of a triphenylphosphonio group on electrophilic substitution of aromatic systems is shown by the fact that even though the nitration of tetraphenylphosphonium bromide in concentrated sulphuric acid led to *o*-, *m*- and *p*-nitro-substituted compounds[855], electrophilic substitution (sulphonation, bromination, nitration or acylation) in (1-naphthyl)triphenylphosphonium bromide occurred exclusively in the 5-position of the naphthalene nucleus[259], regardless of the reactant ratio.

2,4-Pyridinediylbisphosphonium salts react with methanol in the presence of ammonium peroxodisulphate to give the 5-hydroxymethylated compound; when radical conditions were absent, nucleophiles gave the 4-substituted products selectively[746].

Alkenyl-substituted arylphosphonium salts, heated in polyphosphoric acid gave good yields of benzophosphorinium salts[369–398]. Vinyl triphenylphosphonium bromide has shown good radicophilicity[311]; however radical conditions are able to induce intermolecular or intramolecular polymerization of unsaturated substrates[856,857]. It should be mentioned that polyenic phosphonium salts can be prepared as complexes[858]. Finally, molecular hydrogen on neutral Raney nickel reduces $\beta\,\gamma$-ethylenic bisphosphonium salts at 20 °C (55 bar) in quantitative yields[368], and the catalytic hydrogenation of phosphonium salts containing a triple bond (conjugated with a double bond) towards the (*Z*)-ethylenic product has been performed in the presence of a transition metal (20 catalysts used)[859].

F. Phase-transfer Catalysis

The catalysis of reactions between components of immiscible aqueous and organic solutions (phase-transfer catalysis) by quaternary phosphonium salts was first realised by Starks[860], who used hexadecyltributylphosphonium bromide as a catalyst for nucleophilic displacement reactions of an alkyl halide (n-$C_8H_{17}Cl + NaCN$). Then other nucleophilic displacements (S_N2 processes with inversion of configuration at carbon[861]), such as halogen exchanges including fluorination, thio etherification and cyanation, became very common[860–864]. The most frequently used alkyl halides were chlorides and bromide; for sterically hindered substrates, an $E2$ elimination may occur[860]. Such reactions are thought to occur entirely in the organic phase[865], the active nucleophile being the ion pair; however, that is usually not the case when the catalyst is insoluble in the aqueous phase[866] (Scheme 11).

$$\begin{array}{llll} RX + & R'_4P^+\ Y^- & \rightarrow RY + & R'_4\ P^+\ X^- \quad \text{organic phase} \\ \hline & \quad\upharpoonleft\downharpoonright & & \quad\upharpoonleft\downharpoonright \\ MX + & R'_4P^+\ Y^- & \rightleftharpoons MY + & R'_4\ P^+\ X^- \quad \text{aqueous phase} \end{array}$$

$$RX + Y^- \xrightarrow{R'_4P^+} RY + X^- \quad \text{overall organic phase}$$

$$X = Br, Cl, OSO_2Me; \quad Y = F, Cl, Br, I, CN, RS, NO_2, \text{etc.}$$

SCHEME 11

As the concentration of the phosphonium ion remains constant in both the organic and aqueous phases, its effect is really catalytic. The high nucleophilic power and lipophilicity of iodides may block the cationic catalyst in the organic phase, which explains why alkyl iodides are less used. Association phenomena of quaternary phosphonium salts with solvents such as benzene or chloroform have been thought essential for the catalysis of test reactions in the two phase system (reaction 265). The more the water, the less is the association and the slower is the rate[867]. For soft anions, phosphonium salts seem more efficient than ammonium salts, owing to their large size and polarizability, which promote stability of the ion pairs[868]; in the case of hard anions, phosphonium salts can give covalent compounds and this weakens the anion's reactivity[866]. However, phosphonium salts sometimes give better yields than commonly used ammonium salts in phase-transfer reactions[869], but their main advantage is their thermal stability (150–170 °C versus 100–120 °C for ammonium salts), which permits reactions to be carried out at relatively high temperatures; this is a exemplified by the KF fluorination of alkyl or aromatic chlorides[870,871] (reactions 266 and 267). As with ammonium salts, the catalytic efficiency of phosphonium salts increases with the length of chains (at least four carbon atoms); alkyl chains are the most recommended[866,868,872]. Very often used are $n\text{-Bu}_4\overset{+}{P}\,X^-$, $n\text{-Bu}_3\overset{+}{P}CH_2Ph\,X^-$ and $n\text{-Bu}_3\overset{+}{P}C_mH_{2m+1}$($X$ = Cl, Br; m = 16, 18).

$$n\text{-}C_8H_{17}OSO_2Me + KBr \xrightarrow{n\text{-Bu}_4P^+} n\text{-}C_8H_{17}Br + KSO_3Me \tag{265}$$

$t_{1/2}$(min): $CHCl_3$ 165 $CHCl_3$–H_2O(8:1) 648
PhH 17 PhH–H_2O(8:1) 150

$$n\text{-}C_8H_{17}Cl + KF \xrightarrow[160^\circ C/7h]{C_{16}H_{33}\overset{+}{P}Bu_3{}^n\,Br^-} n\text{-}C_8H_{17}F + KCl \tag{266}$$

80%

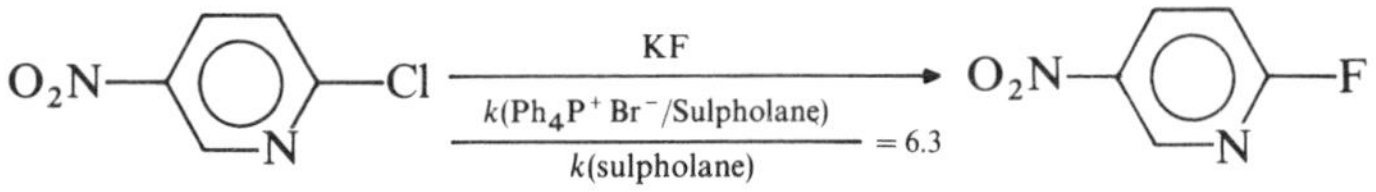

$$O_2N\text{-}C_5H_3N\text{-}Cl \xrightarrow{KF} O_2N\text{-}C_5H_3N\text{-}F \tag{267}$$

$$\frac{k(Ph_4P^+\,Br^-/\text{Sulpholane})}{k(\text{sulpholane})} = 6.3$$

Even moisture-sensitive compounds can be used as substrates, although with a lesser yield in phase-transfer catalysis conditions than in stoichiometric transformations[536] (reaction 268).

$$PhCOCl + KCN \xrightarrow[CH_2Cl_2-H_2O(1:1)]{Ph_3P{=}N{=}PPh_3{}^{7+}Cl^-} PhCOCN + Ph\text{—}C(CN)_2\text{—}OCOPh \tag{268}$$

3 : 1

Good results have been achieved in phosphonio-catalysed alkylation of active methylene compounds and imides which may be steroselective[873] (equation 269). Aqueous sodium hydroxide deprotonation of the phosphonium salt itself in view of a Wittig reaction is also very efficient[315,874–879]. Yields were better using $RCH_2\overset{+}{P}Ph_3\,X^-$ (R = H, Ar, SR) in methylene chloride. The mechanism is thought to be different than in S_N2 reactions: proton abstraction would take place here at the aqueous/organic interface. However, it is noteworthy that the stability of quaternary phosphonium salts under phase-transfer conditions in the presence of aqueous alkaline solutions diminishes dramatically. For a given R_4P^+, the stability decreases in the order $I^- > Br^- >> Cl^-$. It increases strongly either by diminishing the concentration of the base in the aqueous phase or by adding a molar excess of the corresponding inorganic salt. Degradation reactions proceed in the organic phase via extraction of HO^- as $R_4\overset{+}{P}\,HO^-$. Consequently, the extractability and/or reactivity of HO^- in the organic phase have to be minimized in order to attain a pure qantitative reaction[727,730] (Table 26). For less acidic hydrogen atoms, deuteration using NaOD has also been made possible[880].

$$PhCH_2COMe + Bu^nBr \xrightarrow[\text{aq. NaOH}/80^\circ C]{C_{16}H_{33}\overset{+}{P}Bu_3{}^n Br^-} PhCH(Bu^n)COMe \tag{269}$$

TABLE 26. Half-life periods $t_{1/2}$ for the degradation of quaternary phosphonium salts $R_4P^+ Y^-$ in a chlorobenzene–50% aqueous NaOH two-phase system under phase-transfer catalysis in the absence (A)[a] and presence (B)[b] of a molar excess of the corresponding inorganic salt (NaY)[727]

R_4P^+	Y^-	T(°C)	$t_{1/2}$(h) A	$t_{1/2}$(h) B
Bu_4P^+	Cl^-	25	0.03	0.3
	Br^-	25	3.5	220
		60	0.06	5.5
$Bu_3\overset{+}{P}C_{16}H_{33}$	Cl^-	25	0.03	0.8
	Br^-	25	2	220
		60	0.25	
			1.5[c]	
	I^-	25	—[d]	Stable[e]

[a]A PhCl solution (40 ml) of $R_4P^+ Y^-$ (0.02–0.04 M) and 50% aqueous NaOH (40 ml).
[b]As above, in the presence of a molar excess (30 mol/mol $R_4P^+ Y^-$) of NaY, partially as solid phase.
[c]Reaction carried on in the ground NaOH–PhCl solid–liquid system: a PhCl solution (40 ml) of $R_4P^+ Y^-$ (0.02–0.04 M) and 50 mmol of solid NaOH.
[d]8% decomposition after 3 days.
[e]No decomposition was observed after 6 days.

At the end of reactions with phase transfer catalysis removal of the catalyst is sometimes hampered by the formation of stable emulsions or the low solubility of the catalyst in water. This problem can be overcome by the use of a phosphonium salt which incorporates a 1,3-diene unit in the hydrophobic chain as catalyst; during the work-up, a ready Diels–Alder reaction with a dienophile-derivatized silica gel is carried out[15] (reaction 270). Particular experiments should be mentioned, such as hydrolysis of *p*-nitroaryl acetate or sulphonate by a phenolic quaternary phosphonium surfactant[684b], unusual halide substitution of halosilanes under phasetransfer conditions with tetra-*n*-butylphosphonium chloride[881], transfer of permanganate anion from aqueous solutions into benzene by phosphinimino cations $[Bu^n_3PN{=}PR_3]^+$ ($R = NMe_2, NEt_2$)[534,882] and tetrahydroborate reduction[883–885].

$$\text{CH}_3\text{CH=CHCH=CH}(\text{CH}_2)_7\text{—}\overset{+}{\text{P}}\text{Bu}_3\ {}^-\text{OSO}_2\text{Me} \ +$$

$$\text{(Si)—}(\text{CH}_2)_3\text{—NHSO}_2\text{—}\text{C}_6\text{H}_2(i\text{-Pr})_2\text{—triazolinedione} \xrightarrow{\text{Diels–Alder}} \qquad (270)$$

The catalytic activity of polyethylene glycol (PEG) phosphonium salts has been evaluated, in phase-transfer dehydrohalogenation reactions, as slightly better than that of the corresponding PEG ammonium compounds[886] (reaction 271). By comparison with these, the most efficient compound would be that in which only one hydroxyl group is replaced by the quaternary function.

$$\text{Br-}C_6H_4\text{-}CH_2CH_2X \xrightarrow[\text{toluene, PTC}]{\text{KOH 60\%}} \text{Br-}C_6H_4\text{-}CH{=}CH_2 \qquad (271)$$

$X = Cl, Br$

$$PTC = HO - PEG - {}^{+}PR_3\ Br^{-},\ Br^{-}\ R_3P^{+} - PEG - {}^{+}PR_3\ Br^{-}$$

$$[PEG: (CH_2CH_2O)_n CH_2CH_2]$$

Some recent reviews have devoted attention to polymeric phosphonium salts as catalysts ('polymer-supported phase-transfer catalysis') in preparative organic chemistry[887], mainly in S_N reactions such as halogen exchange (including fluorinations in the aromatic series[888]), reaction of haloalkanes with aqueous sodium cyanide, acetate, thiolate, phenoxide, sulphocyanide, etc.[889–891], *C*-alkylation[892] and borohydride reduction[891]. The reactions proceed by a mechanism similar to that of liquid–liquid phase-transfer catalysis[893–896]. The literature focuses on the many empirical parameters which may affect the activity of the catalyst in the triphase process[897,898] (among them, spacer chain, percentage ring substitution[899–901], solvent effects and swelling of the catalyst[902] have particularly attracted attention). Most investigations have used cross-linked polystyrenes with either conventional, isoporous or macroporous[903] structures as support polymers or copolymers. Polymeric phosphonium ions coated on powdered alumina[379] or based on poly(methyl methacrylate) resins[330] have also been investigated. One drawback to all these catalysts is that their activity may be substantially lower than that of soluble catalysts (due to diffusional processes)[330,379]. However, there are immediate advantages in the simplified reaction product work-up and easy and quantitative catalyst recovery. The activity of the catalyst increases with increasing spacer-chain length (for a given ring substitution[545]) (Figure 13).

Polymer-supported 'multi-site' phase-transfer catalysis seems to require the use of less material in order to provide activity comparable to others[253] (Table 27). Quaternary phosphonium ions on polystyrene latices, the particles of which are two orders of magnitude smaller than usual, were shown to be capable of higher activity; coagulation of the catalysts under reaction conditions was minimized by specific treatment[904]. The spacers may also contain ether linkages.

G. Anionic Reactivity

Apart from reactions in which anionic counterparts of phosphonium cations are essentially implicated in a phase-transfer catalysis process (polymer-supported or soluble catalysts; see above), some kinds of chemical transformations in which the anion's reactivity is involved have been studied. There are two major advantages, one being experimental and the other the regenerating capability of the reagent, in monomer- or polymer-supported form. The anionic counterparts of phosphonium salts can have an influence on their own stability or structure (the formation of betaines[163] and allylphosphonium–vinylphosphonium isomerization, for example[275,278]).

Thermally stable tetraphenylphosphonium hydrogen difluoride can act as a powerful *in situ* source of F^{-} in various reactions with organic substrates: as a nucleophile in

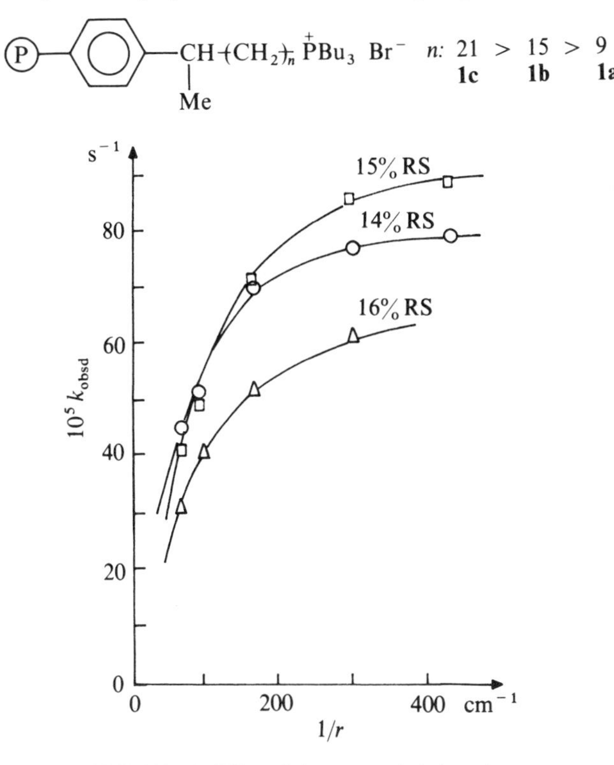

FIGURE 13. Effect of the spacer-chain length on k_{obsd} for the reaction of $C_8H_{17}Br$ with NaCN in the presence of catalysts **1a–c** with 14–17% ring substitution (% RS). (△) Catalyst **1a** with 16% RS; (○) catalyst **1b** with 14% RS; (□) catalyst **1c** with 15% Rs. *r*-Mean particle size (radius)

TABLE 27. Reaction of PhO^- (3) with 1-bromopentane (1) under various phase-transfer catalytic (0.01) conditions (water; 110 °C)

Catalyst	Yield (%) (dist.)
None	7
Ⓟ—$(CH_2)_3$—$\overset{+}{P}Bu_3$ Br^-	81
Ⓟ—CH_2—C(Me)($CH_2\overset{+}{P}Bu_3$)($CH_2\overset{+}{P}Bu_3$) 2 Br^-	85–95

halogen exchange and fluorodenitration; as a catalyst for fluorocarbon oligomerization or as a base or hydrogen bond electron donor in alkylations and Michael addition reactions[106,905]; it may be used either in solution (excellent solubility characteristics) or neat.

Because of its hard electrophilic character and its empty d orbitals, the phosphorus atom in phosphonium salts is able to interact strongly with both the reactive anion itself and the hard nucleophilic centres of any substrate; this may influence the chemio- and regio-selectivities of the substrate–anion reaction. Thus, whilst phosphonium tribromides were observed to give dibromination at the double bond of alkenes[254], selective α-bromination of methyl ketones and their acetals occurred, despite the presence of other functional groups usually reactive towards bromine (activated aromatic ring, double bond, free enolic position)[546]; the results differ from those obtained when molecular bromine was used as brominating agent, and the selectivity appears to be better than with ammonium tribromides. (It is worth noting that methyltriphenylphosphonium tribromide also acts as an efficient mild and highly selective dethioacetalization reagent[906]).

The complex catalyst $Bu_3SnI–Bu_4PI$ provides a reagent for the carbonylation of unreactive or acid-sensitive oxiranes to carbonates; it has been suggested that the iodide anion plays an effective role in nucleophilic cleavage of oxirane rings[907].

The use of molten salts as nucleophile sources has been shown to have a profound effect (rates independent of the anion and badly dependent of substrate, transition state resembling S_N2) on aromatic substitution reactions, which has been attributed to the interaction of the solvent and the molten salt[908] (reaction 272).

$$\text{Y}-\text{C}_6\text{H}_4-\text{OTs} + n\text{-C}_{12}\text{H}_{25}\overset{+}{\text{P}}\text{Bu}^n_3\ \text{I}^- \xrightarrow[\text{neat}]{100\,^\circ\text{C}} \text{Y}-\text{C}_6\text{H}_4-\text{I}$$

$$\text{Y}(\log k) = \text{NO}_2(-5.75), \text{CO}_2\text{Et}(-6.75), \text{H}(-7.27), \text{OMe}(-7.39) \qquad (272)$$

Polymeric phosphonium salt-bound carboxylate, benzenesulphinate and phenoxide anions have been used in nucleophilic substitution reactions for the synthesis of carboxylic acid esters, sulphones and C/O alkylation of phenols from alkyl halides. The polymeric reagent seems to increase the nucleophilicity of the anions[376] and the yields are higher than those for corresponding polymer phase-transfer catalysis (reaction 273).

$$\text{R}^1\text{X} + (\text{P})-\text{C}_6\text{H}_4-\text{CH}_2\overset{+}{\text{P}}\text{Ph}_3\ \text{Y}^- \longrightarrow \text{R}^1\text{Y} + (\text{P})-\text{C}_6\text{H}_4-\text{CH}_2\overset{+}{\text{P}}\text{Ph}_3\ \text{X}^-$$

$$\text{Y} = \text{RCOO}\ (73–95\%), \text{PhSO}_2^-\ (55–80\%), \text{ArO}^-\ (50–90\%) \qquad (273)$$

Long-chain alkylphosphonium azides and thiocyanates have been used as a source of anionic nucleophiles in the reaction of *n*-octyl halides and esters (*n*-octX:X = Cl, Br, I, OTs, OMs); a substitution process is able to occur even in non-polar solvents[909].

Tetraphenylphosphonium nitrite has been found to exist as either an anionic nitrite or a coordinated species, both of which give rise to the covalent phosphorane (oxidizable in nitrate in the presence of air); 1-haloalkane substitution gave a mixture of 1-nitroalkane and alkyl nitrite[528]. Bis(triphenylphosphine)nitrogen (1 +) nitrite $[N(PPh_3)_2]^+ NO_2^-$ reacted extensively with dichloromethane to produce a high yield of a potent nitrosating agent; nucleophilic attack via the oxygen rather than the nitrogen of the ambidentate nitrite ion to yield α-chloromethylated nitrite is postulated[910].

The strong nucleophilicity of aryl thiolate and aryl selenolate anions of tetraphenylphos-

TABLE 28. Quaternary anchored phosphonium chlorometallates in phenylacetylene hydrosilylation with triethylsilane at 80 °C

Isomers: α = PhC(SiEt$_3$)=CH$_2$; β-*cis* = Ph(H)C=C(S:Et$_3$)(H); β-*trans* = Ph(H)C=C(H)(SiEt$_3$)

Anchored complex	Isomer	Product yield (%) (GC data)					
		Reaction time (h)					
		1	2	3	4	5	7
$[P—CH_2\overset{+}{P}Bu_3][RhCl_4]^-$	α	10	20				
	β-*cis*	20	30				
	β-*trans*	35	50				
$[P—CH_2\overset{+}{P}Bu_3]_2[OsCl_6]^{2-}$	β-*cis*	5	7.5	10	12	14	17
	β-*trans*	39	46	50	56	60	66
$[P—CH_2\overset{+}{P}Bu_3]_2[PtCl_6]^{2-}$	α	33	34				
	β-*trans*	52	61				

phonium salts enables them to react with halocarbon solvents, which make these solvents inappropriate for synthetic handling purposes. Moreover, thiolate solutions are sensitive to hydrolysis, which converts the compound into a thiol, and selenolate solutions are extremely sensitive to oxidation, which transforms the compound into diselenide, which limits their use[136].

Introduction of the cyano group through anhydrous $[Ph_3P=N=PPh_3]^+ CN^-$ proceeds smoothly and quantitatively on alkyl and acyl halides (phase-transfer catalysis reactions required higher temperatures and gave by-products)[536].

Oxidizing anionic counterparts were used in organic transformations. Polymeric phosphonium dichromate[107] gave selective oxidation of alcohols. The bisphosponium dichromate $[Ph_3PCH_2PPh_3]^{2+}$ $Cr_2O_7^{2-}$ appears to be particularly mild and selective in the oxidation of alcohols; no acidic by-products are formed and no double bond migration or isomerization occurs[538].

In any application using phosphonium permanganates, care is essential to avoid violent thermal decomposition[137].

Recently, quaternary phosphonium chlorometallates [Ⓟ-$CH_2PBu_3]^+ m\, MCl_n^{m-}$ (non-noble metals included) were used as anchored ionic metal complexes for the hydrosilylation of alkenes and alkynes. Only catalysts containing platinum group metals were found to be fairly active. Heterogenization of $[PtCl_6]^{2-}$, $[RhCl_4]^-$ or $[OsCl_6]^{2-}$ did not influence the regio- and stereo-selectivity[543] (Table 28). Finally, a low-capacity quaternary phosphonium resin such as P-p-$C_6H_4CH_2PBu_3{}^+$ A^- may be more suitable than its ammonium equivalent for anion chromatography[375].

IV. REFERENCES

1. P. Beck, in *Organic Phosphorus Compounds* (Eds G. M. Kosolapoff and L. Maier), Vol. **2**, Wiley–Interscience, New York, 1972, pp. 189–508.
2. K. Jödden, in *Organische Phosphor Verbindungen I* (*Houben-Weyl, Methoden der Organischen Chemie*) (Ed. M. Regitz), Band E1, Georg Thieme Stuttgart, 1982, (a) pp. 491–580; (b) pp. 556–567 (c) pp. 495–516; (d) pp. 516–518; (e) pp. 518–523; (f) 559.

3. J. Emsley and D. Hall, *The Chemistry of phosphorus*, Harper and Row, London, 1976, pp. 254–274.
4. E. Zbiral, in *Organophosphorus Reagents in Organic Synthesis* (Ed. J.I.G. Cadogan), Academic Press, London, 1979, pp. 250–268.
5. D. J. H. Smith, in *Comprehensive Organic Chemistry* (Ed. D. Barton and W. D. Ollis), Vol. **2**, Pergamon Press, Oxford, 1979, pp. 1160–1175.
6. H. R. Hudson, *Top. Phosphorus Chem.*, **11,** 339 (1983).
7. R. Engel, *Synthesis of Carbon–Phosphorus Bonds*, CRC Press, Boca Raton, FL, 1988.
8. T. Minami and I. Yamamoto, in *Reviews on Heteroatom Chemistry* (Ed. S. Oae), Vol. **5**, Myu, Tokyo, 1991, p. 270.
9. S. Trippett, in *Organophosphorus Chemistry* (*Specialist Periodical Reports* Vol. **1**), Chemical Society, London, 1970, pp. 21–23.
10. D. J. H. Smith, in *Organophosphorus Chemistry* (*Specialist Periodical Reports*, Vols **2**–**7**), Chemical Society London, 1971–78.
11. D. W. Allen, in *Organophosphorus Chemistry* (*Specialist Periodical Reports*, Vols **8–22**), Chemical Society London, 1979–91.
12. H. Schindlbaur and F. Mitterhofer, *Fresenius, Z. Anal. Chem.*, **221**, 394 (1966).
13. C. C. Hanstock and J. C. Tebby, *Phosphorus Sulfur*, **15**, 239 (1983).
14. K. L. Duffin and K. L. Busch, *J. Planar Chromatogr.*, **1**, 249 (1988).
15. J. F. W. Keana and D. D. Ward, *Synth. Commun.*, **13**, 729 (1983).
16. C. Larpent, G. Meignan and H. Patin, *Tetrahedron Lett.*, **32**, 2615 (1991).
17. P. Jandik, U. Deschler and H. Schmidbaur, *Fresenius' Z. Anal. Chem.*, **305**, 347 (1981).
18. B. Buglio and V. S. Venturella, *J. Chromatogr. Sci.*, **22**, 276 (1984).
19. G. Aced, H. J. Möckel and S. F. Nelsen, *J. Liq. Chromatogr.*, **12**, 3201 (1989).
20. M. M. Ahmad and M.R. Bryce, *Mol. Cryst. Liq. Cryst.*, **120**, 361 (1985).
21. K. Takahashi, Y. Demura and S. Oota, *Jpn. Pat.*, 63 119 491; *Chem. Abstr.*, **109**, 93329w (1988).
22. S. Ota and K. Takahashi, *Jpn. Pat.*, 020 96 584; *Chem. Abstr.*, **113**, 161216s (1990).
23. A. W. Frank and G. L. Drake, *Text. Res. J.*, **43**, 633 (1973).
24. K. Ramarajan, M. D. Herd and K. D. Berlin, *Phosphorus Sulfur*, **11**, 199 (1981).
25. E. Vedejs, G. P. Meier and K. A. J. Snoble, *J. Am. Chem. Soc.*, **103**, 2823 (1981).
26. G. A. Bowmaker, R. Herr and H. Schmidbaur, *Chem. Ber.*, **116**, 3567 (1983).
27. K. B. Dillon, M. P. Nisbet and T. C. Waddington, *Polyhedron*, **1**, 123 (1982).
28. R. Breslow and L. A. Deuring, *Tetrahedron Lett.*, **25**, 1345 (1984).
29. D. J. Evans, G. J. Leigh and C. J. Macdonald, *Magn. Reson. Chem.*, **28**, 711 (1990).
30. K. S. Fongers, H. Hogeveen and R. F. Kingma, *Tetrahedron Lett.*, **24**, 643 (1983).
31. E. Fluck and J. Lorenz, *Z. Naturforsch, Teil B*, **22**, 1095 (1967); S. O. Grim, E. F. Davidoff and T. J. Marks, *Z. Naturforsch.*, *Teil B*, **26**, 184 (1971).
32. M. M. Crutchfield, C. H. Dungan, J. H. Lechter, V. Mark and J. R. Van Wazer, *Top. Phosphorus Chem.*, **5**, 1 (1967).
33. H. R. Hudson, K. B. Dillon and B. J. Walker in *CRC Handbook of Phosphorus-31 Nuclear Magnetic Resonance Data*, CRC Press, Boca Raton, FL, 1991, p. 181.
34. J. G. Verkade and L. D. Quin, Phosphorus-31 NMR Spectroscopy in Stereochemical Analysis in *Methods in Stereochemical Analysis*, Vol. **8**, VCH, Weinheim, 1987.
35. L. D. Quin and J. J. Breen, *Org. Magn. Reson.*, **5**, 17 (1973).
36. T. A. Albright, W. J. Freeman and E. E. Schweizer, *J. Am. Chem. Soc.*, **97**, 2942 (1975).
37. L. K. Krannich, R. K. Kanjolia and C. L. Watkins, *Inorg. Chim. Acta*, **103**, 1 (1985).
38. H. Goetz, H. Juds and F. Marschner, *Phosphorus*, **1**, 217 (1972).
39. T. A. Albright, W. J. Freeman and E. E. Schweizer, *J. Am. Chem. Soc.*, **97**, 2946 (1975).
40. G. Singh, *Phosphorus Sulfur*, **30**, 279 (1987).
41. I. M. Aladzheva, P. V. Petrovskii, T. A. Mastryukova and M. I. Kabachnik., *Zh. Obshch. Khim.*, **50**, 1442 (1980).
42. K. S. Narayanan and K. D. Berlin, *J. Org. Chem.*, **45**, 2240 (1980).
43. S. M. Cairns and W. E. McEwen, *Heteroatom Chem.*, **1**, 9 (1990).
44. E. M. Briggs, G. W. Brown, P. M. Cairns, J. Jiricny and M. F. Meidine, *Org. Magn. Reson.*, **13**, 306 (1980).
45. I. S. Sigal and F. H. Westheimer, *J. Am. Chem. Soc.*, **101**, 5329 (1979).
46. N. Lowther, P. D. Beer and C. D. Hall, *Phosphorus Sulfur*, **35**, 133 (1988).
47. K. B. Dillon, R. N. Reeve and T. C. Waddington, *J. Chem. Soc., Dalton Trans.*, 2382 (1977); 1318 (1978).

48. R. M. K. Deng and K. B. Dillon, *J. Chem. Soc., Dalton Trans.*, 1911 (1984).
49. J. E. Richman and R. B. Flay, *J. Am. Chem. Soc.*, **103**, 5265 (1981).
50. R. Bartsch, O. Stelzer and R. Schmutzler, *Z. Naturforsch., Teil B*, **36**, 1349 (1981).
51. G. V. Ratovskii, S. L. Belaya, V. I. Donskikh, A. A. Tolmachev and S. P. Ivonin, *Zh. Obshch. Khim.*, **60**, 2698 (1990).
52. K. B. Dillon and T. P. Straw, *J. Chem. Soc., Chem. Commun.*, 234 (1991).
53. M. Brownstein and R. Schmutzler, *J. Chem. Soc., Chem. Commun.*, 278 (1975).
54. M. J. Van Hamme, D. J. Burton and P. E. Greenlimb III, *Org. Magn. Reson.*, **11**, 275 (1978).
55. L. Riesel, H. Vogt and V. Kolleck, *Z. Anorg. Allg. Chem.*, **574**, 143 (1989).
56. D. Schomburg, G. Bettermann, L. Ernst and R. Schmutzler, *Angew. Chem., Int. Ed. Engl.*, **24**, 975 (1985).
57. L. D. Quin, K. A. Mesch and W. L. Orton, *Phosphorus Sulfur*, **12**, 161 (1982).
58. H. J. Cristau, G. Duc and H. Christol, *Synthesis*, 374 (1983).
59. E. Fluck and K. Issleib, *Chem. Ber.*, **98**, 2674 (1965).
60. C. W. Schultz and R. W. Parry, *Inorg. Chem.*, **15**, 3046 (1976).
61. S. Lochschmidt, G. Müller, B. Huber and A. Schmidpeter, *Z. Naturforsch. Teil B*, **41**, 444 (1986).
62. H. Schmidbaur, S. Strunk and C. E. Zybill, *Chem. Ber.*, **116**, 3559 (1983).
63. A. Schmidpeter and S. Lochschmidt, *Angew. Chem., Int. Ed. Engl.*, **25**, 253 (1986).
64. A. Schmidpeter, S. Lochschmidt, K. Karaghiosoff and W. S. Sheldrick, *J. Chem. Soc., Chem. Commun.*, 1447 (1985).
65. N. Weferling, R. Schmutzler and W. S. Scheldrick, *Liebigs Ann. Chem.*, 167 (1982).
66. N. Gurusamy, K. D. Berlin, D. van-der Helm and M. B. Hossain, *J. Am. Chem. Soc.*, **104**, 3107 (1982).
67. H. Schmidbaur and U. Deschler, *Chem. Ber.*, **116**, 1386 (1983).
68. A. Wohlleben and H. Schmidbaur, *Angew. Chem., Int. Ed. Engl.*, **16**, 417 (1977).
69. A. Schmidpeter, S. Lochschmidt and W. S. Sheldrick, *Angew. Chem., Int. Ed. Engl.*, **24**, 226 (1985).
70. S. Lochschmidt and A. Schmidpeter, *Z. Naturforsch., Teil B*, **40**, 765 (1985).
71. R. Appel and G. Erbelding, *Tetrahedron Lett.*, 2689 (1978).
72. R. Uson, J. R. Fornies, R. Navarro, M. A. Uson, M. P. Garcia and A. J. Welch, *J. Chem. Soc., Dalton Trans.*, 345 (1984).
73. H. J. Cristau, A. Chêne and H. Christol, *Synthesis*, 551 (1980).
74. H. H. Karsch, *Phosphorus Sulfur*, **12**, 217 (1982).
75. R. G. Kostyanovskii, Y. I. El'natanov, Sh. M. Shrikhaliev, S. M. Ignatov, I. I. Chervin, S. S. Nasibov, A. B. Zolotoi, O. A. D'yachenko and L. O. Atovmyan, *Izv. Akad. Nauk SSSR, Ser. Khim.*, 2354 (1982).
76. H. J. Bestmann, G. Schmid, H. Oechsner and P. Ermann, *Chem. Ber.*, **117**, 1561 (1984).
77. J. Gonda, P. Kristian and J. Imrich, *Collect Czech. Chem. Commun.*, **52**, 2508 (1987).
78. K. B. Dillon, T. C. Waddington and D. Younger, *J. Chem. Soc., Dalton Trans.*, 790 (1975).
79. S. O. Grim and A. W. Yankowsky, *J. Org. Chem.*, **42**, 1236 (1977).
80. R. Kohle, W. Kuchen and W. Peters. *Z. Anorg. Allg. Chem.*, **551**, 179 (1987).
81. M. J. Gallagher, *Aust. J. Chem.*, **21**, 1197 (1968).
82. V. S. Brovko, N. Skvortsov, A. Yu. Ivanov and V. O. Reikhsfel'd, *Zh. Obshch. Khim.*, **53**, 1648 (1983).
83. J. J. Brophy and M. J. Gallagher, *Aust. J. Chem.*, **20**, 503 (1967).
84. G. A. Bowmaker, C. Dörzbach and H. Schmidbaur, *Z. Naturforsch., Teil B*, **39**, 618 (1984).
85. Ch. M. Angelov and D. D. Enchev, *Phosphorus Sulfur*, **37**, 125 (1988).
86. J. Skolimowski and R. Skowronski, *Phosphorus Sulfur*, **19**, 159 (1984).
87. E. A. Berdnikov, V. L. Polushina, F. R. Tantasheva and E. G. Kataev, *Zh. Obsch. Khim.*, **50**, 993 (1980).
88. H. Christol, H. J. Cristau and J. P. Joubert, *Bull. Soc. Chim. Fr.*, 1421 and 2975 (1974).
89. M. J. Gallagher and I. D. Jenkins, *Top. Stereochem.*, **3**, 48 (1968).
90. C. J. Devlin and B. J. Walker, *Tetrahedron*, **28**, 3501 (1972).
91. H. J. Cristau, D. Bottaro, F. Plénat, F. Pietrasanta and H. Christol, *Phosphorus Sulfur*, **14**, 63 (1982).
92. A. Y. Platonov, E. D. Maiorova, G. S. Akimova and V. N. Chistokletov, *Zh. Obshch. Khim.*, **52**, 451 (1982).
93. G. A. Gray, *J. Am. Chem. Soc.*, **95**, 7736 (1973).

94. W. McFarlane, *Proc. R. Soc. London, Ser. A.*, **306**, 185 (1968).
95. H. J. Bestmann and H. Oechsner, *Z. Naturforsch., Teil B*, **38**, 861 (1983).
96. J. Font and P. de March, *Tetrahedron*, **37**, 2391 (1981).
97. H. J. Bestmann and L. Kisielowski, *Chem. Ber.*, **116**, 1320 (1983).
98. M. I. K. Amer, B. L. Booth and P. Bitrus, *J. Chem. Soc., Perkin Trans. 1*, 1673 (1991).
99. T. Nakajima, H. Kawasaki and T. Koyashiki, Japans Pat. 011 4 267 (1989); *Chem. Abstr.*, **111**, 155120y (1989).
100. H. Schmidbaur and A. Schier, *Chem. Ber.*, **114**, 3385 (1981).
101. D. W. Allen, I. W Nowell and L. A. March, *J. Chem. Soc., Perkin Trans. 1*, 2523 (1984).
102. D. W. Allen, P. E. Cropper, P. G. Smithurst, P. R. Ashton and B. F. Taylor, *J. Chem. Soc., Perkin Trans. 1*, 1989 (1986).
103. D. E. C. Corbridge, *Top. Phosphorus Chem.*, **6**, 235 (1969).
104. E. Paetzold, G. Oehme and H. Pracejus, *J. Prakt. Chem.*, **325**, 429 (1983).
105. S. T. Brown and J. H. Clark, *J. Chem. Soc., Chem. Commun.*, 1256 (1983).
106. S. T. Brown and J. H. Clark, *J. Chem. Soc., Chem. Commun.*, 672 (1985).
107. M. Hassanein, *Eur. Polym. J.*, **27**, 217 (1991).
108. G. P. Schiemenz and P. Nielsen, *Phosphorus Sulfur*, **21**, 267 (1985).
109. R. Gompper, E. Kujath and H. U. Wagner, *Angew. Chem., Int. Ed. Engl.*, **21**, 543 (1982).
110. A. S. Shawali, A. O. Abdelhamid, H. M. Hassaneen and C. Parkanyi, *Phosphorus Sulfur*, **12**, 377 (1982).
111. M. I. Shevchuk, O. M. Bukachuk and D. A. Koshman, *Zh. Obshch. Khim.*, **54**, 1770 (1984).
112. A. J. Bellamy and I. S. MacKirdy, *J. Chem. Soc., Perkin Trans. 2*, 1093 (1980).
113. J. R. Chapman, in *Organophosphorus Chemistry*. (*Specialist Periodical Reports*, Vol. 14), Royal Society of Chemistry, London, 1983, p 278.
114. A. L. Yergey and R. J. Cotter, *Biomed. Mass Spectrom.*, **9**, 286 (1982).
115. M. A. Calcagno and E. E. Schweizer, *J. Org. Chem.*, **43**, 4207 (1978).
116. D. A. McCrery, D. A. Peake and M. L. Gross, *Anal. Chem.*, **57**, 1181 (1985).
117. K. J. Kroha and K. L. Busch, *Org. Mass Spectrom.*, **21**, 507 (1986).
118. H. Mestdagh, N. Morin and C. Rolando, *Org. Mass. Spectrom.*, **23**, 246 (1988).
119. H. Schmidbaur, A. Schier, C.M.F. Frazâo and G. Müller, *J. Am. Chem. Soc.*, **108**, 976 (1986).
120. H. Schmidbaur, G. Blaschke, B. Zimmer-Gasser and U. Schubert, *Chem. Ber.*, **113**, 1612 (1980).
121. S. W. Lee and W. C. Trogler, *J. Org. Chem.*, **55**, 2644 (1990).
122. M. Ul-Haque, W. Horne, S. E. Cremer, P. W. Kremer and P.K. Kafarski, *J. Chem. Soc., Perkin Trans. 2*, 1138 (1981).
123. C. Moret and L. M. Trefonas, *J. Am. Chem. Soc.*, **91**, 2255 (1969).
124. N. D. Gomelya, N. G. Feshchenko, A. N. Chernega, M. Yu. Antipin, Yu. T. Struchkov and I. E. Boldeskul, *Zh. Obshch. Khim.*, **55**, 1733 (1985).
125. R. O. Day, S. Husebye, J. A. Deiters and R. R. Holmes, *J. Am. Chem. Soc.*, **102**, 4387 (1980).
126. A. H. Cowley, R. A. Kemp, J. G. Lasch, N. C. Norman and C. A. Stewart, *J. Am. Chem. Soc.*, **105**, 7444 (1983).
127. A. H. Cowley, C. A. Stewart, B. R. Whittlesey and T. C. Wright, *Tetrahedron Lett.* **25**, 815 (1984).
128. D. W. Allen, I. W. Nowell, A. C. Oades and P. E. Walker, *J. Chem. Soc., Perkin Trans. 1*, 98 (1978).
129. J. D. Gallucci and R. H. Holmes, *J. Am. Chem. Soc.*, **102**, 4379 (1980).
130. H. Schmidbaur, R. Herr and J. Riede, *Angew. Chem., Int. Ed. Engl.*, **23**, 247 (1984).
131. P. Batail, L. Ouahab, J. F. Halet, J. Padiou, M. Lequan and R. M. Lequan, *Synth. Met.*, **10**, 415 (1985).
132. M. Ul-Haque, W. Horne, S. E. Cremer, P. W. Kremer, and J. T. Most, *J. Chem. Soc., Perkin Trans. 2*, 1467 (1980).
133. D. W. Allen, P. E. Cropper and I. W. Nowell, *J. Chem. Res. (S)*, 298 (1987).
134. S. J. Archer, T. A. Modro and L. R. Nassimbeni, *Phosphorus Sulfur*, **11**, 101 (1981).
135. G. Moggi, J. C. J. Bart, F. Cariati and R. Psaro, *Inorg. Chim. Acta*, **60**, 135 (1982).
136. J. M. Ball, P. M. Boorman, J. F. Fait, A. S. Hinman and P. J. Lundmark, *Can. J. Chem.*, **67**, 751 (1989).
137. H. Karaman, R. J. Barton, B. E. Robertson and D. G. Lee, *J. Org. Chem.*, **49**, 4509 (1984).

138. A. J. Bellamy, R. O. Gould and M. D. Walkinshaw, *J. Chem. Soc., Perkin Trans. 2*, 1099 (1991).
139. W. Kunz and L. Maier, *Eur. Pat.* 60222 (1982); *Chem. Abstr.*, **98**, 53908g (1983).
140. H. Schmidbaur, R. Herr, T. Pollok, A. Schier, G. Müller and J. Riede, *Chem. Ber.*, **118**, 3105 (1985).
141. R. Weiss, H. Wolf, U. Schubert and T. Clark, *J. Am. Chem. Soc.*, **103**, 6142 (1981).
142. C. Bianchini, A. Meli and A. Orlandini, *Phosphorus Sulfur*, **11**, 335 (1981).
143. L. Chiche and H. Christol, *J. Chem. Soc., Perkin Trans. 2*, 753 (1984).
144. H. W. Roesky, B. Mainz, M. Noltemeyer and G. M. Sheldrick, *Z. Naturforsch., Teil B*, **43**, 941 (1988).
145. M. R. Bryce, M. M. Ahmad, R. H. Friend, D. Obertelli, S. A. Fairhurst and J. N. Winter, *J. Chem. Soc., Perkin Trans. 2*, 1151 (1988).
146. R. D. Wilson and R. Bau, *J. Am. Chem. Soc.*, **96**, 7601 (1974).
147. D. Mootz, W. Poll, H. Wunderlich and H. G. Wussow, *Chem. Ber.*, **114**, 3499 (1981).
148. S. Pohl, W. Saak and D. Haase, *Z. Naturforsch., Teil B*, 1493 (1987).
149. M. Lequan, R. M. Lequan and F. Robert, *Synth. Met.*, **40**, 49 (1991).
150. G. Helgesson and S. Jagner, *J. Chem. Soc., Dalton Trans.*, 2117 (1988).
151. J. C. Huffman and R. C. Haushaltter, *Z. Anorg. Chem.*, **518**, 203 (1984).
152. W. E. Swartz Jr, and D. M. Hercules, *Anal. Chem.*, **43**, 1066 (1971).
153. A. W. Frank, *Phosphorus Sulfur*, **10**, 147 (1981).
154. H. Zimmer, M. Jayawant, A. Amer and B. S. Ault, *Z. Naturforsch., Teil B*, **38**, 103 (1983).
155. S. W. Ng and J. J. Zuckerman, *J. Organomet. Chem.*, **234**, 257 (1982).
156. Y. Kondo, A. Zanka and S. Kusabayashi, *J. Chem. Soc., Perkin Trans. 2*, 827 (1985).
157. H. Christol, D. Grelet, M. R. Darvich, F. Fallouh, F. Plénat and H. J. Cristau, *Bull. Soc. Chim. Fr.*, 477 (1989).
158. M. S. Chattha and A. M. Aguiar, *J. Org. Chem.*, **38**, 1611 (1973).
159. A. Hercouet and M. Le Corre, *Synthesis*, 157 (1988).
160. W. Ding, W. Cai, J. Pu, W. Shen and J. Dai, *Huaxue Xuebao*, **43**, 53 (1985); *Chem. Abstr.*, **103**, 6435n (1985).
161. H. J. Bestmann, H. Hartung and I. Pils, *Angew. Chem.*, **77**, 1011 (1965).
162. W. M. Abdou, N. M. A. El-Rahman and N. A. F. Ganoub, *Phosphorus Sulfur Silicon*, **61**, 283 (1991).
163. G. Singh, *Phosphorus Sulfur Silicon*, **49–50**, 159 (1990).
164. R. G Ferrillo and A. Granzow, *Thermochim. Acta*, **45**, 177 (1981).
165. S. J. Abraham and W. J. Criddle, *J. Anal. Appl. Pyrol.*, **7**, 337 (1985).
166. J. B. Plumb and C. E. Griffin, *J. Org. Chem.*, **27**, 4711 (1962).
167. D. Asai, A. Okada, S. Kondo and K. Tsuda, *J. Macromol. Sci., Chem.*, **18**, 1011 (1982).
168. H. J. Timpe, U. Oertel and B. Hundhammer, *Z. Chem.*, **25**, 144 (1985).
169. V. N. Listvan, *Zh. Obshch. Khim.*, **55**, 2629 (1985).
170. E. O. Alonso, L. J. Johnston, J. C. Scaiano and V. G. Toscano, *J. Am. Chem. Soc.*, **112**, 1270 (1990).
171. D. T. Breslin and F. D. Saeva, *J. Org. Chem.*, **53**, 713 (1988).
172. R. N. Gedye, Y. N. Sadana and R. Eng., *J. Org. Chem.*, **45**, 3721 (1980).
173. L. Horner and H. Lund, in *Organic Electrochemistry* (Ed. M. Baizer) Marcel Dekker, New York, 1973, p. 729.
174. J. H. P. Utley and A. Webber, *J. Chem. Soc., Perkin Trans. 1*, 1154 (1980).
175. J. H. Wagenknecht and M. M. Baizer, *J. Org. Chem.* **31**, 3885 (1966).
176. V. L. Pardini, L. Roullier, J. H. P. Utley and A. Webber, *J. Chem. Soc., Perkin Trans. 2*, 1520 (1981).
177. E. A. H. Hall and L. Horner, *Phosphorus Sulfur*, **9**, 231 (1980).
178. L. Horner and K. Dickerhof, *Phosphorus Sulfur*, **15**, 331 (1983).
179. L. Horner, H. Winkler, A. Rapp, A. Mentrup, H. Hoffmann and P. Beck, *Tetrahedron Lett.*, 161 (1961). L. Horner, H. Fuchs, H. Winkler and A. Rapp, *Tetrahedron Lett.*, 965 (1963).
180. S. Ling-Chung, K. D. Sales and J. H. P. Utley, *J. Chem. Soc., Chem. Commun.*, 662 (1990).
181. A. Christodoulou, A. Anastopoulos and D. Jannakoudakis, *Can. J. Chem.*, **64**, 2279 (1986).
182. H. J. Cristau, B. Chabaud, L. Labaudinière and H. Christol, *Electrochim. Acta*, **29**, 381 (1984).
183. H. J. Cristau, B. Chabaud and C. Niangoran, *J. Org. Chem.*, **48**, 1527 (1983).
184. M. M. Ahmad, M. R. Bryce, J. Halfpenny and L. Weiler, *Tetrahedron Lett.*, **25**, 4275 (1984).

185. M. M. Ahmad and M. R. Bryce, *Mol. Cryst. Liq. Cryst.*, **120**, 361 (1985).
186. M. I. Shevchuk, V. P. Rudi, I. V. Megera and F. A. Nadtochii, *Zh. Obshch. Khim.*, **51**, 1024 (1980).
187. W. E. Mc Ewen, *Top. Phosphorus Chem.*, **2**, 1 (1965); M. J. Gallagher and I. D. Jenkins, *Top. Stereochem.*, **3**, 1 (1968).
188. L. Horner, *Pure Appl. Chem.*, **9**, 225 (1964); L. Horner, J. P. Bercz and C. V. Bercz, *Tetrahedron Lett.*, 5783 (1966).
189. N. Gurusamy and K. D. Berlin, *J. Am. Chem. Soc.*, **104**, 3114 (1982).
190. H. J. Bestmann, J. Lienert and E. Heid, *Chem. Ber.*, **115**, 3875 (1982).
191. P. Graczyk and M. Mikolajczyk, *Phosphorus Sulfur Silicicon*, **59**, 211 (1991).
192. N. A. Nesmeyanov, V. G. Kharitonov, O. A. Rebrova, P. V. Petrovskii and O. A. Reutov, *Izv. Akad. Nauk SSSR, Ser. Khim.*, 456 (1988).
193. H. J. Cristau, F. Plénat, G. Duc and A. Bennamara, *Tetrahedron*, **41**, 2717 (1985).
194. J. E. Stephenson and J. R. Jackson, *Dangerous Prop. Ind. Mat. Rep.*, **7**, 2 (1987).
195. L. F. Kasukhin, V. S. Brovarets, O. B. Smolii, V. V. Kurg, L. V. Budnik and B. S. Drach, *Zh. Obshch. Khim.*, **61**, 2679 (1991); *Chem. Abstr.*, **117**, 48707x (1992).
196. P. R. McAllister, M. J. Dotson, S. O. Grim and G. R. Hillman, *J. Med. Chem.*, **23**, 862 (1980).
197. R. Houssin, J. P. Hénichart, M. Foulon and F. Baert, *J. Pharm. Sci.*, **69**, 888 (1980).
198. G. I. Van'Kin, N. V. Lukoyanov, A. A. Chaikovskaya, T. N. Kudrya, L. M. Pacheva, T. G. Galenko, T. A. Ivanova, O. A. Raevskii, V. V. Malygin and A. M. Pinchuk, *Khim.-Farm. Zh.*, **25**, 17 (1991); *Chem. Abstr.*, **116**, 106264t (1992).
199. R. A. Mueller, *US Pat.*, 4 297 487; *Chem. Abstr.*, **96**, 20264z (1982).
200. T. Tammer, G. Heubach and B. Sachse, *Ger. Offen.*, DE 3 332 716; *Chem. Abstr.*, **103**, 88078k (1985).
201. I. V. Megera, V. L. Vlad and I. I. Sidorchuk, *Zh. Obshch. Khim.*, **52** 2252 (1982).
202. I. V. Megera, O. B. Smolii, V. K. Patratii, N. G. Prodanchuk and I. I. Sidorchuk, *Khim.-Farm. Zh.*, **16**, 791 (1982); *Chem. Abstr.*, **97**, 182547s (1982).
203. R. Fink, D. Van Der Helm and K. D. Berlin, *Phosphorus Sulfur*, **8**, 325 (1980).
204. F. H. Walker, *Eur. Pat.*, 126 561 (1984), *Chem. Abstr.*, **102**, 95826k (1985).
205. P. Ten. Haken, T. W. Naisby and A. C. Gray, *Br. Pat.*, 2 136 433, *Chem. Abstr.*, **102**, 95820d (1985).
206. W. Wehner and R. Grade, *Eur. Pat.*, 105 843 (1984), *Chem. Abstr.*, **101**, 111199 (1984).
207. J. Curtze, A. Guido, C. Drandarevski and A. A. Ramsey, *Eur. Pat.*, 300 574; *Chem. Abstr.*, **110**, 231888b (1989).
208. A. Wakabayashi, S. Takita, M. Sugiyama, M. Umenó, T. Wada, T. Shibata, Y. Kidó and N. Oyama, *Jpn. Pat.*, 03 261 790 (1991); *Chem. Abstr.*, **116**, 152014j (1992).
209. D. K. Donofrio and W. K. Whitekettle, *US Pat.*, 4 916 123 (1990); *Chem. Abstr.*, **113**, 54317d (1990).
210. Y. Shimano, S. Murakami, Y. Ogawa and J. Hirose, *Jpn. Pat.*, 63 190 080 (1988); *Chem. Abstr.*, **110**, 59498m (1989).
211. J. H. Hansen, *Ger. Pat.*, 08 900 217 (1989); *Chem. Abstr.*, **111**, 79852w (1989).
212. M.W. Klett, and B. Das, *Eur. Pat. Appl.*, 287 289 (1988); *Chem. Abstr.*, **111**, 176297b (1989).
213. Y. Kikuta, E. Miyazaki, T. Saito and Y. Hasegawa, *Jpn. Pat.*, 01 139 654 (1989); *Chem. Abstr.*, **113**, 193542m (1990).
214. K. M. Smith and D. H. Condlyffe, *Br. Pat.*, 2 200 363 (1988); *Chem. Abstr.*, **110**, 9961m (1989).
215. O. Sugimoto, *Eur. Pat.*, 348 194 (1989); *Chem. Abstr.*, **113**, 61086f (1990).
216. T. Yokoyama and M. Ogata, *Jpn. Pat.*, 01 059 873 (1989); *Chem. Abstr.*, **111**, 155442e (1989).
217. S. Yamada, S. Owaki and M. Suzuki, *Jpn. Pat.*, 01 192 875 (1989); *Chem. Abstr.*, **112**, 141179h (1990).
218. M. I. Iwato and M. Katsukawa, *Jpn. Pat.*, 01 272 657 (1989); *Chem. Abstr.*, **112**, 180682m (1990).
219. W. Urano and Y. Sugano, *Jpn. Pat.*, 01 198 620 (1989); *Chem. Abstr.*, **112**, 140474p (1990).
220. B. Hess, H. P. Mueller, H. Heine, B. Brassat, W. Kloeker and W. Uerfingen, *Eur. Pat.*, 296 450 (1988); *Chem. Abstr.*, **110**, 232577m (1989).
221. H. Uno and T. Endo, *J. Polym. Sci.*, **26**, 453 (1988).
222. W. S. Rees and D. Seyferth, *Chem. Abstr.*, **112**, 144325p (1990).
223. M. Kawabe, M. Kimura and H. Inoe, *Jpn. Pat.*, 01 138 231 (1989); *Chem. Abstr.*, **111**, 233935j (1989).

224. K. Harada and M. Fukuhira, *Jpn. Pat.*, 63 270 693 (1988); *Chem. Abstr.*, **110**, 213605c (1989).
225. A. R. Tsyganov, *Izv. Vyssh. Uchebn. Zaved., Khim. Khim. Tekhnol.*, **31**, 128 (1988); *Chem. Abstr.*, **110**, 214744j (1989).
226. A. R. Tsyganov, *Izv. Vyssh. Uchebn. Zaved., Khim. Khim. Tekhnol.*, **31**, 43 (1988); *Chem. Abstr.*, **110**, 200052v (1989).
227. A. Ohki, K. Dohtsu and M. Takagi, *Bunseki Kagaku*, **33**, E187 (1984); *Chem Abstr.*, **101**, 130777y (1984).
228. A. Ohki, H. Dohmitsu and S. Maeda, *Bull. Chem. Soc. Jpn.*, **64**, 476 (1991).
229. P. K. Tse and E. P. Horwitz, *Solvent Extraction Ion exchange*, **8**, 353 (1990).
230. T. Morimoto, K. Hiratsuka, Y. Sanada and H. Ariga, *Jpn. Pat.*, 63 213 914 (1988); *Chem. Abstr.*, **110**, 241289w (1989).
231. T. Samura and N. Kashiwabara, *Jpn. Pat.*, 02 135 321 (1990); *Chem. Abstr.*, **113**, 221419x (1990).
232. T. Muranaka, M. Okamoto, T. Yamagishi and H. Yoneda, *Jpn. Pat.*, 02 248 025 (1990); *Chem. Abstr.*, **114**, 113451p (1991).
233. H. T. Macholdt, A. Sieber, C. Godau and A. Mans, *Eur. Pat.*, 315 082 (1989); *Chem. Abstr.*, **111**, 155982f (1989).
234. M. Akstinat and H. Menzel, *Ger. Pat.*, 4 008 988 (1991); *Chem. Abstr.*, **115**, 213712p (1991).
235. P. Davies, N. B. O'Bryan Jr, and J. B. Philip Jr, *Eur. Pat.*, 328 391 (1989); *Chem. Abstr.*, **112**, 226689z (1990).
236. J. B. Onopchenko, G. M. Singerman and R. T. Sebulski, *US Pat.*, 4 462 935 (1984); *Chem. Abstr.*, **101**, 130899p (1984).
237. D. N. Kramer and P. A. Snow, *US Pat.*, 4 941 989 (1990); *Chem. Abstr.*, **114**, 30213m (1991).
238. B. Escoula, N. Hajjaji, I. Rico and A. Lattes, *J. Chem. Soc., Chem. Commun.*, 1233 (1984).
239. A. Hercouet and M. Le Corre, *Phosphorus Sulfur*, **29**, 111 (1986).
240. H. Schmidbaur and C. E. Zybill, *Chem. Ber.*, **114**, 3589 (1981).
241. M. L. Wang, A. H. Liu and J. J. Jwo, *Ind. Eng. Chem. Res.*, **27**, 555 (1988).
242. W. G. Dauben, J. M. Gerdes and R. A. Bunce, *J. Org. Chem.*, **49**, 4293 (1984).
243. N. Hamanaka, S. Kosuge and S. Iguchi, *Synlett*, 139 (1990).
244. S. Mori, K. Ida and M. Ue, *Eur. Pat.*, 291 074 (1988).
245. S. A. Zalinyan, R. A. Khachatryan and M. G. Indzhikyan, *Arm. Khim. Zh.*, **39**, 457 (1986); *Chem. Abstr.*, **107**, 236832b (1987).
246. A. Salzer and A. Hafner, *Helv. Chim. Acta*, **66**, 1774 (1983).
247. T. Takanami, A. Abe, K. Suda, H. Ohmori and M. Masui, *Chem. Pharm. Bull.*, **38**, 2698 (1990).
248. G. W. J. Fleet, and T. K. M. Shing, *Tetrahedron Lett.*, **24**, 3657 (1983).
249. H. J. Cristau and Y. Ribeill, *Synthesis*, 911 (1988).
250. U. Deschler, N. Holy and H. Schmidbaur, *Chem. Ber.*, **115**, 1379 (1982).
251. R. K. Lulukyan, M. Z. Ovakimyan, G. A. Panosyan and M. G. Indzhikyan, *Arm. Khim. Zh.*, **34**, 474 (1981); *Chem. Abstr.*, **95**, 169300v (1981).
252. M. Vincens, J. T. Grimaldo Moron, R. Pasqualini and M. Vidal, *Tetrahedron Lett.*, **28**, 1259 (1987).
253. J. P. Idoux, R. Wysocki, S. Young, J. Turcot, C. Ohlman and R. Leonard, *Synth. Commun.*, **13**, 139 (1983).
254. A. Akelah, M. Hassanein and F. Abdel-Galil, *Eur. Polym. J.*, **20**, 221 (1984).
255. M. I. Shevchuk, I. N. Chernyuk, E. M. Volynskaya, V. V. Shelest and P. I. Yagodinets, *Zh. Obshch. Khim.*, **50**, 1978 (1980).
256. I. N. Zhmurova, I. M. Kosinskaya and A. M. Pinchuk, *Zh. Obshch. Khim.*, **51**, 1538 (1981).
257. D. A. Koshman, O. M. Bukachuk and M. I. Shevchuk, *Zh. Obshch. Khim.*, **55**, 1738 (1985).
258. M. I. Shevchuk, O. M. Bukachuk and T. A. Zinzyuk, *Zh. Obshch. Khim.*, **55**, 346 (1985).
259. O. M. Bukachuk, D. A. Koshman, V. S. Vorokh and M. I. Shevchuk, *Zh. Obshch. Khim.*, **56**, 1789 (1986).
260. E. Anders and F. Markus, *Tetrahedron Lett.*, **28**, 2675 (1987).
261. E. Anders and F. Markus, *Chem. Ber.*, **122**, 113 (1989).
262. D. W. Allen, I. W. Nowell and L. A. March, *Tetrahedron Lett.*, **23**, 5479 (1982).
263. D. W. Allen and P. E. Cropper, *Polyhedron*, **1**, 129 (1990).
264. T. Migita, T. Nagai, K. Kiuchi and M. Kosugi, *Bull. Chem. Soc. Jpn.*, **56**, 2869 (1983).
265. A. M. Torgomyan, A. S. Pogosyan, M. Zh. Ovakimyan and M. G. Indzhikyan, *Arm. Khim. Zh.*, **33**, 408 (1980); *Chem. Abstr.*, **93**, 186471c (1980).

266. I. L. Rodionov, Yu. N. Luzikov, M. A. Kazankova and I. F. Lutsenko, *Zh. Obshch. Khim.*, **54**, 466 (1984).
267. N. G. Pavlenko, E. I. Sagina and V. P. Kukhar, *Zh. Obshch. Khim.*, **50**, 2379 (1980).
268. M. I. Shevchuk, M. K. Bratenko, I. N. Chernyuk and M. G. Moseichuk, *Zh. Obshch. Khim.*, **57**, 1059 (1987).
269. S.E. Fry and N. J. Pienta, *J. Org. Chem.*, **49**, 4877 (1984).
270. J. A. Kampmeier, S. H. Harris and R. M. Rodehorst, *J. Am. Chem. Soc.*, **103**, 1478 (1981).
271. H. Ohmori, T. Takanami and M. Masui, *Tetrahedron Lett.*, **26**, 2199 (1985).
272. H. G. Viehe and E. Franchimont, *Chem. Ber.*, **95**, 319 (1962).
273. S. I. Miller and C. E. Orzech, C. A. Welch, G. R. Ziegler and J. I. Dickstein, *J. Am. Chem. Soc.*, **84**, 2020 (1962).
274. J. I. Dickstein and S. I. Miller, *J. Org. Chem.*, **37**, 2168 (1972).
275. J. M. McIntosh, H. B. Goobrand and G. M. Masse, *J. Org. Chem.*, **39**, 202 (1974).
276. J. M. McIntosh and R. S. Steevensz, *Can. J. Chem.*, **52**, 1934 (1974).
277. J. M. Swan and S. H. B. Wright, *Aust. J. Chem.*, **24**, 777 (1971).
278. R. A. Khachartryan, A. M. Torgomyan, M. Zh. Ovakimyan and M. G. Indzhikyan, *Arm. Khim. Zh.*, **27**, 682 (1974); *Chem. Abstr.*, **83**, 28334z (1975).
279. E. E. Schweizer, S. De Voe Goff and W. P. Murray, *J. Org. Chem.*, **42**, 200 (1977).
280. J. A. Kampmeier, S. H. Harris and R. M. Roderhost, *J. Am. Chem. Soc.*, **103**, 1478 (1981).
281. R. K. Lulukyan, M. Zh. Ovakimyan and M. G. Indzhikyan, *Arm. Khim. Zh.*, **34**, 563 (1981); *Chem. Abstr.*, **96**, 6803t (1982).
282. R. A. Khachatryan, G. A. Mkrtchyan, F. S. Kinoyan and M. G. Indzhikyan, *Zh. Obshch. Khim.*, **56**, 231 (1986).
283. M. Zh Ovakimyan, R. K. Lulukyan and M. G. Indzhikyan, *Russ. Pat.*, 763 353 (1980); *Chem. Abstr.*, **94**, 84302b (1981).
284. M. Nikhandker, A. Ahmad and M. G. Hossain, *Indian J. Chem.*, **26B**, 773 (1987).
285. J. Barluenga, I. Merino and F. Palacios, *Tetrahedron Lett.*, **31**, 6713 (1990).
286. E. E. Schweizer, J. E. Hayes, A. Rheingold and X. Wei, *J. Org. Chem.*, **52**, 1810 (1987).
287. H. J. Cristau, D. Bottaro, F. Plénat, F. Pietrasanta and H. Christol, *Phosphorus Sulfur*, **14**, 73 (1982).
288. R. K. Lulukayan, M. Zh. Ovakimyan and M. G. Indehikyan, *Arm. Khim. Zh.*, **34**, 995 (1981); *Chem. Abstr.*, **96**, 199800w (1982).
289. R. J. Pariza and P. L. Fuchs, *J. Org. Chem.*, **50**, 4252 (1985).
290. T. Minami, H. Sako, T. Ikehira, T. Hanamoto and I. Hirao, *J. Org. Chem.*, **48**, 2569 (1983).
291. T. Minami, S. Shikita, S. So, M. Nakayama and I. Yamamoto, *J. Org. Chem.*, **53**, 2937 (1988).
292. H. J. Bestmann and L. Kisielowsky, *Tetrahedron Lett.*, **31**, 3301 (1990).
293. A. G. Cameron and A. T. Hewson, *J. Chem. Soc., Perkin Trans. 1*, 2929 (1983).
294. F. Plénat, *Tetrahedron Lett.*, **22**, 4705 (1981).
295. D. J. Burton and D. G. Cox, *J. Am. Chem. Soc.*, **105**, 650 (1983).
296. A. T. Zaslona and C. D. Hall, *J. Chem. Soc., Perkin Trans. 1*, 3059 (1981).
297. J. Skolimowski and M. Simalty, *Tetrahedron Lett.*, **21**, 3037 (1980).
298. W. W. Epstein and M. Garrossian, *Phosphorus Sulfur*, **35**, 349 (1988).
299. C. G. Kruse, E. K. Poels and A. Van Der Gen., *J. Org. Chem.*, **44**, 2911 (1979).
300. S. V. Leig and B. Lygo, *Tetrahedon Lett.* **25**, 113 (1984).
301. S. M. Albonico and M. T. Pizzorno, *An. Asoc. Quim. Argent.*, **70**, 271 (1982); *Chem. Abstr.*, **97**, 6423q (1982).
302. J. B. Ousset, C. Mioskowski, Y. L. Yang and J. R. Falck, *Tetrahedron Lett.*, **25**, 5903 (1984).
303. J. Godoy, S. V. Ley and B. Lygo, *J. Chem. Soc., Chem. Commun.*, 1381 (1984).
304. D. A. Oparin, T. G. Melent'eva and L. A. Pavlova, *Zh. Org. Khim.*, **19**, 1106 (1983).
305. W. Tückmantel, K. Oshima, and K. Utimoto, *Tetrahedron Lett.*, **27**, 5617 (1986).
306. E. Anders, T. Gassner and A. Stankowiak, *Chem. Ber.*, **118**, 124 (1985).
307. E. Anders and T. Gassner, *Chem. Ber.*, **117**, 1034 (1984).
308. J. P. Marino and M. P. Ferro, *J. Org. Chem.*, **46**, 1828 (981).
309. T. Minami, I. Hirao, *Jpn. Pat.*, 60 152 493; *Chem. Abstr.*, **104**, 19679r (1986).
310. M. Ishida, T. Iwata, M. Yokoi, K. Kaga and S. Kato *Synthesis*, 632 (1985).
311. D. H. R. Barton, H. Tago and S. Z. Zard, *Tetrahedron Lett.*, **26**, 6349 (1985).
312. G. V. Romanov, A. A. Lapin and A. N. Pudovik, *Izv. Akad. Nauk SSSR, Ser. Khim.*, 719 (1984).

313. E. A. Suvalova, T. I. Chudakova, P. P. Onys'ko and A. D. Sinitsa, *Zh. Obshch. Khim.*, **57**, 1514 (1987).
314. P. P. Onys'ko, T. V. Kim., E. I. Kiseleva, V. P. Prokopenko and A. D. Sinitsa, *Zh. Obshch. Khim.*, **58**, 35 (1988).
315. K. S. Narayanan and P. D. Taylor, *US Pat.*, 4 997 952 (1991); *Chem. Abstr.*, **115**, 29636r (1991).
316. X. Y. Li and J. S. Hu, *Tetrahedron Lett.*, **28**, 6317 (1987).
317. D. J. Burton and D. M. Wiemers, *J. Fluorine Chem.*, **27**, 85 (1985).
318. D. J. Burton, *J. Fluorine Chem.*, **23**, 339 (1983).
319. D. G. Cox and D. J. Burton, *J. Org. Chem.*, **53**, 366 (1988).
320. L. Yu. Ukhin, Z. S. Morkovnik, D. S. Yufit and Yu. T. Struchkov, *Zh. Obshch. Khim.*, **55**, 2701 (1985).
321. A. A. Prishchenko, M. V. Livantsov, S. A. Moshnikov and I. F. Lutsenko, *Zh. Obshch. Khim.*, **57**, 1664 (1987).
322. S. Tanimoto, S. Jo and T. Sugimoto, *Synthesis*, 53 (1981).
323. A. J. Moore and M. R. Bryce, *Synthesis*, 26 (1991).
324. H. Kunz and G. Schaumlöffel, *Liebigs Ann. Chem.*, 1784 (1985).
325. R. Vilceanu and A. Venczel, *Rev. Chim.*, **31**, 1062 (1980); *Chem. Abstr.*, **94**, 157025a (1981).
326. S. Valverde, B. Herradon, R. M. Rabanal and M. Martin Lomas, *Can J. Chem.*, **65**, 332 (1987).
327. T. N. de Castro Dantas, J. P. Laval and A. Lattes, *Phosphorus Sulfur*, **13**, 97 (1982).
328. M. I. Shevchuk, E. M. Volynskya and A. V. Dombrovskii, *Zh. Obshch. Khim.*, **41**, 1999 (1971); O. M. Bukachuk and M. I. Shevchuk, *Zh. Obshch. Khim.*, **55**, 752 (1985).
329. R. K. Bansal, S. Mathur, J. K. Jain and D. Sharma, *J. Indian Chem. Soc.*, **5**, 134 (1988).
330. A. Akelah, M. Hassanein, M. El-Sakran, E. R. Kenawy and F. Abdel-Galil, *Acta Polym.*, **40**, 129 (1989).
331. G. P. Pavlov, V. A. Kukhtin, V. V. Kormachev and A. V. Kazymov, *Othr. Izobret.*, **110** (1981); *Chem. Abstr.*, **95**, 81214j (1981).
332. G. P. Pavlov, V. V. Kormachev, M. A. Soikova, N. L. Burtseva and A. V. Kazymov, *Zh. Obshch. Khim.*, **57**, 1955 (1987).
333. N. A. Nesmeyanov, V. G. Kharitonov, O. A. Rebrova, P. V. Petrovskii and O. A. Reutov, *Izv. Akad. Nauk SSSR, Ser. Khim.*, 456 (1988); *Chem. Abstr.*, **110**, 114938g (1989).
334. R. J. Linderman and A. I. Meyers, *Tetrahedron Lett.*, **24**, 3043 (1983).
335. F. Plénat, A. Bennamara, L. Chiche and H. Christol, *Phosphorus Sulfur*, **26**, 39 (1986).
336. B. E. Maryanoff and A. B. Reitz, *Phosphorus Sulfur*, **27**, 167 (1986).
337. B. E. Maryanoff, A. B. Reitz, M. S. Mutter, R. R. Inners, H. R. Almond, R. R. Whittle and R. A. Olofson, *J. Am. Chem. Soc.*, **108**, 7664 (1986).
338. A. D. Abell, J. O. Trent and B. I. Whittington, *J. Org. Chem.*, **54**, 2762 (1989).
339. A. D. Abell, K. B. Morris and J. C. Litten, *J. Org. Chem.*, **55**, 5217 (1990).
340. H. J. Cristau and Y. Ribeill, *J. Organomet. Chem.*, **352**, C 51 (1988).
341. M. Schlosser, H. Tuong, J. Respondek and B. Schaub, *Chimia*, **37**, 10 (1983).
342. S. Yamamoto, K. Okuma and H. Ohta, *Bull. Chem. Soc. Jpn.*, **61**, 4476 (1988).
343. M. Le Corre, A. Hercouet and H. Le Baron, *J. Chem. Soc., Chem. Commun.*, 14 (1981).
344. R. A. Michelin, G. Facchin, D. Braga and P. Sabatino, *Organometallics*, **5**, 2265 (1986).
345. I. V. Megera, L. B. Lebedev and M. I. Shevchuk, *Zh. Obshch. Khim.*, **51**, 54 (1981).
346. H. J. Cristau, J. P. Vors and H. Christol, *Synthesis*, 538 (1979).
347. H. J. Cristau, Y. Beziat, C. E. Niangoran and H. Christol, *Synthesis*, 648 (1987).
348. H. J. Cristau, J. Viala and H. Christol, *Bull. Soc. Chim. Fr.*, 980 (1985).
349. J. Viala and M. Santelli, *Synthesis*, 395 (1988).
350. F. R. Benn, J. C. Briggs and McAuliffe, *J. Chem. Soc., Dalton Trans.*, 293 (1984).
351. J. C. Briggs and G. Dyer, *Chem. Ind. (London)*, 163 (1982).
352. J. C. Briggs, Mc Auliffe, W. E. Hill, D. M. A. Minahan, J. G. Taylor and G. Dyer, *Inorg. Chem.*, **21**, 4204 (1982).
353. H. J. Cristau, L. Chiche, F. Fallouh, P. Hullot, G. Renard and H. Christol, *Nouv. J. Chim.*, **8**, 191 (1984).
354. A. Ohki, S. Maeda, T. Takeshita and M. Takagi, *Chem. Lett.*, 1349 (1987).
355. K. Rengan and R. Engel, *J. Chem. Soc., Chem. Commun.*, 1084 (1990).
356. K. Rengan and R. Engel, *J. Chem. Soc., Perkin Trans. 1*, 987 (1991).
357. N. S. Zefirov, T. S. Kuznetsova, S. I. Kozhushkov, K. A. Potekhin, A. V. Maleev and Y. T. Struchkov, *Zh. Org. Khim.*, **10**, 2120 (1987).

358. G. M. Bukachuk, I.V. Megera and M. I. Shevchuk, *Zh. Obshch. Khim.*, **50**, 1730 (1980).
359. H. Christol, H. J. Cristau and J. P. Joubert, *Bull. Soc. Chim. Fr.*, 1421 (1974).
360. H. Christol, H. J. Cristau and J. P. Joubert, *Bull. Soc. Chim. Fr.*, 2263 (1974).
361. H. Christol. H. J. Cristau and J. P. Joubert, *Bull. Soc. Chim. Fr.*, 2975 (1974).
362. H. J. Cristau, L. Labaudiniere and H. Christol, *Phosphorus Sulfur*, **15**, 359 (1983).
363. E. G. Kataev, F. R. Tantasheva, E. A. Berdnikov and B. Ya-Marguus, *Zh. Org. Khim.*, **10**, 1050 (1974).
364. H. H. Karsch, *Z. Naturforsch., Teil B*, **34**, 31 (1979).
365. H. Schmidbaur, C. Paschalidis, O. Steigel Mann and G. Müller, *Angew. Chem., Int. Ed. Engl.*, **28**, 1700 (1989).
366. H. Schmidbaur and T. Costa, *Z. Naturforsch., Teil B*, **37**, 677 (1982).
367. H. Schmidbaur, T. Costa and B. Milewski-Mahrla, *Chem. Ber.*, **114**, 1428 (1981).
368. M. Vincens, J. T. Grimaldo-Moron, R. Pasqualini and M. Vidal, *Tetrahedron Lett.*, **28**, 1259 (1987).
369. L. Horner, H. Kunz and P. Walach, *Phosphorus*, **6**, 63 (1975).
370. L. Horner, P. Walach and H. Kunz, *Phosphorus Sulfur*, **5**, 171 (1978).
371. H. J. Bestmann and L. Kisielowski, *Tetrahedron Lett.*, **31**, 3301 (1990).
372. H. Takeuchi, Y. Miwa, S. Morita and J. Okada, *Chem. Pharm. Bull.*, **32**, 409 (1984).
373. M. Bernard, W. T. Ford and H. W. Taylor, *Polym. Prep.*, 170 (1984).
374. M. Tomoi, E. Nakamura, Y. Hosokawa and H. Kakiuchi, *Makromol. Chem. Rapid. Commun.*, **5**, 281 (1984).
375. L. M. Warth, R. S. Copper and J. S. Fritz, *J. Chromatogr.*, **479**, 401 (1989).
376. M. Hassanein, A. Akelah and F. Abdel Galil, *Eur. Polym. J.*, **21**, 475 (1985).
377. S. M. Khashimova, A. T. Dzhalilov and A. B. Mukhamedgaliev, *Uzb. Khim. Zh.*, 49 (1990); *Chem. Abstr.*, **115**, 15989k (1991).
378. B. A. Mukhamedgaliev, A. V. Safaev, A. T. Dzhalilov and S. M. Khashimova, *Dokl. Akad. Nauk USSR*, 40 (1989) *Chem. Abstr.*, **114**, 123109x (1991).
379. S. Kobayashi, M. Suzuki, T. Sakaya and T. Saegusa, *Makromol. Chem.*, **188**, 457 (1987).
380. S. D. Venkataramu, G. D. Mac Donell, W. R. Purdum, M. El-Deek and K. D. Berlin, *Chem. Rev.*, **77**, 121 (1977).
381. K. L. Marsi, D. M. Lynch and G. D. Homer, *J. Heterocycl. Chem.*, **9**, 331 (1972).
382. M. Ul-Haque, W. Horne, S. E. Cremer and J. T. Most, *J. Chem. Soc., Perkin Trans. 2*, 1000 (1981).
383. H. Schmidbaur and A. Mörtl, *J. Organomet. Chem.*, **250**, 171 (1983).
384. C. H. Chen and K. D. Berlin, *J. Org. Chem.*, **36**, 2791 (1971).
385. C. H. Chen and K. D. Berlin, *Phosphorus*, **1**, 49 (1971).
386. A. M. Aguiar and H. Aguiar, *J. Am. Chem. Soc.*, **88**, 4090 (1966).
387. A. M. Aguiar, K. C. Hansen and J. T. Mague, *J. Org. Chem.*, **32**, 2383 (1967).
388. L. Horner, P. Walach and H. Kunz, *Phosphorus Sulfur*, **5**, 171 (1978).
389. H. Christol, H. J. Cristau, F. Fallouh and P. Hullot, *Tetrahedron Lett.*, 2377 (1979).
390. H. J. Cristau, L. Chiche, F. Fallouh, P. Hullot, G. Renard and H. Christol, *Nouv. J. Chim.*, **8**, 191 (1984).
391. S. D. Venkataramu, M. El-Deek and K. D. Berlin, *Tetrahedron Lett.*, 3365 (1976).
392. A. Bond, M. Green and S. C. Pearson, *J. Chem. Soc. B*, 929 (1968).
393. K. S. Fongers, H. Hogeveen and R. F. Kingma, *Tetrahedron Lett.*, **24**, 643 (1983).
394. G. Märkl and K. H. Heier, *Angew. Chem., Int. Ed. Engl.*, **11**, 1016 (1972).
395. H. Schmidbaur and A. Mörtl, *Z. Naturforsch., Teil B*, **35**, 990 (1980).
396. W. R. Purdum, G. A. Dilbeck and K. D. Berlin, *J. Org. Chem.*, **40**, 3763 (1975).
397. G. A. Dilbeck, D. L. Morris and K. D. Berlin, *J. Org. Chem.*, **40**, 1150 (1975).
398. R. Fink, D. Van Der Helm and K. D. Berlin, *Phosphorus Sulfur*, **8**, 325 (1980).
399. A. S. Radhakrishna, K. D. Berlin and D. Van der Helm, *Pol. J. Chem.*, **54**, 495 (1980).
400. G. D. Macdonell and K. D. Berlin, unpublished results, cf. reference 380 (p. 145).
401. A. G. Rowley, in *Organophosphorus Reagents is Organic Syntheses* (Eds. J. I. G. Cadogan), Academic Press, London, 1979, p. 327.
402. O. Mitsunobu, *Synthesis*, 1 (1981).
403. B. R. Castro, *Org. React.*, **29**, 1 (1983).
404. D. L. Hughes, *Org. React.* **42**, 335 (1992).

405. D. Hellwinkel, in *Organic Phosphorus Compounds* (Eds G. M. Kosolapoff and L. Maier), Vol. **3**, Wiley–Interscience, New York, 1972, (a) p. 197; (b) p. 202; (c) p. 204; (d) p. 205; (e) p. 206.
406. R. Luckenbach in *Organische Phosphor-Verbindungen (Houben-Weyl, E_2)* (Ed. M. Regitz), Vol. II, Georg Thieme, Stuttgart, 1982, p. 873.
407. J. Gloede, *Z. Chem.*, **28**, 352 (1988).
408. D. B. Denney, D. Z. Denney and B. C. Chang, *J. Am. Chem. Soc.*, **90**, 6332 (1968).
409. R. Appel and M. Halstenberg, in *Organophosphorus Reagents in Organic Synthesis* (Ed. J.I.G. Cadogan), Academic Press, London, 1979, pp. 387–432.
410. F. Seez and H. J. Bassler, *Z. Anorg. Allg. Chem.*, **418**, 263 (1975).
411. J. Svara and E. Fluck, *Z. Anorg. Allg. Chem.*, **529**, 137 (1985).
412. S. V. Fridland and M. N. Miftakhov, *Zh. Obshch. Khim.*, **58**, 41 (1988).
413. J. Michalski, J. Mikolajczak and A. Skowronska, *J. Am. Chem. Soc.*, **100**, 5386 (1978).
414. A. Skowronska, M. Pakulski, J. Michalski, D. Cooper and S. Trippett, *Tetrahedron Lett.*, 321 (1980).
415. J. Michalski, M. Pakulski and A. Skowronska, *J. Chem. Soc., Perkin Trans. 1*, 833 (1980).
416. J. Michalski, M. Pakulski, A. Skowronska, J. Gloede and H. Gross, *J. Org. Chem.*, **45**, 3122 (1980).
417. C. K. Tseng, *J. Org. Chem.*, **44**, 2793 (1979).
418. J. Coste, M. N. Dufour, A. Pantaloni and B. Castro, *Tetrahedron Lett.*, **31**, 669 (1990).
419. D. B. Denney and R. R. Dileone, *J. Am. Chem. Soc.*, **84**, 4737 (1962).
420. D. B. Denney and B. H. Garth, J. W. Hanifin, Jr, and H. M. Relles, *Phosphorus Sulfur*, **8**, 275 (1980).
421. J. S. Cohen, *Tetrahedron Lett.*, 3491 (1965).
422. J. S. Cohen, *J. Am. Chem. Soc.*, **89**, 2543 (1967).
423. H. H. Sisler, A. Sarkis, H. S. Ahuja, R. J. Drago and N. L. Smith, *J. Am. Chem. Soc.*, **81**, 2982 (1959).
424. H. H. Sisler, H. S. Ahuja and N. L. Smith, *J. Org. Chem.*, **26**, 1819 (1961).
425. W. A. Hart and H. H. Sisler, *Inorg. Chem.*, **3**, 617 (1964).
426. D. F. Clemens and H. H. Sisler, *Inorg. Chem.*, **4**, 1222 (1965).
427. S. R. Jain, L. K. Krannich, R. E. High Smith and H. H. Sisler, *Inorg. Chem.*, **6**, 1058 (1967).
428. H. H. Sisler and S. R. Jain, *Inorg. Chem.*, **7**, 104 (1968).
429. R. M. Kren and H. H. Sisler, *Inorg. Chem.*, **10**, 2630 (1971).
430. S. E. Frazier and H. H. Sisler, *Inorg. Chem.*, **11**, 1431 (1972).
431. D. F. Clemens, W. Woodford, E. Dellinger and Z. Tyndall, *Inorg. Chem.*, **8**, 998 (1969).
432. A. Hessing and G. Imsiecke, *Chem. Ber.*, **107**, 1536 (1974).
433. H. Ohmori, H. Maeda, K. Konomoto, K. Sakai, and M. Masui, *Chem. Pharm. Bull.*, **35**, 4473 (1987).
434. M. Ohmori, S. Nakai, H. Miyasaka and M. Masui, *Chem. Pharm. Bull.*, **30**, 4192 (1982).
435. J. P. Schaefer and D. S. Weingerg, *J. Org. Chem.*, **30**, 2635 (1965).
436. R. G. Weiss and E. I. Snyder, *J. Org. Chem.*, **35**, 1627 (1970).
437. R. Aneja, A. P. Davies and J. A. Knaggs, *Tetrahedron Lett.*, 67 (1974).
438. S. J. Cristol, R. M. Strom and D. P. Stull, *J. Org. Chem.*, **43**, 1150 (1978).
439. L. A. Jones, C. E. Sumer Jr, B. Franzus, T. T. S. Huang and E. I. Snyder, *J. Org. Chem.*, **43**, 2821 (1978).
440. L. Horner and H. Oediger, *Justus Liebigs Ann. Chem.*, **627**, 142 (1959).
441. H. Zimmer and G. Singh, *J. Org. Chem.*, **28**, 483 (1963).
442. H. J. Cristau and C. Garcia, *Synthesis*, 315 (1990).
443. K. Fukui and R. Sudo, *Bull. Chem. Soc. Jpn.*, **43**, 1160 (1970).
444. G. Singh and H. Zimmer, *J. Org. Chem.*, **30**, 417 (1965).
445. R. Appel and R. Schöllhorn, *Angew. Chem.*, **76**, 991 (1964).
446. H. J. Cristau, E. Manginot and E. Torreilles, *Tetrahedron Lett.*, **32**, 347 (1991).
447. E. Zbiral and L. Berner-Fenz, *Monatsh. Chem.*, **98**, 666 (1967).
448. M. Mc Guiness and H. Shechter, *Tetrahedron Lett.*, **31**, 4987 (1990).
449. H. Teichmann, M. Jatkowski and G. Hilgetag, *J. Prakt. Chem.*, **314**, 129 (1972).
450. R. Appel, R. Kleinstück, K. D. Ziehn and F. Knoll, *Chem. Ber.*, **103**, 3631 (1970).
451. J. M. Downie, H. Heaney and G. Kemp, *Tetrahedron Lett.*, **44**, 2619 (1988).
452. B. Castro and C. Selve, *Bull. Soc. Chim. Fr.*, 2296 (1971).

453. P. Simon, J. C. Ziegler and B. Gross, *Synthesis*, 951 (1979).
454. I. S. Zal'tsman, A. P. Markenko, A. A. Kudryavtsev and A. M. Pinchuk, *Zh. Obshch. Khim.*, **57**, 2272 (1987).
455. J. R. Dormoy and B. Castro, *Tetrahedron Lett.*, 3321 (1979).
456. J. Coste, D. Le-Nguyen and B. Castro, *Tetrahedron Lett.*, **31**, 205 (1990).
457. N. J. De'Ath, J. A. Miller and M. J. Nunn, *Tetrahedron Lett.*, 5191 (1973).
458. A. Kruse, G. Siegemund, A. Schumann and I. W. Ruppert, *Ger. Pat.*, 3 801 248; *Chem. Abstr.*, **112**, 77537u (1990).
459. H. Hoffmann and H. J. Diehr, *Angew. Chem.*, **76**, 944 (1964).
460. P. A. Chopard and R. F. Hudson, *J. Chem. Soc. B*, 1089 (1966).
461. I. J. Borowitz, K. C. Kirby Jr, and R. Virkhaus, *J. Org. Chem.*, **31**, 4031 (1966).
462. H. Hoffmann and H. J. Diehr, *Tetrahedron Lett.*, 583 (1962).
463. A. D. Sinitsa, V. S. Krishtal, V. I. Kal'chenko and L. N. Markovskii, *Zh. Obshch. Khim.*, **50**, 2413 (1980).
464. H. Kunz and P. Schmidt, *Z. Naturforsch., Teil B*, **33**, 1009 (1978).
465. H. Kunz and P. Schmidt, *Chem. Ber.*, **112**, 3866 (1979).
466. S. Ramos and W. Rosen, *Tetrahedron Lett.*, **22**, 35 (1981).
467. J. B. Hendrickson and M. S. Hussoin, *J. Org. Chem.*, **52**, 4139 (1987).
468. K. Dimroth and A. Nurrenbach, *Angew. Chem.*, **70**, 26 (1958).
469. S. F. Fry and N. J. Pienta, *J. Org. Chem.*, **49**, 4877 (1984).
470. K. S. Colle and E. S. Lewis, *J. Org. Chem.*, **43**, 571 (1978).
471. L. V. Nesterov and N. A. Aleksandrova, *Izv. Akad. Nauk SSSR, Ser. Khim.*, **348**, 415 (1971).
472. H. R. Hudson, R. G. Rees and J. E. Weekes, *Chem. Commun.*, 1297 (1971).
473. M. I. Kabachnik, V. A. Gilyarov and M. M. Yusupov, *Dokl. Akad. Nauk SSSR*, **164**, 812 (1965).
474. E. S. Lewis and K. S. Colle, *J. Org. Chem.*, **46**, 4369 (1981).
475. E. S. Lewis and D. Hamp, *J. Org. Chem.*, **48**, 2025 (1983).
476. R. F. Hudson and P. A. Chopard, *Helv. Chim. Acta*, **45**, 1137 (1962).
477. D. I. Phillips, I. Szele, and F. H. Westheimer, *J. Am. Chem. Soc.*, **98**, 184 (1976).
478. H. H. Sisler and N. L. Smith, *J. Org. Chem.*, **26**, 4733 (1961).
479. N. L. Smith and H. H. Sisler, *J. Org. Chem.*, **28**, 272 (1963).
480. J. Von Ellermann, M. Lietz and K. Geibel, *Z. Anorg. Allg. Chem.*, **492**, 122 (1982).
481. B. E. Ivanov, S. E. Krokkina, T. V. Chichkanova and E. M. Kosacheva, *Izv. Akad. Nauk. SSSR, Ser. Khim.*, 173 (1985).
482. L. I. Mizrakh, L. Y. Polonskaya and T. M. Ivanova, *Zh. Obshch. Khim.*, **49**, 2394 (1979).
483. H. J. Cristau, B. Chabaud, A. Chêne and H. Christol, *Phosphorus Sulfur*, **11**, 55 (1981).
484. J. C. Wilburn and R. H. Neilson, *Inorg. Chem.*, **18**, 347 (1979).
485. R. H. Neilson and D. W. Goebel, *J. Chem. Soc., Chem. Commun.*, 769 (1979).
486. R. H. Neilson and P. Wisian-Neilson, *Inorg. Chem.*, **19**, 1875 (1980).
487. W. D. Morton and R. H. Neilson, *Organometallics*, **I**, 623 (1982).
488. W. Ried and H. Appel, *Justus Liebigs Ann. Chem.*, **679**, 51 (1964).
489. L. D. Quin, in *1,4-Cycloaddition Reactions* (Ed. J. Hamer), Academic Press, New York, 1967, pp. 47–96.
490. L. D. Quin, *The Heterocyclic Chemistry of Phosphorus*, Wiley–Interscience, New York, 1981, Chapter 2.
491. H. Teichmann, M. Jatkowski and Hilgetag, *Angew. Chem., Int. Ed. Engl.*, **6**, 372 (1967).
492. L. V. Nesterev and R. I. Mutalopova, *Tetrahedron Lett.*, 51 (1968).
493. J. H. Finley, D. Z. Denney and D. B. Denney, *J. Am. Chem. Soc.*, **91**, 5826 (1969).
494. T. S. Mikhailova, N. S. Skvorstsov, V. M. Ignat'ev, B. I. Ionin and A. A. Petrov, *Dokl. Akad. Nauk SSSR*, **241**, 1095 (1978).
495. C. M. Angelov and D. D. Enchev, *Phosphorus Sulfur*, **37**, 125 (1988).
496. H. Binder and E. Fluck, *Z. Anorg. Allg. Chem.*, **365**, 170 (1969).
497. S. Poignant, J. R. Gauvreau and G. J. Martin, *Can. J. Chem.*, **58**, 946 (1980).
498. H. Schmidbaur and R. Seeber, *Chem. Ber.*, **107**, 1731 (1974).
499. H. Teichmann, C. Auerswald and G. Engelhardt, *J. Prakt. Chem.*, **321**, 835 (1979).
500. J. B. Hendrickson and S. M. Schwartzman, *Tetrahedron Lett.*, 277 (1975).
501. A. Aaberg, T. Gramstad and S. Hesebye, *Tetrahedron Lett.*, 2263 (1979).
502. T. Mukaiyama and S. Suda, *Chem. Lett.*, 1143 (1990).

503. J. B. Hendrickson and M. S. Hussoin, *Synlett*, 423 (1990).
504. J. B. Hendrickson and M. S. Hussoin, *J. Org. Chem.*, **54**, 1144 (1989); Synthesis, 217 (1989).
505. L. Horner and H. Winkler, *Tetrahedron Lett.*, 175 (1964).
506. D. H. R. Barton, D. P. Manly and D. A. Widdowson, *J. Chem. Soc., Perkin Trans. 1*, 1568 (1975).
507. C. Glidewell and E. J. Leslie, *J. Chem. Soc., Dalton Trans.*, 527 (1977).
508. J. Omelanczuk and M. Mikolajczyk, *J. Am. Chem. Soc.*, **101**, 7292 (1979).
509. L. Horner and M. Jordan, *Phosphorus Sulfur*, **8**, 209 (1980).
510. L. Horner and M. Jordan, *Phosphorus Sulfur*, **8**, 215 (1980).
511. J. Omelanczuk and M. Mikolajczyk, *Tetrahedron Lett.*, 2493 (1984).
512. N. Farschtschi, D. G. Gorenstein, T. Fanni and K. Taira, *Phosphorus Sulfur Silicon*, **47**, 93 (1990).
513. M. Wada, M. Kanzaki, M. Fujiwara, K. Kajihara and T. Erabi, *Bull. Chem. Soc. Jpn.*, **64**, 1782 (1991).
514. J. Omelanczuk and M. Mikolajczyk, *Phosphours Sulfur*, **15**, 321 (1983).
515. A. Schmidpeter and C. Weingand, *Angew. Chem.*, **80**, 234 (1968).
516. I. A. Nuretdinov, N. A. Buina and N. P. Grechkin, *Izv. Akad. Nauk SSSR, Ser. Khim.*, 169 (1969).
517. N. Kuhn and H. Schumann, *Phosphorus Sulfur*, **26**, 199 (1986).
518. H. Zimmer, M. Jayawant and P. Gutsch, *J. Org. Chem.*, **35**, 2826 (1970).
519. H. Zimmer and G. Singh, *J. Org. Chem.*, **29**, 1579 (1964).
520. H. J. Cristau, C. Garcia, J. Kadoura and E. Torreilles, *Phosphorus Sulfur Silicon*, **49–50**, 151 (1990).
521. H. J. Cristau, J. Kadoura, L. Chiche and E. Torreilles, *Bull. Soc. Chim. Fr.*, 515 (1989).
522. H. J. Cristau, L. Chiche, J. Kadoura and E. Torreilles, *Tetrahedron Lett.*, **29**, 3931 (1988).
523. H. J. Cristau, E. Manginot and E. Torreilles, *Synthesis*, 382 (1991).
524. M. Takeishi and N. Shiozawa, *Bull. Chem. Soc. Jpn.*, **62**, 4063 (1989).
525. K. Sasse, in *Organischen Phosphorverbindungen* (*Houben-Weyl Methoden der Organischen Chemic*) (Ed. E. Müller) Band XII/I Georg Thieme, Stuttgart, 1961, p. 104.
526. P. Beck, in *Organic Phosphorus Compounds* (Eds G. M. Kosolapoff and L. Maier), Vol. **2**, Wiley–Interscience, New York, 1972, pp. 200–202.
527. H. U. Kibbel and U. Liebelt, *Sulfur Lett.*, **12**, 229 (1991).
528. J. H. Clark and D. J. Macquarrie, *J. Chem. Soc., Chem. Commun.*, 229 (1988).
529. H. P. S. Chauhan, G. Srivastava and R. C. Mehrotra, *Synth. React. Inorg. Met. Org. Chem.*, **12**, 593 (1982).
530. G. A. Doorakian and W. S. Smith, *US Pat.*, 4 340 761 (1982); *Chem. Abstr.*, **98**, 34762p (1983).
531. M. De Giorgi, D. Landini, A. Maia and M. Penso, *Synth. Commun.*, **17**, 521 (1987).
532. A. W. Frank, *Phosphorus Sulfur*, **5**, 19 (1978).
533. A. W. Frank, *Phosphorus Sulfur*, **10**, 207 (1981).
534. R. Bartsch, O. Stelzer and R. Schmutzler, *Z. Naturforsch, Teil B*, **37**, 267 (1982).
535. A. Martinsen and J. Songstad, *Acta Chem. Scand., Ser. A*, **31**, 645 (1977).
536. K. B. Dillon, M. Hodgson and D. Parker, *Synth. Commun.*, **15**, 849 (1985).
537. Y. Chapleur, B. Castro and B. Gross, *Synthesis*, 447 (1983).
538. H. J. Cristau, E. Torreilles, P. Morand and H. Christol, *Tetrahedron Lett.*, **27**, 1775 (1986).
539. S. J. Brown, J. H. Clark and D. J. Macquarrie, *J. Chem. Soc., Dalton Trans.*, 277 (1988).
540. A. W. Frank, D. J. Daigle and S. L. Vail, *Text. Res. J.*, **52**, 678 (1982).
541. S. Ota, S. Shimura, K. Takahashi and T. Sugiya, *Jpn. Pat.*, 02 091 089 (1990); *Chem. Abstr.*, **113**, 172346k (1990).
542. H. Koyama and S. Yokota, *Jpn. Pat.*, 63 190 893 (1988); *Chem. Abstr.*, **109**, 231301j (1988).
543. I. Lovel, Y. Goldberg, M. Shymanska and E. Lukevics, *Appl. Organomet. Chem.*, **1**, 371 (1987).
544. W. T. Ford and M. Tomoi, *Adv. Polym. Sci.*, **55**, 41 (1984).
545. M. Tomoi, N. Kori and H. Kakiuchi, *Makromol. Chem.*, **187**, 2753 (1986).
546. H. J. Cristau, E. Torreilles, P. Morand and H. Christol, *Phosphorus Sulfur*, **25**, 357 (1985).
547. U. Müller, *Z. Naturforsch Teil B.*, **34**, 1064 (1979).
548. B. N. Kozhushko, E. B. Silina, A. V. Gumenyuk, A. V. Turov and V. A. Shokol, *Zh. Obshch. Khim.*, **50**, 2210 (1980).
549. A. T. Mohammed and U. Müller, *Z. Naturforsch, Teil B*, **40**, 562 (1985).
550. V. D. Nefedov, M. A. Toropova, N. E. Shchepina, V. V. Avrorin, V. V. Shchepin and V. E. Zhuravlev, *Radiokhimiya*, **29**, 237 (1987).

551. I. Buscaglioni, C. M. Stables and H. Sutcliffe, *Inorg. Chim. Acta*, **146**, 33 (1988).
552. N. Petragnan, L. T. Castellanos, J. J. Wynne and W. Maxwell, *J. Organomet. Chem.*, **55**, 295 (1973).
553. M. Witt, H. W. Roesky, M. Noltemeyer and A. Schmidpeter, *New J. Chem.*, **13**, 403 (1989).
554. B. Krebs and K. Büscher, *Z. Anorg. Allg. Chem.*, **463**, 56 (1980).
555. M. Lequan, R. M. Lequan, G. Jaouen and P. Delhaes, *Tetrahedron Lett.*, **25**, 4121 (1984).
556. M. Lequan, R. M. Lequan, P. Batail, J. F. Halet and L. Ouahab, *Tetrahedron Lett.*, **24**, 3107 (1983).
557. H. J. Bestmann, *Chem. Ber.*, **95**, 58 (1962).
558. K. Issleib and R. Lindner, *Justus Liebigs Ann. Chem.*, **707**, 112 (1967).
559. H. J. Bestmann and R. Zimmermann, in *Organic Phosphorus Compounds* (Eds G. M. Kosolapoff and L. Maier), Vol. **5**, Wiley-Interscience, New York, 1972, p. 3.
560. H. J. Bestmann and R. Zimmermann, in *Organische Phosphor-Verbindungen* (*Houben-Weyl, E_2*), (Ed. M. Regitz), Georg Thieme, Stuttgart, 1982, (a) p. 625; (b) p. 630; (c) p. 635.
561. I. Gosnly and A. G. Rowley, in *Organophosphorus Reagents in Organic Synthesis* (Ed. J. I. G. Cadogan), Academic Press, London, 1979, p. 20.
562. T. A. Mastryukova, I. M. Alajeva, H. A. Suerbayev, Ye. I. Matrosov and P. V. Petrovsky, *Phosphorus*, **1**, 159 (1971).
563. G. Aksnes and J. Songstad, *Acta Chem. Scand.*, **18**, 655 (1964).
564. V. A. Chauzov, S. V. Agafonov and N. Yu. Lebedeva, *Zh. Obshch. Khim*, **53**, 364 (1983).
565. S. Alunni, *J. Chem. Res* (*S*), 231 (1986).
566. G. Aksnes and H. Haugen, *Phosphorus*, **2**, 155 (1972).
567. A. J. Speziale and K. W. Ratts, *J. Am. Chem. Soc.*, **85**, 2790 (1963).
568. S. Fliszar, R. F. Hudson and G. Salvador, *Helv. Chim. Acta*, **46**, 1580 (1963).
569. R. Luckenbach, in *Organische Phosphor-Verbindungen* (*Houben-Weyl, E2*) (Ed. M. Regitz), Band I, Georg Thieme, Stuttgart, 1982, (a) p. 889; (b) p. 903.
570. D. Seyferth, W. B. Hughes and J. K. Heeren, *J. Am. Chem. Soc.*, **87**, 2847 (1965).
571. D. Seyferth, W. B. Hughes and J. K. Heeren, *J. Am. Chem. Soc.*, **87**, 3467 (1965).
572. G. M. Pilling and F. Sondheimer, *J. Am. Chem. Soc.*, **93**, 1970 (1971).
573. M. Miyano and M. A. Stealey, *J. Org. Chem.*, **40**, 2840 (1975).
574. H. J. Cristau, J. Coste, A. Truchon and H. Christol, *J. Organomet. Chem.*, **241**, C1 (1983).
575. (a) A. Cahours and A. W. Hoffmann, *Justus Liebigs Ann. Chem.*, **104**, 32 (1857).
(b) A. Michaelis and H. V. Soden, *Justus Liebigs Ann. Chem.*, **229**, 316 (1885).
576. H. R. Hays and D. J. Peterson, in *Organic Phosphorus Compounds* (Eds G. M. Kosolapoff and L. Maier), Vol. 3, Wiley–Interscience, New York, 1972, pp. 349–354.
577. R. Luckenbach, *Dynamic Stereochemistry of Pentacoordinated Phosphorus and Related Elements*, Georg Thieme, Stuttgart, 1973; (a) p. 109–124; (b) 141–179.
578. R. R. Holmes, *Pentacoordinated Phosphorus* (ACS Monograph, No. 176), Vol. **II**, American Chemical Society, Washington, DC, 1980, pp. 118–163.
579. H. Heydt and M. Regitz, in *Organische Phosphor-Verbindungen* (*Houben-Weyl, E2*) (Ed. M. Regitz), Georg Thieme Stuttgart, 1982, pp. 51–62.
580. M. A. Battiste and C. T. Sprouse, *Tetrahedron Lett.*, 3164 (1969).
581. P. A. Chopard, *Chimia*, **20**, 172 (1966).
582. D. G. Gilheany, N. T. Thompson and B. J. Walker, *Tetrahedron Lett.*, **28**, 3843 (1987).
583. M. Zanger, C. A. VanderWerf and W. E. McEwen, *J. Am. Chem. Soc.*, **81**, 3806 (1959).
584. W. E. McEwen, K. F. Kumli, A. Blade-Font, M. Zanger and C. A. VanderWerf, *J. Am. Chem. Soc.*, **86**, 2378 (1964).
585. G. W. Fenton and C. K. Ingold, *J. Chem. Soc.*, 2342 (1929).
586. W. E. McEwen, G. Axelrad, M. Zanger and C. A. VanDerWerf, *J. Am. Chem. Soc.*, **87**, 3948 (1965).
587. H. Hoffmann, *Justus Liebigs Ann. Chem.*, **634**, 1 (1960).
588. J. Meisenheimer, J. Casper, M. Höring, W. Lauter, L. Lichtenstadt and W. Samuel, *Justus Liebigs Ann. Chem.*, **449**, 213 (1926).
589. D. Valentine, J. John, F. Blount and K. Toth, *J. Org. Chem.*, **45**, 3691 (1980).
590. J. Bourson, T. G. Guillon and S. Jugé, *Phosphorus Sulfur*, **14**, 347 (1983).
591. J. R. Corfield, N. J. De'Ath and S. Trippett, *J. Chem. Soc. C*, 1930 (1971).
592. H. R. Hays and R. G. Laughlin, *J. Org. Chem.*, **32**, 1060 (1962).

593. K. L. Marsi and J. E. Oberlander, *J. Am. Chem. Soc.*, **95**, 200 (1973).
594. L. Horner, H. Hoffmann, H. G. Wippel and G. Hassel, *Chem. Ber.*, **91**, 52 (1958).
595. G. J. Reilly and W. E. McEwen, *Tetrahedron Lett.*, 1231 (1968).
596. A. W. Smalley, C. E. Sullivan and W. E. McEwen, *Chem. Commun.*, 5 (1967).
597. D. W. Allen, *J. Chem. Soc. B*, 1490 (1970).
598. D. W. Allen, S. J. Grayson, I. Harness, B. G. Hutley and I. W. Mowat, *J. Chem. Soc., Perkin Trans. 2*, 1912 (1973).
599. D. W. Allen, B. G. Hutley and M. T. J. Mellor, *J. Chem. Soc., Perkin Trans. 2*, 63 (1972).
600. Y. Uchida, K. Onoue, N. Tada, F. Nagao, H. Kozawa and S. Oae, *Heteroatom Chem.*, **1**, 295 (1990).
601. D. W. Allen and B. G. Hutley, *J. Chem. Soc., Perkin Trans. 2*, 67 (1972).
602. O. M. Bukachuk, M. G. Nikula, V. N. Belenkov and M. I. Shevchuk, *Zh. Obshch. Khim.*, **56**, 343 (1986).
603. K. E. DeBruin and J. R. Petersen, *J. Org. Chem.*, **37**, 2272 (1972).
604. N. J. De'ath, K. Ellis, D. J. H. Smith and S. Trippett, *Chem. Commun.*, 714 (1971).
605. K. E. DeBruin and D. M. Johnson, *J. Am. Chem. Soc.*, **95**, 4675 (1973).
606. S. M. Cairns and W. E. Mc Ewen, *J. Org. Chem.*, **52**, 4829 (1987).
607. G. L. Keldsen and W. E. Mc Ewen, *J. Am. Chem. Soc.*, **100**, 7312 (1978).
608. G. Aksnes and K. Bergesen, *Acta Chem. Scand.*, **19**, 931 (1965).
609. F. Y. Khalil and G. Aksnes, *Acta Chem. Scand.*, **27**, 3832 (1973).
610. J. G. Dawber, J. C. Tebby and A. A. C. Waite, *J. Chem. Soc. Perkin Trans. 2*, 1923 (1983).
611. J. G. Dawber, J. C. Tebby and A. A. C. Waite, *Phosphorus Sulfur*, **19**, 99 (1984).
612. F. Y. Khalil, M. T. Hanna, F. M. Abdel-Halim and M. El Batouti, *Rev. Roum. Chim.*, **30**, 571 (1985).
613. M. T. Hanna, F. Y. Khalil and S. M. Beder, *Bull. Soc. Chim. Belge*, **96**, 27 (1987).
614. N. S. Isaacs and O. H. Abed, *Tetrahedron Lett.*, **27**, 1209 (1986).
615. G. Aksnes and J. Songstad, *Acta Chem. Scand.*, **16**, 1426 (1962).
616. K. Bergesen, *Acta Chem. Scand.*, **20**, 899 (1966).
617. G. Aksnes and L. J. Brudvick, *Acta Chem. Scand.*, **17**, 1616 (1963).
618. G. Aksness and A. I. Eide, *Phosphorus*, **4**, 209 (1974).
619. D. W. Allen and I. T. Millar, *J. Chem. Soc. B*, 263 (1969).
620. S. E. Cremer, B. C. Trivedi and F. L. Weitl, *J. Org. Chem.*, **36**, 3226 (1971).
621. W. E. Mc Even, A. W. Smalley and C. E. Sullivan, *Phosphorus*, **1**, 259 (1972).
622. A. Schnell and J. C. Tebby, *J. Chem. Soc., Chem. Commun.*, 134 (1975).
623. F. Y. Khalil, F. M. Abdel-Halim, M. T. Hanna and M. El Batouti, *Bull. Soc. Chim. Belge*, **94**, 6 (1985).
624. T. Minami, T. Chikugo and T. Hanamoto, *J. Org. Chem.*, **51**, 2210 (1986).
625. N. J. De'ath and S. Trippett, *Chem. Commun*, 172 (1969).
626. R. S. Berry, *J. Chem. Phys.*, **32**, 933 (1960).
627. F. Ramirez and I. Ugi, *Adv. Phys. Org. Chem.*, **9**, 25 (1971).
628. K. Mislow, *Acc. Chem. Res.*, **3**, 321 (1970).
629. K. E. DeBruin, K. Nauman, G. Zon and K. Mislow, *J. Am. Chem. Soc.*, **91**, 7031 (1969).
630. K. E. DeBruin and K. Mislow, *J. Am. Chem. Soc.*, **91**, 7393 (1969).
631. L. Horner, H. Winkler, A. Rapp, A. Mentrup, H. Hoffmann and P. Beck, *Tetrahedron Lett.*, 161 (1961).
632. K. F. Kumli, W. E. McEwen and C. A. VanderWerf., *J. Am. Chem. Soc.*, **81**, 3805 (1959).
633. A. Blade-Font, C. A. VanderWerf. and W. E. McEwen, *J. Am. Chem. Soc.*, **82**, 2396 (1960).
634. W. E. McEwen, *Top. Phosphorus Chem.*, **2**, 1 (1965).
635. R. Luckenbach, *Phosphorus*, **1**, 223 (1972).
636. R. Luckenbach, *Phosphorus*, **1**, 229 (1972).
637. R. Luckenbach, *Phosphorus*, **1**, 293 (1972).
638. L. Horner, and R. Luckenbach, *Phosphorus*, **1**, 73 (1971).
639. R. A. Lewis, K. Naumann, K. E. DeBruin and K. Mislow, *Chem. Commun.*, 1010 (1969).
640. Mazhar-UI-Hague and C. N. Caughlan, *Chem. Commun.*, 1228 (1968).
641. Mazhar-UI-Hague, *J. Chem. Soc. B*, 934, 938 (1970).
642. D. D. Swank and C. N. Caughlan, *Chem. Commun.*, 1051 (1968).
643. E. Alver and B. H. Holtedahl, *Acta Chem Scand.*, **21**, 359 (1967).

644. S. E. Cremer, R. J. Chorvat and B. C. Trevedi, *Chem. Commun.*, 769 (1969).
645. K. E. DeBruin, G. Zon, K. Naumann and K. Mislow, *J. Am. Chem. Soc.*, **91**, 7027 (1969).
646. K. L. Marsi, *J. Am. Chem. Soc.*, **91**, 4724 (1969).
647. K. L. Marsi, F. B. Burns and R. T. Clark, *J. Org. Chem.*, **37**, 238 (1972).
648. K. L. Marsi, *Chem. Commun.*, 846 (1968).
649. W. Egan, G. Chauvière, K. Mislow, R. T. Clark and K. L. Marsi, *Chem. Commun.*, 733 (1970).
650. K. L. Marsi, *J. Org. Chem.*, **40**, 1779 (1975).
651. K. L. Marsi, and R. T. Clark, *J. Am. Chem. Soc.*, **92**, 3791 (1970).
652. K. L. Marsi, *J. Am. Chem. Soc.*, **93**, 6341 (1971).
653. B. Deslongchamps, *Stereoelectronic effects in Organic Chemistry*, Pergamon Press, Oxford 1983.
654. A. J. Kirby, *The Anomeric Effect and Related Stereoelectronic Effects at Oxygen*, Springer, Berlin, 1983.
655. D. G. Gorenstein, *Chem. Rev.*, **87**, 1047 (1987).
656. L. Horner, H. Hoffmann, W. Klink, H. Ertel and V. G. Toscano, *Chem. Ber.*, **95**, 581 (1962).
657. J. J. Brophy and M. J. Gallagher, *Aust. J. Chem.*, **22**, 1385 (1969).
658. J. J. Brophy, K. L. Freeman and M. J. Gallagher, *J. Chem. Soc. C*, 2760 (1968).
659. A. M. Aguiar, H. Aguiar and D. Daigle, *J. Am. Chem. Soc.*, **87**, 671 (1965).
660. A. M. Aguiar, K. C. Hansen and G. S. Reddy, *J. Am. Chem. Soc.*, **89**, 3067 (1967).
661. J. J. Brophy and M. J. Gallagher, *Chem. Commun.*, 344 (1967).
662. M. J. Gallagher and I. D. Jenkins, *Top. Stereochem.*, **3**, 1 (1968).
663. G. Aksnes and K. Bergesen, *Acta Chem. Scand.*, **20**, 2508 (1966).
664. G. E. Driver and M. J. Gallagher, *Chem. Commun.*, 150 (1970).
665. A. M. Aguiar and M. G. R. Nair, *J. Org. Chem.*, **33**, 579 (1968).
666. M. Davis and F. G. Mann, *Chem. Ind.* (*London*), 1539 (1962).
667. M. Davis and F. G. Mann, *J. Chem. Soc.*, 3770 (1964).
668. D. W. Allen, I. T. Millar and F. G. Mann, *J. Chem. Soc. C*, 1869 (1967).
669. H. Hellmann and J. Bader, *Tetrahedron Lett.*, 724 (1961).
670. M. Schlosser, *Angew. Chem.*, **74**, 291 (1962).
671. D. W. Allen and I. T. Millar, *J. Chem. Soc. C*, 252 (1969).
672. D. W. Allen and I. T. Millar, *Chem. Ind.* (*London*), 2178 (1967).
673. J. R. Corfield, M. J. P. Harger, J. R. Shutt and S. Trippett, *J. Chem. Soc. C*, 1855 (1970).
674. H. Abou-El-Seoud Aly, D. J. H. Smith and S. Trippett, *Phosphorus*, **4**, 205 (1974).
675. E. M. Richards and J. C. Tebby, *J. Chem. Soc. C*, 1059 (1971).
676. D. W. Allen, P. Heathey, B. G. Hutley and M. T. Mellor, *Tetrahedron Lett.*, 1787 (1974).
677. S. Trippett and B. J. Walker, *J. Chem. Soc. C*, 887 (1966).
678. E. M. Richards and J. C. Tebby, *J. Chem. Soc. D*, 494 (1969).
679. H. Christol, H. J. Cristau and M. Soleiman, *Bull. Soc. Chim. Fr.*, 161 (1976).
680. H. J. Cristau, F. Plénat and F. Guida-Pietrasanta, *Phosphorus Sulfur*, **34**, 75 (1987).
681. E. Zbiral and E. Werner, *Justus Liebigs Ann. Chem.*, **707**, 130 (1967).
682. E. M. Richards and J. C. Tebby, *J. Chem. Soc. C*, 1064 (1971).
683. J. R. Shutt and S. Trippett, *J. Chem. Soc. C*, 2038 (1969).
684. (a) D. A. Jaeger and D. Bokilal, *J. Org. Chem.*, **51**, 1350 (1986); (b) *J. Org. Chem.*, **51**, 1352 (1986).
685. E. E. Schweizer and A. T. Wehman, *J. Chem. Soc. C*, 1901 (1970).
686. M. P. Savage and S. Trippett, *J. Chem. Soc. C*, 591 (1968).
687. E. E. Schweizer, T. Minami and D. M. Crouse, *J. Org. Chem.*, **36**, 4029 (1971).
688. H. J. Bestmann, R. Engler and H. Hartung, *Angew. Chem. Int. Ed. Engl.*, **5**, 1040 (1966).
689. D. J. Burton, S. Shin-Ya and R. D. Howells, *J. Fluorine Chem.*, **15**, 543 (1980).
690. H. Bredereck, G. Simchen and W. Griebenow, *Chem. Ber.*, **106**, 3722 (1973).
691. D. W. Allen and J. C. Tebby, *Tetrahedron*, **23**, 2795 (1967).
692. M. M. Rauhut, I. Hechenblekner, H. A. Currier, F. C. Schaeffer and V. P. Wystrach, *J. Am. Chem. Soc.*, **81**, 1103 (1959).
693. S. Trippett, *J. Chem. Soc.*, 2813 (1961).
694. A. Hoffmann, *J. Am. Chem. Soc.*, **43**, 1684 (1921).
695. A. Hoffmann, *J. Am. Chem. Soc.*, **52**, 2995 (1930).
696. W. J. Vullo, *J. Org. Chem.*, **33**, 3665 (1968).
697. M. Grayson, *J. Am. Chem. Soc.*, **85**, 79 (1963).
698. G. Wittig, H. Braun and H. J. Cristau, *Justus Liebigs Ann. Chem.*, **751**, 17 (1971).

699. H. J. Bestmann, H. Häberlein and I. Pilz, *Tetrahedron*, **20**, 2079 (1964).
700. M. Grayson, P. T. Keough and G. A. Johnson, *J. Am. Chem. Soc.*, **81**, 4803 (1959).
701. G. P. Schiemenz, *Chem. Ber.*, **99**, 514 (1966).
702. S. Trippett, *Proc. Chem. Soc.*, **19** (1963).
703. C. E. Griffin and G. Witschard, *J. Org. Chem.*, **29**, 1001 (1964).
704. H. J. Bestmann, F. Seng and H. Schulz, *Chem. Ber.*, **96**, 465 (1963).
705. R. P. Welcher and N. E. Day, *J. Org. Chem.*, **27**, 1824 (1962).
706. H. J. Bestmann, *Angew. Chem. Int. Ed. Engl.*, **4**, 645 (1965).
707. G. W. Fenton and C. K. Ingold, *J. Chem. Soc.*, 2342 (1929).
708. H. Hoffmann, *Chem. Ber.*, **94**, 1331 (1961).
709. H. J. Bestmann and W. Both, *Chem. Ber.*, **107**, 2926 (1974).
710. J. R. Corfield and S. Trippett, *Chem. Commun*, 1267 (1970).
711. P. T. Keough and M. Grayson, *J. Org. Chem.*, **29**, 631 (1964).
712. E. E. Schweizer and R. D. Bach, *J. Org. Chem.*, **29**, 1746 (1964).
713. J. J. Brophy and M. J. Gallahger, *Aust. J. Chem.*, **22**, 1405 (1969).
714. H. Christol, H. J. Cristau, J. P. Joubert and M. Soleiman, *C. R. Acad. Sci. Ser. C*, **279**, 167 (1974).
715. H. Kunz, *Justus Liebigs Ann. Chem.*, 2001 (1973).
716. H. Kunz, *Phosphorus*, **3**, 273 (1974).
717. H. Kunz, *Chem. Ber.*, **109**, 2670 (1976).
718. H. Christol, H. J. Cristau and M. Soleiman, *Tetrahedron Lett.*, 1381 (1975).
719. H. Christol, H. J. Cristau and M. Soleiman, *Tetrahedron Lett.*, 1385 (1975).
720. H. J. Cristau, H. Christol and M. Soleiman, *Phosphorus Sulfur*, **4**, 287 (1978).
721. V. S. Brovarets and B. S. Drach, *Zh. Obshch. Khim.*, **56**, 321 (1986).
722. V. N. Listvan and A. P. Stasyuk, *Zh. Org. Khim.*, **21**, 392 (1985).
723. E. Zbiral and E. Öhler, *Chem. Ber.*, **113**, 2852 (1980).
724. V. V. Kurg, V. S. Brovarets and B. S. Drach, *Dokl. Akad. Nauk Ukr. SSR, Ser. B*, 53 (1990); *Chem. Abstr.*, **114**, 238519 (1911).
725. A. G. Abatoglou and L. A. Kapicak (Union Carbide Corp.), *Eur. Pat. Appl.*, EP 72 560 (1983); *Chem. Abstr.*, **98**, 198 452 p. (1983).
726. M. Halpern, D. Feldman, Y. Sasson and M. Rabinovitz, *Angew. Chem., Int. Ed. Engl.*, **23**, 54 (1984).
627. H. J. Cristau, A. Long and H. Christol, *Tetrahedron Lett.*, 349 (1979).
628. D. Landini, A. Maia and G. Podda, *J. Org. Chem.*, **47**, 2264 (1982).
729. F. Montanari and P. Tundo, *J. Org. Chem.*, **47**, 1298 (1982).
730. D. Landini, A. Maia and A. Rampoldi, *J. Org. Chem.*, **51**, 3187 (1986).
731. W. J. Bailey and S. A. Buckler, *J. Am. Chem. Soc.*, **79**, 3567 (1957).
732. S. T. D. Gough and S. Trippett, *J. Chem. Soc.*, 4263 (1961).
733. L. Maier, in *Organic Phosphorus Compounds* (Eds G. M. Kosolapoff and L. Maier), Vol. **1**, Wiley–Interscience, New York, 1972, p. 54.
734. D. Hellwinkel, *Angew. Chem.* **78**, 985 (1966).
735. P. D. Henson, K. Nauman and K. Mislow, *J. Am. Chem. Soc.*, **91**, 5645 (1969).
736. L. Horner and M. Ernst, *Chem. Ber.*, **103**, 318 (1970).
737. V. D. Makhaev and A. P. Borisov, *Zh. Obshch. Khim.*, **54**, 2550 (1984).
738. E. E. Schweizer, J. G. Thompson and T. A. Ulrich, *J. Org. Chem.*, **33**, 3082 (1968).
739. L. Maier, *Prog. Inorg. Chem.*, **5**, 27 (1963).
740. V. D. Makhaev and A. P. Borisov, *Russ. Pat.*, 1 046 248; *Chem. Abstr.*, **100**, 68529 (1984).
741. W. Eilenberg, *Eur. Pat. Appl.*, EP 364 046 (1990); *Chem. Abstr.*, **113**, 132507g (1990).
742. R. C. Hinton and F. G. Mann, *J. Chem. Soc.*, 2835 (1959).
743. J. J. Brophy and M. J. Gallagher, *Aust. J. Chem.*, **22**, 1399 (1969).
744. E. Lindner and C. Scheytt, *Z. Naturforsch., Teil B*, **41**, 10 (1986).
745. E. Anders and F. Markus, *Chem. Ber.*, **122**, 119 (1989).
746. A. R. Hands and A. J. H. Mercer, *J. Chem. Soc. C*, (a) 1099 (1967); (b) 1331 (1968).
747. L. Horner, *Pure Appl. Chem.*, **52**, 843 (1980).
748. L. Z. Avila and J. W. Frost, *J. Am. Chem. Soc.*, **111**, 8969 (1989).
749. L. Horner and H. Holfmann, in *Neuere Methoden der Präparativen Organische Chemie*, Vol. **II**, Verlag Chemie, Weinheim, 1960, p. 132.
750. L. Horner and K. Dickerhof, *Phosphorus Sulfur*, **15**, 213 (1983).

751. L. Horner, P. Beck and R. Luckenbach, *Chem. Ber.*, **101**, 2899 (1968).
752. H. J. Bestmann and B. Arnason, *Chem. Ber.*, **95**, 1513 (1962).
753. T. Mukaiyama, R. Yoda and I. Kuwajima, *Tetrahedron Lett.*, 23 (1969).
754. L. Horner and H. Lund in *Organic Electrochemistry* (Ed. M. M. Baizer), Marcel Dekker, New York, 1973, p. 735.
755. P. Walach, D. H. Skaletz and L. Horner, *Phosphorus*, **3**, 183 (1973).
756. S. Samaan, *Phosphorus Sulfur*, **7**, 89 (1979).
757. L. Horner and J. Roder, *Phosphorus Sulfur*, **6**, 147 (1976).
758. L. Horner and M. Jordan, *Phosphorus Sulfur*, **6**, 491 (1979); **8**, 209 and 225 (1980).
759. M. Finkelstein, *J. Org. Chem.*, **27**, 4076 (1962).
760. J. P. Hutley and A. Webber, *J. Chem. Soc., Perkin Trans. 1*, 1154 (1980).
761. D. W Allen and L. Ebdon, *Phosphorus Sulfur*, **7**, 161 (1979).
762. J. M. Saveant and S. K. Bingh, *J. Org. Chem.*, **42**, 1242 (1977).
763. E. A. Berdnikov, T. R. Tantasheva, V. I. Morosov, A. V. Il'yasov and A. A. Vafina, *Bull. Acad. Sci. USSR, Chem. Sci.*, **26**, 731 (1977).
764. C. K. White and R. D. Rieke, *J. Org. Chem.*, **43**, 4638 (1978).
765. I. M. Downie, J. B. Lee and M. F. S. Matough, *J. Chem. Soc., Chem. Commun.*, 1350 (1968).
766. B. Castro and C. Selve, *Bull. Soc., Chim. Fr.*, 4368 (1971).
767. B. Castro, Y. Chapleur and B. Gross, *Bull. Soc. Chim. Fr.*, 3034 (1973).
768. A. E. Arbuzov and L. V. Nesterov, *Dokl. Akad. Nauk. SSSR*, **57**, **92** (1953).
769. R. Appel, K. Warning and K. D. Ziehn, *Justus Liebigs Ann. Chem.*, 406 (1975).
770. S. R. Landauer and H. N. Rydon, *J. Chem. Soc.*, 2224 (1953).
771. A. N. Patel, *Zh. Org. Khim.*, **16**, 2621 (1980).
772. L. V. Nesterov and A. Ya. Kessel, *Zh. Obshch. Khim.*, **37**, 728 (1967).
773. H. Hutchins, M. Hutchins and C. Milewski, *J. Org. Chem.*, **37**, 4190 (1972).
774. C. Spangler and T. Hartford, *Synthesis*, 108 (1976).
775. K. Yamada, G. Goto, H. Nagase, Y. Kyotani and H. Hirata, *J. Org. Chem.*, **43**, 2076 (1978).
776. C. Spangler, D. P. Kjell, L. L. Wellander and M. A. Kinsella, *J. Chem, Soc., Perkin Trans. 1*, 2287 (1981).
777. L. V. Nesterov, A. Ya. Kessel and L. I. Maklakov, *Zh. Obshch. Khim.*, **38**, 318 (1968).
778. L. V. Nesterov, A. Ya. Kessel and R. I. Mutalapova, *Zh. Obshch. Khim.*, **39**, 2453 (1969).
779. T. Mukayama, S. Matsui and K. Kashiwagi, *Chem. Lett.*, 993 (1989).
780. A. P. Marchenko, V. A. Kovenya and A. M. Pinchuk, *Zh. Obshch. Khim.*, **48**, 551 (1978).
781. R. K. Haynes and C. Indorato, *Aust. J. Chem.*, **37**, 1183 (1984).
782. A. M. Torgomyan, M. Zh. Ovakimyan and M. G. Indzhikyan, *Arm. Khim. Zh.*, **33**, 63 (1980); *Chem. Abstr.*, **93**, 114639r (1980).
783. A. W. Frank, *Phosphorus Sulfur*, **5**, 197 (1978).
784. J. L. Belletire and M. W. Namie, *Synth. Commun.*, **13**, 87 (1983).
785. Y. Shen and W. Qiu, *Tetrahedron Lett.* **28**, 449 (1987).
786. A. P. Marchenko, G. N. Koidan and A. M. Pinchuk, *Zh. Obshch. Khim.*, **54**, 2691 (1984).
787. D. G. Naae and D. J. Burton, *J. Fluorine Chem.*, **1**, 123 (1971).
788. W. A. Vinson, K. S. Prickett, B. Spahic and P. R. Ortiz de Montellano, *J. Org. Chem.*, **48**, 4661 (1983).
789. D. J. Burton, I. Inouye and J. A. Headly, *J. Am. Chem. Soc.*, **102**, 3980 (1980).
790. H. J. Cristau, M. Fonte and E. Torreilles, *Synthesis*, 301 (1989).
791. J. L. Belletire, D. R. Walley and M. Bast, *J. Synth. Commun.*, **12**, 469 (1982).
792. J. M. McIntosh and F. P. Seguin, *Can. J. Chem.*, **53**, 3526 (1975).
793. W. Flitsch and E. R. Gesing, *Tetrahedron Lett.*, 1997 (1976).
794. E. V. Kuznetsov, T. V. Sorokina and R. K. Valetdinov, *J. Gen. Chem. USSR*, **33**, 2564 (1963).
795. D. Seyferth and J. Fogel, *J. Organomet. Chem.*, **6**, 205 (1966).
796. E. Zbiral and E. Hugl, *Phosphorus*, **2**, 29 (1972).
797. R. L. Webb and J. J. Lewis, *J. Heterocycl. Chem.*, **18**, 1301 (1981).
798. R. L. Webb, B. L. Lam, J. J. Lewis, G. R. Wellman and C. E. Berkoff, *J. Heterocycl. Chem.*, **19**, 639 (1982).
799. E. E. Schweizer and G. J. O'Neill, *J. Org. Chem.*, **30**, 2082 (1965).
800. Y. Okada, T. Minami, S. Yahiro and K. Akinaga, *J. Org. Chem.*, **54**, 974 (1989).
801. J. V. Cooney and W. E. McEwen, *J. Org. Chem.*, **46**, 2570 (1981).

802. E. E. Schweizer and K. K. Light, *J. Org. Chem.*, **31**, 870 (1966); *J. Am. Chem. Soc.*, **86**, 2963 (1964).
803. E. E. Schweizer, *J. Am. Chem. Soc.*, **86**, 2744 (1964).
804. E. E. Schweizer and J. G. Liehr, *J. Org. Chem.*, **33**, 583 (1968).
805. E. E. Schweizer and L. D. Smucker, *J. Org. Chem.*, **31**, 3146 (1967).
806. E. E. Schweizer and C. M. Kopay, *J. Org. Chem.*, **37**, 1561 (1972).
807. G. H. Posner and S. B. Lu, *J. Am. Chem. Soc.*, **107**, 1424 (1985).
808. E. E. Schweizer, A. T. Wehman and D. M. Nycz, *J. Org. Chem.*, **38**, 1583 (1973).
809. J. M. Mc Intosh and H. B. Goodbrand, *Synthesis*, 862 (1974).
810. J. M. Mc Intosh and R. S. Steevensz, *Can. J. Chem.*, **55**, 2442 (1977).
811. A. T. Hewson, *Tetrahedron Lett.*, 3627 (1978).
812. A. G. Cameron, A. T. Hewson and A. H. Wadsworth, *Tetrahedron Lett.*, **23**, 561 (1982).
813. E. Zbiral, *Synthesis*, 792 (1974).
814. V. S. Brovarets, O. P. Lobanov, A. A. Kisilenko, V. N. Kalinin and B. S. Drach, *Zh. Obshch. Khim.*, **56**, 1492 (1986).
815. E. E. Schweizer and S. De Voe, *J. Org. Chem.*, **40**, 144 (1975).
816. E. E. Schweizer and S. De Voe Goff, *J. Org. Chem.*, **43**, 2972 (1978).
817. Zh. A. Aklyan, R. A. Kachatryan and M. G. Indzhikyan, *Arm. Khim. Zh.*, **31**, 194 (1978); *Chem. Abstr.*, **89**, 109799f (1978).
818. R. A. Khachatryan, G. A. Mkrtchyan and M. G. Indzhikyan, *Zh. Obshch. Khim.*, **57**, 813 (1987).
819. E. E. Schweizer, C. J. Berninger and J. G. Thompson, *J. Org. Chem.*, **33**, 336 (1968).
820. E. E. Schweizer, C. J. Berninger, D. M. Crouse, R. A. Davis and R. S. Logothetis, *J. Org. Chem.*, **34**, 207 (1969).
821. P. L. Fuchs, *J. Am. Chem. Soc.*, **96**, 1607 (1974).
822. W. G. Dauben and D. J. Hart, *Tetrahedron Lett.*, 4353 (1975).
823. W. Flitsch and P. Russkamp, *Justus Liebigs Ann. Chem.*, 521 (1983).
824. J. M. Muchowski and P. H. Nelson, *Tetrahedron Lett.*, **21**, 4585 (1980).
825. E. E. Schweizer and C. S. Kim, *J. Org. Chem.*, **36**, 4033 (1971).
826. T. Minami, T. Hanamoto and I. Hirao, *J. Org. Chem.*, **50**, 1278 (1985).
827. E. Zbiral, M. Rasberger and H. Hengstberger, *Justus Liebigs Ann. Chem.*, **725**, 22 (1969).
828. Y. Tanaka and S. Miller, *J. Org. Chem.*, **15**, 2708 (1973).
829. C. Ivancsis and E. Zbiral, *Monatsh. Chem.*, **106**, 839 (1975).
830. N. Gakis, H. Heimgartner and H. Schmid, *Helv. Chim. Acta*, **57**, 1403 (1974).
831. M. K. Tasz, F. Plénat, H. J. Cristau and R. Skowronski, *Phosphorus Sulfur Silicon*, **57**, 143 (1991).
832. K. D. Gundermann and A. Garming, *Chem. Ber.*, **102**, 3023 (1969).
833. R. A. Ruden and R. Bonjouklian, *Tetrahedron Lett.*, 2095 (1974).
834. I. V. Megera, *Zh. Obshch. Khim.*, **54**, 2153 (1984).
835. M. I. Shevchuk and O. M. Bukachuk, *Zh. Obshch. Khim.*, **52**, 830 (1982).
836. R. Houssin, J. L. Bernier and J. P. Hénichart, *J. Heterocycl. Chem.*, **22**, 1185 (1985).
837. E. Zbiral, *Tetrahedron Lett.*, 5107 (1970).
838. P. I. Yagodinets, I. N. Chernyuk and M. I. Shevchuk, *Zh. Obshch. Khim.*, **54**, 2789 (1984).
839. H. J. Cristau, B. Chabaud and H. Christol, *J. Org. Chem.*, **49**, 2023 (1984).
840. M. Regitz, A. E. M. Tawfik and H. Heydt, *Justus Liebigs Ann. Chem.*, 1865 (1981).
841. H. Kunz and H. H. Bechtolsheimer, *Synthesis*, 303 (1982).
842. A. Hercouet and M. Le Corre, *Tetrahedron*, **37**, 2867 (1981).
843. A. Arnoldi and M. Carughi, *Synthesis*, 155 (1988).
844. F. Plenat and H. J. Cristau, unpublished results.
845. A. P. Martnyuk, V. S. Brovarets, O. P. Lobanov and B. S. Drach, *Zh. Obshch. Khim.*, **54**, 2186 (1984).
846. O. V. Smolii, V. S. Brovarets, V. V. Pirozhenko and B. S. Drach, *Zh. Obshch Khim.*, **58**, 2635 (1988).
847. O. V. Smolii, V. S. Brovarets and B. S. Drach, *Zh. Obshch. Khim.*, **57**, 2145 (1987).
848. J. Gosselck, H. Schenk and H. Ahlbrecht, *Angew. Chem., Int. Ed. Engl.*, **6**, 249 (1967).
849. K. Okuma, K. Kojima, S. Yamamoto, H. Takeuchi, K. Yonekura and H. Ohta, *Fukuoka Univ. Sci. Rep.*, **21**, 139 (1991); *Chem. Abstr.*, **115**, 208079h (1991).
850. C. C. Walker and U. Shechter, *Tetrahedron Lett.* 1447 (1965).
851. H. Kunz and H. H. Bechtolsheimer, *Justus Liebigs Ann. Chem.*, 2068 (1982).

852. G. D. Mac Donell, K. D. Berlin, S. E. Ealick and D. Van der Helm, *Phosphorus Sulfur*, **4**, 187 (1978).
853. H. Kunz and H. Kauth, *Z. Naturforsch. Teil B*, **34**, 1737 (1979).
854. G. Märkl, *Angew. Chem.*, **75**, 168 (1963).
855. T. A. Modro and A. Piekos, *Bull. Acad. Pol., Ser. Sci, Chim.*, 585 (1968) and 345 (1970); *Chem. Abstr.*, **71**, 29932g (1969) and **73**, 130429q (1973).
856. R. Rabinowitz and R. Marcus, *J. Polym. Sci., Part A*, **3**, 2063 (1965); *Chem. Abstr.*, **63**, 4401 (1965).
857. S. M. Khashimova, A. T. Dzhalilov and A. B. Mukhamedgaliev, *Uzb. Khim. Zh.*, 49 (1990); *Chem. Abstr.*, **115**, 159895 (1991).
858. P. Pinsard, J. P. Lellouche, J. P. Beaucourt and R. Grée, *Tetrahedron Lett.*, **31**, 1137 (1990).
859. E. A. Broger, *Eur. Pat. Appl.*, EP 100 839 (1984); *Chem. Abstr.*, **101**, 111194q (1984).
860. C. M. Starks, *J. Am. Chem. Soc.*, **93**, 195 (1971).
861. D. Landini and F. Rolla, *Synthesis*, 565 (1974).
862. G. Pantini, *Technol. Chim.*, **5**, 54 (1985), and references cited therein.
863. C. M. Starks and C. Liotta, *Phase-Transfer Catalysis, Principles and Techniques*, Academic Press, New York, 1978.
864. E. V. Dehmlow, *Angew. Chem., Int. Ed. Engl.*, **13**. 170 (1974); E. V. Dehmlow and S. S. Dehmlow, *Phase-Transfer Catalysis* 2nd ed., Verlag Chemie, Weinheim, 1983.
865. D. Landini, A. M. Maia and F. Montanari, *J. Chem. Soc., Chem. Commun.*, 950 (1975).
866. D. Landini, A. M. Maia and F. Montanari, *J. Chem. Soc., Chem. Commun.*, 112 (1977).
867. L. Horner and J. Gerhard, *Justus Liebigs Ann. Chem.*, 838 (1980).
868. A. W. Herriott and D. Picker, *J. Am. Chem. Soc.*, **97**, 2345 (1975).
869. C. Moesch, O. Prioux, C. Raby, *Ann. Pharm. Fr.*, **44**, 243 (1986).
870. J. H. Clark and D. J. Macquarrie, *Tetrahedron Lett.*, **28**, 111 (1987).
871. Y. Kimura, Y. Yoshida and Y. Suzuki, *Jpn. Pat.*, 0283364 (1990); *Chem. Abstr.*, **113**, 77908 (1990).
872. C. M. Starks and R. M. Owens, *J. Am. Chem. Soc.*, **95**, 3613 (1973).
873. T. Sakakibara and R. Sudoh, *J. Org. Chem.*, **40**, 2823 (1975); T. Sakakibara, M. Yamada and R. Sudah, *J. Org. Chem.*, **41**, 736 (1976).
874. G. Märkl and A. Merz, *Synthesis*, 295 (1973).
875. W. Tagaki, I. Inoue, Y. Yano and T. Okonogi, *Tetrahedron Lett.*, 2587 (1974).
876. S. Hunig and J. Stemmler, *Tetrahedron Lett.*, 3151 (1964).
877. R. Bruos and M. Anteunis, *Synth. Commun.*, **6**, 53 (1976).
878. C. Botteghi, G. Gaccia and S. Gladiali, *Synth. Commun.*, **6**, 549 (1976).
879. Y. Yano and W. Tagaki, *Kagaku No Ryoiki Zokan*, 113 101 (1975); *Chem. Abstr.*, **86**, 42534h (1977).
880. W. J. Splillane, H. J. M. Dou and J. Metzger, *Tetrahedron Lett.*, 2269 (1976).
881. R. Damrauer, *J. Organomet. Chem.*, **216**, C1 (1981).
882. F. Yamashita, A. Atsumi and H. Inoue, *Nippon Kagaku Kaishi*, 1102 (1975); *Chem. Abstr.*, **83**, 113378 (1975).
883. M. Cinquini, F. Montanari and P. Tundo, *J. Chem. Soc., Chem. Commun.*, 393 (1975).
884. S. Colonna and R. Fornasier, *Synthesis*, 531 (1975).
885. R. Fornasier, F. Montanari, G. Rodda and P. Tundo, *Tetrahedron Lett.*, 1381 (1976).
886. S. Grinberg and E. Shaubi, *Tetrahedron*, **47**, 2895 (1991).
887. S. L. Regen, *J. Am. Chem. Soc.*, **98**, 6270 (1976).
888. Y. Yoshida, Y. Kimura and M. Tomoi, *Chem. Lett.*, 769 (1970).
889. H. Molinari and F. Montanari, *J. Chem. Soc., Chem. Commun.*, 638 (1977).
890. M. Cinquini, S. Colonna, H. Molinari, F. Montanari and P. Tundo, *J. Chem. Soc., Chem. Commun.*, 394 (1976).
891. P. Tundo, *J. Chem. Soc., Chem. Commun.*, 641 (1977); P. Tundo and P. Venturello, *J. Am. Chem. Soc.*, **101**, 6606 (1979).
892. M. Komeili-Zadeh, H. J. M. Dou and J. Metzger, *J. Org. Chem.*, **13**, 156 (1978).
893. S. Regen and J. J. Besse, *J. Am. Chem. Soc.*, **101**, 4059 (1979).
894. H. Molinari, F. Montanari S. Quici and P. Tundo, *J. Am. Chem. Soc.*, **101**, 3920 (1979).
895. H. Takeuchi, M. Kikuchi, Y. Miwa and J. Okada, *Chem. Pharm. Bull.*, **30**, 3865 (1982).
896. H. Takeuchi, Y. Miwa, S. Morita and J. Okada, *Chem. Pharm. Bull.*, **32**, 409 (1984).
897. E. Chiellini, R. Solaro and S. D'Antone, *Makromol. Chem., Suppl.*, **5**, 82 (1981).
898. W. T. Ford, *Polym. Sci. Technol.*, **24**, 201 (1984).

899. M. Tomoi, E. Ogawa, Y. Hosokama, and H. Kakiuchi, *J. Polym. Sci., Polym. Chem. Ed.*, **20**, 3421 (1982).
900. P. L. Anelli, F. Montanari and S. Quici, *J. Chem. Soc., Perkin Trans. 2*, 1827 (1983).
901. M. Tomoi, S. Shiiki and H. Kakiuchi, *Makromol. Chem.*, **187**, 357 (1986).
902. M. Tomoi, E. Nakamura, Y. Hosokawa, H. Kakiuchi, *J. Polym. Sci. Polym. Chem. Ed.*, **23**, 49 (1985).
903. M. Tomoi, Y. Hosokawa and H. Kakiuchi, *J. Polym. Sci., Polym. Chem. Ed.*, **22**, 1243 (1984).
904. M. Bernard, W. T. Ford and T. W. Taylor, *Macromolecules*, **17**, 1812 (1978).
905. S. J. Brown and J. H. Clark, *J. Fluorine Chem.*, **30**, 251 (1985).
906. H. J. Cristau, A. Bazbouz, P. Morand and E. Torreilles, *Tetrahedron Lett.*, **27**, 2965 (1986).
907. A. BaBa, T. Nozaki and H. Matsuda, *Bull. Chem. Soc. Jpn.*, **60**, 1552 (1987).
908. S. E. Fry and N. J. Pienta, *J. Am. Chem. Soc.*, **107**, 6399 (1985).
909. D. Landini, A. Maia, F. Montanari and F. Rolla, *J. Org. Chem.*, **48**, 3774 (1983).
910. J. C. Fanning and L. K. Keefer, *J. Chem. Soc., Chem. Commun.*, 955 (1987).

CHAPTER **3**

Preparation, properties and reactions of phosphoranes

RAMON BURGADA

Université Pierre et Marie Curie, Laboratoire de Chimie des Organoéléments, CNRS, 4 Place Jussieu, F-75252 Paris Cedex 05, France

and

RALPH SETTON

Centre National de la Recherche Scientifique, Centre de Recherche sur la Matière Divisée, 1B rue de la Férollerie, F-45071 Orléans Cedex 2, France

The chemistry of organophosphorus compounds, Vol. 3
Edited by F. R. Hartley

I. INTRODUCTION

The developments in the chemistry of pentacoordinated phosphorus have been so considerable during the last 35 years as to be truly exceptional in the history of chemistry. This is due to the combination of a number of factors, the first of course being the natural growth of chemical research, resulting in the synthesis by Wittig and Reiber[1] in 1959 of PPh_5, the first organic phosphorane, whose trigonal bipyramid (TBP) structure was only determined 15 years later by Wheatley[2]. The same period also witnessed the synthesis of monocyclic oxyphosphoranes by Ramirez[3] and of acyclic pentaoxyphosphoranes by Denney and Relles[4], in addition to the tremendous acceleration of phosphorane chemistry due to the results obtained using NMR.

Even more fundamental, to our mind, are the following three factors:

1. Phosphoranes represented a new domain of phosphorus chemistry, rich in possibilities from the fundamental and the practical points of view. The discovery of their stereochemical non-rigidity, i.e. the exchange of axial and equatorial positions in a TBP structure without bond cleavage, found to occur in PF_5 at a frequency of 10^8–10^9 Hz, led Berry[5] to suggest the presence of an interconversion mechanism. The fact that this is a general property of phosphoranes immediately set the fascinating and complex challenge of the synthesis of enantiomeric phosphoranes with optical activity depending solely on the P atoms.
2. The equilibria P(III)$\rightleftharpoons$P(V), P(IV)$\rightleftharpoons$P(V) and P(VI)$\rightleftharpoons$P(V) show that phosphoranes constitute 'pivots', allowing pentacoordinated phosphorus to be reached from any other coordination number, as probably postulated for the first time in 1929 by Fenton and Ingold[6], and later observed many times experimentally.
3. Lastly, the finding by Westheimer[7] of considerable differences in the rates of hydrolysis of five-membered cyclic and acyclic phosphates also led him to postulate the existence of a pentacoordinated intermediate likely to undergo the stereomutations described Berry[5]. This threw new light on the mechanism of nucleophilic substitution at tetracoordinated P atoms, which is at the heart of a large number of reactions occurring in living organisms and in more formal chemistry.

II. STRUCTURES AND GEOMETRIES ASSOCIATED WITH THE PENTACOVALENT PHOSPHORUS ATOM

A. The Pentacovalent Phosphorus Atom

Before proceeding to the preparation and reactions of phosphoranes, it is important to recall some of the fundamental properties of the pentacoordinated P atom, but since this

TBP TPa TPb CT

FIGURE 1

subject has been abundantly covered in a number of excellent publications, we shall merely give here an outline of the different points and the literature references particularly apt to provide supplementary details.

According to the valence shell electron pair repulsion (VSEPR) theory of Gillespie[8], the five ligands around a P atom will occupy positions such that they are as far away from each other as possible, so as to minimize their interaction energy. Two geometrical models (Figure 1) are particularly favoured: the trigonal bipyramid (TBP) of point group symmetry D_{3h}, and the tetragonal pyramid (TP) of symmetry C_{4v}.

The TBP has three coplanar bonds, at 120° from each other forming an equatorial plane and, on either side of this plane and perpendicular to it, two colinear axial or apical bonds, longer by a factor $\sqrt{2}$. Obviously, this ideal structure is only strictly applicable in simple cases, when the five ligands are identical and/or are free from steric strains caused, for instance, by their forming small rings in apical–equatorial (*ae*) positions. Differences in the ligands or the presence of strains induce important deformations[9,10] in the TBP, and other geometries must then be envisaged[11], such as the TP (TPa or TPb), or even the 'skew structure' of symmetry C_s[12] (Figure 2)*.

Experimental results and theoretical calculations[8,16,17] both show that the TBP structure is more favourable, with TPa at 1.5 kcal mol^{-1} higher energy than TBP, and TPb, with four ligands in the basal plane, at 33 kcal mol^{-1} above TBP (1 kcal = 4.184 kJ). The TBP is the configuration most often met among phosphoranes, even when the latter involve up to two *ae* rings.

In addition to the VSEPR theory[8] mentioned above, other theoretical or semiempirical approaches have addressed the problem of the positions occupied by various ligands as a function of their nature, for comparison with the numerous experimental results now available: molecular orbital calculations, (four-electron, three-centre model with neglect of the P d orbitals)[18,19]; semiempirical calculations[20,21]; non-empirical calculations[22,23]; and hybrid orbitals[24,25].

These studies revealed the importance of three factors, which will now be reviewed.

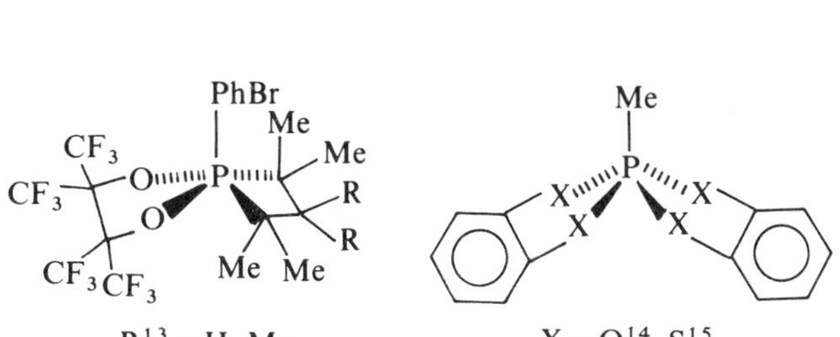

PhBr Me Me R R Me Me CF_3 CF_3 CF_3 CF_3 O O P

R^{13} = H, Me

Me P X X X X

X = O^{14}, S^{15}

FIGURE 2

* Literature references in Figure 2 correspond to structural determinations by X-ray diffraction.

B. Electronic Factors (Electronegativity, Apicophilicity, Polarity Rules)

The study of a large number of (mostly fluorinated) phosphoranes[26,27] showed that, as far as the three equatorial bonds were concerned, it was the P 3s orbital which was most important since the increase in electron density along the equator induces its decrease along the polar axis. According to Bent[25], electropositive substituents will seek maximum s characteristics whereas electronegative substituents will seek maximum p characteristics. Hence, the rule of electronegativities states that the most electronegative of the five groups around a pentacoordinated P atom will be found preferentially in the two apical positions.

Further insight into the role played by electronegativity in determining the two apical substituents was given by the good agreement between theoretical studies of the concept of apicophilicity[23,28] and experimental determinations by dynamic NMR[29] of the free energy of activation and of the coupling constants J_{PC} of a number of compounds involved in some specific equilibria. The main conclusion was that, for a compound with a TBP structure, the apicophilicity of a substituent involves both its electronegativity and its aptness at forming π bonds on passing from an equatorial to an apical position. Apicophilicity is enhanced by increased electronegativity, is diminished by the presence of free electron pairs in the substituent and is favoured by vacant low-lying orbitals.

The empirical relative scale of apicophilicity[29,30,31] F > H > CF_3 > OPh > Cl > SMe > OMe > NMe_2 > Me > Ph compares fairly well with that drawn up after a theoretical study based on the acyclic H_4PX structure[23], Cl > CN > F > C≡CH > H > Me > OH > O^- > S^- > NH_2 > BH_2. However, when these two scales are compared with others drawn up by different authors[29,31], such as H, OMe, $(CF_3)_2C(H)O$ > PhO, PhS > NMe_2 > Me > Ph, Cl or[32] CN, PhO > MeO, MeS > Me_2N, Me, CH_2=CH_2, Ph, it is obvious, from the different positions of Cl or the absence of difference between O and S, that the apicophilicity of a group must be due to the combined influences of many factors, such as electronegativity, $p\pi$–$d\pi$ interaction, steric effects and polarizability. The possibility of setting up another scale[32] based on Hammett's inductive substituent constant σ_1 confirms the fact that the apicophilicity of a group is sensitive to the nature of the ligands in the molecule.

A single equatorial π acceptor preferentially orients its acceptor orbitals perpendicularly to the equatorial plane; conversely, the orbitals of a π donor system are coplanar with this plane (Figure 3). These theoretical findings are strongly supported by experimental results obtained with sulphur[33] and aminophosphoranes[34] (note that, for the latter, the energy of activation for rotation around the P—N bond has been evaluated as 5–12 kcal mol^{-1}). In the TP geometry with a C_{4v} symmetry, π donor ligands will prefer apical positions whereas π acceptors will favour basal sites. If there is only a single π system among the basal substituents, it will be parallel to the basal plane for an acceptor, but parallel to the apical bond for a donor[35].

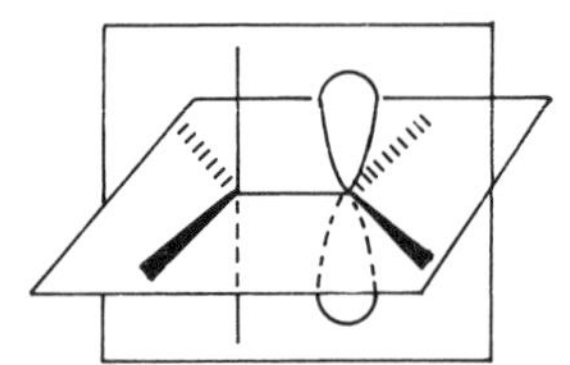

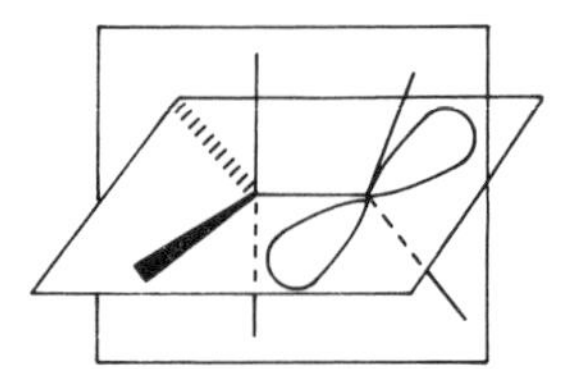

FIGURE 3. Possible orientations of the acceptor π orbital with respect to the equatorial plane of the molecule

C. Steric Constraints

The presence of steric strains implies, *a priori*, at least the presence of a ring in which two out of the five bonds to the P atom are involved. This immediately raises a number of questions: what is the resulting geometry of the TBP, the TP (or of any other structure involved) as a function of the nature and of the number of atoms directly bonded to the P atom, and how will the presence of steric constraints affect the apicophilicity of the ligands?

At present, cyclic phosphoranes have been classified into four main families (Figure 4): monocyclic, bicyclic (spiro and fused), tricyclic and tetracyclic.

Rings with four or five atoms are most often found[29,30,36–39] to have adopted an *ae* configuration. The outcome of conflicts for the occupation of the apical position between the electronegativity of a ligand and strain in the ring depends on the nature of the other ligands and on the final, overall, resulting structures[29,40,41]. There are two means of minimizing important strains due to small or to unsaturated rings coplanar with the P atom: (a) variations of the bond lengths between P and the five ligands, as in **12** (Figure 5)* in which the two P—O bonds of the pentagonal ring are longer than the apical P—O bonds of the hexagonal ring; (b) adoption of variable geometries[66] ranging from the nearly perfect TBP (cf. **6**, Figure 5) to the ideal TP structure (with an angle of 105° between the apical bond and the basal ligands[67]) or to the strongly deformed TP or skew C_s structure. Thus, in the TP, the most electronegative ligands will preferentially occupy basal positions and small rings will be in dibasal position (Figure 2 and 5).

Six-membered rings are likely to occupy either an *ae* or a diequatorial (*ee*) configuration in the TBP, even if the latter is energetically less favourable. The dynamic NMR study by Chang *et al.*[61] of oxyphosphoranes led them to suggest *ae* positions for the dioxaphosphorinane rings in compounds such as **11** (Figure 5), a structure recently confirmed by X-ray diffraction (XRD)[48]. The same *ae* structure was also recently found by XRD[59,60] in the bicyclic pentaoxyphosphorane **23** with $R^1 = t$-Bu, $R^2 = H$, R = $CH(CF_3)_2$. Further, the temperature dependence of the NMR signal led Trippett and coworkers[29] to conclude that in **23**, with $R^1 = R^2 = H$ and R = Ph, the dioxaphosphorinane ring was also *ae*, with a 6 kcal mol^{-1} barrier to the *ae* → *ee* stereomutation, but that with a dioxaphospholane ring, the ΔG^* barrier was 17.4 kcal mol^{-1}.

The XRD analysis of solid pentaoxyphosphorane **12** and the NMR study of its solution also point to an *ae* structure[50]. The same is true for the oxazaphosphorinane rings in **26**[64,65,68], with a twisted boat configuration[60,68], and for the rings in the dithiaphosphorinane **8** and the diazaphosphorinane **10**[47,49]. This, incidentally, points to the fact that one of the N atoms in **10** and one of the S atoms in **8** occupy an apical position, thus forcing the OPh group to occupy an equatorial position in spite of its stronger apicophilicity.

Consider now compounds **27, 28** and **29** in Figure 6. Note first that the steric constraints

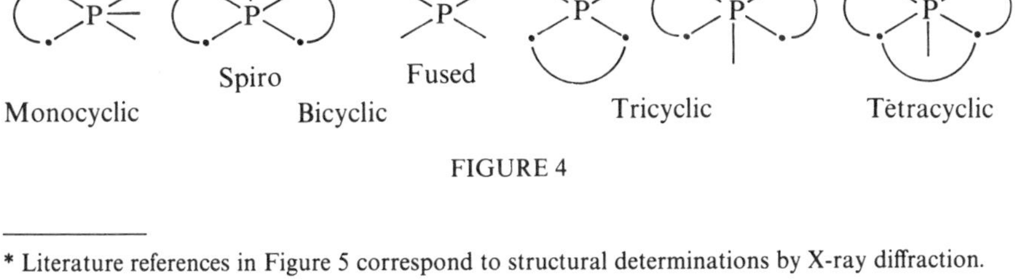

Monocyclic | Spiro | Fused | Bicyclic | Tricyclic | Tetracyclic

FIGURE 4

---

* Literature references in Figure 5 correspond to structural determinations by X-ray diffraction.

(**1**)[42] (**2**)[43] (**3**)[44]

(**4**)[45] (**5**)[45] (**6**)[46]

(**7**)[40] (**8**)[47] (**9**)[48]

(**10**)[49] (**11**)[48] (**12**)[50]

(*continued*)

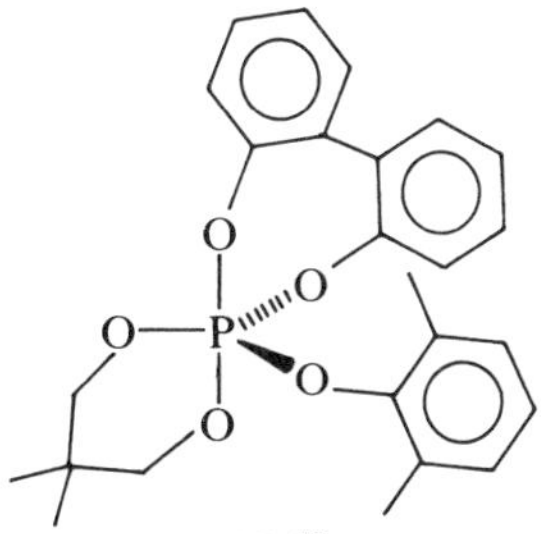

(**13**)[48] 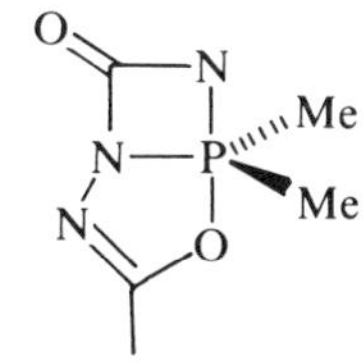 (**14**)[51] 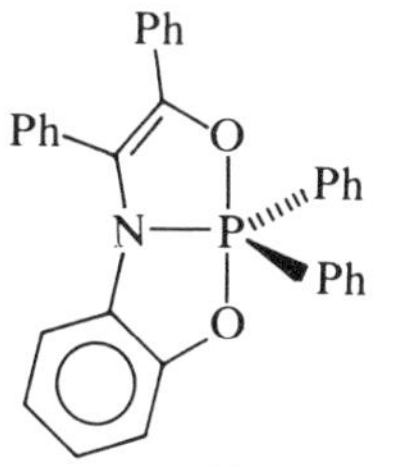 (**15**)[52]

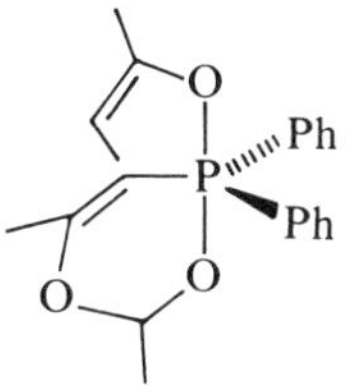

(**16**)[53] 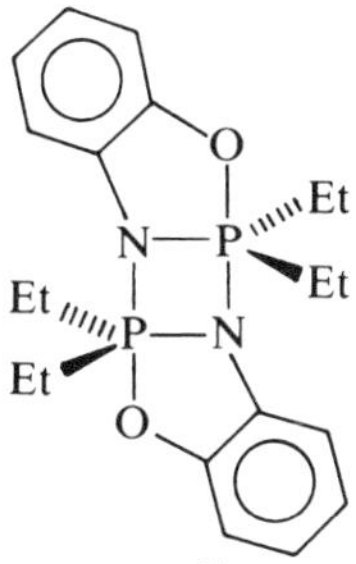 (**17**)[54] 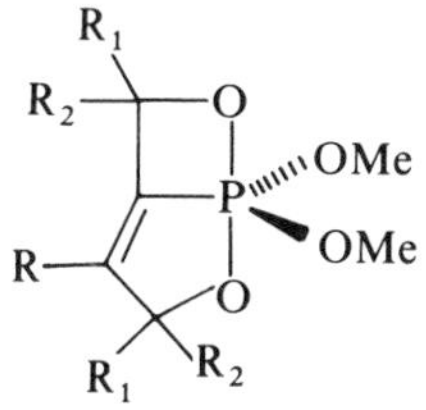 (**18**)[55]

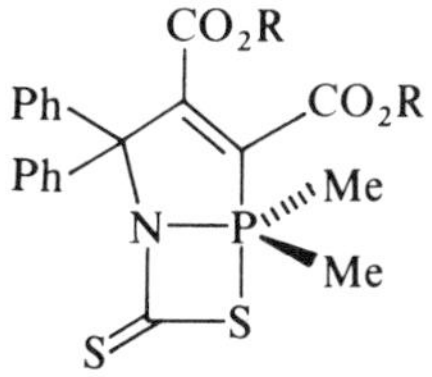

(**19**)[56] 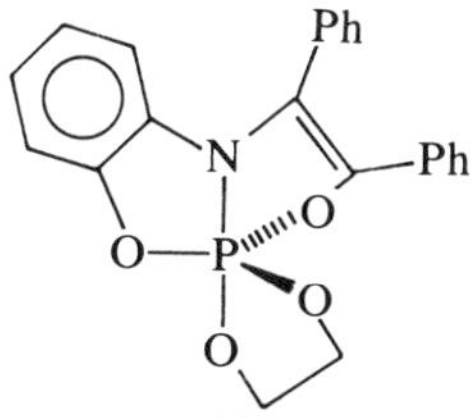 (**20**)[57] 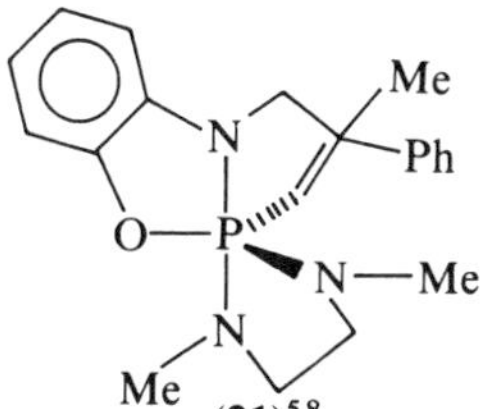 (**21**)[58]

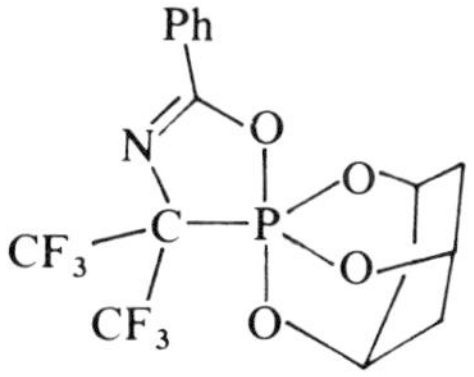

(**22**)[58] 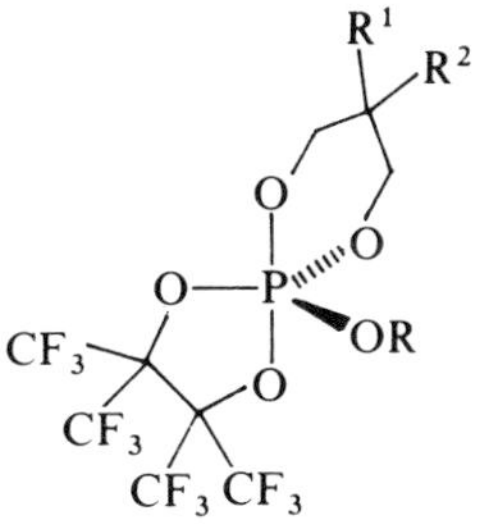 (**23**)[59–61] 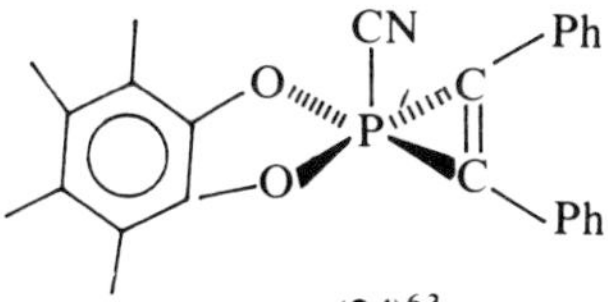 (**24**)[63]

(continued)

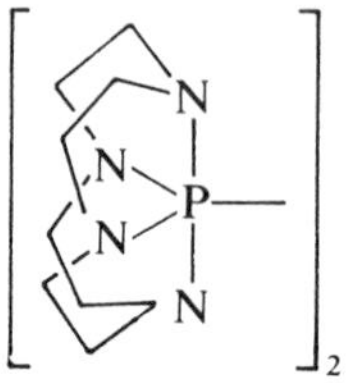

(25)[63]

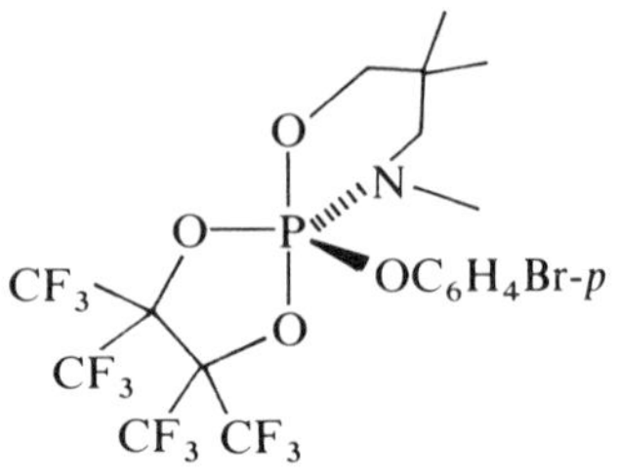

(26)[64]

FIGURE 5

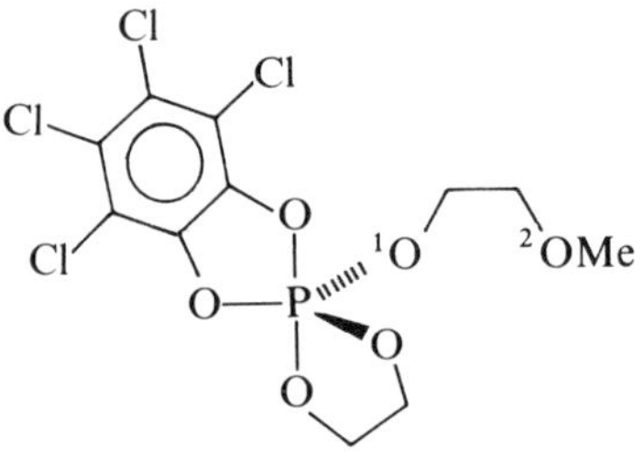

(27)

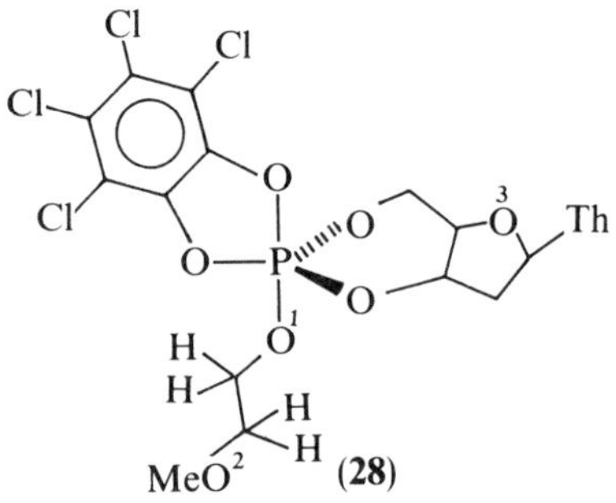

(28)

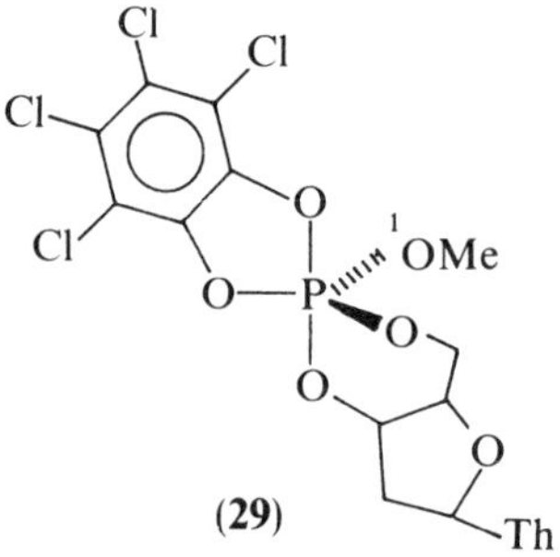

(29)

FIGURE 6

due to the two five-membered rings in **27** impose an equatorial position for the OCH_2CH_2OMe chain, but that the six-membered ring in **28** and **29** can adopt either an *ee* or an *ae* configuration, with a half-chair conformation in the former case or a twisted boat conformation in the latter, in agreement with numerous previous observations on similar structures[50,60,68].

Compounds **28** and **29** are pentacoordinated models of the activated state of 3′,5′-cyclic adenosine monophosphate (cAMP), which plays a dominant role as second messenger in cell metabolism regulation[69]. A 'conformational transmission effect' found by dynamic ^1H and ^{13}C NMR[70] results in 51% of *trans* position for the $O_{(1)}$ and $O_{(2)}$ atoms, but only when

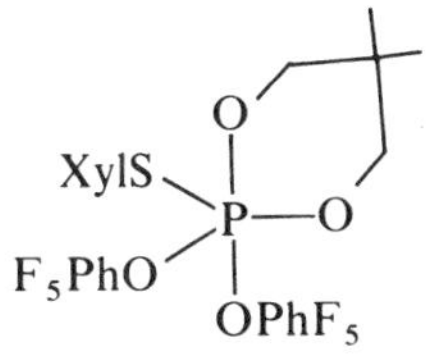

(30a) 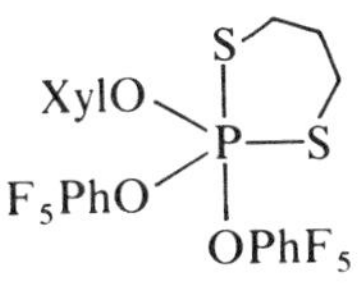 (30b) 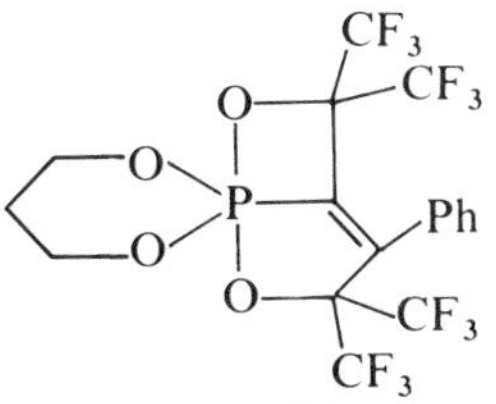 (31)

FIGURE 7

the OCH_2CH_2OMe chain is apical as in **28**, in which case the electrons of the atom $O_{(1)}$ are more easily retained than when this atom is equatorial as in **27**, when only 12% of the molecules have *trans* $O_{(1)}$–$O_{(2)}$ atoms. Even though MNDO calculations[70] on different theoretical isomers of a phosphorane with three OMe groups and a dioxaphosphorinane ring (as in **28** or **29**) showed that the *ae* configuration was energetically more favourable than the *ee* configuration by 3–4 kcal mol^{-1}, the dioxaphosphorinane ring adopts an *ee* configuration in **28**, in contrast with **29**, which follows the general rule. In other words, the OCH_2CH_2O chain is more apicophilic than the OMe or OEt groups.

The conformation of the *ae* dioxaphosphorinane ring in all the examples given above was preferentially boat, or twisted boat, rather than chair, as found by NMR or XRD. The only exception was due to the presence, in the crystal, of intermolecular hydrogen bonds inducing a twisted chair conformation[71]. Two recent results should however be taken into account to complete the picture.

(a) As determined by XRD[72], the dioxaphosphorinane and dithiaphosphorinane rings in compounds **30a** and **30b** (Figure 7) are both *ae*, but the conformation is chair in the former and twisted boat in the latter. Further, **30a** is probably the first example of a TBP with nearly equal-sized apical and equatorial P—O bonds (1.584_8 and 1.588_7 Å, respectively) for two identical ligands in the dioxaphosphorinane ring. The stable chair conformation for the saturated *ae* six-membered phosphorinane ring of the TBP in **30a** proves that a boat conformation is not a necessary requirement, in spite of the contention by Trippett and coworkers[29] that the latter form was given preference because it is the only one allowing the lone pair of electrons of the heteroatom in the equatorial ring to reside in the equatorial plane.

(b) The second example is provided by the synthesis of and the ^{1}H NMR and XRD data for the phosphorane **31**[73] (Figure 7). In this case, the TBP is distorted and the *ee* dioxaphosphorinane ring is in the chair conformation, whereas MNDO calculations predict[70] a half-chair conformation flattened about phosphorus. Moreover, conforma-

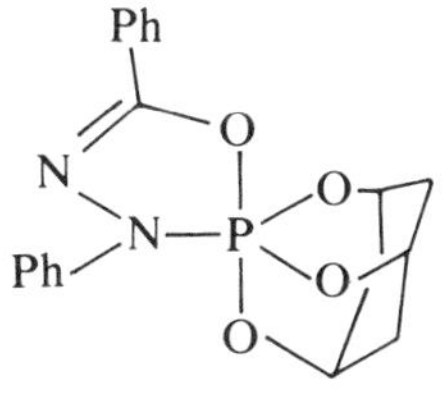

(32) 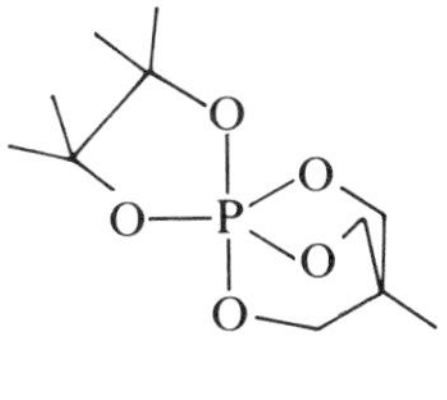 (33)

FIGURE 8

tional analysis by NMR (J_{HH} and J_{HP}) of **31** yielded parameters corresponding to an *ee* chair structure for the dioxaphosphorinane ring; these results conflict with the conclusions drawn[70] from the corresponding parameters obtained on **28**, namely that the preferred structure of the dioxaphosphorinane ring in **28** undergoing stereomutation is *ee*.

We can conclude this discussion of six-membered rings with compounds **22**[58] (Figure 5), **32**[74] and **33**[75] (Figure 8) in which the cage structure imposes an *ee* position for one of the heteroatomic cycles. Structural determinations on **22** and **32** by XRD show that their geometry is relatively different from the ideal TBP.

The oxyphosphoranes **34** (Figure 9), with a ring of seven atoms, and **35**, with eight, have been prepared by Abdou *et al.*[76a]. The temperature variation of their ^{1}H, ^{13}C or ^{31}P NMR spectra led them, on the basis of the usual criteria (δ, J and coalescence of signals), to suggest an *ee* geometry for these rings; the value of the intramolecular barrier to reorganization is too low to be measured when $R^1 = H$, but it is about 13.8–16 kcal mol^{-1} when $R^1 = t$-Bu. More recently, Burton *et al.*[77] studied the same compounds (with $R^1 = H$ and R = 2,6-dimethylphenyl) and other structures similar to **34** and **35** and showed, using XRD, that the seven- and eight-membered rings in these phosphoranes were *ae*, and that this structure was retained in solution, as proved by ^{1}H, ^{13}C and ^{31}P NMR spectroscopy. Similarly, the phosphorane **35a** was synthesized by the reaction of 2,2-methylenebis(4-methyl-6-*tert*-butyl)phenol on $P(OCH_2CF_3)_3$ in the presence of *N*-chlorodiisopropylamine and its structure, as determined by XRD[76b], was found to be TBP with an *ee* eight-membered ring, in agreement with the conclusions of Abdou *et al.*[76a] concerning the conformation of the ring in **34**.

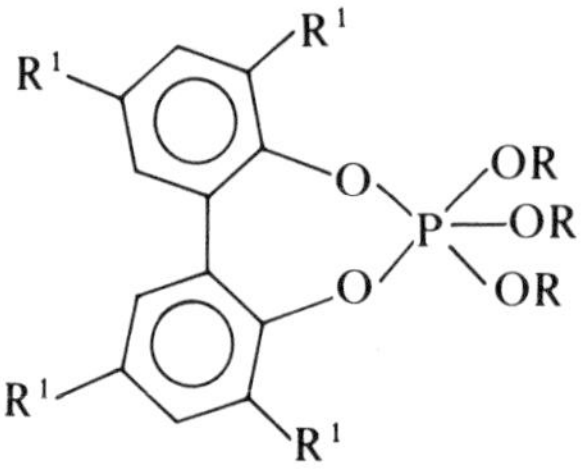

(**34**) 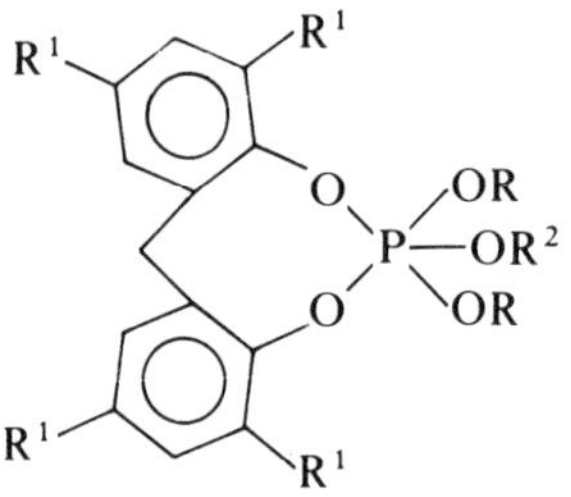 (**35**)

$R = CH_2CF_3$; R^1 = H or T – Bu^t; R^2 = Et or CH_2CF_3

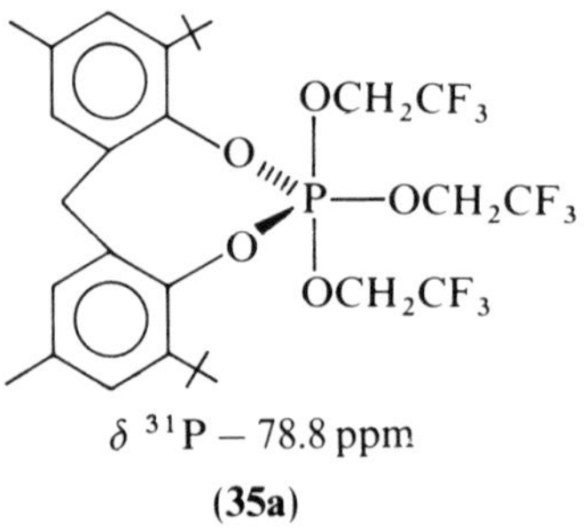

δ ^{31}P – 78.8 ppm

(**35a**)

FIGURE 9

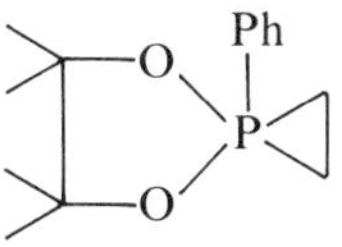

(36) 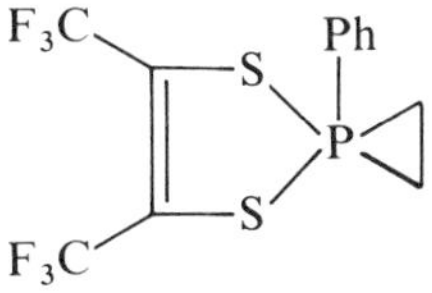 (37) 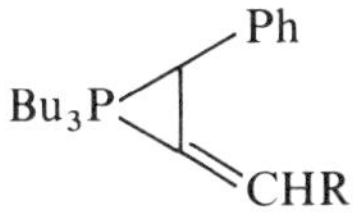 (38)

(a) R = H
(b) R = Ph

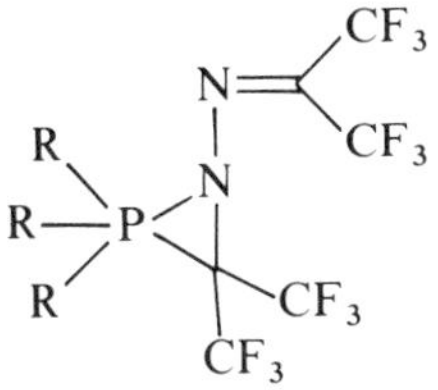

(39) (40)

(a) R = OMe
(b) R = OEt
(c) $R = NMe_2$

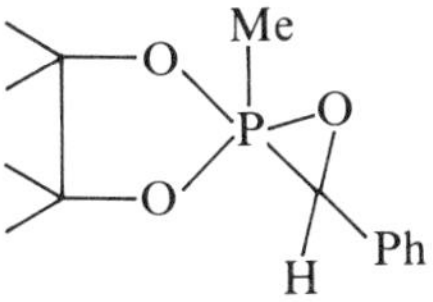

FIGURE 10

Three-membered rings are as rare as seven- and eight-membered rings, but much less stable. Among the compounds **36**–**40** (Figure 10), the phosphorane **36** is stable at −80° C[78] whereas **37** is unstable even at that temperature[79]. The phosphorinane **38** has been described as stable[80], but there are doubts as to its structure[81] as well as to that of **39**[82]. Lastly, the oxaphosphirane **40**[83] is in fact in equilibrium with its dimer. In view of the instability of these compounds, it seems hazardous to ascribe specific geometries to them, except in the case of the stable phosphirene **24** (Figure 5), in which the P atom is pentacoordinated and whose structure, a deformed TP, has been unambiguously determined by XRD[62].

D. Symmetry Factors

The importance of symmetry factors was underlined by Ramirez and coworkers[84]. A semiempirical quantum mechanical CNDO/2 method was used to estimate chemical binding energies, and hence the relative stabilities, in addition to the electron density on the atoms of the isomers with a given set of five substituents. The results of these calculations highlighted the increase in stability obtained by substantial back-donation of electrons from the ligand via the p orbital into the vacant P 3d orbital, in proportions dependent on the relative position and on the nature of the substituents: stability is acquired by decreasing the polarity of the bonds, and the stablest among the positional isomers is the one with the least difference between the electron density on the P atom and the average of the electron density of the ligand set. Since d-orbital interaction is smaller, with concomitantly less back-donation in the apical than in the equatorial position, the most electronegative elements or groups tend to settle at the apices and are found to be at larger distances than the equatorial substituents since bond length varies roughly inversely with the electron density.

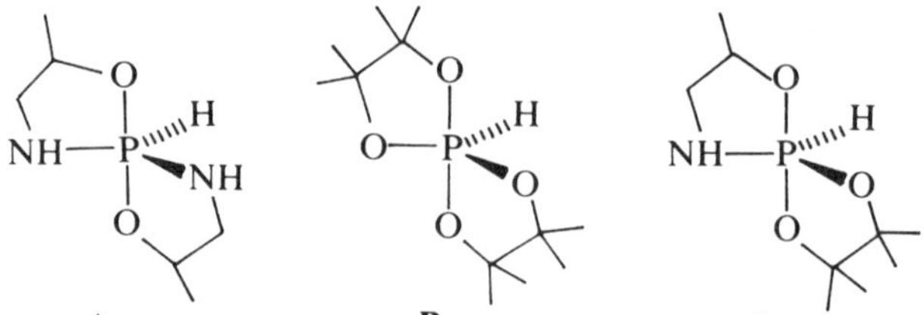

	A	B	C
Apical position	O, O	O, O	O, O
Equatorial position	N, N, H	O, O, H	O, N, H
% P(III) at 100° C	0	0	20

FIGURE 11. Influence of symmetry of the substituents in the equatorial plane on the stability of a P(V) structure

Another result, termed stability by ligand set symmetry and accruing from these calculations, is that stability is decreased if either of the (*eee*) or (*aa*) subsets contains a 'foreign' ligand, such as in ($R^1R^1R^2$) and/or (R^3R^4).

A good illustration of these concepts is provided by the three spirophosphoranes **A**, **B** and **C** in Figure 11. Since all of them are apt to be involved in a P(III) $\rightleftharpoons$ P(V) equilibrium, the proportion of P(III) can be taken as a probe of the stability of the pentacoordinated compound. It was thus found[85] that **C**, with three different substituents in the *e* subset, is the one for which the proportion of P(III) is highest.

III. PERMUTATIONAL ISOMERISM

A. Count of Isomers in a Pentacoordinated, Trigonal Bipyramid (TBP) Structure

Changing one isomer into another by permutation of the ligands implies the necessity of knowing, in each case, the number of isomers formed by these possible permutations, and of designating unambiguously each resulting structure. Ruch and coworkers[86] applied Polya's theorem[87] giving the number of isomers, N, to TBP structures of pentacoordinated phosphorus:

$$N = n!/g$$

where n is the number of different achiral ligands and g is the order of the symmetry sub-group of the polyhedron. For a TBP structure of symmetry D_{3h} with five ligands,

$$g = E, 2C_3, 3C_2$$

hence $g = 6$.

$N = 5!/6 = 20$ isomers or 10 pairs of racemates

For a TP of symmetry C_{4v},

$$g = E, 2C_2, 2C_4$$

hence $g = 5$.

$N = 5!/4 = 30$ isomers or 15 pairs of racemates

Taking the TBP structure as an example, the 20 isomers can be described with reference to the isomer in Figure 12, which itself can be described as having bonds* 1 and 2 in apical

* For the remainder of this section, the numbers 1–5 will be used to characterize both the bonds of pentacovalent P atoms and/or the ligands in these positions linked to the central atom.

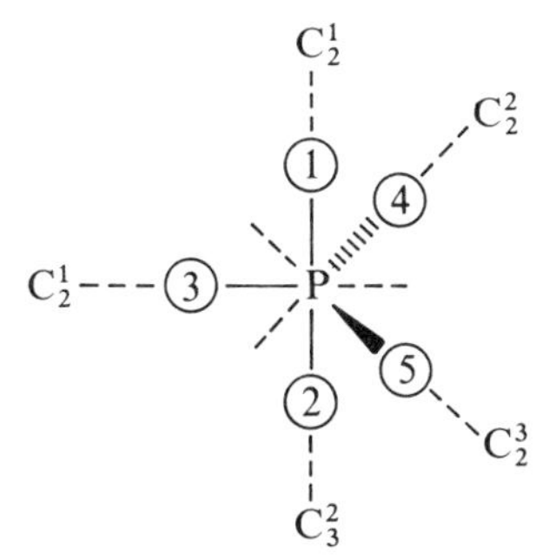

FIGURE 12

positions and bonds 3, 4 and 5 in equatorial positions. A shorthand notation for this description is (12)(345) or, more simply, merely 12.

Applying now the Cahn–Ingold–Prelog rule and viewing 2 from 1 along the C_3^1 axis, we note that the equatorial ligands have been numbered clockwise from 3 to 5. For the other isomers, we can permute the two apical ($a \rightleftharpoons a$) or any two of the equatorial ($e \rightleftharpoons e$) ligands (Figure 13): still looking along the same axis and in the same direction, we find that the only difference is that we have now created the optical isomer of 12, which can be denoted $\overline{12}$. Similarly, by permutation of ligands 3 and 2 ($a \rightleftharpoons e$), a new isomer, denoted 13, is obtained. The different isomers will therefore be:

12, 13, 14, 15	$\overline{12}$, $\overline{13}$, $\overline{14}$, $\overline{15}$
23, 24, 25	$\overline{23}$, $\overline{24}$, $\overline{25}$
34, 35	$\overline{34}$, $\overline{35}$
45	$\overline{45}$

Identities among the ligands will, of course, decrease the number of isomers: thus, two identical five-membered rings (met, for instance, in spirophosphoranes) will reduce the number of isomers to 16 because none of the rings can be diapical and four structures are no longer attainable.

Isomers resulting from permutations of the position of the ligands have been called permutational[25], polytopal isomers[88] or topomers[89]. The interchange of ligands in the same molecule has itself been named polyhedral or permutational isomerization, stereoisomerization, stereomutation or topomerization. Pseudorotation is now reserved for Berry pseudorotation (BPR), a permutation of ligands occurring according to the mechanism described by Berry[5].

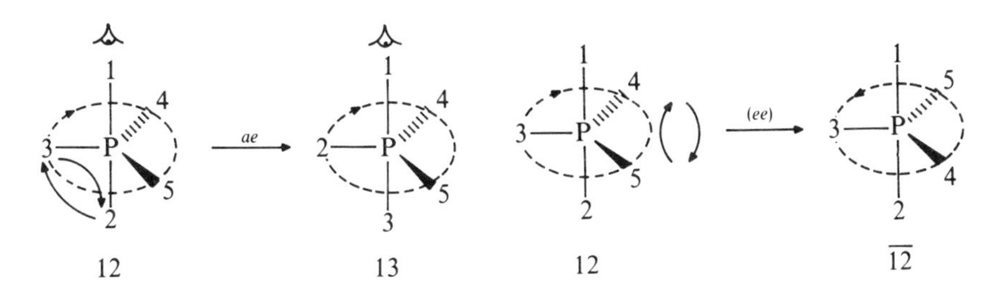

FIGURE 13

The large number of isomers and the complexity associated with their interconversion led a number of workers to propose graphs visualizing the isomers and the interconnecting paths[28,90-97]. Muetterties, for instance, empirically divided all the possible permutations into 6 classes[88,98]. Later, Gielen and van Lautem[99] reached the same conclusion but only considered BPR as the reaction mechanism, as also did Brocas[100], who, nevertheless, believed that differences in the isomers formed resulted from differences in reaction kinetics. The problem was independently re-examined, from a purely theoretical point of view and without reference to any given mechanism of formation, by Hasselbarth and Ruch[101] and by Klemperer[102], and again semiempirically by Musher[103]. They all agreed with the division of the isomerization operators into the following six classes or modes:

Class	*N*(TBP)	Permutations	
C_0	1	*eee*, (*aa.ee*)	the identity mode
C_1	3	*eaea*, (*ae.aee*), *aeae*	
C_2	6	*ae*, *eeea*, *eeaa*, (*ee.aae*)	
C_3	1	*aa*, *ee*, (*aa.eee*)	the racemization mode
C_4	6	*aee*, *aae*, (*ae.ee*), *aaeee*	
C_5	3	(*ae.ae*), *aeaee*	

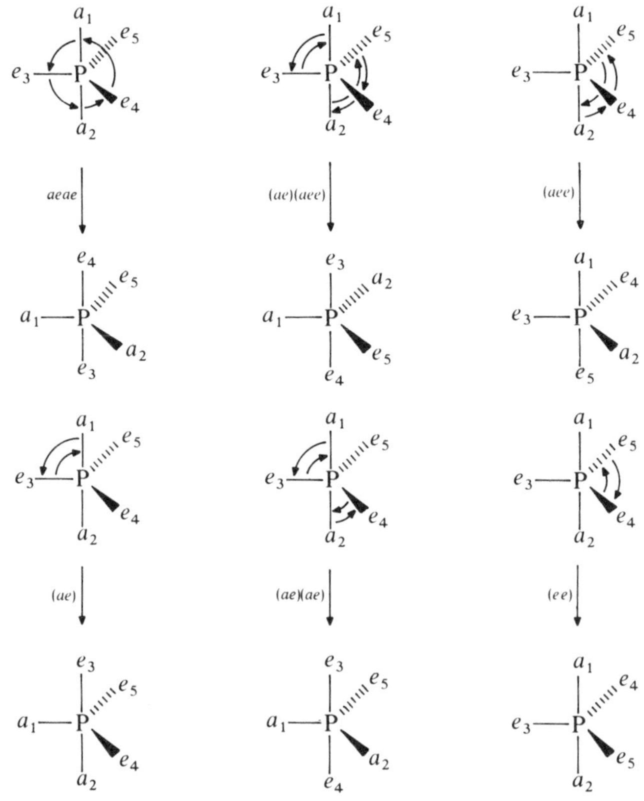

FIGURE 14. Example of the theoretically possible interconversions among isomers of a P(V) structure by permutation of the substituents

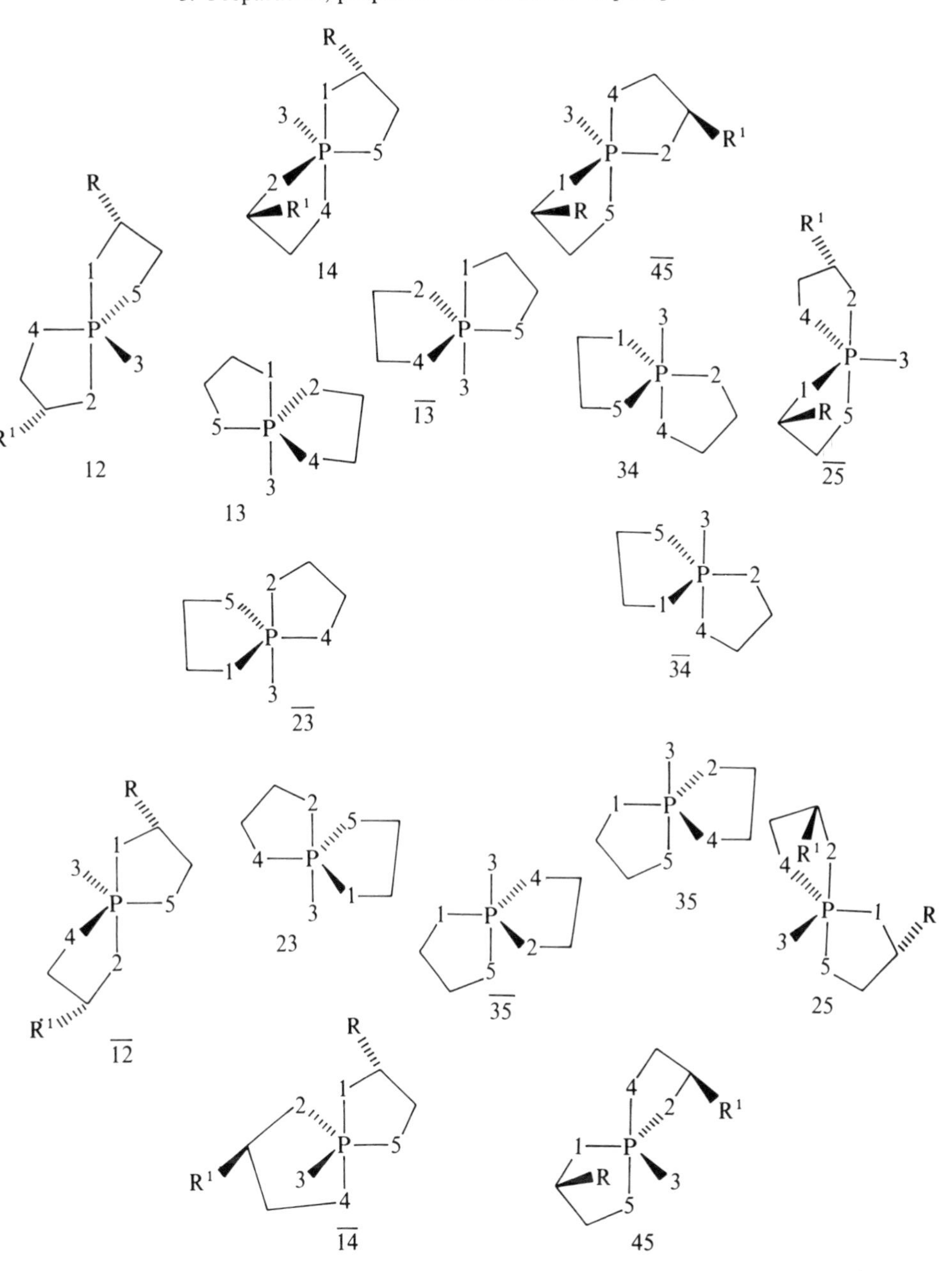

FIGURE 15. The sixteen possible isomers of a spirophosphorane containing a chiral P atom and two chiral C atoms. The chirality of the C atoms, which are in position a to the atoms 1 and 2 respectively, is due to the substituents R and R^1 which they bear, but R and R^1 need not necessarily be different. (For clarity, R and R^1 have not been drawn in the case of the isomers on the inner circle of the Figure.) The isomers on the outer circle are stable, the others, on the inner circle, do not exist or are only metastable

where N(TBP) is the number of trigonal bipyramids which can be obtained. The order of the letters in the operators listed indicates the sequence of permutations; for instance, *eaea* describes the following operation: starting from isomer 12, an equatorial ligand (say 3) replaces an apical ligand (say 1), which itself replaces another equatorial ligand (4), with the latter going to the second apical site (2) vacated by its ligand gone to occupy the empty equatorial site (3) first vacated. Isomer 12 has therefore now become 34, and ligand 5 has not participated in the fourfold ring of permutations. Note also that the notation (*ae.ee*) indicates two separate rings, the first being the permutation of an apical and an equatorial ligand and the second the permutation of the two other equatorial ligands.

It should be remembered that although the number of different TBP is strictly limited to 20 (the sum of the values in the second column of the table above), the different operators of a given mode may each give identical results. This is obviously true for modes C_0 (the identity mode) and C_3 (the racemization mode), but also for all other modes: thus, in the class C_4 and starting from isomer 12, the isomers 13, 14, 15, 23, 24 or 25 can be obtained using either *aee* or *aae*. Starting then from any isomer, successive applications of the six classes of permutations will yield all the 20 isomers.

Figure 14 gives graphical illustrations of some of the permutations in classes C_1–C_3, and Figure 15 shows the 16 isomers of a spirophosphorane with two identically substituted ($R = R^1$) five-membered rings which, in isomer 12, are linked at 1, 5 and 2, 4. The reasons for this choice are that it shows:

(a) the effect of the different permutations on the position of the substituents R and R^1 with respect to the extracyclic ligand; thus, R and R^1 are pseudo-apical and *trans* with respect to 3 in 12, but pseudo-apical and *cis* in $\overline{12}$; the same observation is applicable to the isomers 45 and $\overline{45}$, with R and R^1 in pseudo-equatorial positions; from the positions of R and R^1, the positions of all the other ligands at 1, 5 and 4, 2 can be quickly and visually deduced;
(b) the influence of a cyclic constraint on the number of stable isomers; the eight isomers in the outer circle are stable, whereas those on the inner circle are, energetically, at a disadvantage;
(c) the reason for the elimination of isomers 15 and $\overline{15}$, 24 and $\overline{24}$, due to the fact that each would require a diapical ring;
(d) finally, that an *ee* or *aa* permutation transforms 12 into $\overline{12}$, 45 into $\overline{45}$, etc.

B. Mechanisms of Stereomutations

Mechanisms will be called regular if they do not involve any bond cleavage, as against irregular mechanisms involving bond rupture and recombination, occurring, for instance, in the equilibria between P(V) and other hybridization states of phosphorus [P(III), P(IV), P(VI)]. It is not always easy to exclude with absolute certainty irregular processes, because the latter may be induced by mere solvent effects, and P(IV) may be formed by acid catalysis or P(VI) by basic catalysis. Irregular mechanisms may, however, be detected by the thermal variation of ^{31}P NMR signals; depending on the rate of the interconversion processes, two cases may arise:

(a) if the equilibrium in which the P atom is involved is slow with respect to the characteristic time scale of the method of observation, the chemical shifts observed correspond to two states of hybridization, and the amplitude of each signal varies with the temperature or the solvent;
(b) if the equilibrium is fast, the single chemical shift observed is the weighted mean of the signals due to the different P species present.

In yet a third possibility, the signal corresponding to coupling between P and other atoms, such as H, or F or C, may be present or have disappeared, indicating that the bonds involved have been preserved or broken.

The trigonal bipyramid structure of PF_5, $MePF_4$, $F_3P(CF_3)_2$ and F_3PCl_2, has been proved by various techniques such as electron diffraction[104], IR[105] and Raman spectroscopy[105] and nuclear quadrupole resonance[106]: the apical and equatorial PF bonds are different, resulting in non-equivalence of the F atoms occupying these two positions. The observation by Berry[5] that the fluorine atoms became equivalent above − 100 °C led him to propose a mechanism whereby equivalence was obtained by rapid exchange between apical and equatorial F atoms without bond cleavage. Taking one of the equatorial ligands as a reference, this mechanism implies variations of the angles between the reference and each of the four other ligands in the TBP: the 180° angle between the two apical substituents is reduced to 120°, while the 120° between the two other equatorial ligands widens to 180°. There is no rotation of any group with respect to any other group during the whole process which, starting from a D_{3h} symmetry, passes through an intermediate C_{4v} to end up once again as D_{3h}, as shown in Scheme 1: the reference pivot ligand (3 in Scheme 1) retains its *e* configuration whereas ligands 1 and 2, which originally were both in apical configuration, are both equatorial at the end of the process, while both 4 and 5, which were *e*, are now *a*. The choice of 4 or 5 as pivot would have led to 35 and $\overline{34}$.

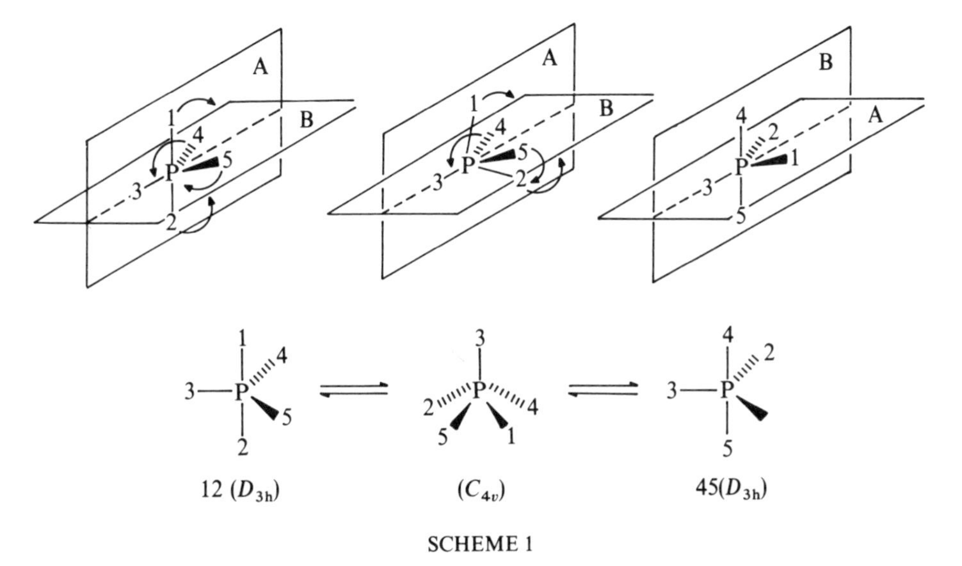

SCHEME 1

The analysis by Whitesides and coworkers[107] of the line widths in the variable-temperature, proton-decoupled ^{31}P NMR spectra of Me_2NPF_4 led them to conclude that, with the Me_2N ligand as pivot and throughout any stereomutation resulting in the interchange of apical and equatorial F atoms, the latter always behaved as inseparable pairs. This is in agreement with the Berry mechanism but excludes five out of all the schemes which can be called upon to explain the stereomutations[98,88], and immediately sets three questions:

(a) whether stereomutations involving original structures fairly different from the ideal TBP occur via the same mechanism;
(b) whether inseparable pairs, such as those mentioned above, are always part of the mechanism;

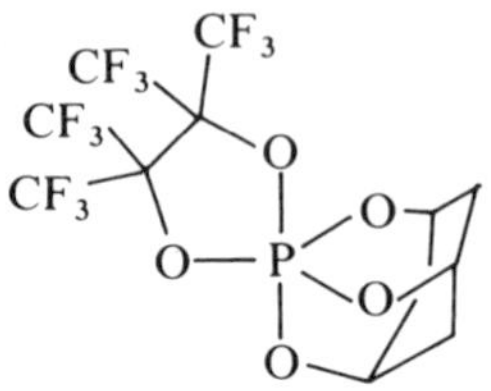

(41) 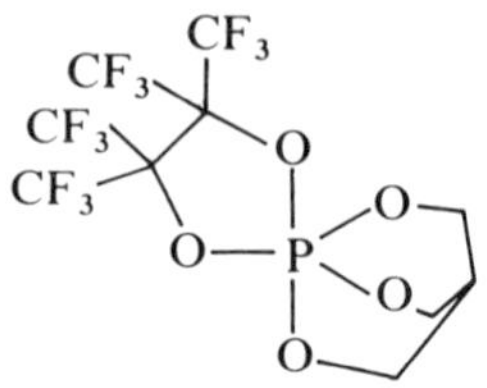 (42) 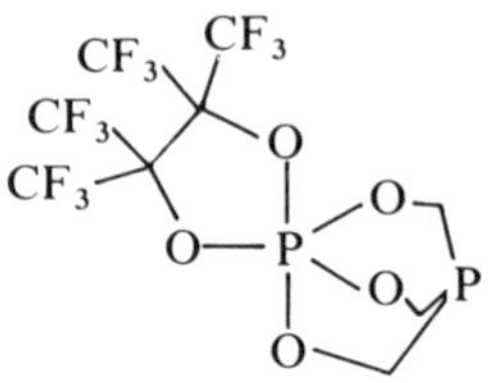 (43)

FIGURE 16

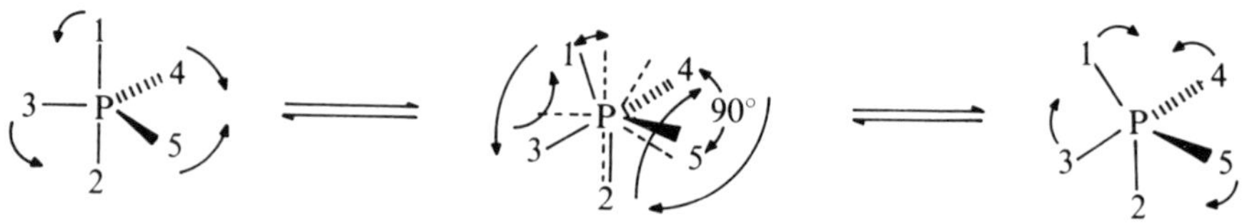

1–3 pair
2–4–5 trio

12 (D_{3h})

$0° \rightarrow 60°$ TR

Local $C_{2v'}$ and $C_{3v'}$ symmetry

35 (D_{3h})

FIGURE 17. The turnstile rotation (TR) mechanism (see text)

(c) whether the process underlying stereomutations in polycyclic phosphoranes (cf. **41**, **42** and **43**, Figure 16), which cannot accept variations of the angles between the substituents of the BPR, is the same.

Ramirez, Ugi and coworkers[28,84] have put forward the turnstile rotation (TR) mechanism, which can explain the fast stereomutations of compounds such as **41**, **42** or **43** without contradicting the conclusions of Whitesides and coworkers. Referring to Figure 17, the process can be defined with reference to three types of movements of the ligands of a TBP structure:

(i) a compression to 90° of the 120° angle between the two *e* ligands in the (4, 5, 2) trio formed by two *e* ligands (here 4 and 5) and one of the *a* ligands (say 2); at that point, all the angles between any two of the P—4, P—5 and P—2 bonds are 90°;

(ii) a concerted swivel by about 9° of the (1, 3) pair in the (P—1, P—3) plane defined by the two bonds, around an axis passing through P and perpendicular to that plane, and in a direction away from the trio so that the bisector of angle 3P1 is now colinear with the axis of the pyramid P245, and ligands 3 and 2 are eclipsed; this configuration is called 0°-TR as no rotation of the pair has occurred with respect to the position of the substituents in the trio;

(iii) a rotation around the axis of the pyramid, and in opposite directions, of 24° for the pair and of 36° for the trio, equivalent to a total rotation of 60° and bringing 3 over 4 (or 5, if the rotations are reversed) and completing the TR process: a reverse tilt by about 9° then occurs, in addition to an expansion to 120° of angle 5P2 (or angle 4P2) restoring the TBP configuration, as in 35 (Figure 17).

In this model, when the total rotation is only 30° (midway through the third process), one of the points of highest energy is reached, corresponding to a 30°-TR barrier. If this energetic barrier is too high or the final (eclipsed) configuration is unfavourable, the rotation may be reversed or, on the contrary, it may continue until a better configuration (say 14, $\overline{23}$ or $\overline{34}$, equivalent to TR, TR^2 or TR^3, respectively) is reached.

Evidently, the same process may start with any one of the (1, 4), (1, 5), (2, 3), (2, 4) or (2, 5) pairs instead of the (1, 3) pair. Similarly, while the molecule is in the 0°-TR configuration, a switch may occur, resulting in a permutation of one of the ligands in the pair and one of the ligands in the trio, resulting in the creation of a new pair and a new trio; however, in polycyclic structures such as **41**, **42** or **43**, the pair must be the five-membered ring and the trio must be the polycyclic ligand.

C. Comparison of the Turnstile Rotation and Berry Pseudorotation Processes

Detailed discussions of these problems have appeared[66,108-111]. As a short summary, one can list the following conclusions. (1) Both the TR and BPR processes respect Whitesides and coworkers' criterion of concerted motion by pairs[107]. (2) It is always possible to find a TR process equivalent to a given BPR mechanism; in fact, up to four different TR processes may lead to the same result so that, other things being equal, TR is more probable than BPR. (3) In BPR, four out of five ligands participate in the motion without relative internal rotation, while all five ligands and internal rotation are involved in TR. (4) TR is applicable to all structures, whereas BPR cannot explain the stereomutation of certain phosphoranes with cage structures.

Both the TR and BPR mechanisms thus seem about equally probable, with a slight edge for TR in the case of cyclic systems, but other mechanisms may be imagined in the future.

IV. CHIRAL PHOSPHORANES[66,112-116]

The synthesis of an optically active chiral phosphorane in which the optical activity is solely connected with the P atom implies, for instance, the preparation of a trigonal bipyramid or a tetragonal pyramid with five different achiral ligands which would, at least theoretically, entail the possible formation in the first case of ten or in the second case of fifteen diastereoisomers requiring separation, a formidable task even in the absence of stereomutation. The problem has nevertheless been solved using three different approaches exploiting the particular properties of spirophosphoranes, as hereafter detailed.

An examination of the isomers depicted in Figure 15 leads to the following conclusions:

(a) The P atom in the spirophosphoranes is an asymmetric centre. If the two rings are chemically and stereochemically identical, the product of a synthesis will be a racemate theoretically separable into its two enantiomers, but a single stereomutation would be sufficient to invert the configuration, voiding all attempts at separation. Structures 15 and $\overline{15}$, 24 and $\overline{24}$ are excluded since five-membered rings cannot be *aa*; likewise, unstable and energetically unfavourable *ee* conformations are excluded.

(b) The presence of an asymmetric centre on each of the rings in addition to that on the P atom creates eight stable and eight unstable racemates subject to interconversion by stereomutation, but in the absence of stereomutations and with $R = R^1$, the number of racemates is reduced to three.

(c) If 12, the reference isomer, has two rings which both have *R* or *S* centres of asymmetry, the eight stable isomers in Figure 15 are not enantiomeric (12 is not the enantiomer of $\overline{12}$) but diastereoisomeric, and the enantiomers of the eight racemates can only be obtained by the presence of two rings of inverse configuration. In other words, 12 and $\overline{12}$ will only be enantiomers if the cyclic moieties bearing R and R^1 in 12 and $\overline{12}$ are optical antipodes.

As an example, the positions occupied by R and R^1 in the eight isomers of Figure 15 are (the chiral atoms bearing R and R^1 are in the same configuration):

Isomer:	12	$\overline{25}$	$\overline{45}$	14	$\overline{12}$	25	45	$\overline{14}$
Position of R and R^1:	*AtAt*	*AtEc*	*EtEt*	*AcEt*	*AcAc*	*AcEt*	*EcEc*	*AtEc*

where

At (or *Ac*) = R in pseudo-apical position *trans* (or *cis*) with respect to the fifth, non-cyclic ligand and

Et (or *Ec*) = R in pseudo-equatorial position *trans* (or *cis*) with the respect to the fifth, non-cyclic ligand,

where 'pseudo' is used to indicate that R and R^1 are directly linked to the atom in an *e* or *a* position.

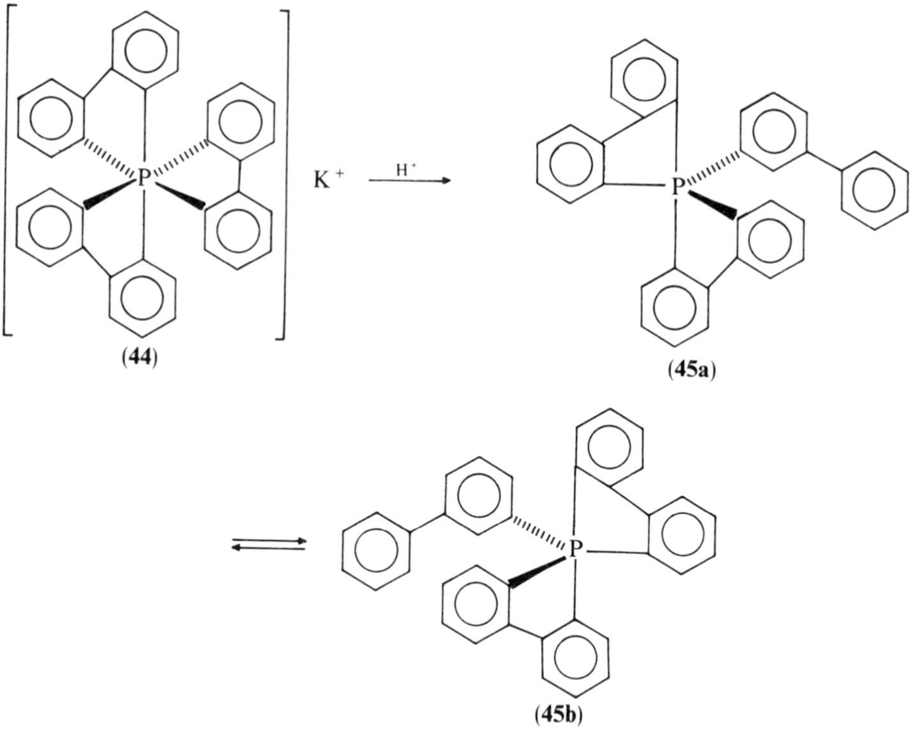

K$^+$ $\xrightarrow{H^+}$

(44) **(45a)**

⇌

(45b)

SCHEME 2

The spirophosphorane **45** (Scheme 2) was synthesized by Hellwinkel[117] by acid cleavage of the optically active hexacoordinated anion **44** previously separated into its two enantiomers. However, and as explained in (a) above, because of the easy stereomutation (1.5 kcal mol^{-1} at room temperature), racemization resulted from the inversion of configuration **45a**⇌**45b**. Using, as an NMR probe, Me substituents symmetrically placed on each aromatic cycle of the biphenylylene groups, Hellwinkel and coworkers[118] examined homologues of **45** in which the fifth ligand was Me, Et, $PhCH_2$, Ph, β-$C_{10}H_7$, etc., and found that their rate of stereomutation at room temperature decreased as the bulk of the fifth ligand increased. Similar results were obtained, at different temperatures, by Whitesides and Bunting[119]. The free enthalpy of the reorganization process has been calculated[114,119] [$\Delta G^* \approx 12$ kcal mol^{-1} (Me, $-60°$ C); $\Delta G^* \approx 17$ kcal mol^{-1} (α-$C_{10}H_7$, $+120$ °C)]. Ultimately, when the extracyclic ligand is an 8-methoxy-1-naphthyl or 8-dimethylamino-1-naphthyl group, the probe Me groups on the Ph rings give four NMR signals and no stereomutation occurs up to 100 °C[120] (ΔG^* has been estimated to be then larger than 23 kcal mol^{-1}) and the optically active form has a chance of existing, as proved by Hellwinkel[117] when he ingeniously isolated the first optically active phosphorane, helped, as was Pasteur in his historic work, by the fortunate occurrence of a spontaneous separation of the racemate: starting from the optically active, hexacoordinate anion **46** (Scheme 3) only differing from **44** (Scheme 2) in that it bears a single substituent on an aromatic ring, Hellwinkel obtained, by a slightly selective acid hydrolysis, the three phosphoranes **47**, **48** and **49** expected. Of the three, **47** is the only one in which stereomutation will not produce racemization, owing to the presence of two non-equivalent rings (cf. 12 and $\overline{12}$ in Figure 15). In fact, **47** is an equilibrium mixture of two diastereoisomers undergoing a fast stereomutation, with $[\alpha]^{24}_{578} = 94 \pm 1°$ (thf).

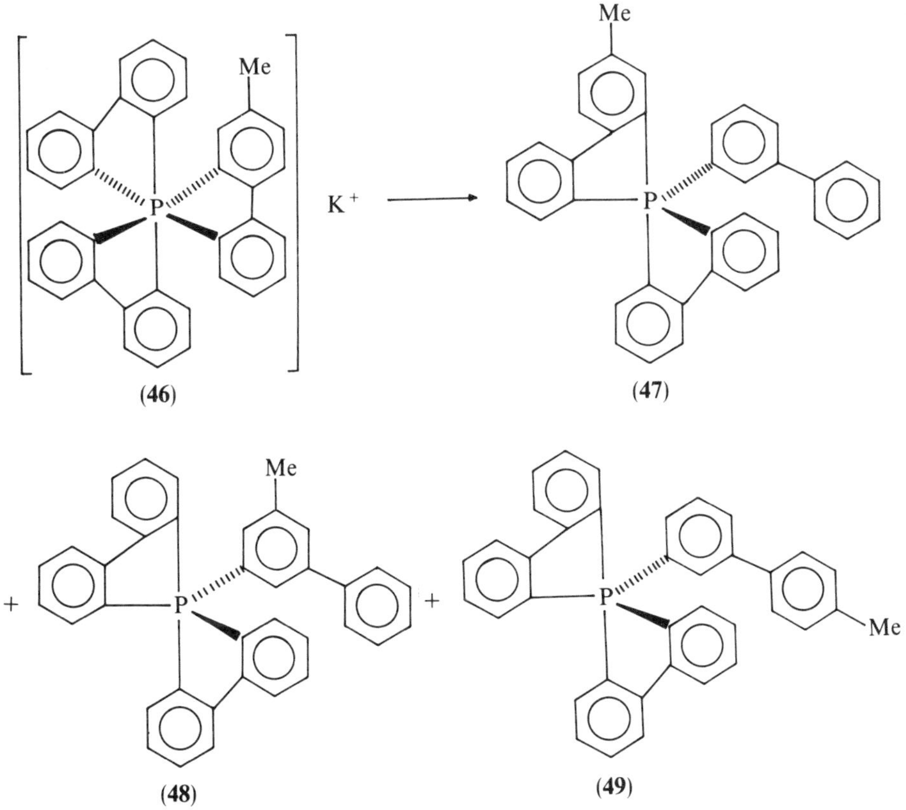

Me Me K$^+$ → P P

(**46**) (**47**)

+ P + P Me Me

(**48**) (**49**)

SCHEME 3

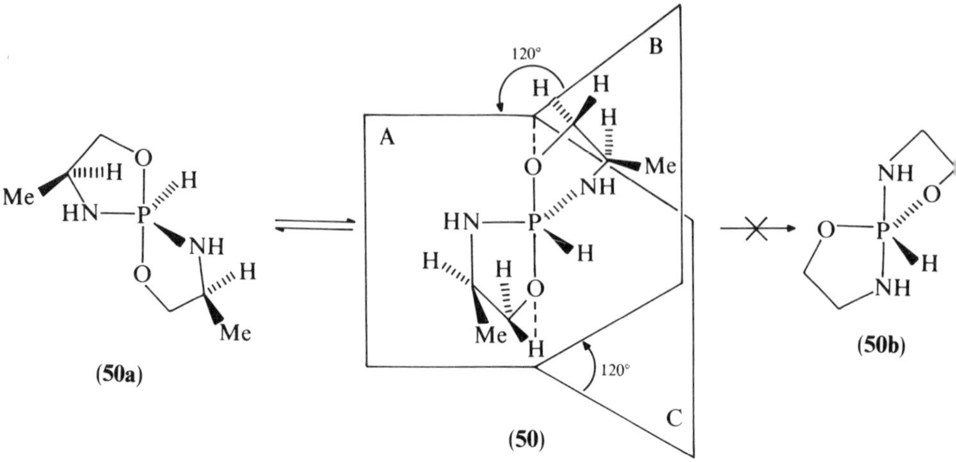

(50a) (50) (50b)

SCHEME 4

The second optically active phosphorane was obtained by Burgada *et al.*[121] via a completely different route: the spirophosphorane **50** (Scheme 4) was synthesized by the reaction of an amino alcohol (alaninol) on trisdimethylaminophosphane [$P(NMe_2)_3$]. The electronegativity rules force an equatorial position for the N atoms and rule out any stereomutation which could lead to the isomer **50b**. The three racemic diastereoisomeric spirophosphoranes expected are formed, but using the optically active L-(+)-alaninol creates the following constraints: (a) the diastereoisomer in which the Me substituent is *cis* with respect to the PH plane on one ring and *trans* on the other cannot be formed; in this structure, the chiral C atom is *R* in one ring and *S* in the other; (b) the only diastereoisomers which can be formed are **50** and **50a**, eventually in stereomutational equilibrium, as found for **47** above.

The compound prepared showed two series of PH and MeCH signals in the NMR spectrum, corresponding to the two diastereoisomers, and $[\alpha]^{25}_{578} = 30.4°$. As seen by the

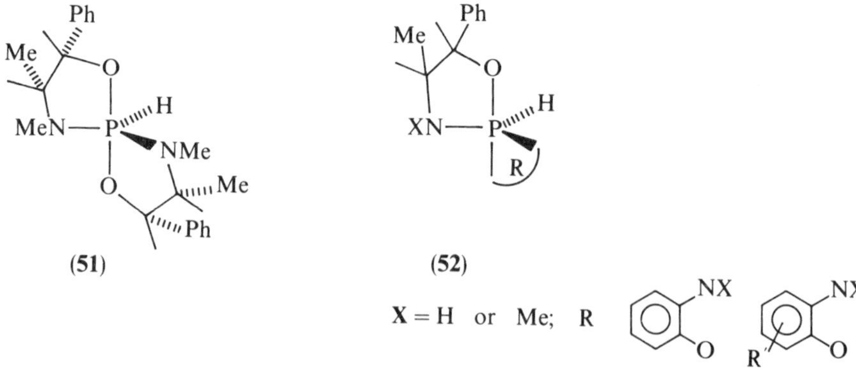

(51) (52)

X = H or Me; R

FIGURE 18

temporal variation of the rotary power[121], the two isomers in solution equilibrate by mutarotation. A single diastereoisomer is obtained by crystallization, and the solvent-dependent rate of epimerization of the P atom is slow at room temperature: in benzene and at 20 °C, a mixture with an initial 80:20 ratio of the two isomers is changed to 60:40 after 30 min, while heating to 100 °C a solution in PhCl which, at −20 °C, contained a single diastereoisomer leads to the 50:50 mixture.

The same principle was later largely developed and applied by Wolf and coworkers[122–127] to the synthesis of a series of optically active spirophosphoranes with a PH bond derived from ephedrine (Figure 18). Problems concerning the stereochemistry, the structure and the stereomutations of the optically active spirophosphoranes with this structure were treated in detail. Among them were the determination by XRD of the structure of the pure epimer of **51**[123] (Figure 18), the demonstration of the chirality of a racemic spirophosphorane by NMR using optically active shift reagents[124], the influence of temperature and of the solvent on the kinetics of epimerization[125] and the fundamental problem of the absolute configuration[123,126,128].

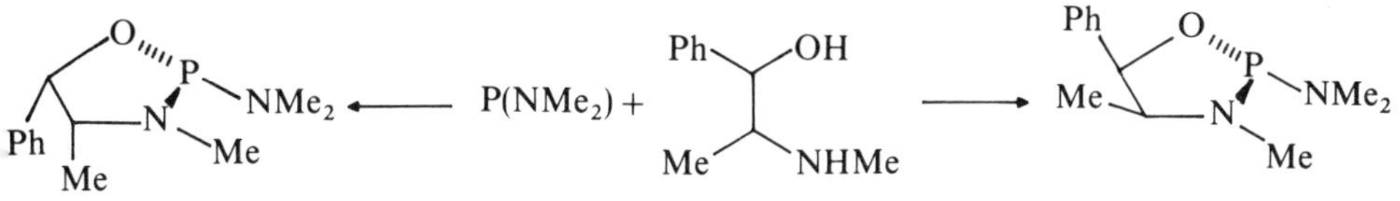

(−)-ephedrine

(**53**)

SCHEME 5

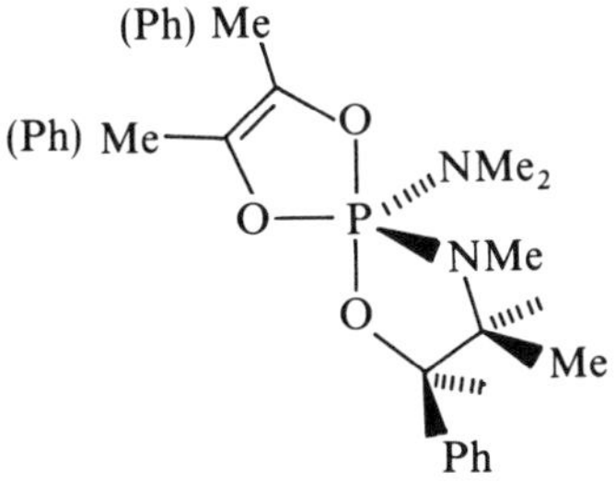

(**54**)

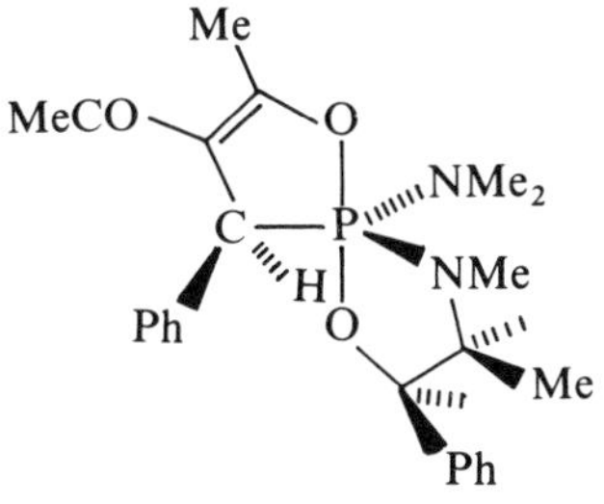

Me $[\alpha]_D^{22}$ −298°
Ph $[\alpha]_D^{22}$ −290°

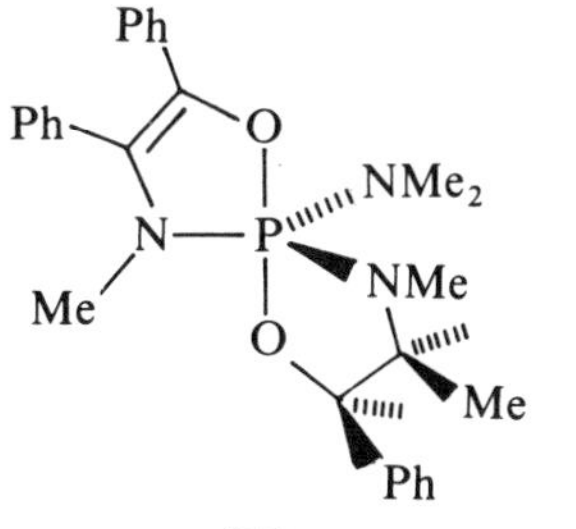

(**55**)

(**56**)

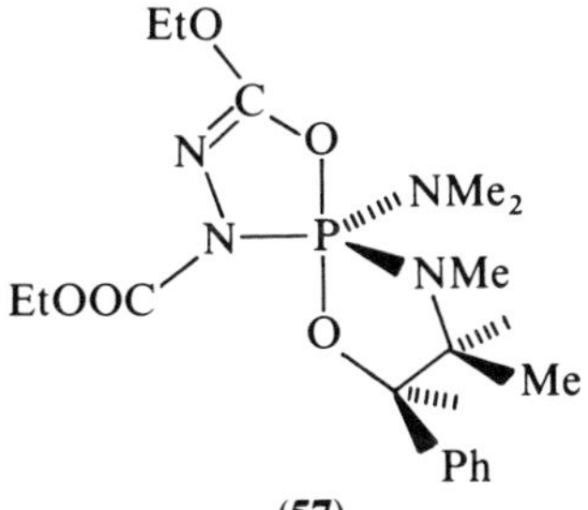

(**57**)

FIGURE 19

Even if some of the optically active spirophosphoranes mentioned above can exist as single species in the solid state, all of them are more or less rapidly transformed, when dissolved, into binary mixtures of diastereoisomers[114–129] such as 12 and $\overline{12}$. However, a third type of synthesis yielded pure, optically active phosphoranes not subject to change when dissolved[129]. This was obtained by the stereospecific synthesis of the oxazaphospholidine **53**[130] (Scheme 5) derived from the −(1*R*, 2*S*)enantiomer of ephedrine by means

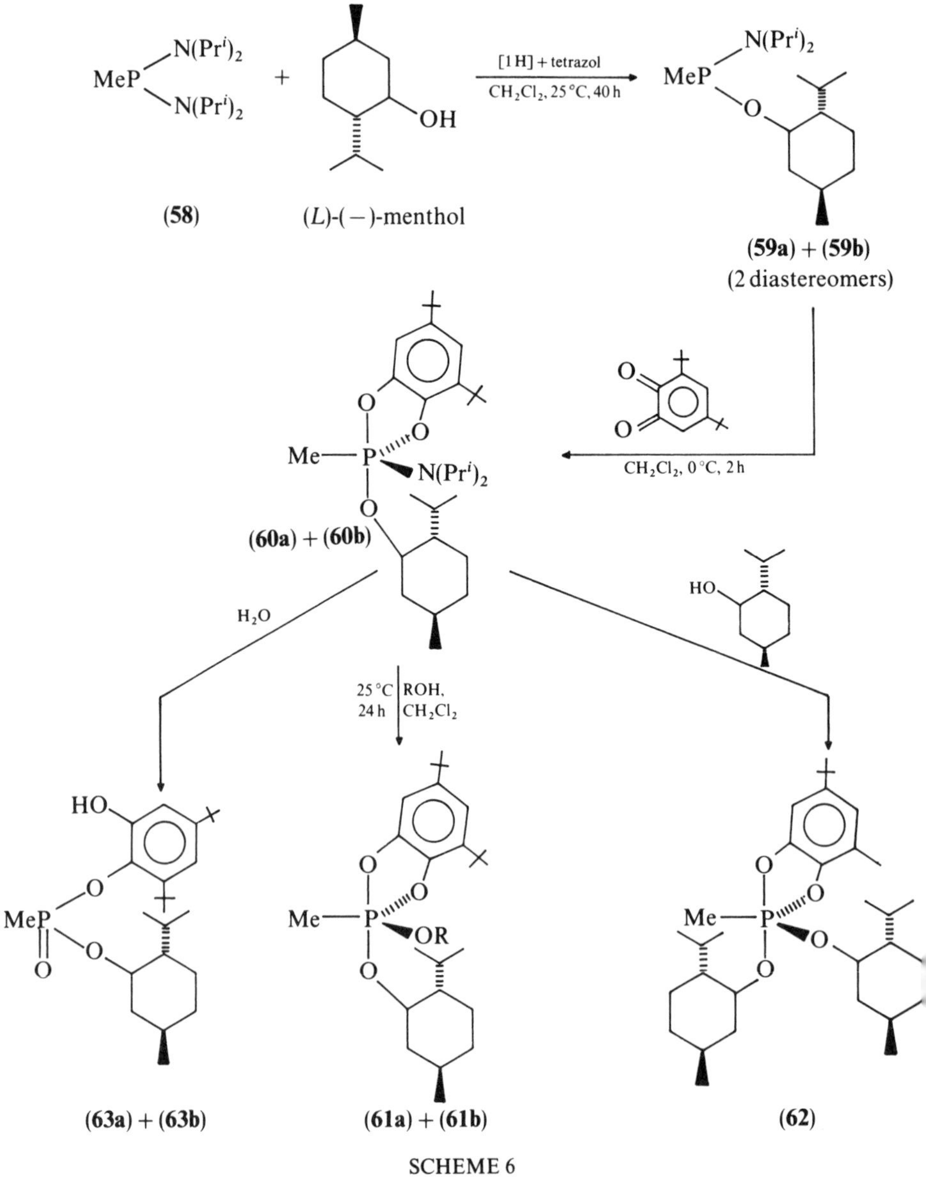

MeP($N(Pr^i)_2$)$_2$ (**58**) + (*L*)-(−)-menthol → [1 H] + tetrazol, CH_2Cl_2, 25 °C, 40 h → (**59a**) + (**59b**) (2 diastereomers)

→ CH_2Cl_2, 0 °C, 2 h → (**60a**) + (**60b**)

H_2O → (**63a**) + (**63b**)

ROH, CH_2Cl_2, 25 °C, 24 h → (**61a**) + (**61b**)

HO-menthyl → (**62**)

SCHEME 6

of a general reaction of this type of compound with α-diketones or ethylenic ketones[131] and also with ketoimines[132], to yield the spirophosphoranes **54, 55** and **56**, while the reaction with ethyl azodicarboxylate[133] yielded the spirophosphorane **57**. It is the stereospecificity of these reactions which is at the root of the formation of only one of the two diastereoisomers possible for each of **54, 56** and **57**, or the four possible for **55** (Figure 19). Heating **54** at 100 °C in PhCl yields a 50:50 equilibrium mixture of the two diastereoisomers, unchanged at room temperature and with the same rotary power. The free energy of activation of stereomutation is certainly larger than 24 kcal mol^{-1}.

There are probably two reasons behind the stereospecificity of the syntheses of phosphoranes **54, 55, 56** and **57**: the exothermicity of the reaction, likely to result in kinetic control, and the geometry of the transition state, close to that of the reagents[129] (Hammond's principle[134]).

To our knowledge, the first examples of asymmetrically substituted monocyclic phosphoranes are **60** and **61**, described by Moriarty *et al.*[135] and involving the reaction of a substituted *o*-benzoquinone[136,137] (Scheme 6) on an aminophosphine (**59**), itself obtained by alcoholysis of **58** with L-(−)-menthol. In contrast to the amino phosphine **53** (Scheme 5), **59** is a mixture of the diastereoisomers **59a** and **b**, and its reaction with 3,5-di-*tert*-butyl-1,2-benzoquinone yields two diastereoisomeric phosphoranes, **60a** and **b**. Finally, alcoholysis of the P(V)—NR_2 bond[138] in **60a** and **b** leads to **61a** and **b** or **62**.

The scheme followed more recently by Moriarty *et al.*[139] again involved a racemic phosphine (**64**, in Scheme 7), used to synthesize a racemic phosphorane **65**, but the optically active ligand was introduced only at that point, to replace NMe_2. As before, two diastereoisomers, **66a** and **b**, were obtained, as proved by the two series of signals in the ^{31}P, ^{1}H and ^{13}C NMR spectra, and of the δ and *J* values which were determined for the

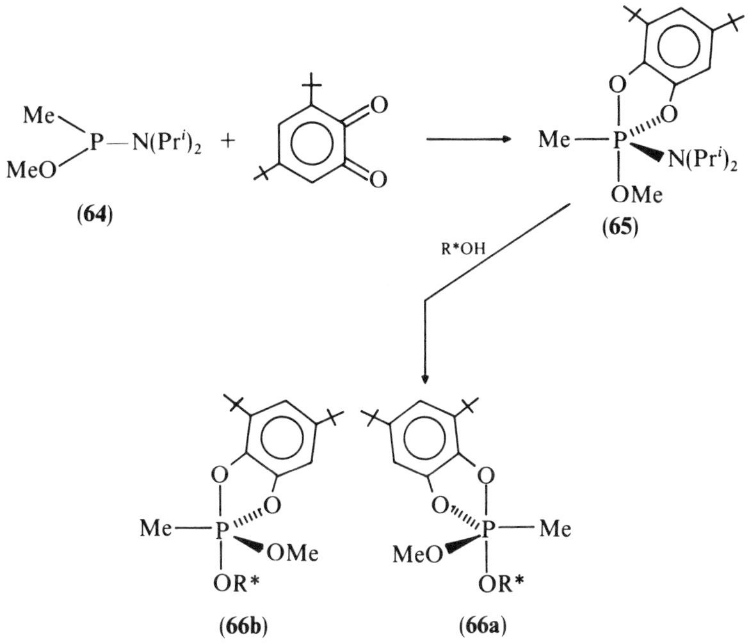

Me
P—N(Pr^i)$_2$ +
MeO
(64)
Me—P—N(Pr^i)$_2$
OMe
(65)
R*OH
Me—P—OMe
OR*
(66b)
MeO—P—Me
OR*
(66a)

SCHEME 7

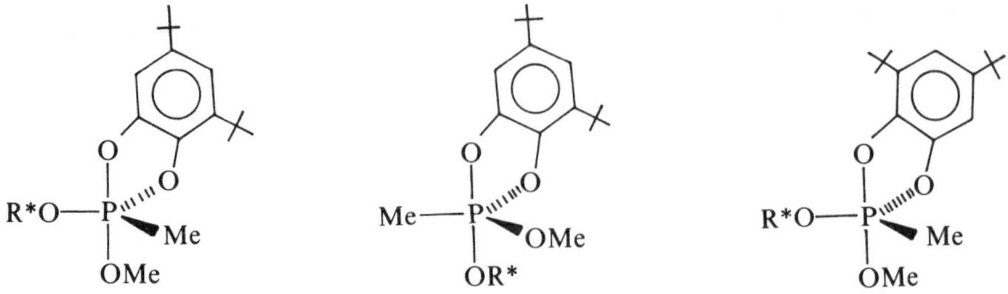

(**66′**) (**66″**) (**66‴**)

FIGURE 20

MeP moiety. The diastereoisomers were separated by chromatography or by crystallization, and the absolute configuration (TBP structure) of the P atom in one of the stereoisomers was determined by XRD. The rotary powers were weak, namely 9.40° for **66a** and 2.94° for **66b**, with R*OH = $MeCH(OH)C(Me)_2CH_2NMeCOC_6H_4(NO_2)_2$.

Epimerization of the isolated isomers was obtained by heating at 90° C, and the energy of activation of the first-order process found to be 27 kcal mol^{-1}. Since the transformation of **66a** into **66b** corresponds to an *ee* permutation (class C_3), three more isomers (**66′**, **66″** and **66‴**) can be envisaged (Figure 20) in which one would still have an *ae* ring and an *e* Me

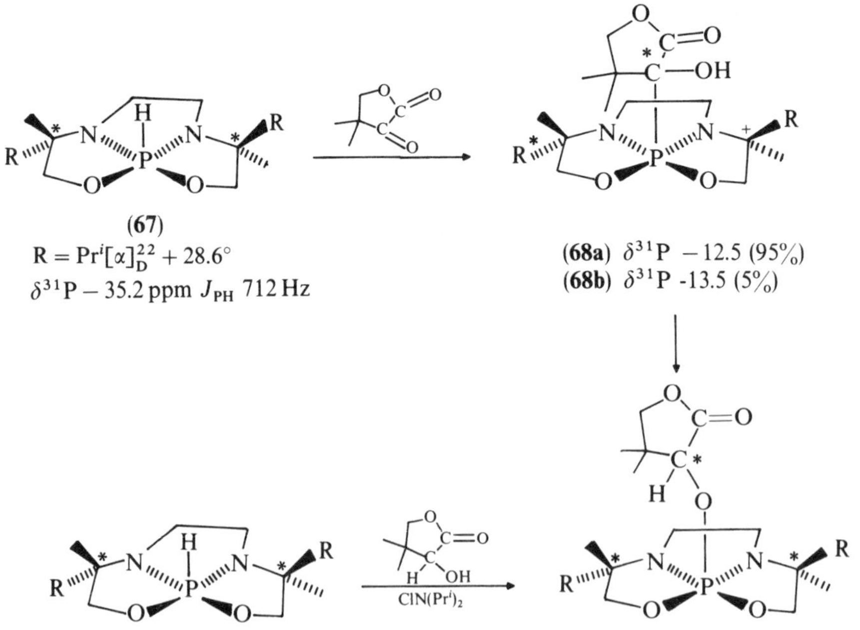

(**67**)
R = $Pr^i [\alpha]_D^{22}$ + 28.6°
$\delta^{31}P$ − 35.2 ppm J_{PH} 712 Hz

(**68a**) $\delta^{31}P$ − 12.5 (95%)
(**68b**) $\delta^{31}P$ -13.5 (5%)

(**67**)
R = Pr^i

$ClN(Pr^i)_2$

(**69a**) $\delta^{31}P$ − 18.5 (75%)
(**69b**) $\delta^{31}P$ -18.0 (25%)

SCHEME 8

group, but the other substituents would be permuted, resulting from fast stereomutations of **66a** and **b**, in strict respect of the electronegativity and steric rules.

We conclude the list of examples with **67**, a tricyclic phosphorane with a PH bond, synthesized[140] from the chiral diaminodiol *N*, *N'*-bis(1-alkyl-2-hydroxyethyl)ethylenediamine and $P(NMe_2)_3$ (Scheme 8). Compound **67**, which gives a single NMR signal at room temperature, is a mixture of two indistinguishable stereoisomers formed by the fast stereomutation leading to epimerization at the P atom. Condensation of **67** with dihydro-4,4-dimethylfuran-2,3-dione quickly forms the diastereoisomeric phosphoranes **68a** and **b** with a PC* bond, the high selectivity of the reaction resulting from the asymmetric induction during the formation of the PC bond. Phosphoranes **68a** and **b** are unstable and are transformed into **69a** and **b**. The reaction of **67** on dihydro-3-hydroxy-4,4-dimethylfuran-2(3*H*)-one in the presence of 2 mol of chlorodiisopropylamine leads directly to the stable diastereoisomer **69a**, which can be isolated (Scheme 8).

V. EQUILIBRIA AMONG DIFFERENTLY COORDINATED PHOSPHORUS SPECIES

A. Introduction

The importance of equilibria in the chemistry of pentacoordinated phosphorus arises from the possibility of passing from a given structure to another, even if the latter has a totally different molecular geometry and distribution of electrons, by a mere shift of hybridization. This has important consequences on the chemical properties of phosphoranes and on the possibilities of interconversion of isomers by bond rupture and recombination processes akin to regular true stereomutations, i.e. without bond cleavage.

The following example is a good illustration of these two facts: the spirophosphorane **70** (Scheme 9), with a TBP structure, is a racemate of the enantiomers **x** and $\bar{\mathbf{x}}$, which can both be in tautomeric equilibrium with **70a**, a derivative of tricoordinated phosphorus. The pentacoordinated P is electrophilic and can therefore undergo a number of reactions with nucleophilic reagents, whereas the tricoordinated P atom of **70a** is nucleophilic or even biphilic, and **70a** itself, with its OH group, has a nucleophilic protic centre with all its specific chemical properties.

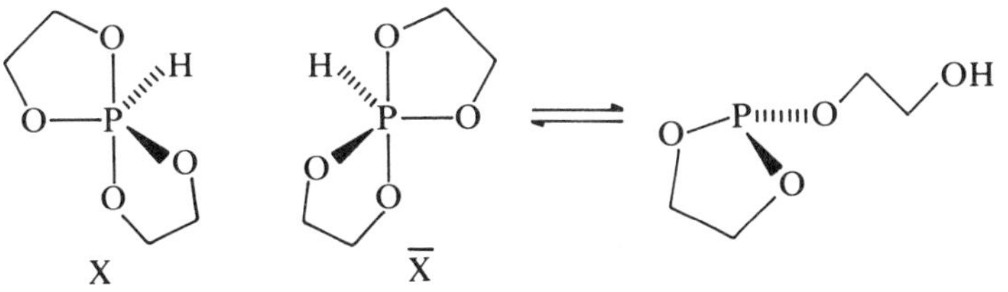

X $\bar{X}$

(70) **(70a)**

SCHEME 9

It is clear that the stereochemistry of the phosphorane is subordinated to the **70**$\rightleftharpoons$**70a** equilibrium resulting in an interconversion between **x** and $\bar{\mathbf{x}}$, and therefore in the automatic racemization of the structure. The same reasoning applies to homologues of **70** obtained by substitution on the C atoms of either ring, even if this substitution produces new centres of asymmetry besides that on the central P atom.

The phosphorane **70** can also undergo stereomutation without bond cleavage by the Berry process (Berry pseudorotation with the PH bond as pivot) or by the Ramirez process (turnstile rotation). The result, in either case, is an inversion of configuration at the P atom, with consequent racemization of **70**.

This example illustrates the importance of the tautomeric P(III)⇌P(V) equilibrium yielding the same result as a stereomutation without bond cleavage. In view of their importance in phosphorane chemistry, we shall therefore study the following equilibria: P(III)⇌P(V), P(IV)⇌P(V), P(V)⇌P(V) and P(VI)⇌P(V).†

B. P(III)⇌P(V) Equilibria

1. Spirophosphoranes and polycyclic phosphoranes with a PH bond

The first tautomeric, P(III)⇌P(V), chain⇌ring equilibria involving the migration of a proton were found during the study of spirophosphoranes with a PH bond, a number of which (**71**, **72** and **73**, Scheme 10) were first described as P(III) derivatives **71a**[141], **72a**[142] and **73a**[141] in agreement with their chemical reactivity, and later as P(V) compounds[143,144], as indicated by the ^{1}H and ^{31}P NMR results, this suggesting an equilibrium[145], confirmed by IR results[146] and the thermal variation of NMR signals[147]. This is a common occurrence in solutions of PH-bearing spirophosphoranes, in which the P(V)/P(III) ratio can vary from 0 to ∞. The rate of interconversion is sufficiently low to permit observation by ^{1}H and ^{31}P NMR. Thus, a solution of **71** in CH_2Cl_2 is 90% P(V) (^{31}P NMR signal at −26 ppm) and 10% **71a** [P(III) signal at 130 ppm].

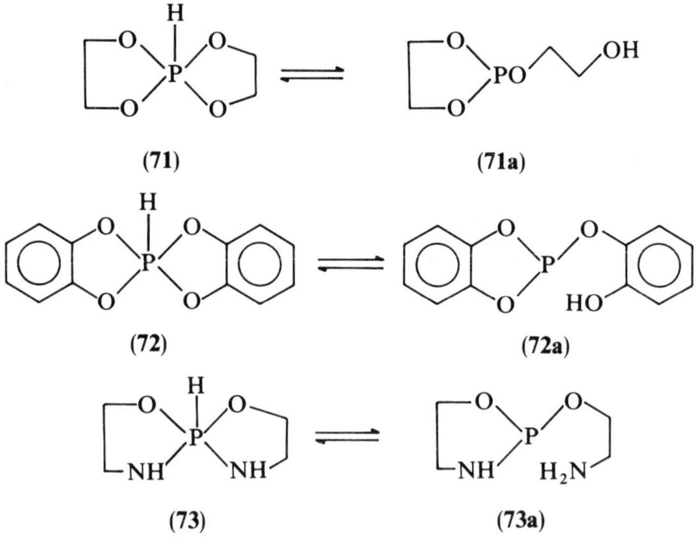

(**71**) (**71a**)

(**72**) (**72a**)

(**73**) (**73a**)

SCHEME 10

The thermal variation of the equilibria (Scheme 11) shown by about 100 spirophosphoranes has clarified the role played by a number of factors[148,149], and the interactions among them have been abundantly discussed[85,150]. These factors are:

† For the remainder of this chapter, the notation P(III), P(IV), P(V) and P(VI) refers to tricoordinated, tetracoordinated, pentacoordinated and hexacoordinated P atoms respectively, not to different valencies.

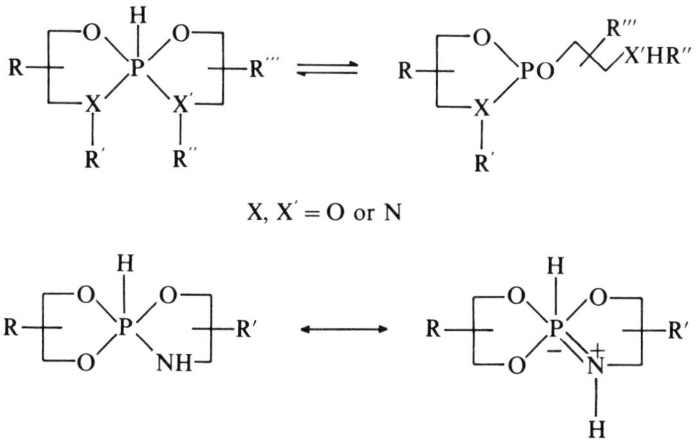

X, X′ = O or N

SCHEME 11

(a) temperature: its increase favours P(III) rather than P(V);
(b) basic solvents also favour P(III): thus, pure **72** is 100% pentacovalent at room temperature, and in equilibrium with 25% **72a** at 100 °C, but its solution in $O{=}P(NMe_2)_3$ contains 100% **72a** even at room temperature;
(c) unsaturation: the presence of aromatic cycles (pyrocatechols, *o*-aminophenols, etc.; cf. **71** and **72**, Scheme 10) favour P(V);
(d) symmetry factors: see Figure 11;
(e) steric factors: the presence of one to eight aromatic or aliphatic substituents on the rings favours P(V);
(f) electronic factors: they refer to the hardness[150] as defined by Pearson, or to the relative basicity[85,149] of P and of one of the O or N atoms (at x or x′, in Scheme 11); when *x* or *x*′ = N, the nature of R′ or R″ also affects the process; further, IR studies[151] have shown that a mesomer could also be involved.

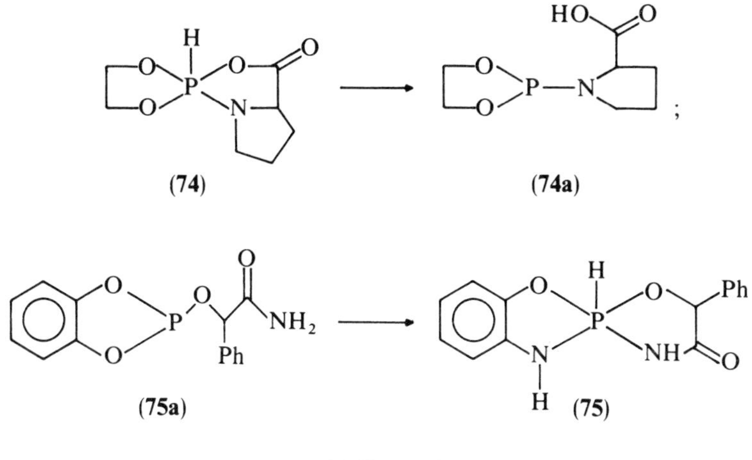

(**74**) (**74a**)

(**75a**) (**75**)

SCHEME 12

The equilibria involving **74** or **75** induce the formation and breaking of a PO bond and the closure and opening of one of the rings (Scheme 12). The equilibrium constants have been determined[152,153] and correspond to $\Delta G^* = 23.8$ and $24\,\text{kcal mol}^{-1}$, respectively.

This tautomeric chain–ring equilibrium can be generalized to a large variety of pentacoordinated structures, with different architectures surrounding the P atom and the four O or N atoms besides the PH bonds, namely **76**, the spirophosphorane derived from *o*-phenylenediamine[154], **77**, from diaminomaleonitrile[155], and the cyclamphosphoranes **79** and **80**, derived from cyclic polyamines[156]. It is interesting to note in the series of tetraaminophosphoranes, prepared by the action of trisdimethylaminophosphane on the macrocyclic amines[157] (Scheme 13), that **78**, with four five-membered rings, is pure P(V), that **81**, with four six-membered rings, is pure P(III), but that the simultaneous presence of

(**76**) ⇌ (**76a**)

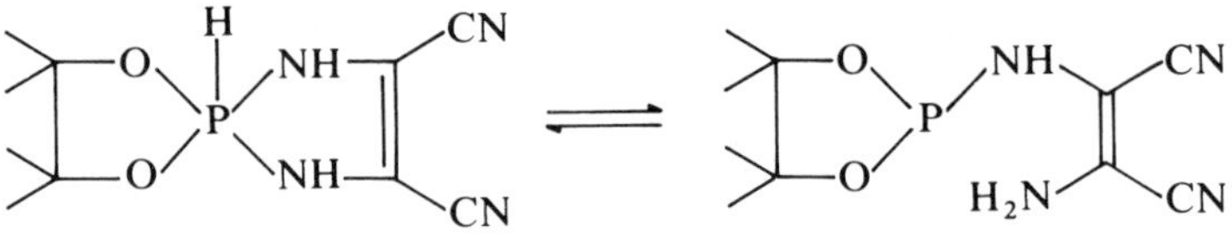

(**77**) ⇌ (**77a**)

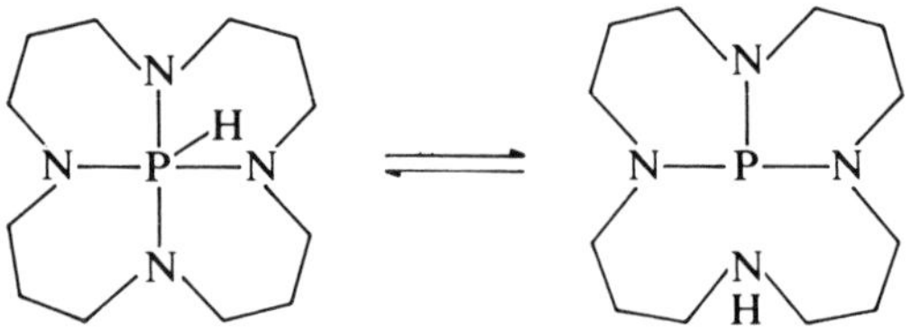

0% (**81**) ⇌ 100% (**81a**)

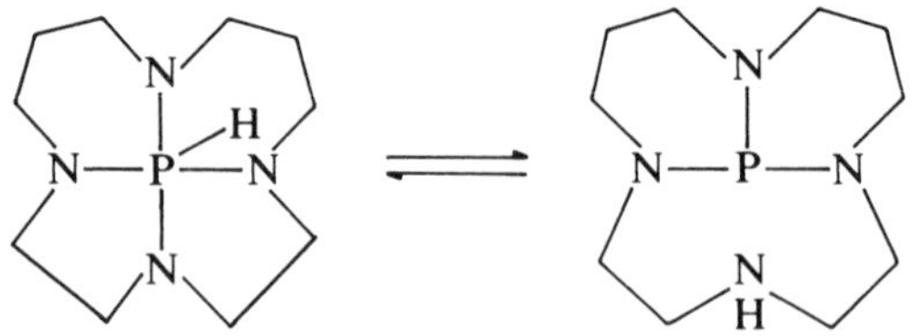

12% (**80**) ⇌ 88% (3 forms) (**80a**)

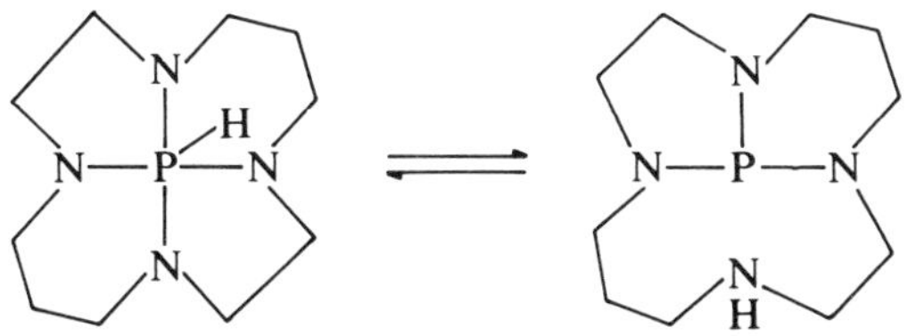

82% (79) ⇌ 18% (79a)

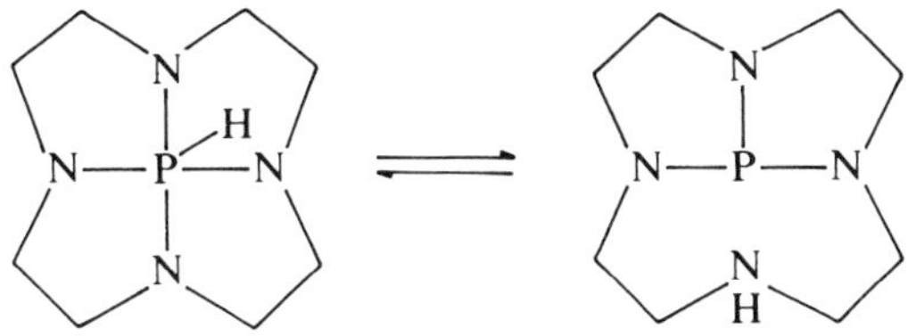

100% (78) ⇌ 0% (78a)

SCHEME 13

two adjacent six-membered and two adjacent five-membered rings in **80** is detrimental to P(V), whereas the latter is favoured when the five and six-membered rings are opposed, as in **79**.

In the spirophosphoranes with a PH bond, the presence of a six-membered ring destabilizes the pentacoordinated form. Thus, the spirophosphorane D in Scheme 14, although identified by ^{31}P NMR in a reaction mixture, could not be isolated[158]. Similarly, Van Lier *et al.*[153] attempted the preparation of spirophosphoranes containing a dioxaphosphorinane ring, such as **A**, **B** or **C** in Scheme 14. The synthesis of **A** starting from the P(III) compounds **E** or **F** showed that the **E**⇌**F** equilibrium existed, but no **A** was detected although its existence as an intermediate between **E** and **F** was deemed highly probable. Conversely, the precursors **G** and **H** of the spirophosphorane **C** are not in equilibrium, a fact which, in principle, excludes the existence of the pentacoordinated **C**. Precursors **E**, **F**, **G** and **H** are, themselves, in equilibrium with the tricoordinated diphosphite formed by the transesterification between two molecules, with the subsequent release of the diol, as for instance in

2**H** ⇌ I + dimethylcyclohexanediol.

2. Monocyclic phosphoranes and fused bicyclic phosphoranes with a PH bond

Phosphoranes devoid of polycyclic or spiran structures but with a PH bond can still be prone to P(III)⇌P(V) equilibria, as shown by the study of the changes in the ^{31}P NMR spectra brought about by the thermal variations to which were submitted the monocyclic phosphoranes[159] **82** (Scheme 15), which are intermediates in the synthesis of the spirophosphoranes **83** and whose structure and properties provide a good explanation for the redistribution of ligands observed in this series[160,161].

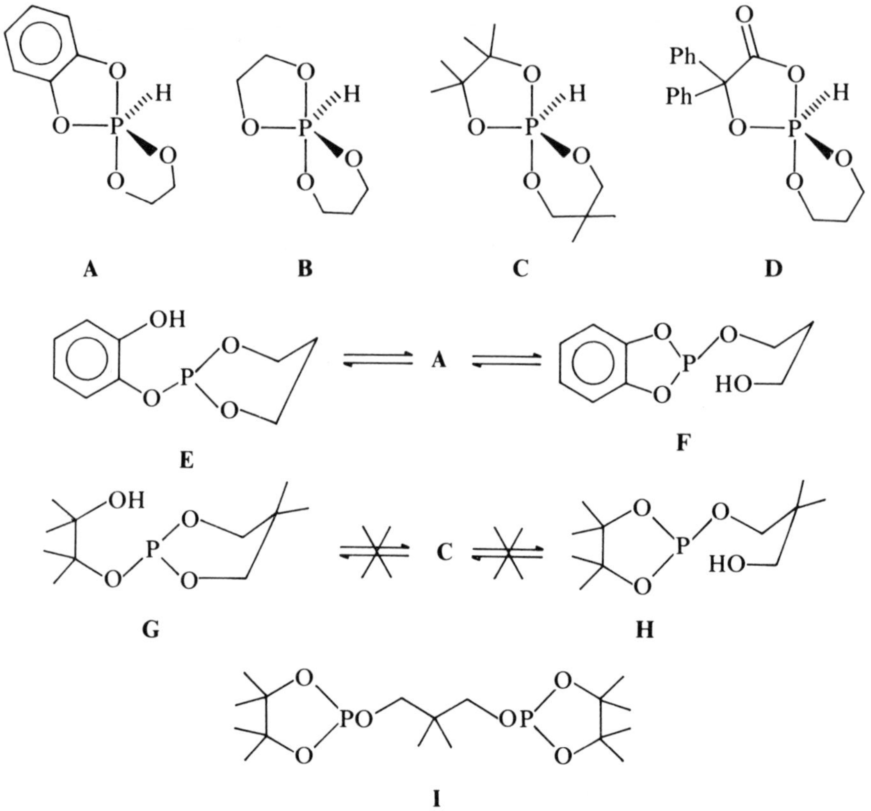

A B C D

E F

G H

I

SCHEME 14

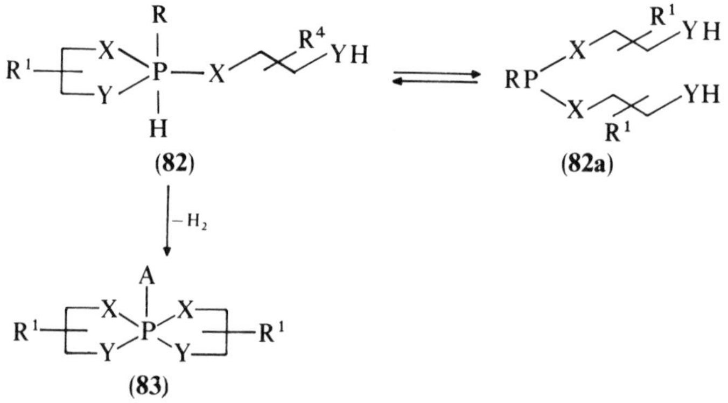

(82) (82a)

$-H_2$

(83)

R = Me, Ph
X = O, Y = O, NH
R^1 = Alkyl or Aryl

SCHEME 15

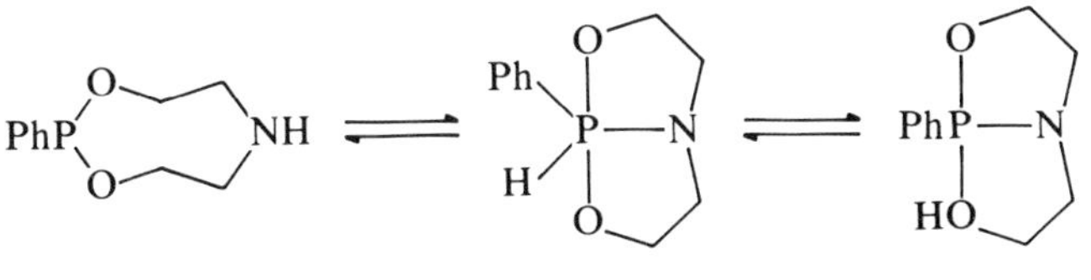

(84b) (84) (84a)

SCHEME 16

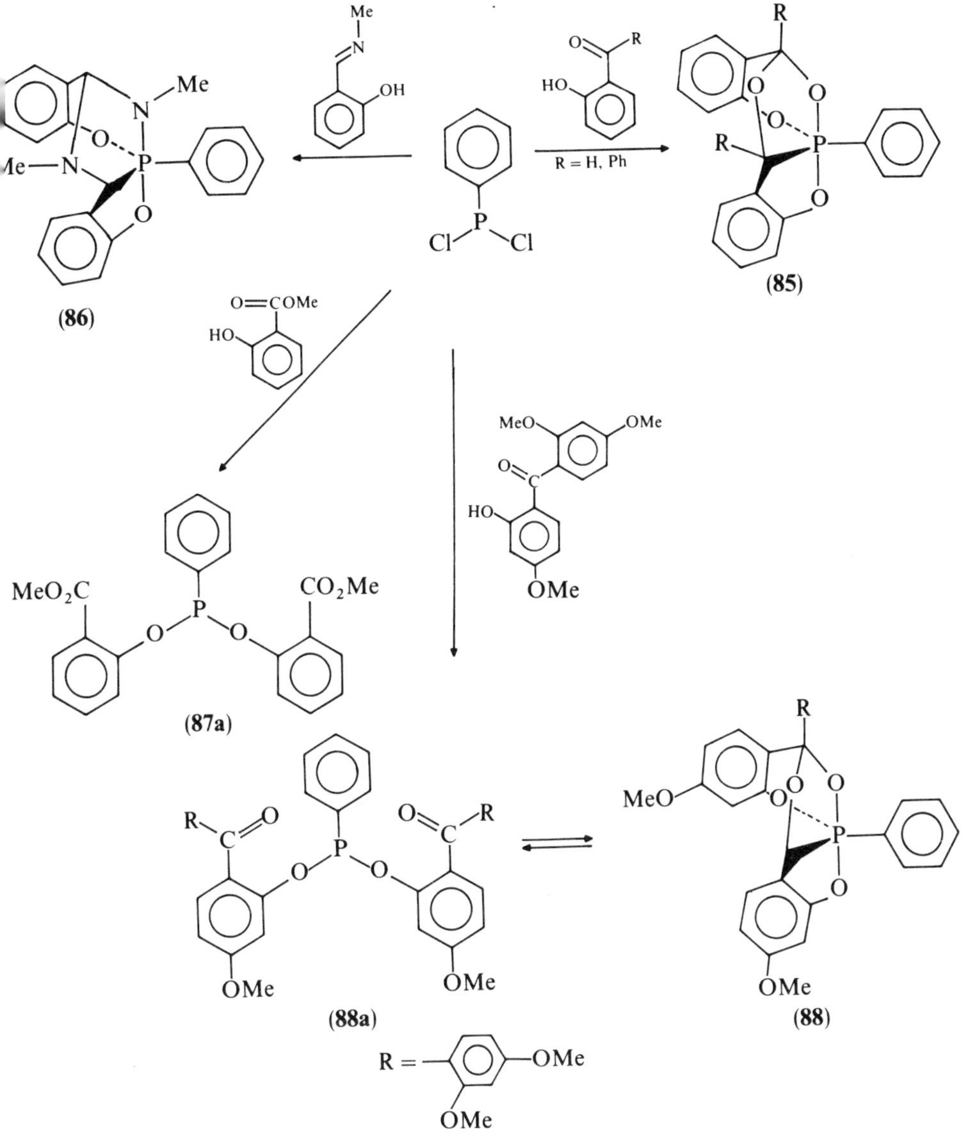

R = H, Ph

(85)

(86)

(87a)

(88a)

(88)

R = 3,4-dimethoxyphenyl (–C$_6$H$_3$(OMe)$_2$)

SCHEME 17

The fused bicyclic structure of **84** is peculiar in that two P(III)⇌P(V) equilibria can be envisaged for its solutions (Scheme 16), but no trace of the tautomeric P(III) derivative **84a** or of the cyclic **84b** could be found. However, the chemical reactivity observed suggests that the equilibrium does exist, but is almost totally displaced towards P(V)[162].

It was shown by potentiometric titration[163] that the acidity of the proton of the PH bond is stronger in spiranic than in the homologous monocyclic phosphoranes.

3. The P(III)⇌P(V) oxidative cyclization equilibria

The reaction of phenylphosphorus dichloride on 2-aldehydephenols[164] or 2-ketophenols[164,165] was found to yield tricyclic phosphoranes such as **85**, or **86** when the reaction was performed with salicylaldehyde methylimine[164] (Scheme 17). As shown by the authors, the formation of **85** follows that of the P(III) derivative **85a**. The reaction can even stop at the P(III) stage, with no cyclization, if steric hindrance is too great (R = MeC_6H_4) or if the electrophilic characteristic of the C=O group on the phenol is too low (R = OMe), so that only **87a** is formed. A judicious choice of steric and electronic parameters allowed the authors[164] to obtain **88** in which the presence of the expected equilibrium with the tricoordinated **88a** was fully demonstrated[166].

C. P(IV)⇌P(V) Equilibria

1. Tautomeric equilibria involving hydroxyphosphoranes

Hydroxy phosphoranes are intermediate forms normally resulting from the attack of a nucleophilic, hydrogen-bearing reagent NuH on a P=O moiety. The pentacoordinated derivative **II** (Scheme 18) results from the attack of NuH on the ($L^1L^2L^3$) face of tetracoordinated **I** (the so-called 'axial entry'), while **III** would be the result of an attack on the (OL^2L^3) face of the tetrahedron. This type of mechanism is found during the hydrolysis of phosphates[7,167]. The return to P(IV) can occur with or without retention of the primitive configuration, depending on a number of abundantly studied factors[36,84], such as the nature of the nucleophilic agent (electronegativity and steric bulk), the apicophilicity of the three L ligands, the presence of cyclic structures and possible stereomutations and the presence of acid-, base- or metal ion-catalysed reactions.

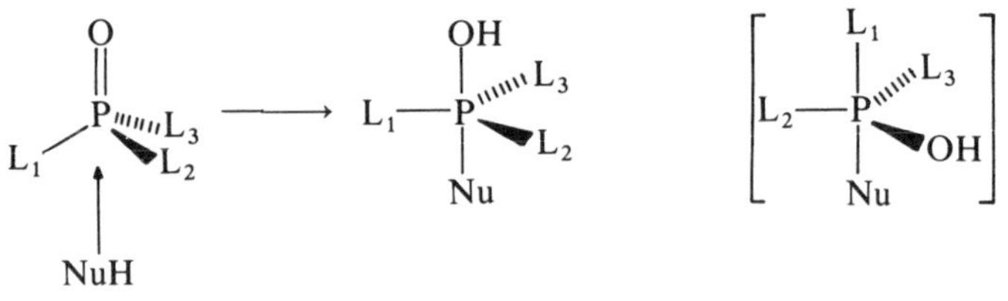

O
P
L_1 L_3 L_2
NuH
I

OH
L_1—P L_3 L_2
Nu
II

[L_1
L_2—P L_3 OH
Nu]
III

SCHEME 18

Hydroxyphosphoranes with a lifetime sufficiently long for them to be studied by classical techniques or to be 'trapped' by a chemical reagent are therefore particularly interesting models for the analysis of $SN_2P(IV)$ reaction mechanisms.† One of the first, if not *the* first, attempts at trapping a hydroxyphosphorane occurred with neopentyl hypophosphite **89**[168] (Scheme 19). The presence of the pentacoordinated **90** was deduced, even though the authors did not present any spectroscopic evidence, by the reaction with diazomethane to yield transiently **91** and, finally, **92**. More recently[169], the particularly stable spirophosphorane **93** was obtained by permanganic oxidation of the phosphine **94**, probably via the phosphine oxide **95** with free carboxylic groups (Scheme 20). (Strictly, the reversible interconversion of **93** and **95** by alkali fusion or acidification is not an equilibrium.)

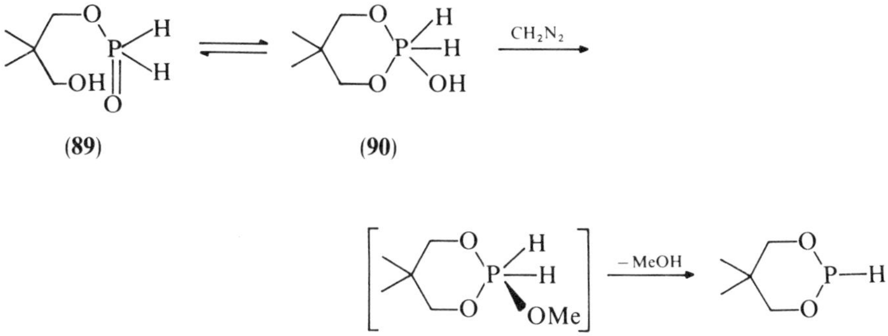

(**89**) (**90**) (**91**) (**92**)

SCHEME 19

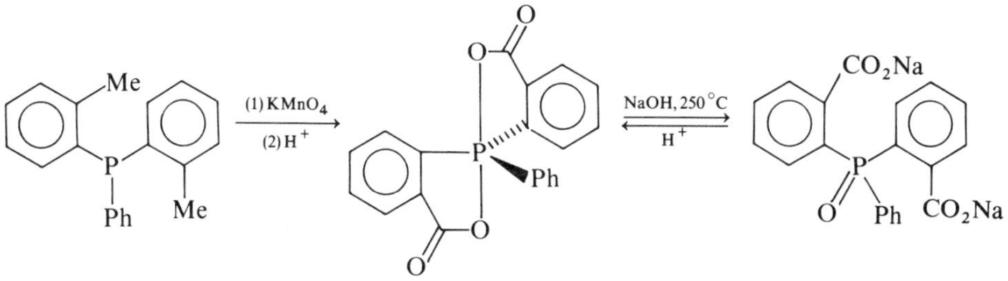

$\delta^{31}P$ −60.8 ppm

(**94**) (**93**) (**95**)

SCHEME 20

Equilibria between hydroxyphosphoranes and phosphoric esters have been claimed to occur as intermediates[170–172], or have been shown to exist by trapping the hydroxy compound with diazomethane as above[173], or have been displaced by acido–basic variations[174], as with **96** and **97** in Scheme 21.

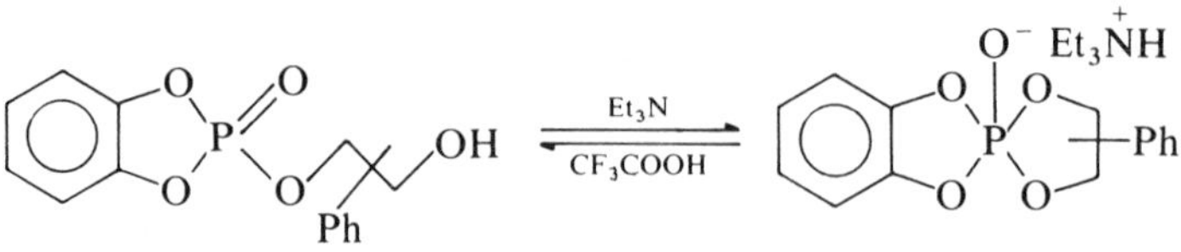

(96) (97)

SCHEME 21

The first spectroscopic evidence of the simultaneous existence of P(IV) and P(V) in a cyclic phosphate ⇌ hydroxyphosphorane equilibrium was given by Ramirez *et al.*[175] in a variable-temperature ^{31}P NMR study of **98** and **99** (Scheme 22): at -39 °C, in acetone-d_6 the signals from the two species are narrow, but they broaden at 32 °C and finally coalesce at 55 °C; the temperature of coalescence is about 10 °C in acetonitrile-d_3. Further, diazomethane yielded the Me ester of **98** and acetyl chloride gave acetylated derivatives of **98** and **99**.

(**98**) δ^{31}P -27 ppm (**99**) δ^{31}P 6.7 ppm

SCHEME 22

A second example (Scheme 23), involving a spiran structure (**100**) in equilibrium with the P(IV) derivative **101**, and with proofs provided by the same experimental techniques, was given by Granoth and Martin[176,177]. The **100** ⇌ **101** interconversion was demonstrated by the fact that the single ^{31}P NMR peak observed at $+54.9$ ppm at 28 °C splits into two narrow peaks separated by 80 ppm at -10 °C. The respective intensities of the two peaks indicate that the proportion of P(V) increases at low temperatures.

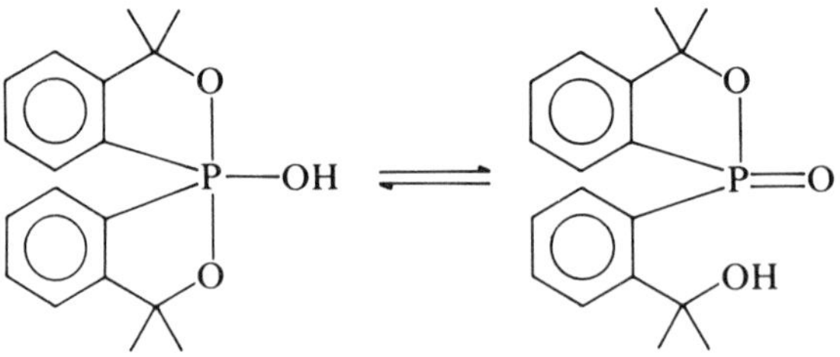

δ^{31} -26.3 ppm (MeOD at -10 °C) δ^{31}P 53.7 ppm (MeOD at -10 °C)

(**100**) (**101**)

SCHEME 23

In addition to the possibility of detecting their presence as intermediates during certain reactions, some hydroxyphosphoranes can be isolated as salts or can be detected as unstable intermediates in the course of a reaction[178]: thus, the phosphorane **102** was prepared[179] and the structure of its triethylammonium salt determined by XRD[180].

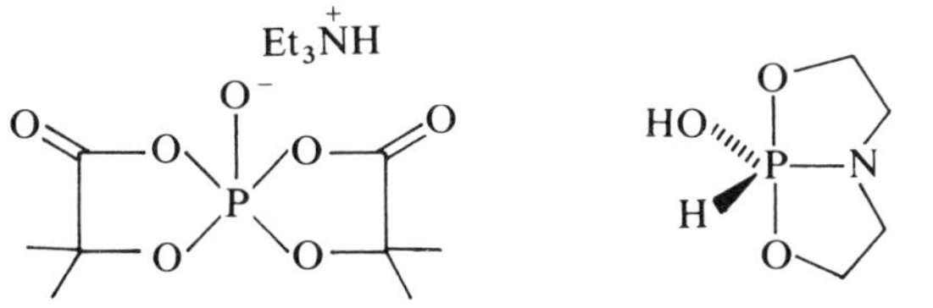

(102) (103)

FIGURE 21

Similarly, the hydroxyphosphorane **103** with a PH bond was detected during the hydrolysis of the dimethyl-6,6-dioxa-2,8-aza-5-phospha(III)-1-bicyclo[3.3.0]octane[181] (Figure 21).

The structure of phosphorane **102** is very close to a TBP[180], whereas that of the hydroxyphosphorane **105** is a tetragonal pyramid[182] owing to the important steric constraint (Scheme 24). The ^{31}P NMR chemical shift found for the phosphorane **105** dissolved in MeCN indicates that the equilibrium with the tetracoordinated form **105a** is nearly totally in favour of P(V). The shift is reversed, however, when another CH_2 link is added to the cyclobutane ring. Both **105** and **105a** are prepared (although in small yields) from the corresponding tetrols and **107**, or by ozonolysis of the phosphite **104a**. As above, the anionic form of **105** can also be alkylated.

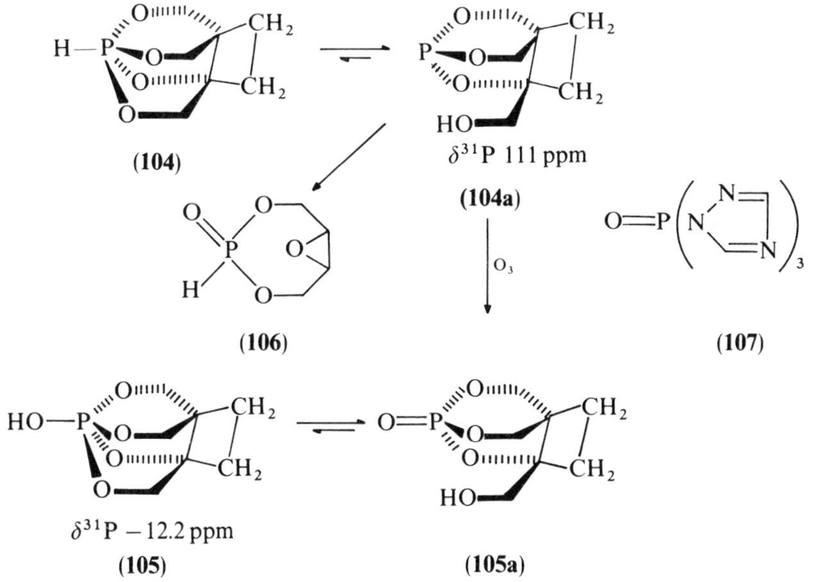

(104) δ^{31}P 111 ppm (104a) (106) (107) δ^{31}P −12.2 ppm (105) (105a)

SCHEME 24

The same tetrols, treated with $P(NMe_2)_3$, yield **104a**, which changes to the heptacyclic hydrogenphosphonate **106**, probably by intramolecular epoxidation. The pentacoordinated form of **104a**, namely **104**, which would result from a P(III)$\rightleftharpoons$P(V) equilibrium, is not detected by ^{31}P NMR. Similarly, no P–H coupling can be seen, although the presence in the IR spectra of the stretching PH vibration at 2300 cm^{-1} seems to indicate a rapid exchange of the proton link to O or P in a system shifted toward **104a**. Finally, replacing the cyclopentane or cyclobutane ring in the tetrols by cyclopropane prevents the-formation of **104** or **105**. This elegant piece of research thus brings much information on equilibria involving six- or seven-membered rings as a function of the geometry of the tetrol, itself controlled by the small ring (cyclopropane → cyclobutane → cyclopentane).

In addition to the proton, as seen in the examples above, other atoms, such as Cl[183] (**108** and **109**, in Scheme 25), or groups, such as N_3[184] (Scheme 26), can also migrate from the P atom on to a C=O group. In the latter case, the phosphorane **110**, prepared from chlorinated **113** and azidotrimethylsilane[185], is in equilibrium with its tetracoordinated form, the acyl azide **111**, both ultimately evolving to **112** by a Curtius rearrangement[186]. The phosphine oxide **111** has not been detected spectroscopically, but it would be a logical step in the formation of **112**. In this respect, it is interesting to compare the pairs (**113**, **113a**) and (**114**, **114a**): in the former, only **113** can be detected spectroscopically whereas, in the latter, only **114a** is detected (Scheme 27).

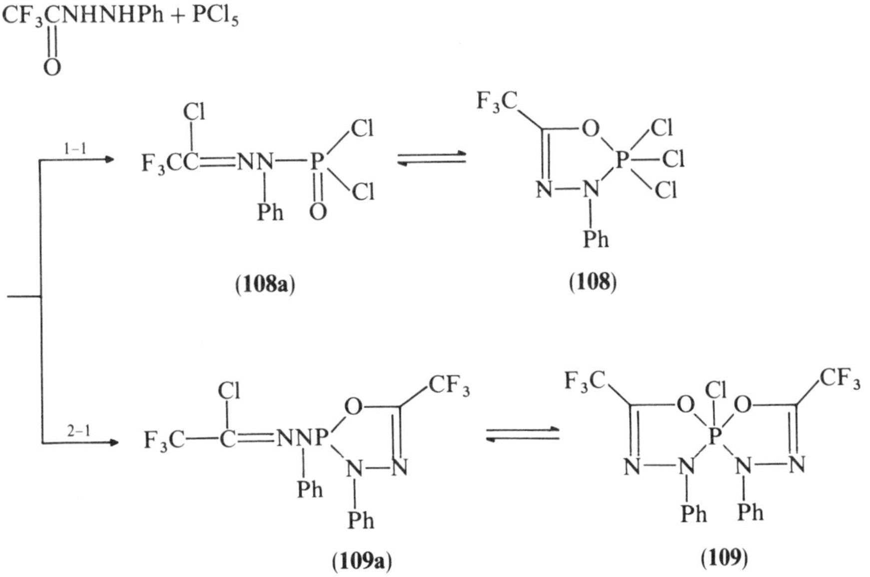

$CF_3C(=O)NHNHPh + PCl_5$

1-1 → $F_3CC(Cl)=NN(Ph)-P(=O)Cl_2$ ⇌ (108)

(**108a**) (**108**)

2-1 → (**109a**) ⇌ (**109**)

SCHEME 25

$\delta^{31}P$ −60 ppm

(**110**) ⇌ (**111**) $\xrightarrow{-N_2}$ (**112**)

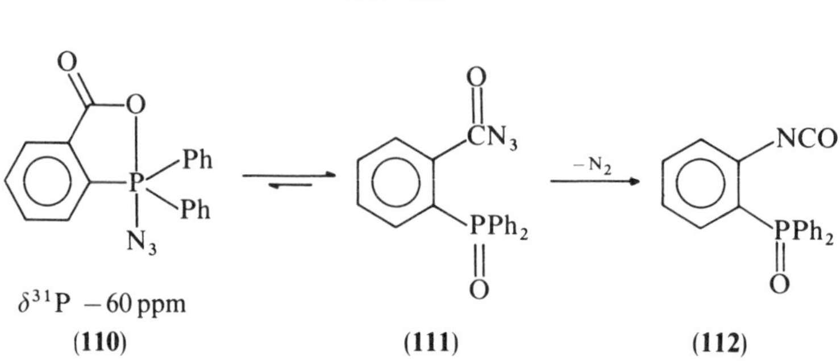

SCHEME 26

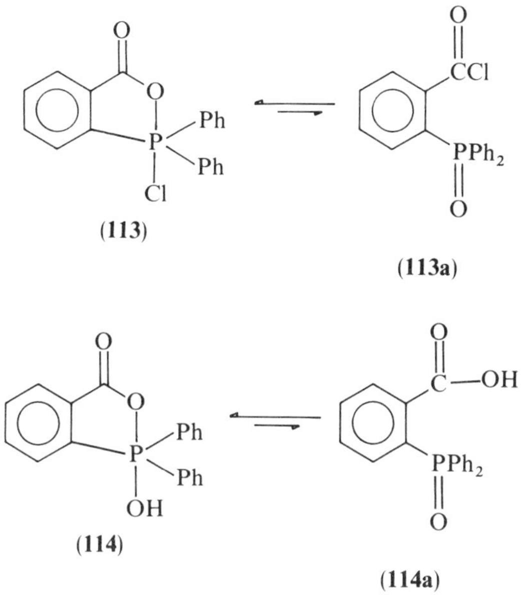

(**113**) ⇌ (**113a**)

(**114**) ⇌ (**114a**)

SCHEME 27

2. *Tautomeric iminophosphane ⇌ aminophosphorane equilibria*

The addition of a molecule X—Z (often with Z = H) to a P=O, P=N, P=C or P=S double bond can always be a possibility. The reaction may occur with a distinct molecule, or it may even be intramolecular when XH is a constitutive part of the phosphorus-bearing molecule. Structures such as **A**, **A′**, **A″**, **C** and **C′** in Scheme 28 are representative of a large proportion of phosphorus compounds: of (mono)phosphazenes, and of their cyclophosphazane dimers (over 1000 literature references!).

The phosphazene bond[187] can be formed via the reaction of tricoordinated phosphorus on azides[188] (the Staudinger–Meyer reaction[185]), or by the reaction of some P(V) derivatives, such as PCl_5, Ph_3PCl_2 and $(PhO)_3PCl_2$, on RNH_2 (the Kirsanov reaction[189]). Other methods also exist which, however, will not be detailed here. Generally, the XZ or XH fragment is borne by a chain linked to the N atom (structure **A**) or to the P atom of the iminophosphane ≡P=N— group (structure **A′**) or it may originate from a different molecule which will add to **A″**. The pentacoordinated P atom of the **B** and **B′** aminophosphoranes in Scheme 28 is part of a five-membered ring; this very common occurrence is, of course, dependent on the nature of the XZ moiety.

The iminophosphanes **A** and **A′** can be involved in two kinds of equilibria*: (a) the first (*a* in Scheme 28) leads to the pentacoordinated **B** or **B′** structures by cyclization, e.g. by nucleophilic attack of X on the P atom, and migration of atom Z, which is most often a proton, to the nitrogen; (b) the second (*b* in Scheme 28) results in the dimerization of the structure and in the formation of the cyclic diphosphazane **C** or **C′**.

* Reactions (such as thermal transpositions) resulting in a rupture of the equilibrium will not be considered here.

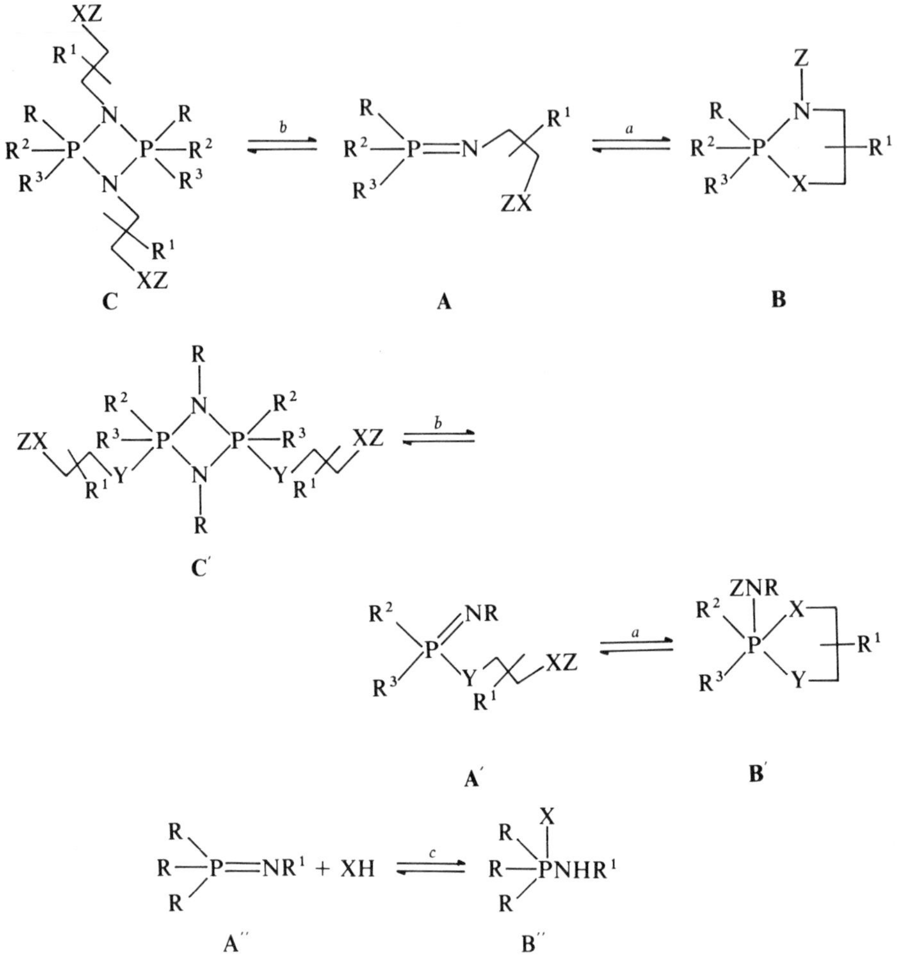

SCHEME 28

As expected, the evolution of the equilibria **A** (or **A**′)$\rightleftharpoons$**B** (or **B**′) and **A** (or **A**′)$\rightleftharpoons$**C** (or **C**′) depends on the temperature and on the nature of the solvent, but more particularly on the structure of the groups linked to the P or N atoms, i.e. on their steric characteristics, their electronegativity, their electron donor or acceptor properties, etc. The sum of these interactions results either in the creation of a single structure (**A**, **B** or **C**), or in the formation of an equilibrium when the difference in the thermodynamic stability of the various species is only slight.

In general, it was found that low temperatures and the presence of a five-membered iminophosphane ring (e.g. an R^2R^3P=dioxaphospholane ring, or an R^2R^3P=oxazaphospholane ring, for **A**′) tend to favour the formation of an aminophosphorane, in accordance with the stability sequence acyclic < monocyclic < spirocyclic. Similarly, one can deduce, from Scheme 28, that if the XZ bond cannot be broken, Z cannot migrate to the N atom and the only possibility will be the equilibrium *b*.

a. Examples of type a equilibria. Using **115** and **115a** (Scheme 29) as models for phosphoranes and the corresponding phosphazenes, respectively, Stegmann *et al.*[190] analysed the role of temperature, of the nature of the solvent and of the substituents, all of them factors found to influence the equilibria investigated.

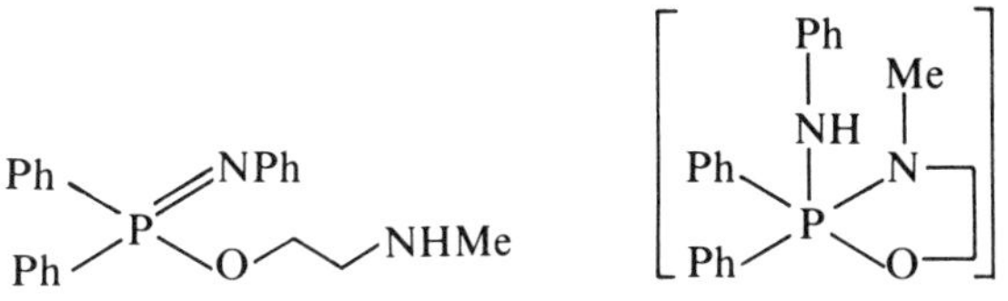

(**116a**) (**116**)

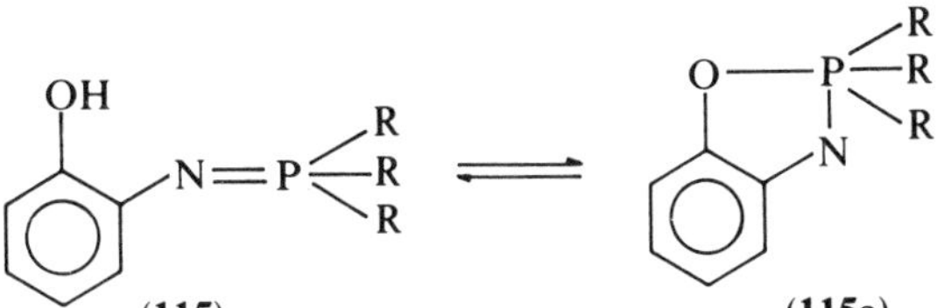

(**115**) (**115a**)

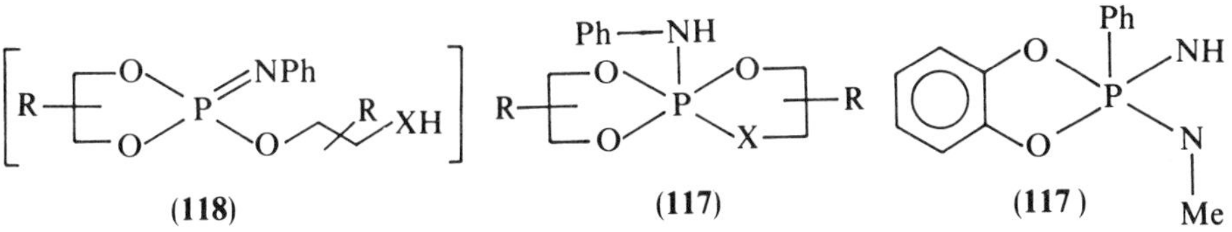

(**118**) (**117**) (**117**)

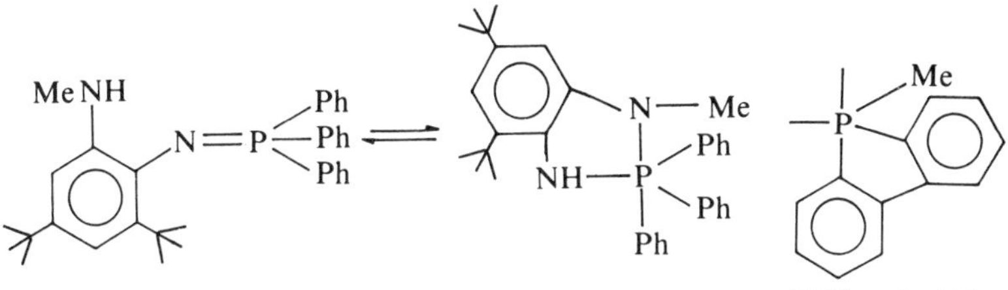

(**119a**) (**119**) (**120**) ⇌ (**120a**)

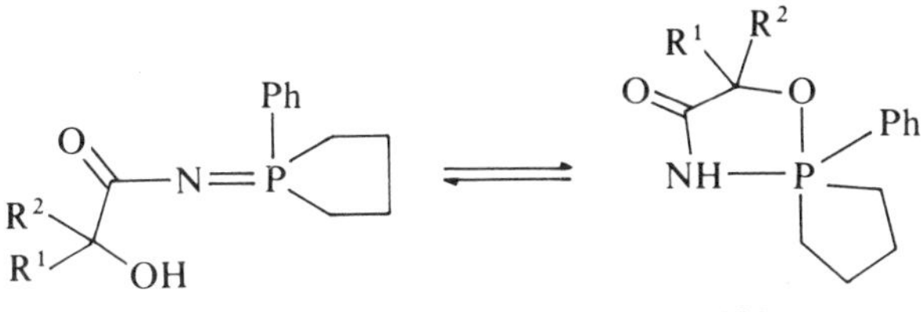

(**121a**) (**121**)

SCHEME 29

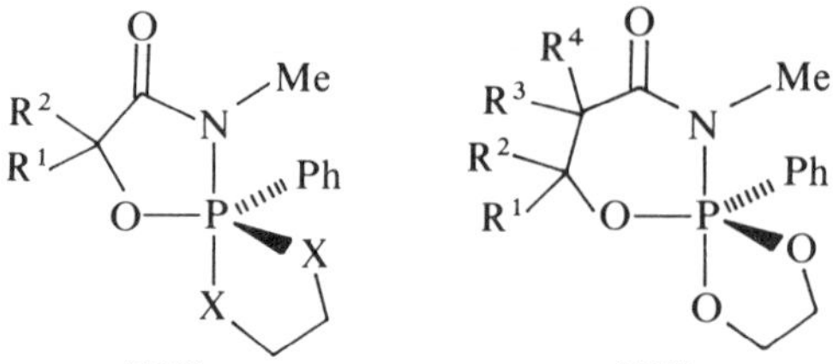

(122) (123)

FIGURE 22

The system aminophosphorane (**116**)–phosphazene (**116a**) (Scheme 29) studied by Sanchez *et al.*[191] was found to behave differently: only **116a** can be detected by NMR, i.e. there is no equilibrium. Exactly the opposite situation was found with the system **117**–**118**, in which the only observable species was the aminophosphorane **117**. Evidently, the increase in thermodynamic stability results from the formation of a spirophosphorane structure. Similar conclusions were reached by Gololobov *et al.*[197] in the course of a study of structures similar to that of **117′**. More recently, Stegmann *et al.*[193] extended the scope of their research to substituted 1,2-aryldiamines in a study of the equilibria **119**⇌**119a** and **120**⇌**120a**. The thermodynamic parameters ΔH, ΔG and ΔS were determined by NMR. Here too, the position of the equilibrium was found to depend on the substituents (steric and electronic effects), on the solvent and on the temperature.

The reaction of 2- or 3-hydrocarboxylic acid azides with 2-phenyl-1,3-dioxaphospholanes either yields spirophosphoranes or results in an equilibrium spirophosphorane (**121**)⇌iminophosphane (**121a**) (Scheme 29)[194]. Deprotonation of the two participants in the equilibrium by strong bases forms spirocyclic anions which, by reaction with methylating agents, finally yield the corresponding *N*-methylated spirophosphoranes **122** or **123** (Figure 22). The abnormally low value of the coupling constant J_{PNCH} of these compounds suggests an apical position for N, in contradiction with the standard electronegativity rules, but confirmed by XRD[194]: thus, **122** (with $R^1 = Ph$, $R^2 = H$, $X = O$) is a trigonal bipyramid, with bond lengths PO = 1.608(5) Å for the equatorial *O* in the heterocycle, 1.629(5) Å and 1.662(5) Å for PX and PN (apical) = 1.797(6) Å.

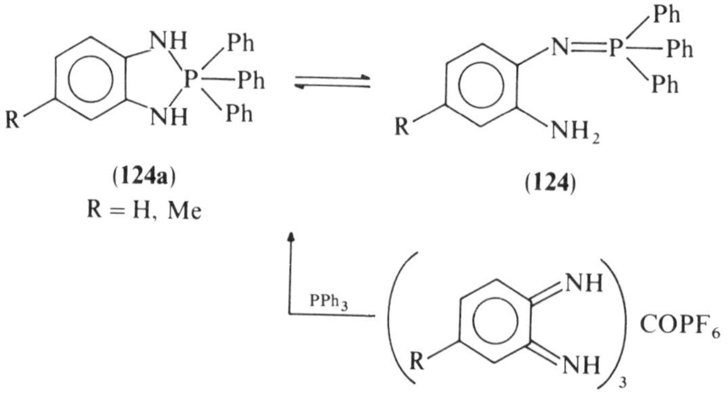

(124a)
R = H, Me

(124)

(125)

SCHEME 30

The authors also prepared other compounds, such as the phosphorane **123** with a seven-membered ring, and homologues of **122** in which X is replaced by a biphenylene moiety similar to that in **120**. When X in **122** is replaced by CH_2, XRD gives the bond lengths as PO = 1.617(4), PN = 1.887(6) Å. The important geometrical modifications brought about by methylation of the N atom are thus revealed.

Lastly, mention should be made of an original scheme of access[195] to an iminophosphane (**124**)$\rightleftharpoons$aminophosphorane (**124a**) equilibrium starting from PPh_3 and an *o*-benzoquinone diimine derivative (**125**) (Scheme 30).

b. Examples of type b equilibria. As above, type *b* equilibria are also conditioned by the usual steric and electronic factors so that, once again, three possibilities arise: the stable structure can be a monomer (**A**, **A′**), or a dimer (**C**, **C′**), or an equilibrium may form. Thus, strong electron-attracting substituents (Cl or F) promote charge separation between N and P ($\equiv P^+—N^-—$) and subsequent dimerization[189,196,197]. For instance, the iminophosphane **125a** (Scheme 31) is transformed into the cyclophosphazane **125** (with possible *cis–trans* isomers for the R substituent).

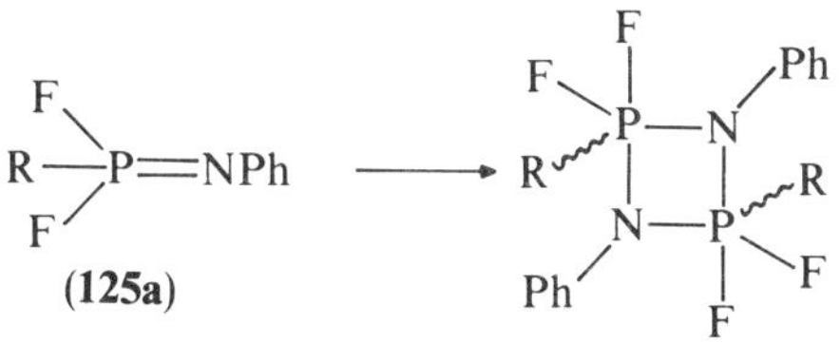

(**125a**)

R = Ph, NEt_2

(**125**)

SCHEME 31

Similarly, the presence of a bulky or of an electronegative substituent on N favours the iminophosphane among the non-cyclic structures. If the P atom is part of a ring (and especially if the latter is five-membered), a new parameter needs to be considered, namely the 'cycle effect' which, by relieving the strain due to the $sp^3 \rightarrow sp^3d$ change of hybridization, promotes pentacoordinated structures. This is particularly well illustrated by the behaviour of the structurally similar compounds **126a, 127a** and **128a** (Scheme 32). Although the N atom of the iminophospholane **126a** bears a bulky trimethylsilyl group, the effect of the electronegativity of the substituents on the P atom is such that dimer **126** is the stable form[198]. Replacing F in **126a** by a trimethylsilyl group inverts the situation, so that the monomer is now the stable form[199]. In **128**, prepared from the azidophosphorane **129**, both the steric hindrance and the electronegativity of the non-cyclic substituent on the P atom are reduced, and only the two isomers of **128** are obtained[184]. The presence of a ring also enhances the chemical reactivity of iminophospholanes, which are generally more reactive than their non-ring homologues[200].

The occurrence of an equilibrium between cognate monomers and dimers was unambiguously proved with the 2-imino-1,3,2-dioxaphospholanes (Scheme 33). The **130a:130** ratio is 62:38 in C_6H_6, 64:36 in $CHCl_3$ and 53:47 in CCl_4[196]. Similarly, the tetramethyldioxaphospholane, which induces electronic effects opposite to those of a benzodioxaphospholane cycle, does not hinder the establishment of the equilibrium **131a**$\rightleftharpoons$**131**[201]. In the latter ring, the attractive inductive effect of the two oxygen atoms is not compensated by a donor mesomeric effect of their p doublets, and conjugating occurs with the aromatic ring rather than with the d orbital of the P atom, thus accentuating the $P^+—N^-$ polarization of the phosphazene bond. Conversely, in the former ring, the attractive

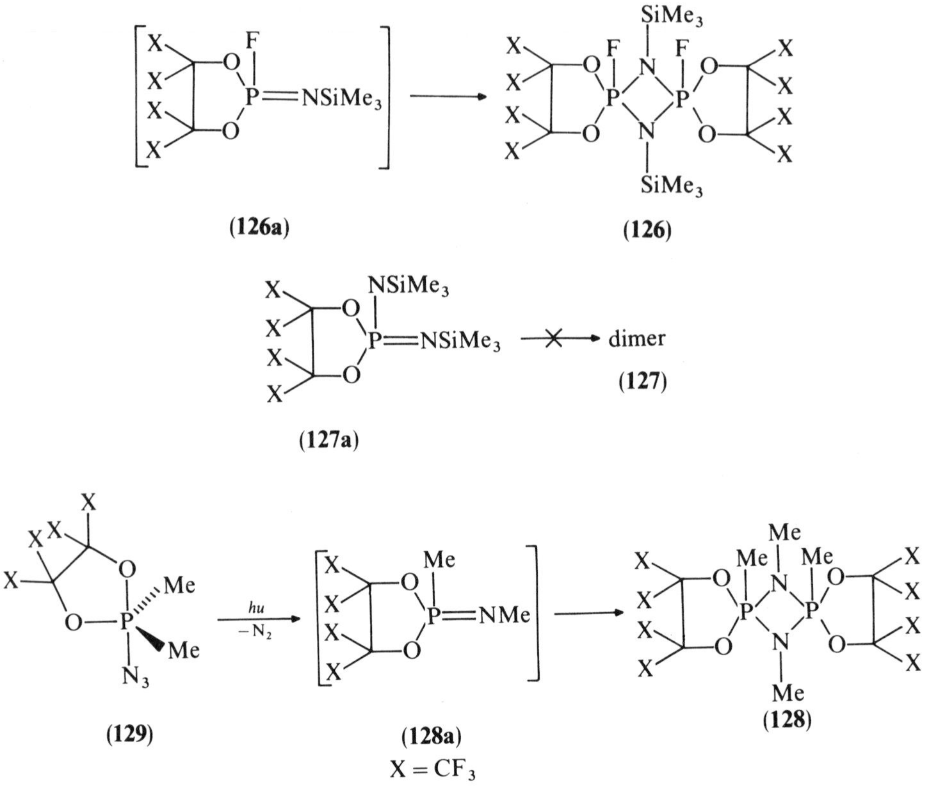

(126a) (126)

(127a) (127)

(129) (128a) (128)

X = CF_3

SCHEME 32

inductive effect of the O atoms is decreased by the donor effect of the Me group and by the π back-donation of the O doublets on the d orbitals of phosphorus[202].

Endocyclic P=N bonds tend to promote dimerization (Scheme 34): the phosphazane **132** is formed[203] from the precursor **132a**, and even if **134a** is the result of the thermolysis of the phosphorane **133**, as proved by the possibility of trapping it with a C=O bond[204], the product isolated is the dimer **134**.

With reference to the equilibria in Scheme 28, note that when Z = H in ZX, it is found that, preferentially, **A**$\rightleftharpoons$**B** and **A′**$\rightleftharpoons$**B′**, i.e. the equilibria are prototropic, but that otherwise **A**$\rightleftharpoons$**C** and **A′**$\rightleftharpoons$**C′**. That this is not a general rule is proved by the **C′** → **A** → **B′** sequence[205] in Scheme 35: heating the stable dimer **135** yields the spirophosphorane **137** via the iminophosphane **136**.

Marre *et al.*[206] compared the results obtained with dioxa-, oxaza- and diazaphospholanes (**138**, **139** and **140**, respectively, in Figure 23), all of them containing a P=N bond of which the N atom bore groups such as $SO_2C_6H_4Me$, Ph or Et, with a decreasing inductive electron-attractive effect: the dimerization is prevented by the diazaphospholane ring, whatever the substituent R′ on the imidic N of **140**. The ‘ring effect’, which usually promotes P(V) structures, seems to be inoperative in this case, possibly because the possible dimer **141**, which is not formed, would require either two apical N atoms or a diequatorial diazaphospholane ring, in contradiction to the steric and electronegativity rules, even if some exceptions to these rules do exist.

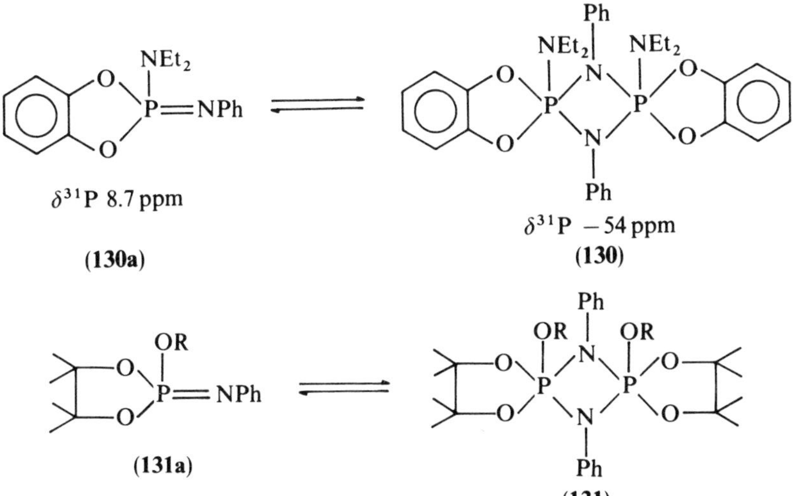

$\delta^{31}P$ 8.7 ppm

(**130a**)

$\delta^{31}P$ −54 ppm

(**130**)

(**131a**)

(**131**)

SCHEME 33

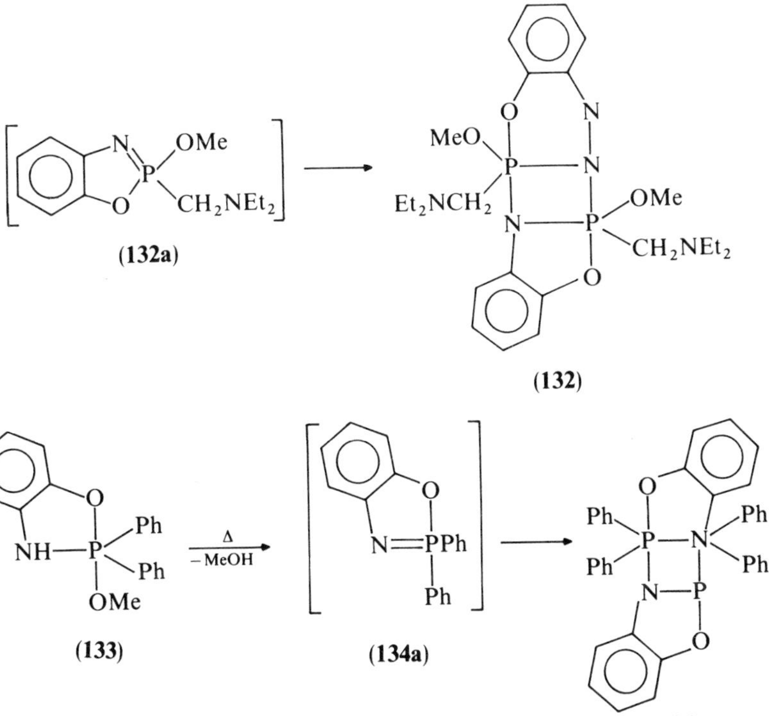

(**132a**)

(**132**)

(**133**)

Δ
− MeOH

(**134a**)

(**134**)

SCHEME 34

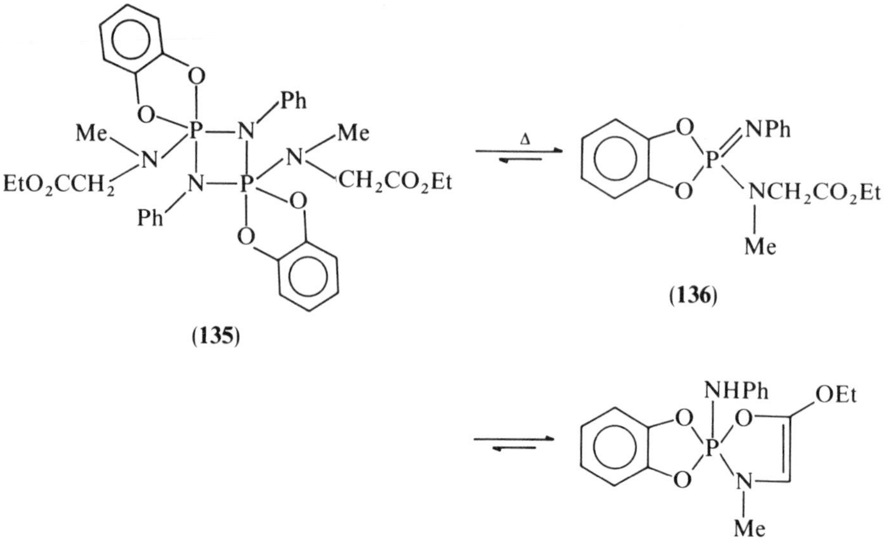

(135)

(136)

(137)

SCHEME 35

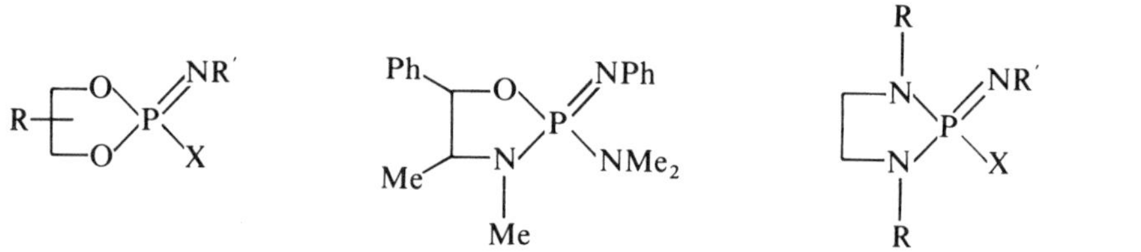

(138)

R = H, Me

X = OMe or NMe_2

(139)

(140)

R = Me or Ph

X = OMe or NMe_2

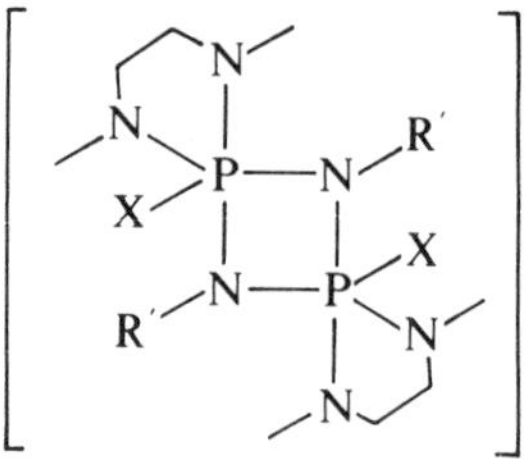

(141) not formed

FIGURE 23

The same authors[206] also studied the reaction of methanol on iminophospholanes similar to **138** and **140** in Figure 23. the result (Scheme 36) was that the low-temperature (−40 °C) addition of MeOH to **142** formed the adduct **143**, visible by ^{1}H and ^{31}P NMR, and finally changed to the thermodynamically stable product of the reaction **144**, but the occurrence of the equilibrium **142**$\rightleftharpoons$**143** could not be demonstrated. As regards the diazaiminophosphane, it only formed hydrogen bonds with MeOH, detected by a slight change $\Delta\delta = 7$–10 ppm in the δ values of the ^{31}P NMR signal.

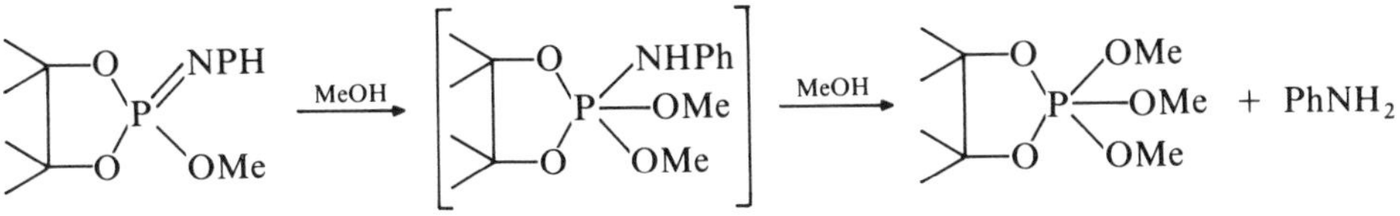

$\delta^{31}P$ 5.8 – 6.4 ppm (**142**) —MeOH→ [$\delta^{31}P$ −57 ppm (**143**)] —MeOH→ $\delta^{31}P$ − 55 ppm (**144**) + $PhNH_2$

SCHEME 36

3. Ylide$\rightleftharpoons$phosphorane equilibria

Ylides are isoelectronic with the iminophosphanes (or phosphazenes) just considered and, consequently, share with them much of their chemical reactivity[207–210] (Scheme 37), hence many of the equilibria involving the phosphazene bond and illustrated in Scheme 28 can be transposed, with the appropriate modifications, to the ylides. Thus, the ylide (**145**)$\rightleftharpoons$phosphorane (**146**) equilibrium described by Schmidbauer and Stuhler[211] and involving the addition or elimination of MeOH (Scheme 38) is comparable to the reaction described in Scheme 36 and the equilibrium *c* in Scheme 28. Similarly, ylides can be looked upon as intermediate kinetic products which will eventually change to a thermodynamically stable phosphorane[212–214], as in Scheme 39. The ylide **149**, visible by ^{1}H and ^{31}P NMR, is quantitatively formed between 0 and 20 °C and trapped by reaction of the 1,3-dipole with phenol. It is, however, relatively unstable and progressively changes, again quantitatively, to the well characterized phosphorane **150**.

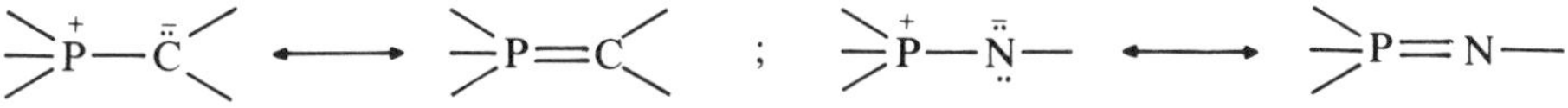

$$\overset{+}{P}-\overset{-}{C} \longleftrightarrow P{=}C \quad ; \quad \overset{+}{P}-\overset{-}{N}- \longleftrightarrow P{=}N-$$

SCHEME 37

$$Me_3P{=}CH_2 \underset{-MeOH}{\overset{+MeOH}{\rightleftharpoons}} Me_4POMe$$

(**145**) (**146**)

SCHEME 38

The inverse reaction to that described above and leading to the formation of an ylide by transformation of a phosphorane by thermodynamic evolution has also been observed[212–214] (Scheme 40): the phosphorane **151**, formed at −50 °C by addition of $P(OMe)_3$ to methyl acetylenedicarboxylate in the presence of MeOH to trap the 1,3-dipole, is rapidly transformed, at −20 °C, into the ylide **152**.

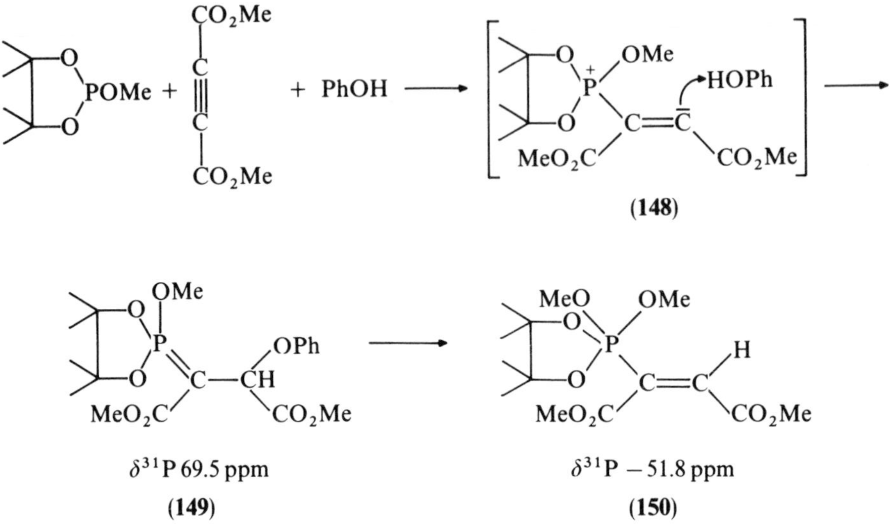

δ^{31}P 69.5 ppm

(**149**)

δ^{31}P −51.8 ppm

(**150**)

SCHEME 39

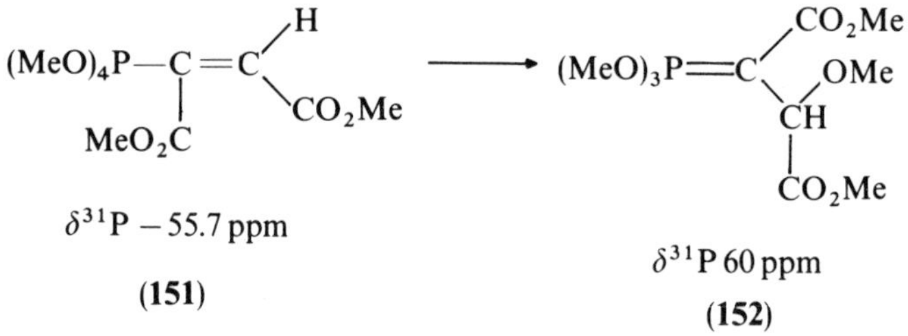

δ^{31}P −55.7 ppm

(**151**)

δ^{31}P 60 ppm

(**152**)

SCHEME 40

It is clear, therefore, that the line between the sp^3 and sp^3d types of hybridization is affected by steric and electronic factors. The interpretation of the **149** → **150** and **151** → **152** sequences as due solely to the 'ring effect' promoting P(V) structures would undoubtedly be naive. The migration of PhO, MeO, R_2N, etc., groups in these isomerization reactions is governed by their own acido–basic and steric properties. The migration of a PhO group from a C atom of the ylide **149** on to the P atom of the phosphorane **150** was interpreted by the HSAB theory and the symbiotic effect as a competition between the two sites[215,216]. On the basis of these results, it is evident that as regards the three-component [P(III), activated acetylenic bond, trap for the 1,3-dipole] reactions in Schemes 39 and 40, a judicious choice of reagents can allow access to structures in which the energy barrier between the ylide and the phosphorane is sufficiently low for the equilibrium to become possible. This has indeed been achieved[212–217] by using an aromatic acid as trap for the 1,3-dipole **148** in Scheme 39.

The equilibrium **153** ⇌ **154** (Scheme 41) was demonstrated by the reversible variations of the proportion of the constituents with changes of the solvent, the **153**:**154** ratio being

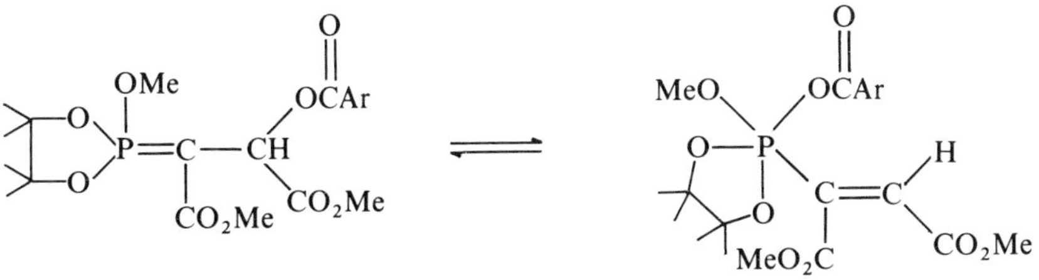

Ar = Ph $\delta^{31}P$ 69 ppm

(**153**)

Ar = Ph $\delta^{31}P$ −48.9 ppm

(**154**)

SCHEME 41

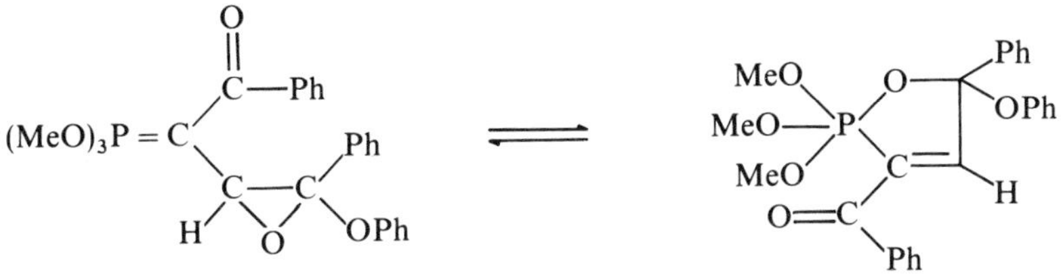

$\delta^{31}P$ 53 ppm

(**155**)

δ^{31} −33 ppm

(**156**)

SCHEME 42

41.5:58.5 in CH_2Cl_2 at 20 °C and 18:82 in CCl_4. The system shown in the sequence is stable for many hours.

Another example of ylide–phosphorane equilibrium is found in the reaction of trimethyl phosphite with acetylenic ketones, in the presence of phenol as trapping reagent[216] (Scheme 42). Here, too, the **155:156** ratio depends on the nature of the solvents, being 37:63 in CH_2Cl_2 and 74.5:25.5 in CCl_4.

Scheme 43 shows the details of the different steps involved in the equilibrium. The nucleophilic attack of the P(III) derivative on the acetylenic bond yields a 1,3-dipole which, after a fast protonation, frees a Z^- ion. If the subsequent addition of this ion occurs on the P atom (reaction *a*), a P(V) phosphorane is formed, but the addition of Z^- on the ethylenic C atom (reaction *b*) results in the formation of an ylide. Both of these reactions occur under kinetic control and, in both cases, X is always an OR group from the initial acetylene dicarboxylic ester. When the acetylenic compound is a diketone and X is an alkyl or aryl moiety, the C=O group is much more electrophilic and the attack by the Z^- ion produces an alcoholate (reaction *c*), a new intermediate which can cyclize on to the P^+ to form a phosphorane, or attack the α-C atom to form an ylide as in Scheme 42. Hence, reactions *a* and *c* can coexist, and are strongly dependent on the nature of the trapping reagent and of the P compound, but reaction *b* is blocked, whatever the reagent. This is well illustrated by the reaction of the 2-methoxytetramethylphospholane **147** on dibenzoylacetylene in the presence of methanol as trapping reagent. The proportions of the vinylphosphorane **157** and spirophosphorane **158** formed (Figure 24) are 13% and 84%, respectively.

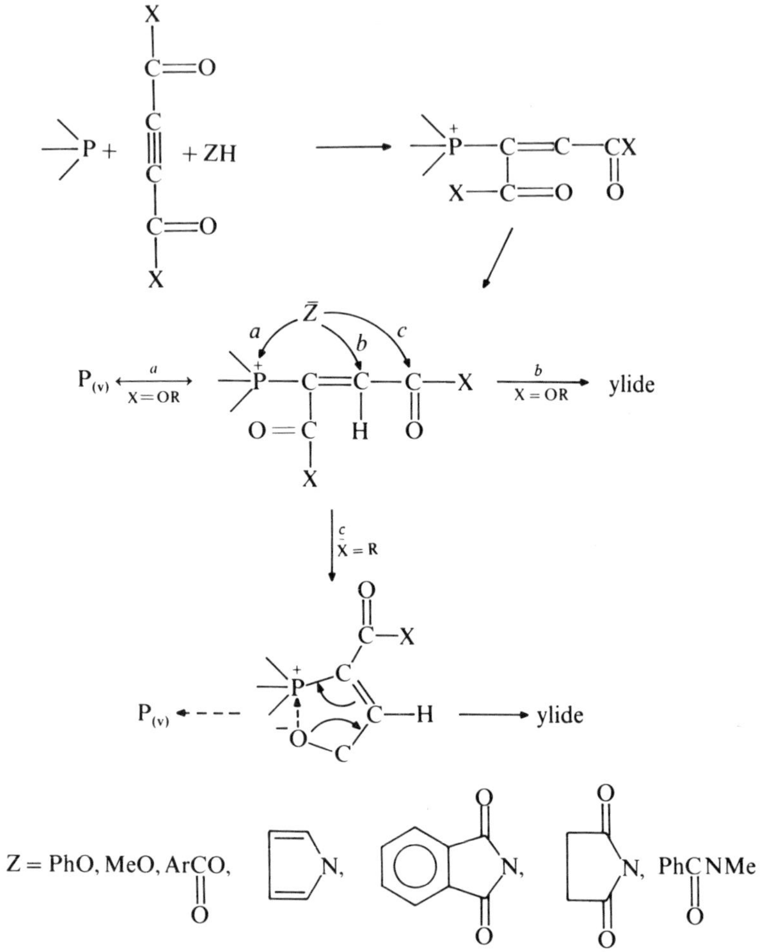

X
C=O
C
C
C=O
X

P + ... + ZH ⟶ P—C=C—CX, X—C=O, O

$P_{(v)}$ ←(a, X=OR)→ P—C=C—C—X, O=C, H, O, X —(b, X = OR)→ ylide

Z̄: a, b, c

(c, X = R)

O, C—X, P, C, C—H, ⁻O, C

$P_{(v)}$ ← - - - ⟶ ylide

Z = PhO, MeO, ArCO, N, N, N, PhCNMe

SCHEME 43

MeO, O, OMe, P, O, C=C, H, PhC, CPh, O, O

OMe, O, P, O, Ph, OMe, C=, H, O=C, Ph

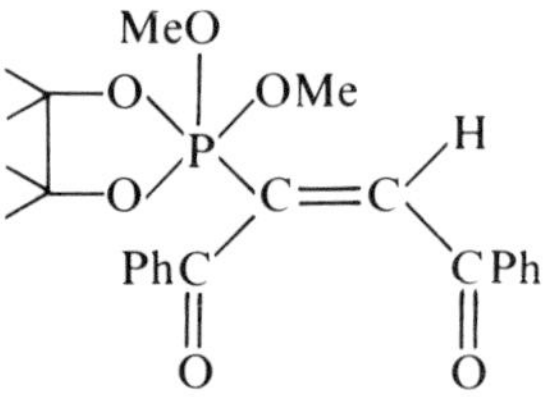

$\delta^{31}P - 50$ ppm

(**157**)

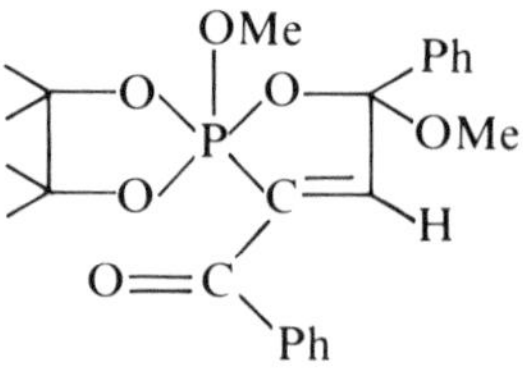

$\delta^{31}P - 26$ and -30 ppm (2 diastereomers)

(**158**)

FIGURE 24

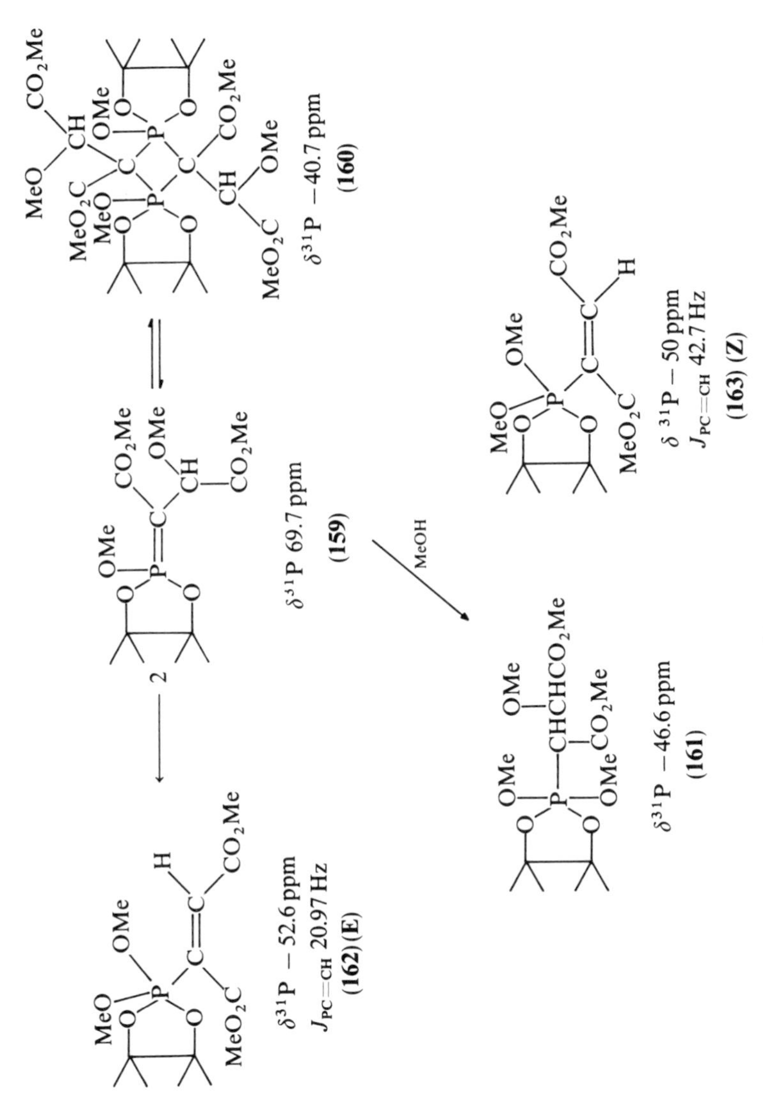
MeOH
δ31P 69.7 ppm
(159)
δ31P −40.7 ppm
(160)
δ31P −46.6 ppm
(161)
δ31P − 52.6 ppm
JPC=CH 20.97 Hz
(162) (E)
δ 31P − 50 ppm
JPC=CH 42.7 Hz
(163) (Z)

SCHEME 44

Section V.B.2 gave a few examples of dimerization of the phosphazene bond, and it was also stated therein that dimerization was considerably promoted by the presence of a strongly polarized P=N bond. A similar occurrence in ≡P=C was only observed[213] with the ylide **159** (Scheme 44), obtained by a series of steps analogous to those in Scheme 39, with methanol as trapping reagent. Proper control of the amount of MeOH used is essential: if the amount is (or is nearly) stoichiometric, only the ylide **159** is formed; a large excess of it yields the phosphorane **163**, which is stable under these conditions. (The mechanism of its formation is certainly a *trans*-addition to the triple bond, analogous to the addition of amines to activated acetylene derivatives described by Huisgen *et al.*[218], without formation of an intermediate 1,3-dipole or of the ylide **159**.) Similarly, there is no formation of the phosphorane **161** when MeOH is in excess, and the ylide **159** is therefore the result of a protonation of the 1,3-dipole followed by the addition of MeO^- on the ethylenic C atom (reaction *b*, Scheme 43). Under these conditions, the ylide **160** was also detected, although not isolated. The addition of MeOH to the mixture of **159** and **160** shifts the system to the formation of the saturated phosphorane **161**.

The reactions in Schemes 39, 40, 43 and 44 involve stabilized ylides. The preparation of ylides using Ph_3P in a three-component reaction, with methyl acetylenedicarboxylate and an alcohol as trapping reagent, was first described by Wilson and Tebby[219] (Scheme 45). The ylide ⇌ phosphorane equilibria described so far involve the intramolecular rearrangement of quasi-phosphonium species, but Granoth *et al.*[220] have shown that phosphoranes can also be in equilibrium with phosphonium ylides: the deprotonation of the phosphorane **165** creates the ion **166** in equilibrium with a charged ylide, as shown in Scheme 46. This equilibrium has been studied by variable-temperature ^{31}P NMR, and the addition of MeI results in methylation of **166** (δ $^{31}P = -54.9$ ppm).

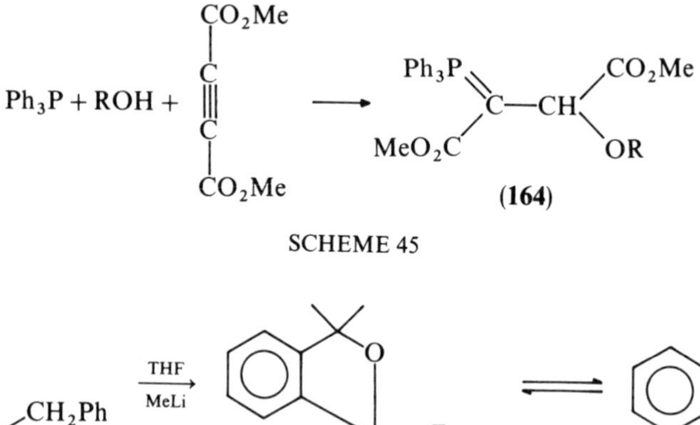

$Ph_3P + ROH + MeO_2C{-}C{\equiv}C{-}CO_2Me \longrightarrow Ph_3P{=}C(CO_2Me){-}CH(CO_2Me)(OR)$

(164)

SCHEME 45

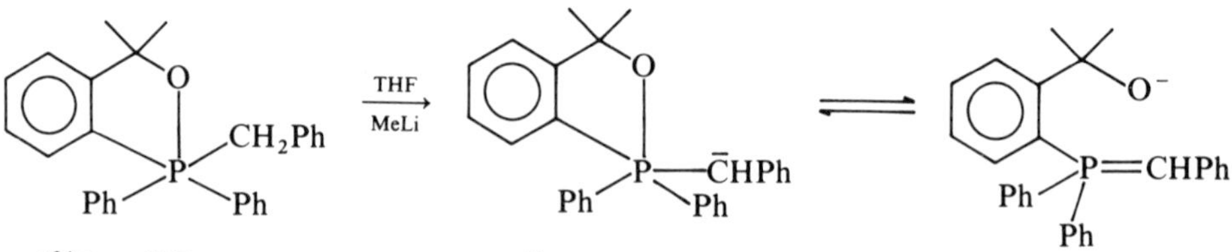

$\delta^{31}P$ −60.7 ppm **(165)** $\xrightarrow[\text{MeLi}]{\text{THF}}$ $\delta^{31}P$ −39.8 ppm **(166)** ⇌ $\delta^{31}P$ 16.2 ppm **(167)**

SCHEME 46

4. *Phosphonium salt ⇌ phosphorane equilibria*

The oldest known example of a pentacoordinated structure in equilibrium with an ionic tetra- or hexacoordinated structure[221] is probably the inorganic compound PCl_5:

$$2\,PCl_5 \rightleftharpoons (PCl_4)^+ + (PCl_6)^-$$

and

$$PCl_5 \rightleftharpoons (PCl_4)^+ Cl^-$$

A similar example, more in line with organic chemistry, was recently described by Dillon and Straw[222]: the phosphorane **168** is found in solutions, whereas the ionic species **169** and **170** make up the solid. In many cases, the redistribution of substituents between two phosphorane molecules can be ascribed to the existence of a P(IV)$\rightleftharpoons$P(V) equilibrium involving a phosphonium salt[223], as with **171**$\rightleftharpoons$**172**. Similarly, the study of the Mitsunobu reaction[224] resulted in bringing to light successive equilibria such as **173**$\rightleftharpoons$**174**$\rightleftharpoons$**175** (Scheme 47).

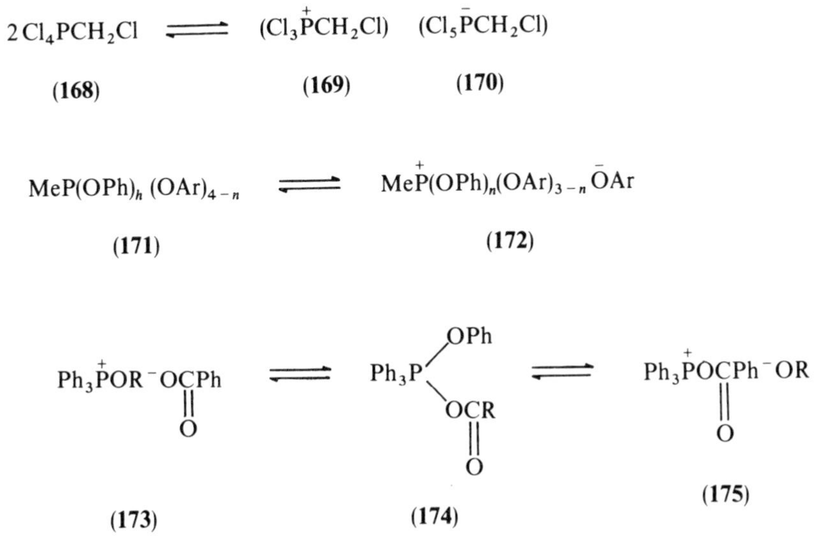

$$2\,Cl_4PCH_2Cl \rightleftharpoons (Cl_3\overset{+}{P}CH_2Cl)\ (Cl_5\overset{-}{P}CH_2Cl)$$

(**168**) (**169**) (**170**)

$$MeP(OPh)_n(OAr)_{4-n} \rightleftharpoons Me\overset{+}{P}(OPh)_n(OAr)_{3-n}\,\overset{-}{O}Ar$$

(**171**) (**172**)

$$Ph_3\overset{+}{P}OR\ {}^-OC(=O)Ph \rightleftharpoons Ph_3P(OPh)(OC(=O)R) \rightleftharpoons Ph_3\overset{+}{P}OC(=O)Ph\ {}^-OR$$

(**173**) (**174**) (**175**)

SCHEME 47

(**176**) + $MeOSO_2CF_3$ $\rightleftharpoons$ (**177**) $CF_3SO_2O^-$

(**178**) $\underset{}{\overset{H^+}{\rightleftharpoons}}$ (**179**)

Ph_3P (**180**) $\overset{\Delta}{\rightleftharpoons}$ $Ph_3\overset{+}{P}$ (**181**)

$\delta^{31}P$ −36.5 ppm

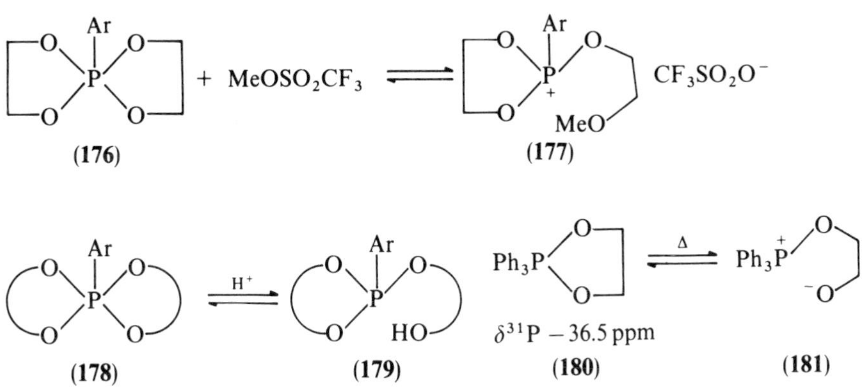

SCHEME 48

The equilibria **176**⇌**177** and **178**⇌**179** in Scheme 48 correspond to yet another possibility, namely the addition or removal of certain species. They correspond to the results of a study of the hydrolysis mechanisms of spirophosphoranes with penta- or hexa-atomic cycles presented by McClelland and coworkers[225]. Slightly different is the equilibrium **180**⇌**181** examined by Kelly and Evans[226] and involving the phosphorylation of 1,2-diols (and triols) and diethoxytriphenylphosphorane. This interesting cyclodehydration provides a means of access to oxiranes starting from the corresponding diols: thus, butane-1,2,4-triol yields 3,4-epoxybutan-1-ol (81%) and tetrahydrofuran-3-ol (19%) by reaction with diethoxytriphenylphosphorane at 40 °C. The contribution of dialkoxytriphenylphosphoranes to the Mitsunobu reaction has also been studied[227], as well as the extension of the reaction to mercaptoalcohols[226] and to methyl *α-D*-glucopyranoside[229].

The reactivity and use in organic synthesis of 1, 3, 2, λ^5-dioxaphospholanes has been presented[230]. Further details on various topics can be found in the following reviews: Gloede (1988)[344]; Markovskii *et al.*[342], work prior to 1987 on linear equilibria between phosphoranes and phosphonium salts; and Polezhaeva and Cherkasov[340], results obtained prior to 1982 with cyclic derivatives.

Up to this point, the criterion used to ascertain the presence of an equilibrium, i.e. the simultaneous presence of two well defined entities, was mainly based on ^{1}H, ^{13}C or ^{31}P NMR determinations at different temperatures or in various solvents. However, it turns out that the chemical shift in ^{31}P NMR spectra (usually negative with respect to the H_3PO_4 reference) is often inadequate to permit a choice among various possibilities, namely a true phosphorane, a phosphonium salt or an equilibrium between these two forms.

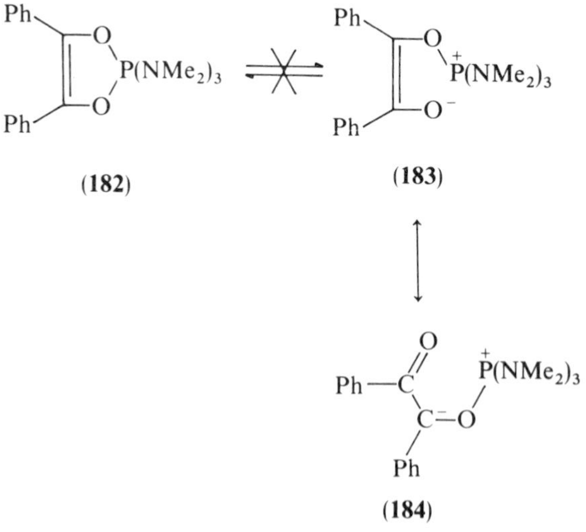

(**182**) (**183**) (**184**)

SCHEME 49

A case in point is the derivative **182** (Scheme 49) obtained by the reaction of $P(NMe_2)_3$ on benzil. First prepared by Burgada[231], it was described as a P(V) compound on the basis of the -25 ppm ^{31}P chemical shift in C_6H_6. Shortly after, on the basis of the variations of the chemical shift in various solvents (-29.9 ppm in C_6H_6; -13.1 ppm in CH_2Cl_2) and on the isolation of the two forms by crystallization from hexane, Ramirez *et al.*[232] declared that **182** was in equilibrium with the betaine **183**. These are convincing arguments, all the more so since they agree with the results of a study[233] performed using reagents chosen to detect these forms (activated alkenes, Me_3SiCl, sulphur) which led to the conclusion that **183** and **184**, both of them ionized, existed in the solution. However, more recently, in the course of an investigation on the reactions of cyclic, polycylic and linear aminophosphines, Denney and coworkers[234,235] refuted the existence of the ionized forms **183**, as deduced from the lack of difference among the olefinic C atoms and among the *ipso* C atoms of the benzene ring in the ^{13}C NMR spectrum, with the values $\delta = 133.05$ ppm and $J_{POCC} = 11.6$ Hz: the *o*-, *m*- and *p*-C atoms were seen as a single resonance, and the Me groups on the nitrogen atom gave a single doublet. These facts are characteristic of a single pentacoordinated structure for **182** undergoing a fast stereomutation, even at -75 °C. Further, the ^{13}C NMR spectrum in CD_2Cl_2 (a solvent causing important shifts of the ^{31}P NMR δ) yields values similar to those above, namely $\delta = 135.35$ ppm and $J_{POCC} =$ 8.5 Hz. This was followed by work[235] on model compounds analogous to **182**, in which the aromatic ring bore, on the *p*-C atom, substituents such as Cl, F, Me, MeO or NO_2. Only the *p*-nitro compound was found to ionize, all the others remaining true phosphoranes. However, lowering the temperature of the solution of the *p*-Cl or *p*-F derivatives was found to abolish the POCC coupling; this was attributed to a fast ionization of the phosphoranes caused by the temperature dependence of the solvent dielectric constant. The authors concluded that the variations of δ caused by the changes in the physical characteristics of the solvent, as observed with **182** and similar compounds, could be explained by a participation of canonic forms, in proportions dependent on the polarity of the solvent.

Another example of the help provided by ^{13}C NMR in the determination of the structural characteristics of phosphoranes, when the information based on δ values from ^{31}P NMR spectra is uncertain or inconclusive, was given by Lowther *et al.*[236], who showed that **185** (with $\delta = 33.6$ ppm) and **186** ($\delta = 16.8$ ppm) in C_6H_6 are both pentacoordinated in spite of the positive δ values (Figure 25).

All these results cast doubt on the validity of the conclusions regarding some of the equilibria presented, for instance as regards the structures of **187** and **188** in Scheme 50 (cf. the review by Polezhaeva and Cherkasov[340], and the references therein). Thus, Aganov *et al.*[237] pointed out that another kind of equilibrium was also possible, featuring intramolecular permutation, as with the groups R^1 and RC=O in **189** and **190**. Both forms are detected by NMR whereas **191**, which would constitute a logical intermediate in this migration, cannot be observed.

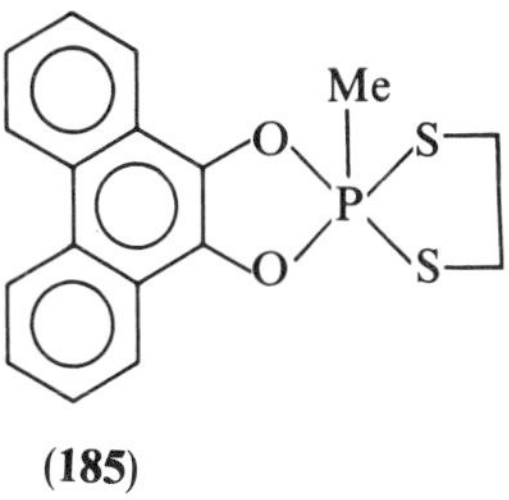

Me O O P S S

(185)

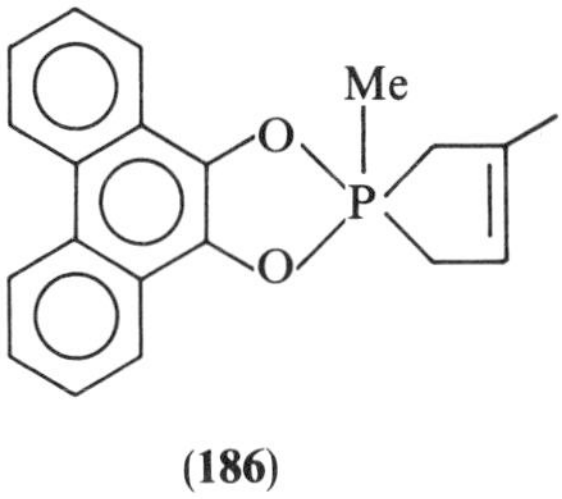

Me O O P

(186)

FIGURE 25

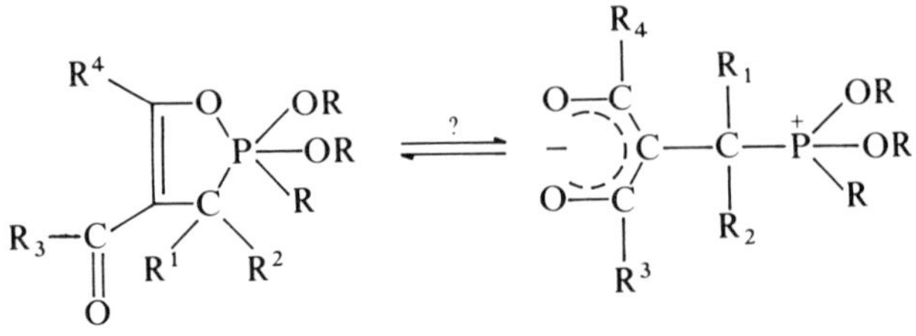

(187) (188)

SCHEME 50

D. P(V)$\rightleftharpoons$P(V) Equilibria

Three sets of phenomena are grouped under this heading, with the possibility that the first two coexist:

(a) Equilibria among pentacoordinated structures with the same substituents differing only by their stereogeometry, e.g. stereoisomers. (Stereomutations were examined in detail in Section III and many of the reviews at the end of the reference list give full details.)
(b) Equilibria among pentacoordinated structures with rearrangements in the distribution of the substituents, i.e. with migration of a functional group. The equilibrium between the phosphoranes **189** and **190** (Scheme 51) is a typical example, with the undetectable phosphonium salt **191** as the intermediate form.

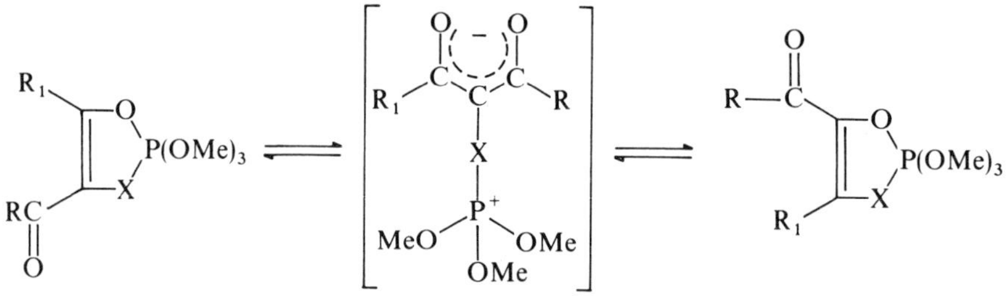

(189) (191) (190)

X = O, CHR_2, NPh

SCHEME 51

(c) The third type of equilibria are true, reversible chemical reactions between different pentacoordinated species, as in Scheme 52[85,238], in which all species are present simultaneously and are detectable. In this **A**$\rightleftharpoons$**B** type of equilibrium, the same final composition can be reached by starting either with the mixture **A** + acid or with the mixture **B** + base, and this final composition can be modified by changing the initial molecular ratio: with 2 mol of base (or acid), the equilibrium can be driven to completion because of the subsequent neutralization of the acid (or base) formed, and **A** (or **B**) can be isolated.

The equilibrium in Scheme 52 evolves, more or less rapidly and depending on the characteristics of R^2 and of the phosphorane, towards the formation of R^2CONMe_2 and

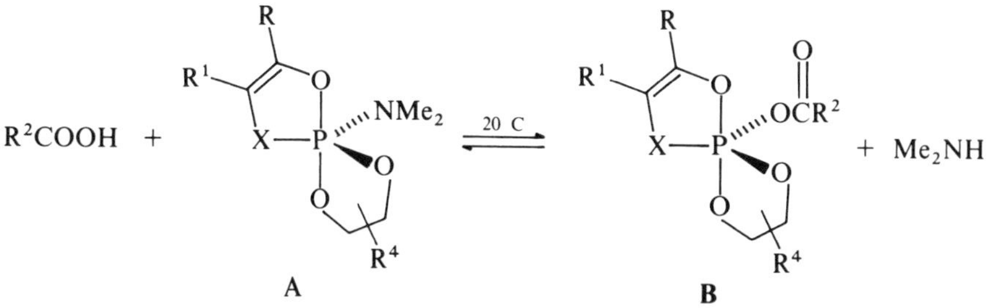

R²COOH +
NMe₂
20 C
OCR²
+ Me₂NH
A
B

X = O, NMe, CHR^3

SCHEME 52

of a cyclic phosphate. Acyloxyspirophosphoranes (such as **B** in Scheme 52) are very reactive: even at 20 °C, they are, for instance, quantitatively and rapidly alcoholysed to the P(V)—OR compound. Although the corresponding hexacoordinated intermediate between **A** and **B** can be imagined, its presence has never been detected.

Another example of this type of equilibrium is provided[239] by the reaction of the cyclenphosphorane **78** (already met in Section V.B.1) with zinc or cadmium chloride (Scheme 53). A single peak is found in the ^{31}P NMR spectrum of the mixture of **78** + $ZnCl_2$ (molar ratio 1:0.5), at −51.2 ppm and with $J_{PH} = 653$ Hz. These values lie between those given by **78** and **192**, pointing to a bond-rupture ⇌ bond-formation equilibrium for N—Zn. Complexes such as **192** readily form salts such as **193** with strong acids, and the progressive change of δ and J values in the sequence **78** → **192** → **193** is instructive, especially as the structure of **193** was confirmed by XRD. The interconversion **194** + acid ⇌ **195** + base (Scheme 53), similar to that just mentioned, was demonstrated by the same authors[240] and the structure of **195** was also determined by XRD.

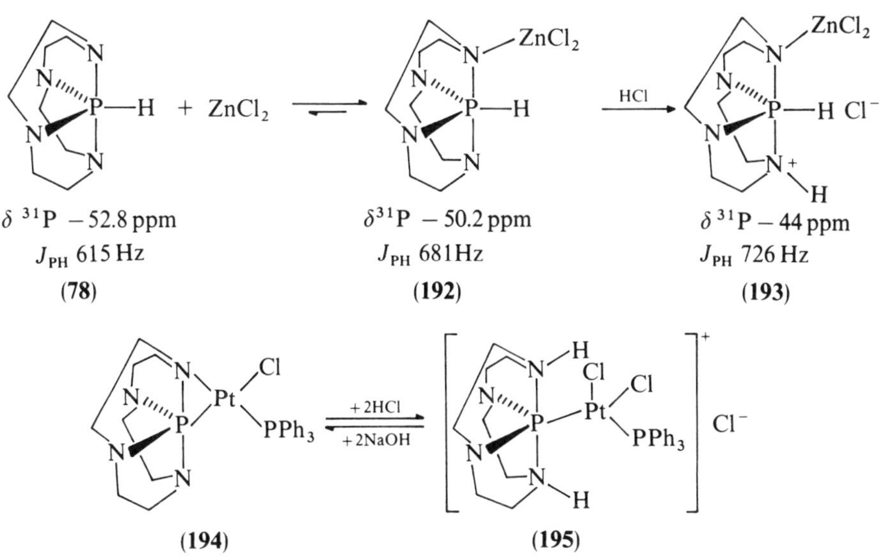

+ ZnCl₂
ZnCl₂
HCl
ZnCl₂
H Cl⁻
δ ³¹P −52.8 ppm
J_PH 615 Hz
(78)
δ³¹P − 50.2 ppm
J_PH 681Hz
(192)
δ ³¹P − 44 ppm
J_PH 726 Hz
(193)
Cl
PPh₃
+2HCl
+2NaOH
Cl
PPh₃
Cl⁻
(194)
(195)

SCHEME 53

E. P(V) ⇌ P(VI) Equilibria

Work on this subject prior to 1966 is reviewed by Webster[335]; the review by Cherkasov and Polezhaeva[341] covers the synthesis and reactivity of hexacoordinated phosphorus derivatives up to 1987. As early as 1911, Werner[241] suggested the octahedral structure now known to be generally adopted by hexacoordinated phosphorus derivatives. The P(VI) compounds give a ^{31}P NMR signal upfield with respect to analogous pentacoordinated structures, usually between -60 and -300 ppm from the H_3PO_4 reference.

One of the interesting points connected with this type of coordination springs from the fact that nucleophilic substitution on phosphoranes probably occurs via the formation of an octahedral complex[242–245]. Further, it should be recalled that the mechanism of irregular stereomutations of pentacoordinated structures in basic media may also involve a hexacoordinated intermediate. Figure 26 shows just a few of the numerous and diverse structures known (**196**–**203**) but, even then, some general observations can be made.

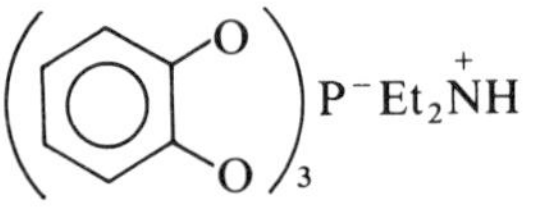

$P^- Et_2\overset{+}{N}H$

δ −82 ppm (1963)[246]

(196)

$(F_3C)_2\bar{P}F_4$ Cs^+

(1968)[248]

(197)

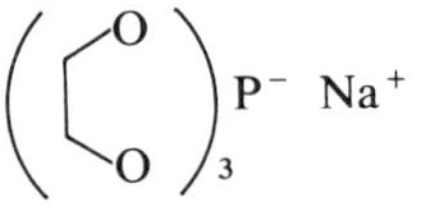

P^- Na^+

δ −89 ppm (1971)[247]

(198)

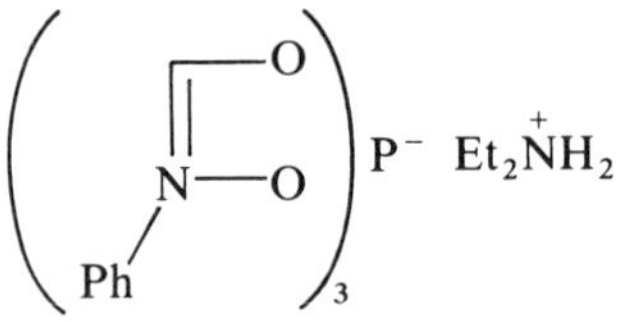

P^- $Et_2\overset{+}{N}H_2$

δ −77 ppm (1977)[250]

(199)

$Et_2\overset{+}{N}H_2$

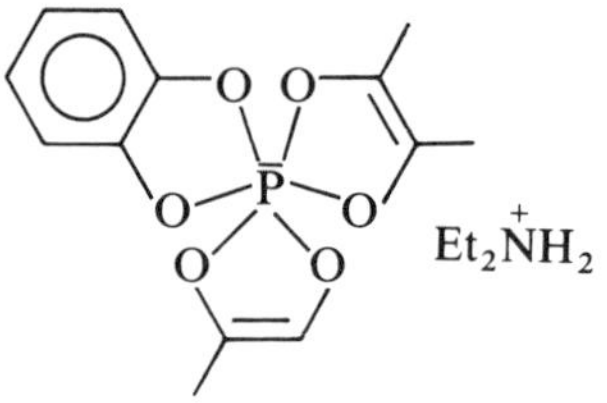

(1977)[251]

(200)

$(MeO)_6P^-K^+$

δ −145 ppm (1976)[649]

(201)

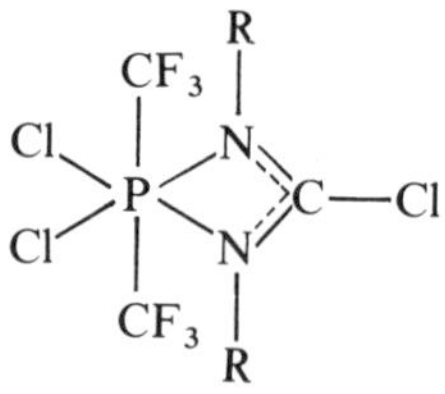

(1990)[252]

(202)

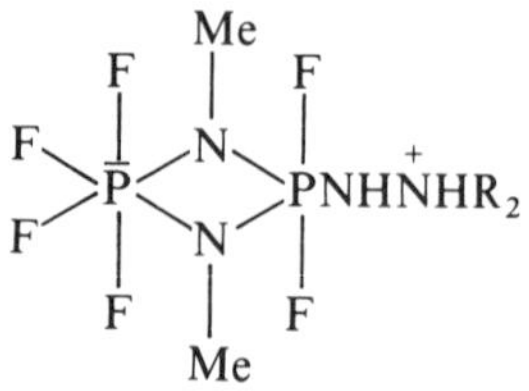

$PNH\overset{+}{N}HR_2$

(1988)[253]

(203)

FIGURE 26

Tricyclic compounds, and especially those with a five-membered ring, are particularly stable, as is, for instance, the typical tricatecholate[246] **196**, whereas linear compounds such as **201** cannot be isolated even when shown to exist in a solution[249], an exception being **197**, which, although linear, is extremely stable[248]. Also noteworthy is the ^{31}P NMR chemical shift displaced towards high fields as the structure changes from tricyclic to linear. Numerous structural determinations by XRD have proved the octahedral structure of these molecules, e.g. of **196**[254] and **202**[252] and their homologues.

The catecholate structure seems particularly favourable to the formation of hexacoordinated compounds, especially in basic media. Thus, the reaction of pyrocatechol on dimethylaminobenzodioxaphospholane (Scheme 54) allowed Lopez *et al.*[255] to obtain **204**, the first hexacoordinated bicyclic derivative with a PH bond (rather than the spirophosphorane **206**). Moreover, heating **204** yields the triscatecholate **205**, identical with **196** (Figure 26). Conversely, **206** is obtained quantitatively by alcoholysis of the P—pyrrole bond in neutral media[256] (Scheme 54). The reaction of an aryl or alkyl benzodioxaphospholane **207** (Figure 27) with pyrocatechol in the presence of NEt_3 yields a hexacoordinated biscatecholate with a PH bond (**208**), which can be isolated and which,

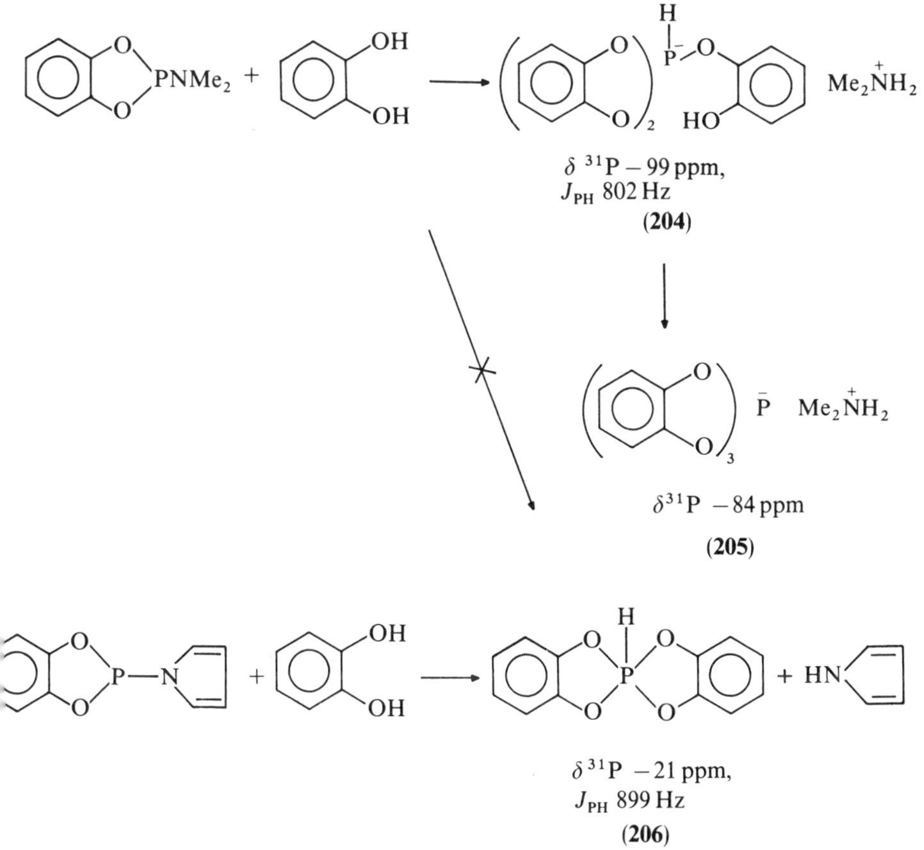

PNMe$_2$ + OH OH → H P O HO $Me_2\overset{+}{N}H_2$

δ ^{31}P -99 ppm,
J_{PH} 802 Hz
(**204**)

$\bar{P}$ $Me_2\overset{+}{N}H_2$

$\delta^{31}P$ -84 ppm
(**205**)

P—N + OH OH → H P + HN

δ ^{31}P -21 ppm,
J_{PH} 899 Hz
(**206**)

SCHEME 54

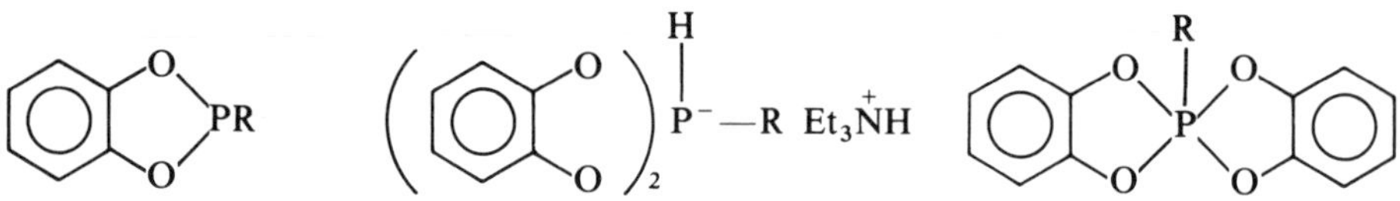
PR
H
P⁻—R Et$_3\overset{+}{N}$H
R
P

R = Me δ −113.5 ppm

(207) **(208)** **(209)**

FIGURE 27

when heated to 83 °C (R = Me), gives the spirophosphorane **209** with evolution of 1 mol of H_2 and regeneration of the amine[257]. This reaction, which was also observed by Pudovik *et al.*[258] on analogous compounds, is specific to catechol derivatives. It provided the first evidence for the presence of P(V)⇌P(VI) equilibrium under acido–basic control[259] (Scheme 55); when put in the presence of methanol, phenol, pyrocatechol or pyrrolidine in basic media (i.e. with NEt_3 added in the case of ROH compounds), the spirophosphorane **206** forms the hexacoordinated complex **210**, which, in turn, can regenerate **206** if a sufficiently strong acid, such as CF_3COOH, is added. Since CH_3COOH is not strong enough to perform this reversion, the basicity of **210** must be intermediate between that of CF_3COO^- and of CH_3COO^-. Compound **206** can also react with amino alcohols to form an intermolecular complex such as **211**.

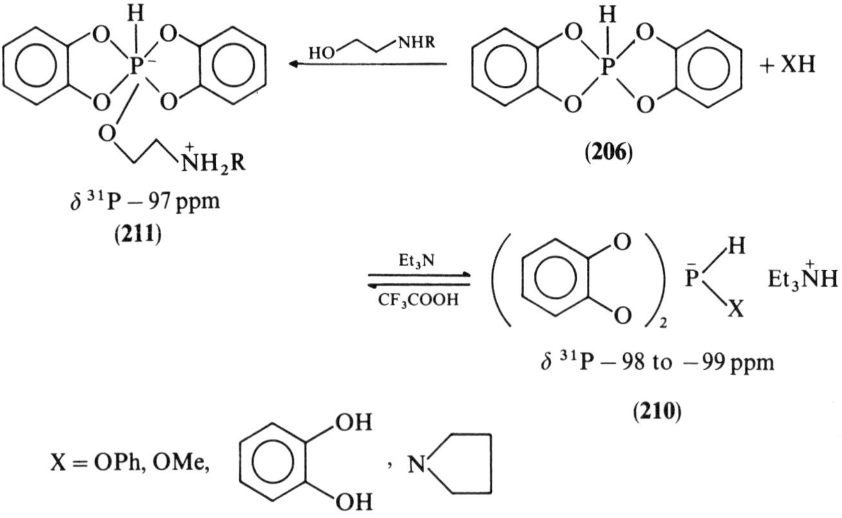
HO⁀NHR
+ XH
$\overset{+}{N}H_2R$
δ ^{31}P − 97 ppm
(211)
(206)
Et$_3$N
CF$_3$COOH
Et$_3\overset{+}{N}$H
δ ^{31}P − 98 to −99 ppm
(210)
X = OPh, OMe,
OH
OH

SCHEME 55

All these observations led Bernard[260] to a study of the behaviour of the spirophosphorane **206** in the presence of trimethylamine or pyridine (Scheme 56); an insoluble complex (**212**) rapidly precipitates with the former, whereas the addition of the latter progressively shifts the PH NMR doublet of **206** to high field, with $\delta = -94$ ppm (for one equivalent added) or $\delta = -69$ ppm (for two equivalents). This was interpreted as resulting from the formation of the hexacoordinated complex **213** in rapid equilibrium with **206**. Similar results were reported by Muñoz and coworkers[261] and by Schmidtpeter *et al.*[262],

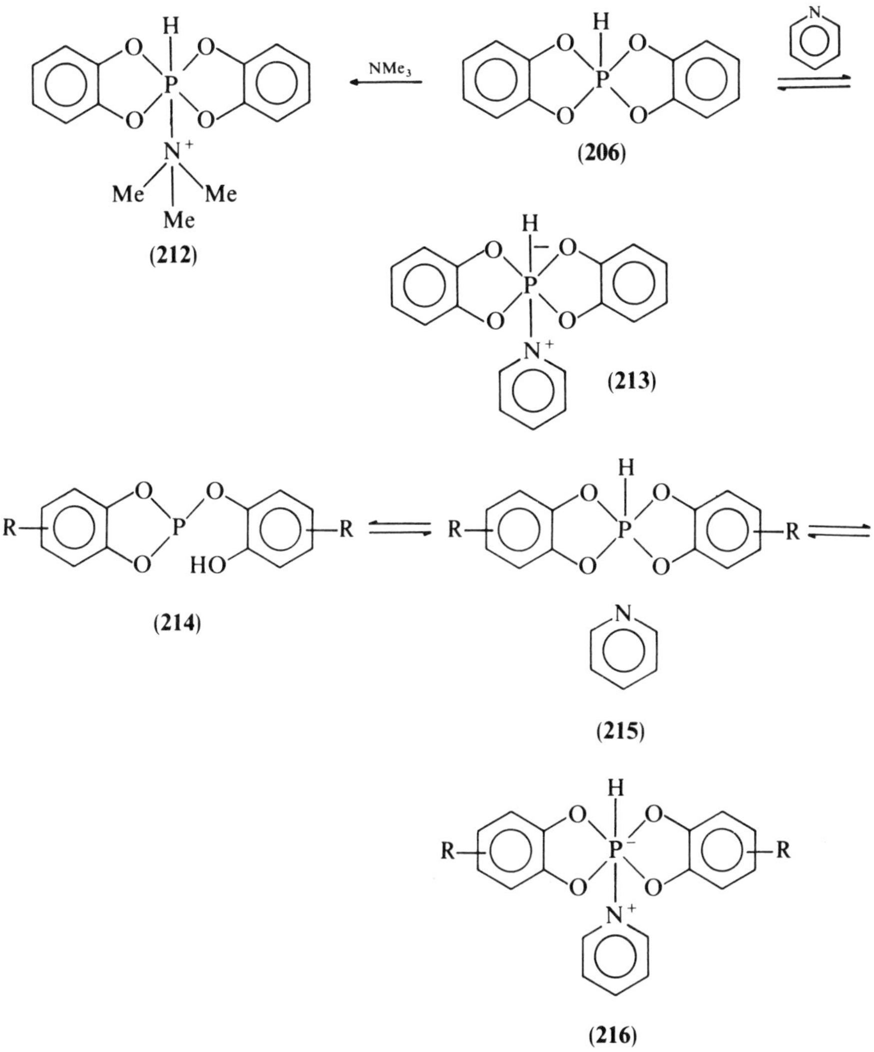

(212) (206) (213) (214) (215) (216)

SCHEME 56

both groups working on spirophosphoranes of substituted pyrocatechols. In his treatment of the quantitative determination of the equilibrium thermodynamic parameters, Koenig[263] studied the influence of various solvents and of the temperature (from -45 to 100 °C) and demonstrated the presence of two successive equilibria, namely P(III)$\rightleftharpoons$P(V)$\rightleftharpoons$P(VI) (Scheme 56), with P(VI) favoured at low temperatures. The reversibility of the equilibria is therefore a proof of the succession **214**$\rightleftharpoons$**215**$\rightleftharpoons$**216**.

Starting from the spirophosphorane **217** with two different rings, Ramirez *et al.*[264] obtained the octahedral complex **218** in the presence of pyridine and also deduced that a fast P(V)$\rightleftharpoons$P(VI) equilibrium was present. The same **217** with phenol and NEt_3 yields the

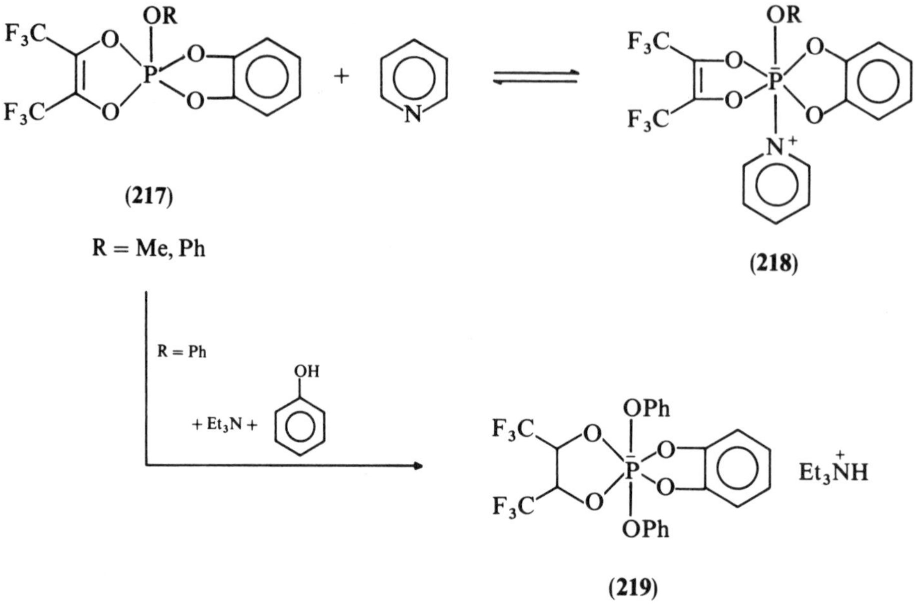

(217)

R = Me, Ph

(218)

R = Ph

+ Et_3N +

(219)

SCHEME 57

complex **219**, and it would appear that this ion pair is stabler, as a solid or in solution, than the zwitterion **218** (Scheme 57). Two geometries are possible for **219** if an octahedral structure is assumed: in **219a**, the two phenolic moieties are adjacent (*cis*), but they are colinear (*trans*) in **219b**. Two quadruplets are found in the ^{19}F NMR spectrum, at − 13.3 and − 14.1 ppm, with $J_{FF} = 10$ Hz. The non-equivalence of the two CF_3 groups points to **219a** (Figure 28), in agreement with the conclusions later reached by Minaev and Minkin[265], who concluded that the addition of a nucleophilic reagent to a pentacoordinated structure is easiest with a *cis* conformation. An NMR study by FontFreide and Tripett[266] of a number of hexacoordinated compounds similar to **219c** (Figure 28) also led

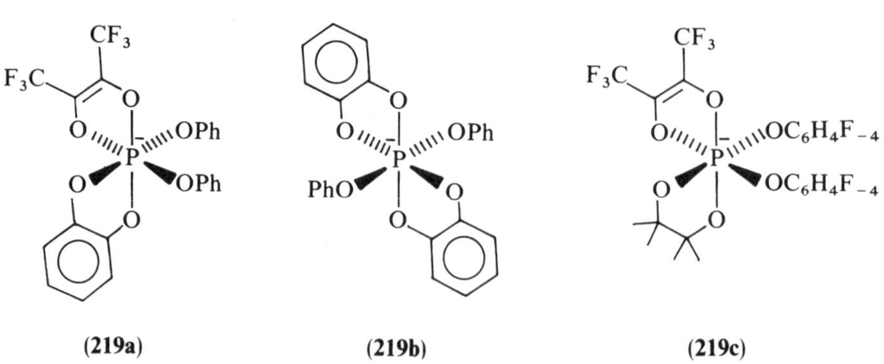

(219a) (219b) (219c)

FIGURE 28

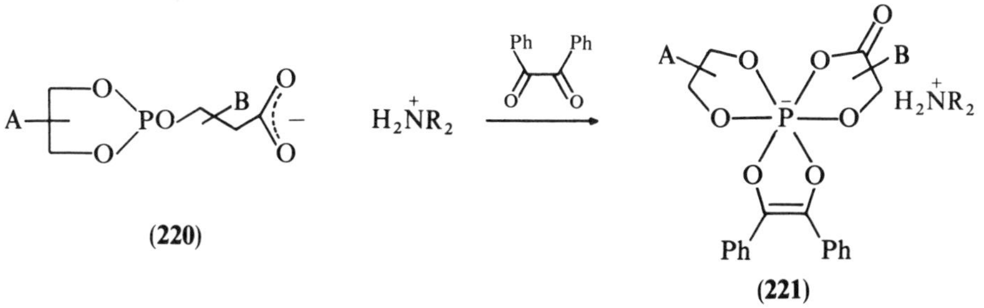

A, O, PO, O, B, O, O, − $H_2\overset{+}{N}R_2$ (Ph, Ph, O, O) → A, O, O, B, O, O, P, O, O, O, Ph, Ph, $H_2\overset{+}{N}R_2$

(220)

(221)

A = CH_2CH_2, B = CHPhCH, R = Et: $\delta^{31}P$ −93 ppm
A = CH_2CH_2, B = CHMeCH, R = Et: $\delta^{31}P$ −95 ppm
A = Me_2C—CMe_2, B = CHMeCH, R = Me: $\delta^{31}P$ −97 ppm

SCHEME 58

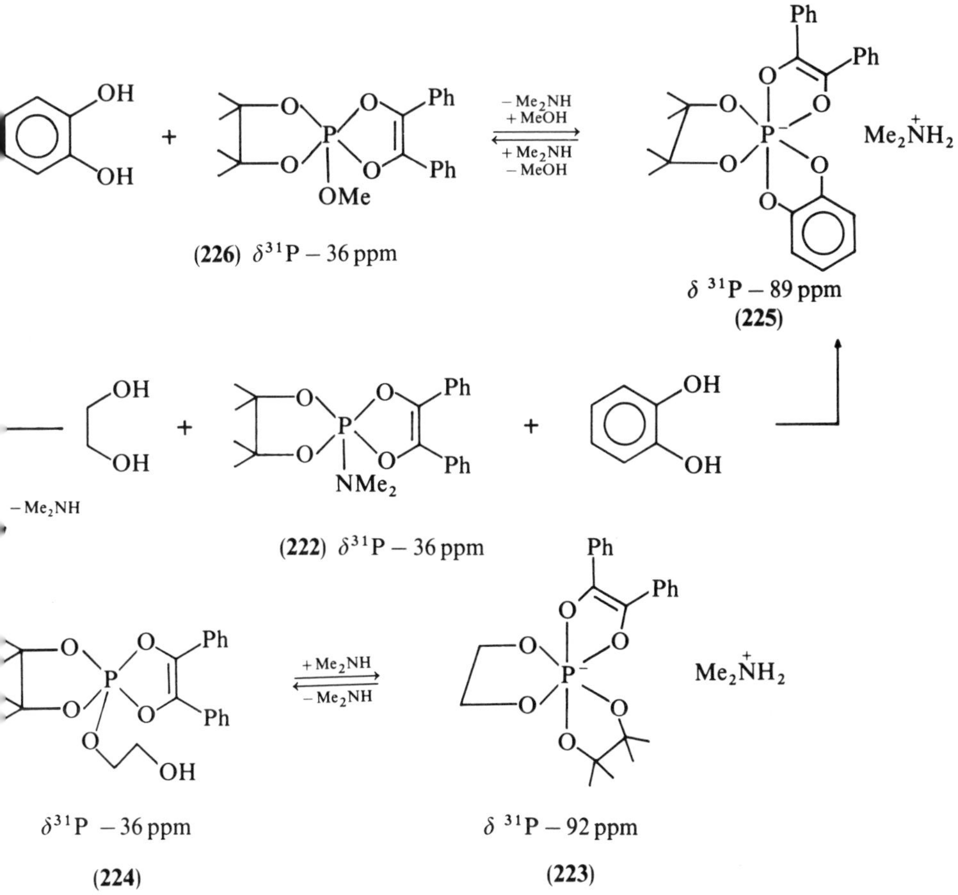

OH, OH + O, O, P, O, O, Ph, Ph, OMe ($-Me_2NH$, $+MeOH$ ⇌ $+Me_2NH$, $-MeOH$) Ph, Ph, O, O, O, P, O, O, O, O, $Me_2\overset{+}{N}H_2$

(226) $\delta^{31}P$ − 36 ppm

δ ^{31}P − 89 ppm
(225)

OH, OH + O, O, P, O, O, Ph, Ph, NMe_2 + OH, OH

$-Me_2NH$

(222) $\delta^{31}P$ − 36 ppm

O, O, P, O, O, Ph, Ph, O, OH ($+Me_2NH$ ⇌ $-Me_2NH$) Ph, Ph, O, O, O, P, O, O, O, O, $Me_2\overset{+}{N}H_2$

$\delta^{31}P$ −36 ppm

(224)

δ ^{31}P − 92 ppm

(223)

SCHEME 59

them to conclude that the *cis*-form was predominant at room temperature. One should keep in mind, however, that, like their parent P(V) derivatives, the hexacoordinated phosphorus compounds are likely to undergo regular stereomutations.

The first chelate-type hexacoordinated compounds were prepared by Koenig *et al.*[267] (Scheme 58) by allowing an aminodioxaphospholane to react first with a hydroxyacid, thus obtaining the tricoordinated compound **220**, and then on benzil. Varying the substitution on the ring or the hydroxyacid yields the species **221a–c**. The same scheme was observed by Bernard and Burgada[268] during a study of the alcoholysis of the P—NR_2 bond in spirophosphoranes (Scheme 59); the alcoholysis of **222** by pyrocatechol at room temperature gives a quantitative yield of **225**, whereas slight heating of **222** in the presence of glycol gives the spirophosphorane **224** with elimination of dimethylamine. If the latter is replaced in the reacting medium, the system evolves towards the complex **223**. Similarly, the reversible alcoholysis of **225** to **226** shows that the equilibria **225**⇌**226** and **223**⇌**226** constitute proof of the formation of hexacoordinated intermediates during the nucleophilic sustitution of phosphoranes in basic media.

The presence of an electron-accepting group ($R = CF_3$ or Cl) in compounds such as **227** or **228** gave the first examples of a P(V)⇌P(VI) equilibrium accompanied by migration of a chlorine atom (Scheme 60). The equilibrium is displaced towards the P(V) form when the attracting power of R increases, the polarizing power of the solvent decreases or the temperature increases[269]. Nebegretskii and coworkers[270] also studied structures akin to **229** with different groups linked to the P atom, among them a spirophosphorane with Ph and F replaced by a catechol ring, and found that the P(VI) compound could be an intermediate in equilibrium with the two phosphoranes **230** and **231**, thus providing a new example of irregular permutational isomerism.

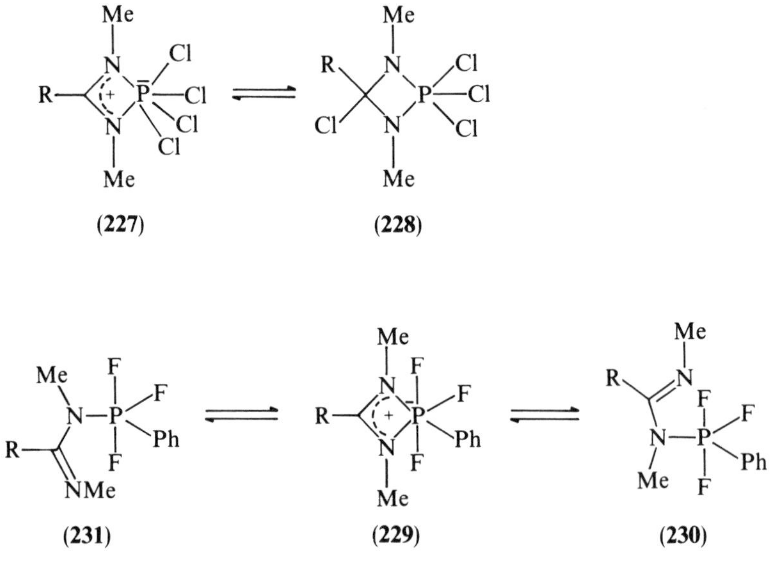

(227) ⇌ **(228)**

(231) ⇌ **(229)** ⇌ **(230)**

SCHEME 60

To summarize, the P(V)⇌P(VI) equilibria show that the stability of pentacoordinated and hexacoordinated structures depend on the same factors, namely the nature of the solvent and on the basic character of the reacting medium, the temperature, and/or the

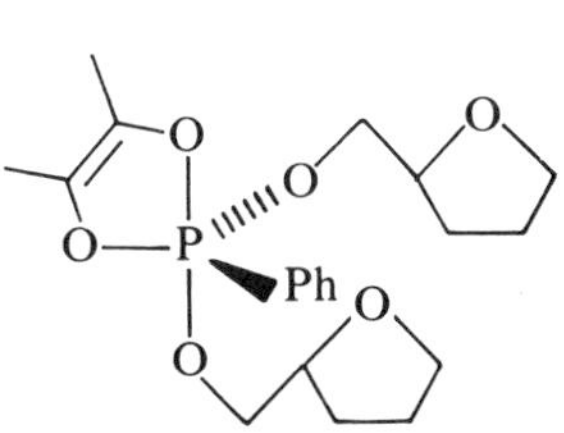

(2 diastereomers)

(232)

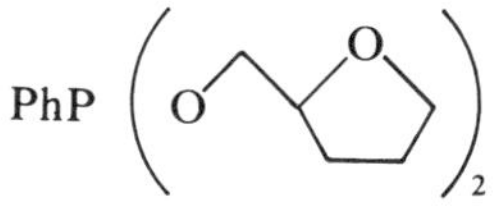

PhP(O-tetrahydrofurfuryl)$_2$

2 *meso* forms + **DL** pair

(233)

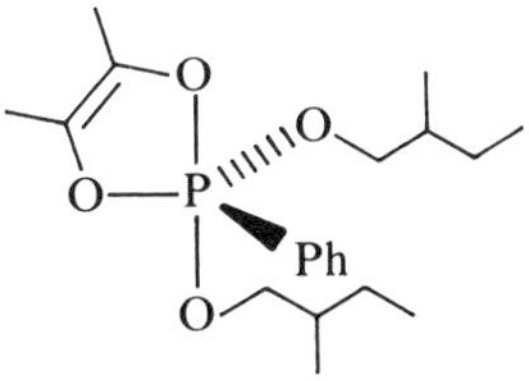

$PhP(OCH_2CH(Me)CH_2Me)_2$

(234)

(3 diastereomers)

(235)

FIGURE 29

steric and electronic characteristics of the substituents. Moreover, a fast P(V)⇌P(VI) equilibrium strongly displaced in favour of P(V) is often not easily detected, as illustrated by the following example. Starting from **233** (Figure 29), a mixture of the two *meso* forms and of the racemate of phenylbis(tetrahydrofurfuryloxy)phosphine, the phosphorane **232** was prepared by Koole *et al.*[271]. The phosphorane, which should normally have been a mixture of three diastereoisomers, gave only two signals, at $\delta = -42.1$ and -42.09 ppm in the ^{31}P NMR spectrum, from which the authors deduced that, even at -70 °C, two *meso* forms were rapidly interconverting by stereomutation. Since this implies a passage by an intermediate in which the ring occupies a diequatorial conformation in the Berry mechanism, there is a violation of the ring strain rule. This result should be compared to the synthesis by Denney *et al.*[272] of the phosphorane **235** from phenylbis(2-methyl-1-butoxy)phosphine **234** and the observation of the three signals expected for the diastereoisomers ($\delta = -34.49$, -34.52 and -34.56) in the ^{31}P NMR spectrum. The only difference between **232** and **235** is the presence in the former of the O atoms of the tetrahydrofuran rings which can form a zwitterion-type bond with the P atom, thus forcing a regular stereomutation mechanism for the hexacoordinated structure. This could perhaps also be the reason for the difference $\Delta\delta = 7.5$ ppm observed in the spectra of the two phosphoranes in spite of the apparently strictly identical environment of the phosphorus atoms.

Ramirez *et al.*[273] also used the possible formation of a hexacoordinated intermediate to account for the OMe-OPh ligand exchange between $P(OPh)_5$ and 2,2,2-trimethoxy-2,2-dihydro-1,3,2-trioxaphospholene.

The following example shows that the results of some reactions may be unexpected: by allowing $P(NMe_2)_3$ to react with *p-tert*-butylcalix[4]arene, Khasnis *et al.*[274] obtained the hexacoordinated compound **236** instead of the expected phosphorane with a PH bond, **238** (Scheme 61). This result is, however, consistent with the particular reactivity of the pyrocatechols in basic media mentioned above (Schemes 54 and 55). Conversely, submitting **238** to heat or to the action of CF_3COOH yields nearly quantitatively the P(III) derivative **237** rather than **238**. The equilibrium P(III) (**237**)⇌P(V) (**238**), if it exists, must be totally shifted to the tricoordinated form. The action of LiBu on **236** does not affect the PH bond but rather results in the deprotonation of the NH bond and the formation of a lithium salt whose structure, as determined by XRD[275], places the P atom nearly in the plane of the four O atoms, the H atom of the PH bond within the aromatic 'cone' and the N atom at the apex of the structure.

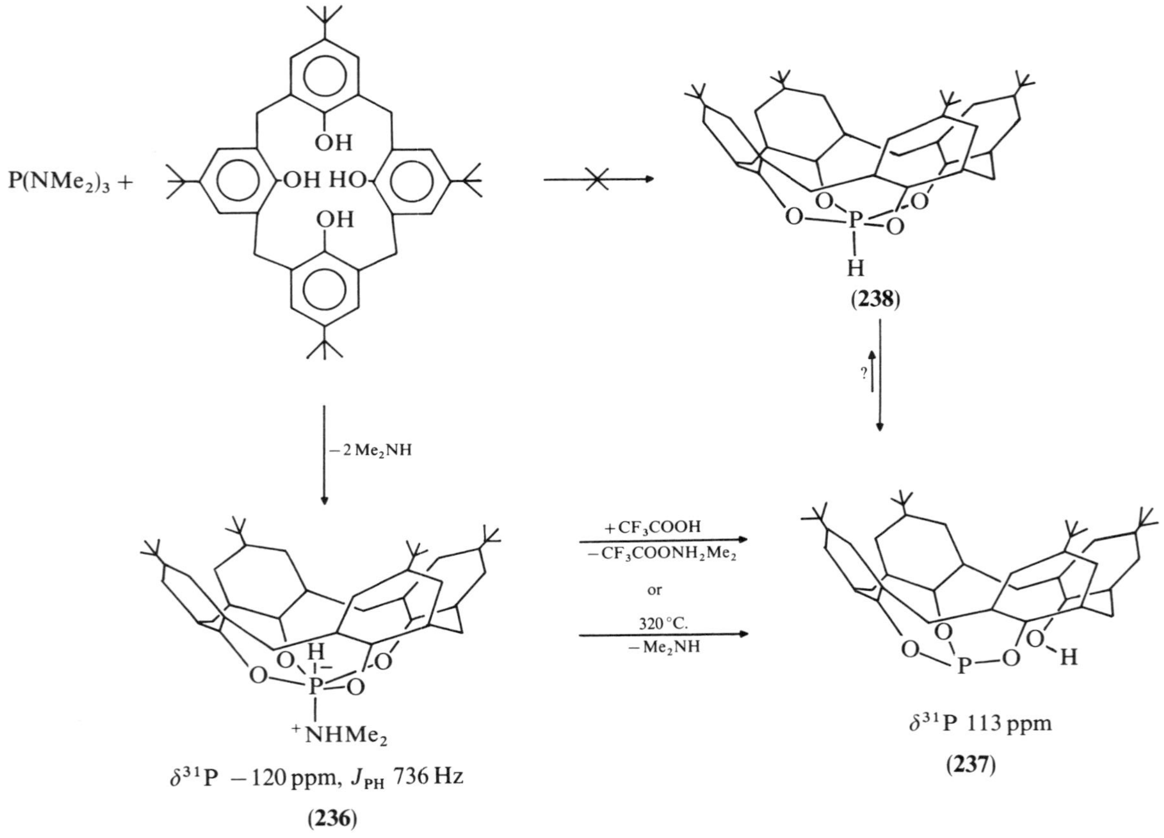

$P(NMe_2)_3$ +

$-2\,Me_2NH$

(**238**)

?

$+CF_3COOH$

$-CF_3COONH_2Me_2$

or

320 °C.

$-Me_2NH$

$^{+}NHMe_2$

$\delta^{31}P$ −120 ppm, J_{PH} 736 Hz

(**236**)

$\delta^{31}P$ 113 ppm

(**237**)

SCHEME 61

VI. SYNTHESIS OF PHOSPHORANES

As indicated in Section I, pentacoordinated structures can be reached from virtually all the other coordination states of phosphorus, but a survey of the methods used places those starting from P(III) derivatives as the most popular, followed by the methods using P(V) compounds, with synthesis from P(IV) structures coming last. As it seems impossible to detail all the reactions used, we shall limit this section to a few general examples and results, especially since many of the methods have already been reviewed in the preceding sections.

A. Syntheses of Phosphoranes Based on P(III) Derivatives (Table 1)

Many P(V) compounds can be obtained by simple addition of a halogen or a halogenated molecule (Table 1, reactions 1–7). The phosphorane **242** produced by reaction 4 (4-iodo-2-oxo-1,3,5,5,8-pentamethyl-1,3,8-triaza-5-azonia-4 λ^5-phosphaspiro-[3,4]octanyl iodide) and whose structure was ascertained by ^{13}C, ^{15}N, and ^{31}P NMR of the solid, is the first example of a compound with a P—I bond. Some of these reactions, e.g. No. 3, yielding **241**, can give rise to P(V)$\rightleftharpoons$phosphonium salt equilibria.

Another series of reactions (Nos 8–15) show different possibilities for the addition of conjugated double bonds on a P(III) atom. Although this type of reaction has been stated to be a concerted six-electron dissociative process, and therefore stereoselective, its scope is limited by the fact that it cannot be generalized to all P(III) compounds as some of them are fairly unreactive, or yield phosphonium salts or may even be dealkylated.

In contrast, the reactions with peroxides or sulphenates (Nos 16–19) are general and lead to a large number of linear or cyclic phosphoranes such as **256**–**264**, even if some of the latter, especially when they are linear, are sometimes unstable. The reaction with sulphenates (No. 17), which yields as an intermediate a thiophosphorane such as **257**, may be followed by reaction with a further molecule of sulphenate to yield the fully oxygenated derivative **258**. A similar reaction is observed with dithiete. This reaction, which was discovered by Denney and coworkers[292], has led to a very large number of linear, monocyclic, polycyclic and spiranic thiophosphorus derivatives, among which **261** is notable for being the first thiophosphorane with five PS bonds.

Reactions 20 and 21 are interesting in their use of the oxidizing properties of an *N*-chloro derivative. In reaction 22, the two Cl atoms on the P(III) compound play a somewhat similar role, while in reaction 23 oxidizing power is provided by the S—S bond of diphenyl disulphide.

The last two reactions in Table 1 (Nos 24 and 25) show the reaction of a P(III) compound with each of the C=O groups in 2 mol of reagent to yield diverse cyclic compounds.

B. Syntheses of Phosphoranes from P(V) Compounds (Table 2)

The simple halogenated phosphoranes are particularly strong Lewis acids, and one of their most characteristic properties is their high reactivity towards amines, alcohols, thiols or their metallic derivatives, and their ability to form, in these reactions, pentacoordinated substitution products. The P(V)—halogen bond can easily be reduced to P(V)—H (Nos 1 and 11), and the latter is a good starting point for the functionalization of pentacoordinated structures by reaction with halogen derivatives (No. 8): acid chlorides in dimethyl sulphoxide and benzyl bromide in the presence of Me_3N thus yield the compounds **283** to **286**, whereas unsaturated molecules with aldehydic, ethylenic, acetylenic or azo groups, or unsaturated α,γ-diketones, undergo addition to the unsaturated bond (if it is activated) and yield the derivatives **287**–**291** (No. 9).

TABLE 1. Synthesis of some pentacoordinated phosphorus compounds starting from tricoordinated phosphorus derivatives

No.	P(III)	Reagent	P(V)	Ref.
1	$P(CH_2Cl)_3$	Cl_2	$Cl_3P(CCl_3)_2$ **(239)**	276
2	R_2NPF_2	X_2	$R_2NPF_2X_2$ **(240)**	277
3	$P(OAr)_3$	X_2	$(ArO)_3PX_2$ **(241)**	278
4	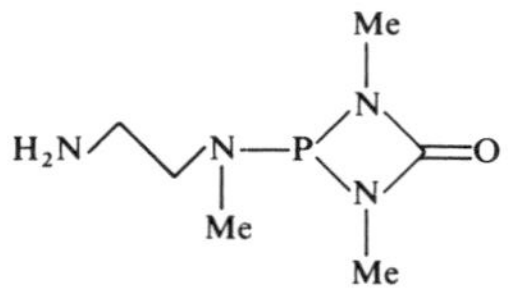 [structure]	I_2	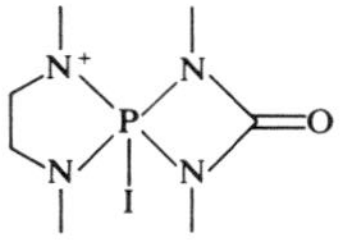 [structure] **(242)**	279
5–6	RPX_2	HF, R^1ZH	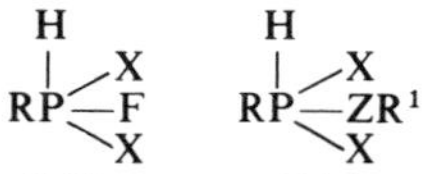 [structure] **(243)** [structure] **(244)**	280
7	PF_3	$(F_3C)_2NCl$	$(F_3C)_2NPF_3Cl$ **(245)**	281
8	PCl_3, $P(OR)_3$, 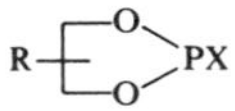 [structure]; X = Br, Cl, F, SR, NMe_2, MeS, OR	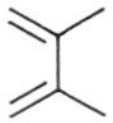 [structure]	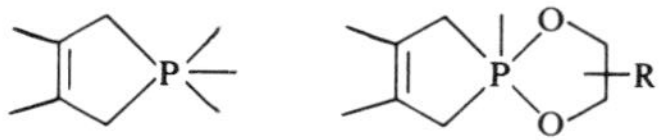 [structure] **(246)** [structure] **(247)**	282, 283

9	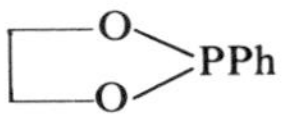 O O PPh	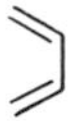	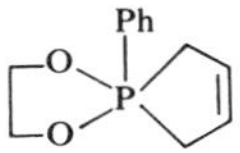 Ph O P O (**248**)	284
10	$(RO)_2PCO_2Me$	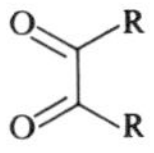 O R O R	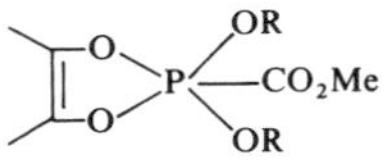 O P O OR CO_2Me OR (**249**)	285–6
11	 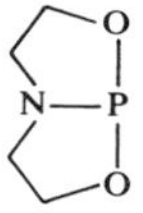 N P N N O N P O	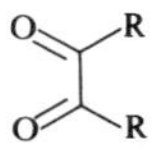 O R O R	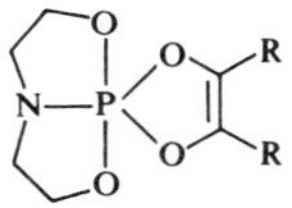 O N P O O R O R N N P N O R O R (**250**) 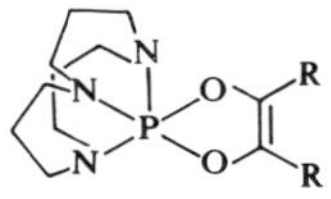 (**251**)	234 290
12	$P(OR)_3$	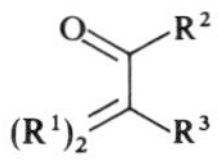 O R^2 $(R^1)_2$ R^3	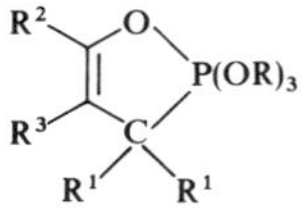 R^2 O $P(OR)_3$ R^3 C R^1 R^1 (**252**)	286
13	$P(OR)_3$	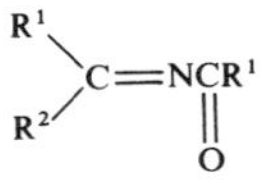 R^1 R^2 C=NCR^1 O	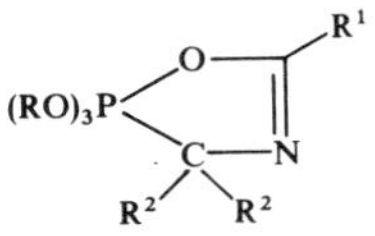 $(RO)_3P$ O R^1 N C R^2 R^2 (**253**)	287

TABLE 1. (*continued*)

No.	P(III)	Reagent	P(V)	Ref.
14	$P(OR)_3$	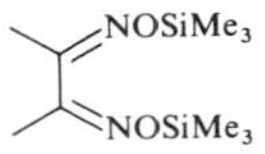 NOSiMe$_3$, NOSiMe$_3$	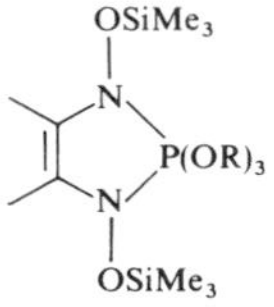 OSiMe$_3$, N, P(OR)$_3$, N, OSiMe$_3$ (254)	288
15	$(Et_2N)_2PC{\equiv}CP(NEt_2)_2$	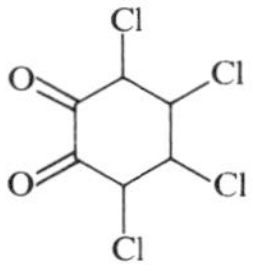 O, O, Cl, Cl, Cl, Cl	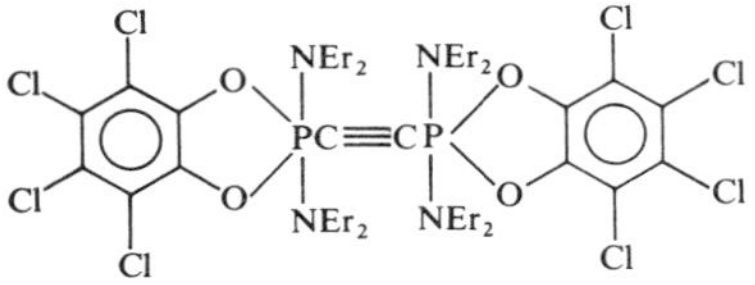 Cl, Cl, Cl, Cl, O, O, NEr$_2$, NEr$_2$, PC≡CP, NEr$_2$, NEr$_2$, O, O, Cl, Cl, Cl, Cl (255)	289
16	$(RO)_3P$, $RP(OR)_2$	R^1OOR^1	$(RO)_3P(OR^1)_2$ (256) $(RO)_2PR(OR^1)_2$ (259)	290
17	R_3P, 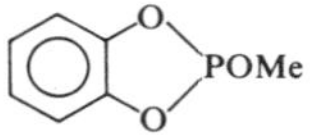 O, O, POMe	$PhSOR^1$	 O, O, P, OMe, SPh, OR^1 (257) ⟶ O, O, P, OR^1, OMe, OR^1 (258)	291
18	R_3P, 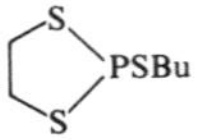 S, S, PSBu	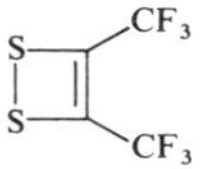 S, S, CF$_3$, CF$_3$	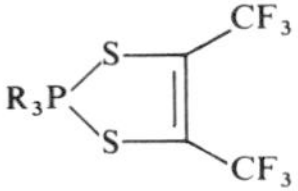 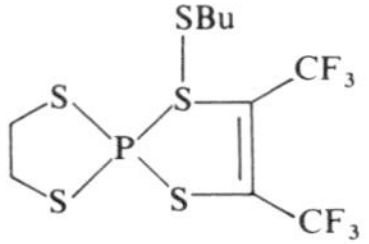 R$_3$P, S, S, CF$_3$, CF$_3$ (260) S, S, P, SBu, S, S, CF$_3$, CF$_3$ (261)	292

19	Ph_3P, $(MeO)_3P$, Ph_2POMe	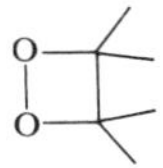	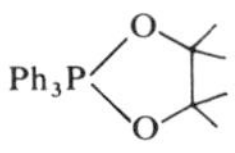 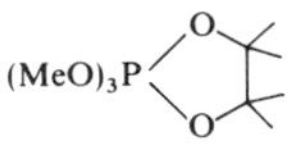 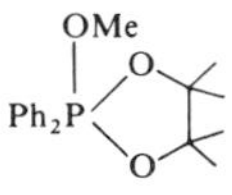 Ph_3P (262) $(MeO)_3P$ (263) Ph_2P OMe (264)	293
20	$P(OR)_3$	$ClN(Pr^i)_2 + R^1$ OH OH 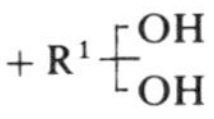	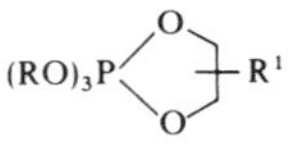 $(RO)_3P$ O O R^1 (265)	294
21	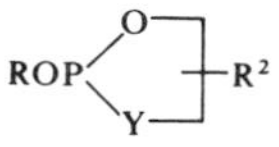 ROP O Y R^2	$ClN(Pr^i)_2 + R^1$ OH OH 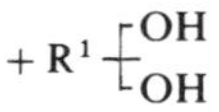	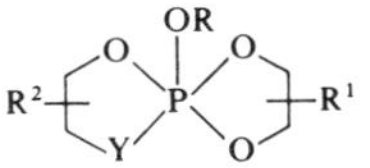 R^2 O Y P OR O O R^1 (Y = O, NMe, S) (266)	294
22	$PhPCl_2$	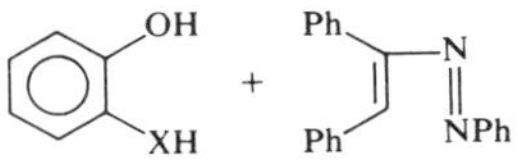 OH XH + Ph Ph N NPh	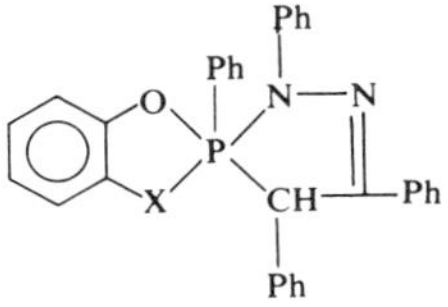 O X P Ph N N Ph CH Ph Ph (267) X = O, NH	295
23	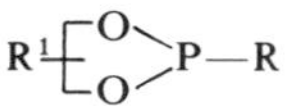 R^1 O O P—R	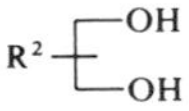 R^2 OH OH + PhSSPh(− 2 PhSH)	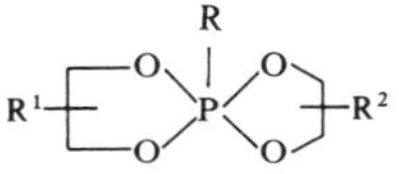 R^1 O O P R O O R^2 (268)	296

(continued)

TABLE 1. (*continued*)

No.	P(III)	Reagent	P(V)	Ref.
24	$(MeO)_2PC{\equiv}CR$	$2R^1C(=O)R^2$, 2 PhCOCN	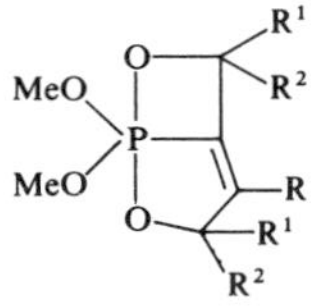 (269) 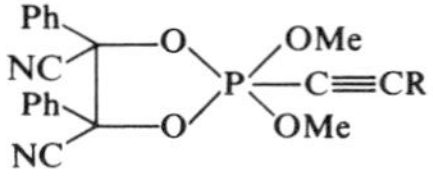 (270)	297
25	$(RO)_3P$, $PhP(OMe)_2$	$2R^1CHO$, $2Me_2C{=}C{=}O$ 2 OHC C_6H_4 NO_2-*p*	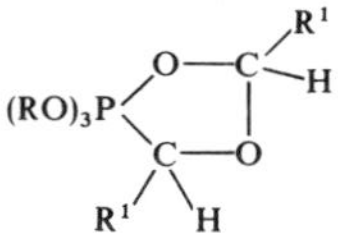 (271) 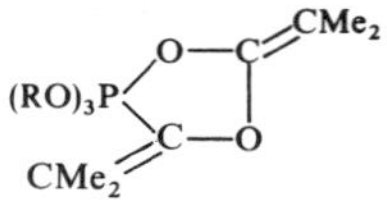 (272) 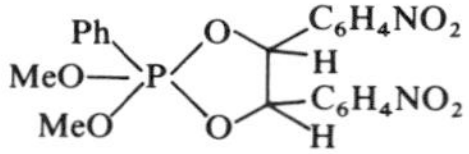 (273)	298

TABLE 2. Synthesis of some pentacoordinated phosphorus compounds starting from other pentacoordinated phosphorus derivatives

No.	P(V)	Reagent	P(V)	Ref.
1	RPX_4	Me_3SnH	$RPHX_3$ **(274)**	288
2	$RPHF_3$	RONa	$RP(H)(F)_2(OR)$ **(275)**	300
3	PCl_5	$R_2C{=}CH_2$	$R_2C{=}CHPCl_4$ **(276)**	301
4	$O{=}C(NMe)_2PCl_3$	$MeNHC({=}O)NHMe$	$[O{=}C(NMe)_2]_2PCl$ **(277)**	302
5	$R^1R^2PF_3$	R^3_2NH	$R^1R^2PF(NR^3_2)_2$ **(278)**	303
		2-amino-4,6-dimethylphenol (OH, NH_2)	$C_6H_4(O)(NH)PR^1R^2F$ **(279)**	304
		$Me_2N(SiMe_3)_2$	$[R^1R^2FP(\mu\text{-}NMe)]_2$ **(280)**	305

(*continued*)

TABLE 2. *(continued)*

No.	P(V)	Reagent	P(V)	Ref.
6	PCl_5	PhOH (phenol)	$P(OPh)_5$ (**281**)	306
7	$ArP(H)F_3$	R^1CHO	$ArPF_3(CHOHR^1)$ (**282**)	307
8	spirophosphorane with P–H, R = CF_3	X_2, $PhCH_2Br-Et_3N$	(**283**) PX; (**284**) PCH_2Ph	308
		MeCOCl, Me_2SO	(**285**) PC(O)Me; (**286**) POH	
9	spirophosphorane with P–H, R = H, Me Y = O, NMe	R^1CHO, $R^1CH{=}NR^2$,	(**287**) $PCHOHR^1$, $\geqslant PCHR^1NHR^2$ (**288**)	309
		$-C(O)-CH{=}CH-C(O)-$	$\geqslant PCH(C(O)-)CH_2C(O)-$ (**289**)	310
		$R^1C{\equiv}CR^2$, $ROC(O)N{=}NC(O)OR$	$\geqslant P(R^1)C{=}C(R^2)-$ (**290**); $\geqslant PN(CO_2R)-NH(CO_2R)$ (**291**)	

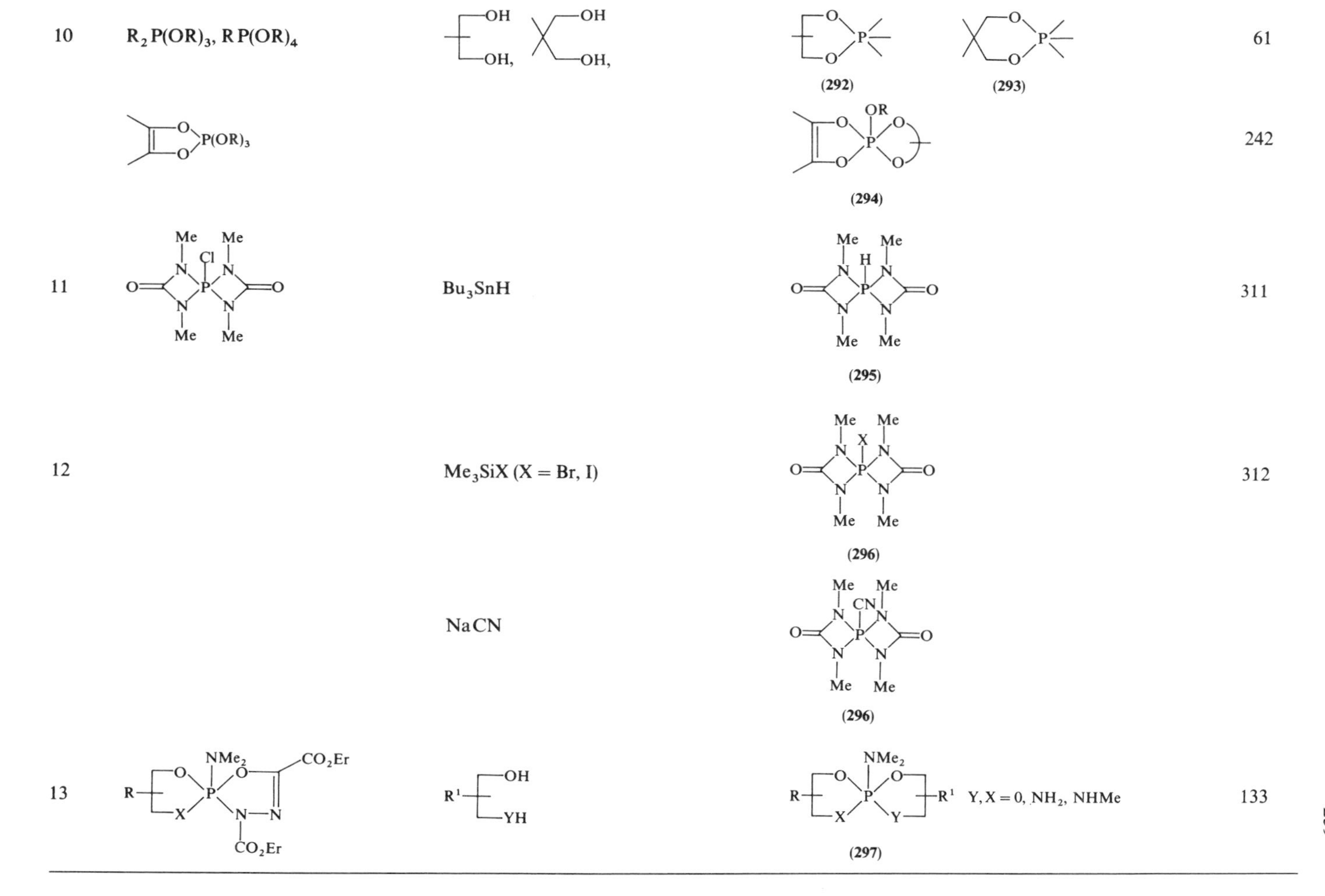

10	$R_2P(OR)_3$, $RP(OR)_4$	OH, OH, OH, OH,	(292) (293)	61
			(294)	242
11		Bu_3SnH	(295)	311
12		Me_3SiX (X = Br, I)	(296)	312
		NaCN	(296)	
13	NMe_2, CO_2Er, CO_2Er	R^1, OH, YH	(297) Y, X = 0, NH_2, NHMe	133

TABLE 3. Synthesis of some pentacoordinated phosphorus compounds starting from tetracoordinated phosphorus derivatives

No.	P(IV)	Reagent	P(V)	Ref.
1	$R_4\overset{+}{P}X^-$	$NaBH_4$, $LiAlH_4$	R_4PH **(283)**	117, 326
2	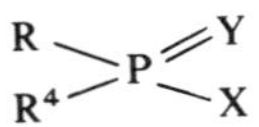 $R R^4 P(=Y)X$; X = O, S; X = Cl, OH, SH	$SbF_3[SF_4, SOCl_2, HF, Cl_2]$	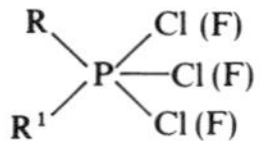 $RR^1P(Cl)_3$ [(F)] **(302)**	327
3	$(F_3C)_3PO$	$(Me_3Si)_2O$	$(F_3C)_3P(OSiMe_3)_2$ **(303)**	328
4	$R_3P=CH_2$	oxetane	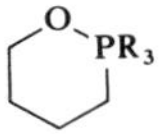 1,2-oxaphosphorinane, PR_3 **(304)**	329
5	$ROP(=O)Cl_2$	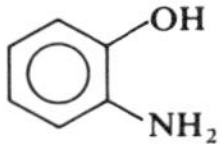 2-aminophenol	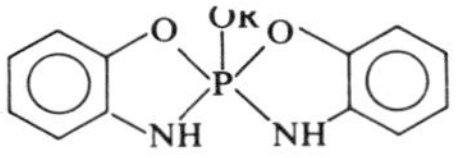 spiro bis(benzoxazaphosphole) P–OR **(305)**	330
6	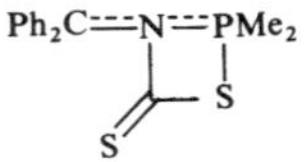 $Ph_2C{=}N{=}PMe_2$ (four-membered ring with C=S, S)	$ROC(=O)C{\equiv}CC(=O)OR$	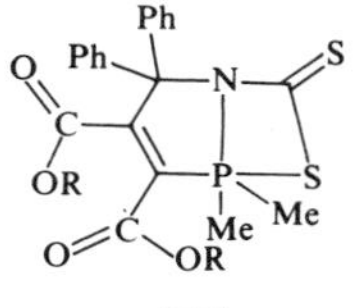 **(306)**	56
7	$RR^1P(=O)Cl$	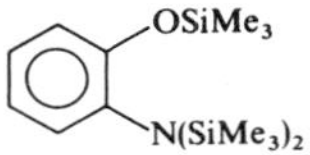 2-(trimethylsiloxy)-N,N-bis(trimethylsilyl)aniline, $OSiMe_3$, $N(SiMe_3)_2$	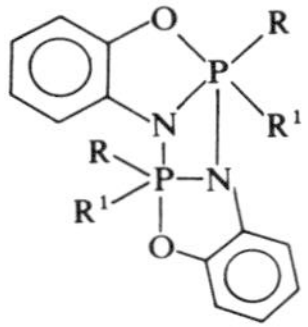 **(307)**	331

This type of reaction has been extended to monocyclic phosphoranes with a PH bond[313]. A comparative study of the reactivity of the cyclamphosphorane **79**⇌**79a** and cyclenephosphorane **78** (cf. Scheme 13) towards trifluoromethyl phenyl ketone evidenced two distinct addition mechanisms[314] (Scheme 62). The first, operative in the presence of the equilibrium, is probably an attack by the P(III) form (**79a**) on the oxygen atom of the ketone; the second, occurring with **78** and yielding **299**, corresponds to a (slow) attack on the C atom of the ketone by the intermediate phosphoranide **78′**. This hypothesis is strongly bolstered by the fact that the formation of the adduct **299** is extremely fast in the presence of LiMe. A similar occurrence has been observed with a tetroxyspirophosphorane with a PH bond whose P(III) form cannot be detected but which adds very fast to aldehydes in the presence of strong bases, probably by means of an intermediate phosphoranide structure[315].

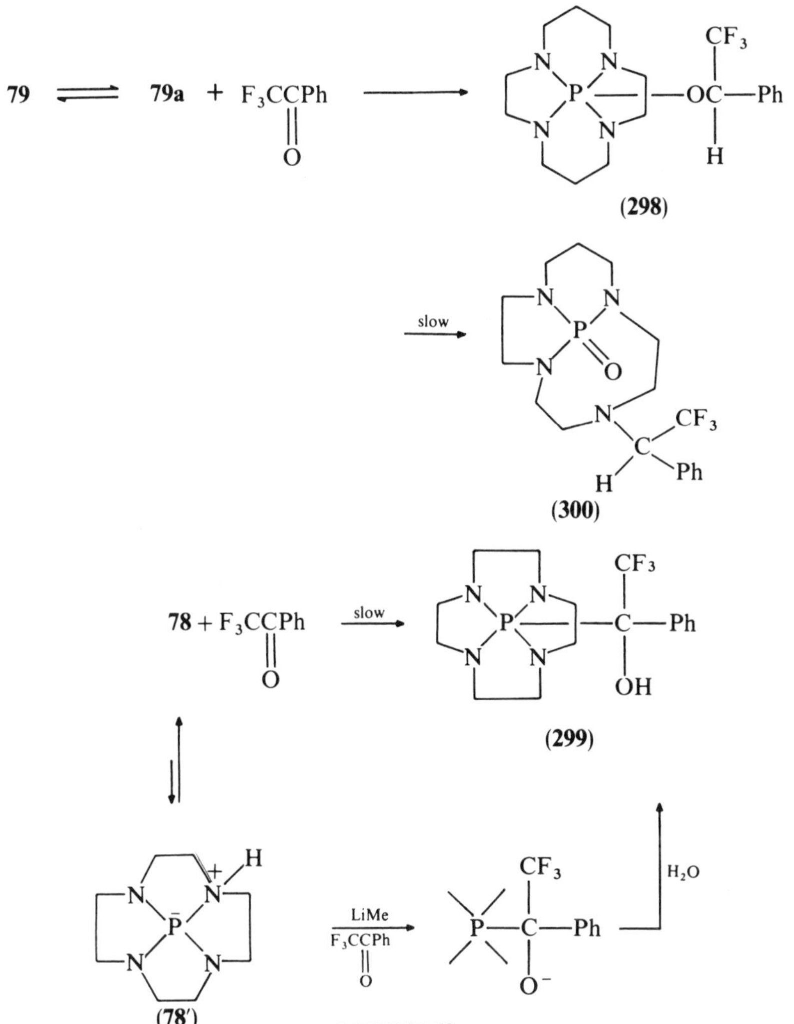

79 ⇌ **79a** + $F_3CC(O)Ph$ → (**298**)

slow → (**300**)

78 + $F_3CC(O)Ph$ —slow→ (**299**)

(**78′**) —LiMe, $F_3CC(O)Ph$→ P—C(CF_3)(Ph)—O^- —H_2O→ (**299**)

SCHEME 62

Treatment of the cyclenephosphorane **78** with LiBu in tetrahydrofuran leads to a polymeric phosphoranide in which the intermolecular links are LiTHF units bonding apical N atoms. As shown by XRD[316], the geometry is that of a deformed trigonal bipyramid, with the electron doublet in an *e* position. (The first structural determinations by XRD are due to Sheldrick *et al.*[317] and Schomberg *et al.*[318].) The particular properties of cyclenphosphoranes are essentially due to the strong nucleophilic characteristics of the apical N atoms[319].

A fairly general method of synthesis of oxyphosphoranes is based on an exchange reaction between two OR groups linked to the P atom, and a diol, an amino alcohol[320] or pyrocatechol (No. 10). The driving force of this reaction, similar, in a way, to transesterification, is the high thermodynamic stability resulting from the change of two linear groups to monocyclic, bicyclic or even spirocyclic substitution. The alcohol (often ethanol) which is liberated does not give rise to the possible reverse reaction. Among other dihydroxy compounds used are ethylene, propylene, neopentyl and styrene glycol, cyclohexanediol and butane-2,3-diol. There are, however, limitations to this reaction: butane-1,4-diol and pentane-1,5-diol are dehydrated to tetrahydrofuran and tetrahydropyran, respectively, while 5-aminopentan-1-ol and ethanolamine yield cyclohexylamine and aziridine, with simultaneous formation of a P(IV) compound[321] such as $OPPh_3$ if the starting phosphorane was $Ph_3P(OEt)_2$, or $OP(OEt)_3$ if it was $P(OEt)_5$[322,323]. Nevertheless, unsaturated cyclic phosphoranes can also undergo exchange reactions with alcohols and diols while retaining their dioxophospholane ring[244,324] and even spirophosphoranes with a particularly labile ring, such as that obtained by reaction of a cyclic P(III) compound with ethyl azodicarboxylate, can turn out to be stable under these conditions (No. 13).

C. Synthesis of Phosphoranes Starting from P(IV) Compounds (Table 3)

Although far from exhaustive, Table 3 shows that tetracoordinated compounds as diverse as phosphonium salts, phosphinates, thiophosphinates, phosphine oxides, phosphonium ylides and halophosphates can all lead to P(V) derivatives by addition, substitution, or oxidoreduction reactions.

Note

The rapid evolution and the multifarious activity in the field of phosphorus chemistry is well illustrated by the proposal (Dupart)[332] of a new systematic classification of free and coordinated phosphorus compounds to complete the systems proposed by Wolf[333] and by Perkins *et al.*[334]. In this new system, a code number *NFE*(*NPL*)*D* could be used to describe the compound, in which *NFE* would be the number of free electrons, *N* the number of valence shell electrons, *L* the number of ligands and *D* the number of electron doublets donated or accepted, which would be negative if the P atom acted as a donor (a Lewis base), or positive if P acted as an acceptor (a Lewis acid).

VII. REFERENCES

1. G. Wittig and M. Reiber, *Justus Liebigs Ann. Chem.*, **562**, 147 (1949).
2. P. J. Wheatley, *J. Chem. Soc.*, 2206 (1964).
3. F. Ramirez, *Pure Appl. Chem.*, **9**, 337 (1964); *Acc. Chem. Res.*, **1**, 168 (1968).
4. D. B. Denney and H. M. Relles, *J. Am. Chem. Soc.*, **86**, 3897 (1964).
5. R. S. Berry, *J. Chem. Phys.*, **32**, 933 (1960).
6. G. W. Fenton and C. K. Ingold, *J. Chem. Soc.*, 2342 (1929).
7. F. H. Westheimer, *Acc. Chem. Res.*, **1**, 70 (1968); *Pure Appl. Chem.*, **49**, 1059 (1977); J. A. Grelt, F. H. Westheimer and J. M. Sturtevant, *J. Biol. Chem.*, **250**, 5059 (1975).

8. R. J. Gillespie, *Angew. Chem., Int. Ed. Engl.*, **6**, 819 (1967); R. J. Gillespie, in *Molecular Geometry*, Van Nostrand, London, 1972; R. J. Gillespie and R. S. Nyholm, *Q. Rev. Chem. Soc.*, **11**, 330 (1957); R. J. Gillespie, *Chem. Soc. Rev.*, 59 (1992).
9. J. Brocas, M. Gielen and R. Willem, *The Permutational Approach to Stereochemistry*, McGraw-Hill, New York, 1983.
10. R. F. Hudson, *Structure and Mechanism in Organo Phosphorus Chemistry*, Academic Press, New York, 1965.
11. G. E. Kimball, *J. Chem. Phys.*, **8**, 188 (1960).
12. R. J. Gillespie, *J. Chem. Soc.*, 1002 (1952).
13. J. A. Howard, D. R. Russell and S. Trippett, *J. Chem. Soc., Chem. Commun.*, 856 (1973).
14. H. Wunderlich, D. Mootz, R. Schmutzler and M. Wieber, *Z. Naturforsch., Teil B*, **29**, 32 (1974).
15. M. Eisenhut, R. Schmutzler and W. Sheldrick, *J. Chem. Soc., Chem. Commun.*, 144 (1973).
16. M. L. Muetterties and R. A. Shum, *Q. Rev. Chem. Soc.*, **20**, 245 (1966).
17. D. Hellwinkel, *Chem. Ber.*, **99**, 3628 (1966).
18. R. E. Rundle, *Rec. Chem. Prog.*, **23**, 195 (1962); *J. Am. Chem. Soc.*, **85**, 112 (1963).
19. J. I. Musher, *Angew. Chem., Int. Ed. Engl.*, **8**, 54 (1969).
20. J. B. Florey and L. C. Cusachs, *J. Am. Chem. Soc.*, **94**, 3040 (1972).
21. R. S. Berry, M. Tamres, C. J. Ballhausen and H. Johansen, *Acta. Chem. Scand.*, **22**, 231 (1968); P. C. Van der Voor and R. S. Drago, *J. Am. Chem. Soc.*, **88**, 3255 (1966); R. D. Brow and J. B. Peel, *Aust. J. Chem.*, **21**, 2605 (1968).
22. A. Rauk, L. C. Allen and K. Mislow, *J. Am. Chem. Soc.*, **94**, 3047 (1972).
23. R. S. McDowell and H. Streitwieser, Jr, *J. Am. Chem. Soc.*, **107**, 5849 (1985).
24. R. B. King, *J. Am. Chem. Soc.*, **91**, 7211 (1969).
25. H. A. Bent, *Chem. Rev.*, 275 (1961).
26. E. L. Muetterties, W. Mahler and R. Schmutzler, *Inorg. Chem.*, **2**, 613 (1963); E. L. Muetterties, W. Mahler, K. J. Packer and R. Schmutzler, *Inorg. Chem.*, **3**, 1298 (1964); W. Mahler and E. L. Muetterties, *Inorg. Chem.*, **4**, 1520 (1965).
27. I. Ugi, D. Marquarding, H. Kusacek, G. Gokel, P. Gillespie and F. Ramirez, *Angew. Chem., Int. Ed. Engl.*, **9**, 703 (1970); R. D. Spatley, W. C. Hamilton and J. Ladell, *J. Am. Chem. Soc.*, **89**, 2272 (1967).
28. P. Gillespie, P. Hoffmann, H. Klusacek, D. Marquarding, S. Pfohl, F. Ramirez, E. A. Tsolis and I. Ugi, *Angew. Chem., Int. Ed. Engl.*, **10**, 687 (1971).
29. A. P. Stewart and S. Trippett, *J. Chem. Soc., Chem. Commun.*, 1279 (1970); R. K. Cram and S. Trippett, *J. Chem. Soc., Perkin Trans. 1*, 1300 (1973); E. Duff, S. Trippett and P. J. Whittle, *J. Chem. Soc., Perkin Trans. 1*, 972 (1973); S. Trippett and P. J. Whittle, *J. Chem. Soc., Chem. commun.*, 2302 (1973); J. I. Dickstein and S. Trippett, *Tetrahedron Lett.*, **24**, 2203 (1973); S. Trippett, *Pure Appl. Chem.*, **40**, 595 (1974); S. A. Bone, S. Trippett, M. W. White and P. J. Whittle, *Tetrahedron Lett.*, **20**, 1975 (1974); S. A. Bone, S. Trippett and P. J. Whittle, *J. Chem. Soc., Perkin Trans. 1*, 80 (1977); J. Brierly, S. Trippett and P. J. Whittle, *J. Chem. Soc., Perkin Trans. 1*, 1977 (1977); S. Trippett and R. E. L. Waddling, *Tetrahedron Lett.*, **2**, 193 (1979); P. B. Kay and S. Trippett, *J. Chem. Res. (S)*, 62 (1986).
30. J. Emsley and D. Hall, *The Chemistry of Phosphorus*, Harper and Row, London, 1976.
31. K. E. DeBruin, A. G. Padilla and M. T. Campbell, *J. Am. Chem. Soc.*, **95**, 4681 (1973).
32. G. Buono and J. R. Llinas, *J. Am. Chem. Soc.*, **103**, 4532 (1981); R. G. Cavell, J. A. Gibson and K. I. The, *J. Am. Chem. Soc.*, **99**, 7841 (1977).
33. S. C. Peake and R. Schmutzler, *J. Chem. Soc. A*, 1049 (1970).
34. M. A. Landau, V. V. Sheluchenko, G. I. Drozd, S. S. Dubov and S. Z. Ivin, *Zh. Strukt. Khim.*, **8**, 1097 (1967); J. J. Harris and B. Rudner, *J. Org. Chem.*, **33**, 1392 (1968); J. S. Harman and D. W. N. Sharp, *Inorg. Chem.*, **10**, 1538 (1971); E. L. Muetterties, P. Meakin and R. Hoffmann, *J. Am. Chem. Soc.*, **94**, 5674 (1972).
35. R. Hoffmann, J. M. Howell and E. L. Muetterties, *J. Am. Chem. Soc.*, **94**, 3047 (1972).
36. P. Gillespie, F. Ramirez, I. Ugi and D. Marquarding, *Angew. Chem., Int. Ed. Engl.*, **12**, 91 (1973).
37. R. Luckenbach, *Dynamic Stereochemistry of Pentacoordinated Phosphorus and Related Elements*, Georg Thieme, Stuttgart, 1973.
38. R. R. Holmes, *Pentacoordinated Phosphorus—Structure and Spectroscopy (ACS Monograph*, No. 175), American Chemical Society, Washington, DC, 1980, pp. 1–48.
39. W. S. Sheldrick, *Top. Curr. Chem.*, **73**, (1978).

40. G. O. Doak and R. Schmutzler, *J. Chem. Soc., Chem. Commun.*, 476 (1970).
41. N. J. De'Ath, D. Z. Denney and D. B. Denney, *J. Chem. Soc., Chem. Commun.*, 272 (1972); N. J. De'Ath and D. B. Denney, *J. Chem. Soc., Chem. Commun*, 395 (1972).
42. W. C. Hamilton, S. J. LaPlaca, F. Ramirez and C. P. Smith, *J. Am. Chem. Soc.*, **89**, 2268 (1967); R. D. Spartley, W. C. Hamilton and J. Ladell, *J. Am. Chem. Soc.*, **89**, 2272 (1967).
43. F. Ramirez, O. P. Madan and R. S. Heller, *J. Am. Chem. Soc.*, **87**, 731 (1965).
44. J. I. C. Cadogan, R. O. Gould, S. E. B. Gould, P. A. Sadler, S. W. Swire and B. S. Tait, *J. Chem. Soc., Perkin Trans. 1*, 2392 (1975).
45. P. Narayanan, H. M. Berman, F. Ramirez, J. F. Marecek, Y. F. Chaw and V. A. Prasad, *J. Am. Chem. Soc.*, **99**, 3336 (1977).
46. M. Willson, F. Mathis, R. Burgada, R. Enjalbert, J. J. Bonnet and J. Galy, *Acta Crystallogr., Sect. B*, **34**, 629 (1978).
47. K. C. Kumara Swamy, J. M. Holmes, R. O. Day and R. R. Holmes, *J. Am. Chem. Soc.*, **112**, 6092 (1990).
48. K. C. Kumara Swamy, R. O. Day, J. M. Holmes and R. R. Holmes, *J. Am. Chem. Soc.*, **112**, 6095 (1990).
49. R. R. Holmes, K. C. Kumara Swamy, J. M. Holmes and R. O. Day, *Inorg. Chem.*, **30**, 1052 (1991).
50. J. Yu, A. M. Arif and W. G. Bentrude, *J. Am. Chem. Soc.*, **112**, 7451 (1990).
51. A. Schmidpeter, J. Luber, D. Schomburg and W. S. Sheldrick, *Chem. Ber.*, **109**, 3581 (1976).
52. W. S. Sheldrick, *Acta Crystallogr., Sect. B*, **32**, 925 (1976).
53. D. D. Swank, C. N. Caughlan, F. Ramirez and J. F. Pilot, *J. Am. Chem. Soc.*, **93**, 5236 (1971).
54. M. I. Kabachnick, V. A. Giliarov, N. A. Tikhonina, A. E. Kalinin, V. G. Adrianov, Yu. T. Stuchkov and G. I. Timofeeva, *Phosphorus*, **5**, 65 (1974).
55. I. V. Konovalova, I. S. Dokuchaeva, Yu G. Trichin, L. Burnaeva, V. N. Chistokletov and A. N. Pudovik, *Zh. Obshch. Khim.*, **59**, 1726 (1989).
56. M. Fulde, W. Ried and J. W. Bats, *Helv. Chim. Acta*, **72**, 139 (1989).
57. W. S. Sheldrick, D. Schomburg and A. Schmidpeter, *Acta Crystallogr., Sect. B*, **36**, 2316 (1980).
58. H. L. Carrel, H. M. Bermann, J. S. Ricci, Jr, W. C. Hamilton, F. Ramirez, J. F. Marecek, L. Kramer and I. Ugi, *J. Am. Chem. Soc.*, **97**, 38 (1975).
59. D. Schomburg, H. Hacklin and G. V. Röschenthaler, *Phosphorus Sulfur*, **35**, 241 (1988).
60. J. H. Yu, A. E. Sopchik, A. M. Arif, W. G. Bentrude and G. V. Röschenthaler, *Heteroatom Chem.*, **2**, 177 (1991).
61. B. C. Chang, W. E. Conrad, D. B. Denney, D. Z. Denney, R. Edelman, R. L. Powell and D. W. White, *J. Am. Chem. Soc.*, **93**, 4004 (1971).
62. M. Ehle, O. Wagner, U. Bergsträber and M. Regitz, *Tetrahedron Lett.* **24**, 3429 (1990).
63. J. E. Richman, R. O. Day and R. R. Holmes, *J. Am. Chem. Soc.*, **102**, 3955 (1980).
64. S. A. Bone, S. Trippett and P. J. Whittle, *J. Chem. Soc., Perkin Trans. 1*, 80 (1977).
65. J. H. Barlow, S. A. Bone, D. R. Russell, S. Trippett and P. J. Whittle, *J. Chem. Soc., Chem. Commun.*, 1031 (1976).
66. P. Lemmen, R. Baumgartner, I. Ugi and F. Ramirez, *Chimi. Scr.*, **28**, 451 (1988).
67. A. Strich and A. Veillard, *J. Am. Chem. Soc.*, **95**, 5574 (1973).
68. J. H. Yu and W. G. Bentrude, *J. Am. Chem. Soc.*, **110**, 7897 (1988); J. H. Yu and W. G. Bentrude, *Tetrahedron Lett.*, **30**, 2195 (1989).
69. See references cited in refs 50, 68 and 70.
70. N. L. H. L. Broeders, L. H. Koole and H. M. Buck, *Heteroatom Chem.*, 2205 (1991); N. L. H. L. Broeders, L. H. Koole and H. M. Buck, *J. Am. Chem. Soc.*, **112**, 7475 (1990).
71. R. O. Day, K. C. Kumara Swamy, L. Fairchild, J. M. Holmes and R. R. Holmes, *J. Am. Chem. Soc.*, **113**, 1627 (1991).
72. J. Hans, R. O. Day, L. Howe and R. R. Holmes, *Inorg. Chem.*, **30**, 3132 (1991).
73. Y. Huang, A. M. Arif and W. G. Bentrude, *J. Am. Chem. Soc.*, **113**, 7800 (1991).
74. W. C. Hamilton, J. S. Ricci, Jr, F. Ramirez, L. Kramer and P. Stern, *J. Am. Chem. Soc.*, **95**, 6335 (1973).
75. D. Bernard and R. Burgada, *C. R. Acad. Soc., Sect. C*, **271**, 418 (1970).
76. (a) W. F. Abdou, D. B. Denney, D. Z. Denney and S. D. Pastor, *Phosphorus Sulfur*, **22**, 99 (1985); (b) T. Prakasha, R. O. Day and R. R. Holmes, *Inorg. Chem.*, **31**, 725 (1992).
77. S. D. Burton, K. C. Kumara Swamy, J. M. Holmes, R. O. Day and R. R. Holmes, *J. Am. Chem. Soc.*, **112**, 6104 (1990).
78. D. B. Denney and L. S. Shih, *J. Am. Chem. Soc.*, **96**, 317 (1974).

79. B. C. Campbell, D. B. Denney, D. Z. Denney and L. S. Shih, *J. Chem. Soc., Chem. Commun.*, 854 (1978).
80. G. T. Gasparyan, M. Z. Ovakinyan, T. A. Ahamyan and M. G. Indzhikyan, *Arm. Khim. Zh.*, **37**, 520 (1984); *Chem. Abstr.*, **102**, 78993v (1985).
81. F. Mathey, *Chem. Rev.*, **90**, 997 (1990).
82. K. Burger, J. Fehn and W. Thenn, *Angew. Chem.*, **85**, 542 (1973).
83. M. T. Boisdon and J. Barrans, *J. Chem. Soc., Chem. Commun.*, 615 (1988).
84. F. Ramirez and I. Ugi, *Advances Physical Organic Chemistry*, Vol. 9 (Ed. V. Gold), Academic Press, London and New York, 1971; F. Ramirez and I. Ugi, *Bull. Soc. Chim. Fr.*, 453 (1974); F. Ramirez, S. Pfohl, E. A. Tsolig, J. F. Pilot, C. P. Smith, I. Ugi, D. Marquarding, P. Gillespie and P. Hoffmann, *Phosphorus*, **1**, 1 (1971); I. Ugi, D. Marquarding, H. Klusacek, P. Gillespie and F. Ramirez, *Acc. Chem. Res.*, **4**, 288 (1971).
85. R. Burgada, *Bull. Soc. chim. Fr.*, 407 (1985).
86. E. Ruch and I. Ugi, *Theor. Chim. Acta*, **4**, 287 (1966); E. Ruch, W. Hasselbarth and B. Richter, *Theor. Chim. Acta*, **19**, 288 (1970).
87. G. Polya, *Acta Math.*, **68**, 145 (1937).
88. E. L. Muetterties, *J. Am. Chem. Soc.*, **91**, 1636, 4115 (1969); *Rec. Chem. Prog.*, **31**, 51 (1970).
89. G. Binsch, E. L. Eliel and H. Kessler, *Angew. Chem., Int. Ed. Engl.*, **10**, 570 (1971).
90. P. C. Lauterbur and F. Ramirez, *J. Am. Chem. Soc.*, **90**, 6722 (1968).
91. K. Mislow, *Acc. Chem. Res.*, **3**, 321 (1970).
92. M. Gielen, in *Chemical Applications of Graph Theory* (Ed. A. T. Balaban), Academic Press, New York, 1972, p. 261.
93. D. Gorenstein and F. H. Westheimer, *J. Am. Chem. Soc.*, **92**, 634 (1970).
94. A. T. Balaban, D. Farcasiu and R. Banica, *Rev. Rouma. Chim.*, **11**, 1205 (1966).
95. J. Dunitz and V. Prelog, *Angew. Chem., Int. Ed. Engl.*, **7**, 726 (1968).
96. M. Gielen and J. Nasielski, *Bull. Soc. Chim. Belg.*, **78**, 339 (1969).
97. D. J. Cram, J. Day, D. R. Rayner, D. M. von Schriltz, D. J. Duchamp and D. C. Garwood, *J. Am. Chem. Soc.*, **92**, 7369 (1970).
98. E. L. Muetterties, *Int. Rev. Sci.*, **9**, 37 (1970); **10**, 1 (1972).
99. M. Gielen and N. van Lautem, *Bull. Soc. chim. Belg.*, **79**, 679 (1970).
100. J. Brocas, *Top. Curr. Chem.*, **32**, 43 (1972).
101. W. Hasselbarth and E. Ruch, *Theor. Chim. Acta*, **29**, 259 (1973).
102. W. G. Klemperer, *J. Am. Chem. Soc.*, **94**, 380, 2105, 6490, 8360 (1972); *Inorg. Chem.*, **11**, 2668 (1972); *J. Chem. Phys.*, **56**, 5478 (1972).
103. J. I. Musher, *J. Am. Chem. Soc.*, **94**, 5662 (1972).
104. K. W. Hasen and L. S. Bartell, *Inorg. Chem.*, **4**, 1775 (1965); L. S. Bartell and K. W. Hansen, *Inorg. Chem.*, **4**, 1777 (1965).
105. J. E. Griffiths, R. P. Carter, Jr, and R. R. Holmes, *J. Chem. Phys.*, **41**, 863 (1964).
106. R. R. Holmes, R. P. Carter, Jr, and G. E. Peterson, *Inorg. Chem.*, **3**, 1748 (1964).
107. G. M. Whitesides and H. L. Mitchell, *J. Am. Chem. Soc.*, **91**, 5384 (1969); M. Eisenhut, H. L. Michell, D. D. Traficante, R. J. Kaufman, J. M. Deutch and G. M. Whitesides, *J. Am. Chem. Soc.*, **96**, 5385 (1974); G. M. Whitesides, M. Eisenhut and W. M. Bunting, *J. Am. Chem. Soc.*, **96**, 5398 (1974).
108. P. Wang, D. R. Agriotis, A. Streitwieser and P. v̇. R. Schleyer, *J. Chem. Soc., Chem. Commun.*, **3**, 201 (1990).
109. A. E. Reed and P. v̇. R. Schleyer, *J. Am. Chem. Soc.*, **112**, 1434 (1990); H. Wasada and K. Hirao, *J. Am. Chem. Soc.*, **114**, 16 (1992).
110. T. P. E. auf der Heyde and H. B. Bürgi, *Inorg. Chem.*, **28**, 3982 (1989).
111. G. Robinet, M. Barthelat, V. Gasmi and J. Devillers, *J. Chem. Soc., Chem. Commun.*, 1103 (1989).
112. M. J. Gallagher in *Stereochemistry of Heterocyclic Compounds, Part II* (Ed. W. L. F. Armarego), Wiley, Chichester, 1977, p. 339.
113. H. Christol and H. J. Cristau, *Ann. Chim. (Paris)*, **6**, 179 (1971).
114. D. Hellwinkel, *Chimia (Zürich)*, **22**, 488 (1968).
115. D. Hellwinkel, *Angew. Chem.*, **78**, 749 (1966).
116. M. J. Gallagher and I. D. Jenkins in *Top. Stereochem.*, **3**, 1 (1968).
117. D. Hellwinkel, *Chem. Ber.*, **99**, 3628, 3642, 3660 (1966); A. Hellwinkel and H. J. Wilfinger, *Tetrahedron Lett.*, 3423 (1969).

118. D. Hellwinkel and H. J. Wilfinger, *Justus Liebigs Ann. Chem.*, **742**, 163 (1970); D. Hellwinkel and M. Bach, *Naturwissenschaften*, **56**, 214 (1969).
119. G. M. Whitesides and M. Bunting, *J. Am. Chem. Soc.*, **89**, 6801 (1967).
120. D. Hellwinkel and H. J. Wilfinger, *Phosphorus*, **2**, 87 (1972); D. Hellwinkel, C. Wunšche and M. Bach, *Phosphorus*, **2**, 167 (1973).
121. R. Burgada, M. Bon and F. Mathis, *C.R. Acad. Sci., Ser. C*, **277**, 1499 (1967).
122. J. Ferekh, J. F. Brazier, A. Muñoz and R. Wolf, *C.R. Acad. Sci., Ser. C*, **270**, 865 (1970).
123. M. G. Newton, J. E. Collier and R. Wolf, *J. Am. Chem. Soc.*, **96**, 6888 (1974).
124. D. Houalla, M. Sanchez and R. Wolf, *Org. Magn. Reson.*, **5**, 451 (1973); D. Houalla, M. Sanchez, R. Wolf, M. Bois, D. Gagnaire and J. B. Robert, *Org. Magn. Reson.*, **6**, 340 (1974).
125. A. Klaebe, A. Cachapuz–Carrelhas, J. F. Brazier, M. R. Marre and R. Wolf, *Tetrahedon Lett.*, 3971 (1974); A. Klaebe, A. Cachapuz–Carrelhas, J. F. Brazier, D. Houalla and R. Wolf, *Phosphorus Sulfur*, **3**, 61 (1977); A. Klaebe, J. F. Brazier, F. Mathis and R. Wolf, *Tetrahedron Lett.*, 4367 (1972).
126. R. Contreras, J. F. Brazier, A. Klaebe and R. Wolf, *Phosphorus*, **2**, 67 (1972); J. F. Brazier, A. Cachapuz–Carrelhas, A. Klaebe and R. Wolf, *C.R. Acad. Sci., Ser. C*, **277**, 183 (1973).
127. A. Muñoz, M. Koenig, G. Gence and R. Wolf, *C.R. Acad. Sci., Ser. C*, **278**, 1353 (1974).
128. A. Klaebe, A. Cachapuz–Carrelhas, J. F. Brazier and R. Wolf, *J. Chem. Soc., Perkin Trans. 2*, 1668 (1974); A. Klaebe, J. F. Brazier, A. Cachapuz–Carrelhas, B. Garrigues and R. Marre, *Tetrahedron*, **38**, 2111 (1982).
129. D. Bernard and R. Burgada, *Phosphorus*, **3**, 187 (1974).
130. R. Burgada, in *Colloque International sur la Chimie du Phosphore*, CNRS, Paris, 1969, p. 240.
131. D. Bernard and R. Burgada, *C.R. Acad. Sci., Ser. C*, **272**, 2077 (1971).
132. D. Bernard and R. Burgada, *Tetrahedron*, **31**, 797 (1975).
133. H. Gonçalves, J. R. Dormoy, Y. Chapleur, B. Castro, H. Fauduet and R. Burgada, *Phosphorus Sulfur*, **8**, 147 (1980).
134. G. S. Hammond, *J. Am. Chem. Soc.*, **77**, 334 (1955).
135. R. M. Moriarty, J. Hiratake and K. Liu, *J. Am. Chem. Soc.*, **112**, 8575 (1990).
136. F. Ramirez and N. B. Desai, *J. Am. Chem. Soc.*, **85**, 3252 (1963).
137. W. M. Abdou, M. R. Mahran and M. M. Sidky, *Phosphorus Sulfur*, **27**, 345 (1986).
138. L. Yu. Sandalova, L. I. Mizrakh and V. P. Evdakov, *Zh. Obshch. Khim.*, **36**, 1451 (1966); D. Bernard and R. Burgada, *C.R. Acad. Sci. Paris, Ser. C*, **277**, 433 (1973); R. Burgada, *Phosphorus Sulfur*, **3**, 187 (1974).
139. R. Moriarty, J. Hiratake, K. Liu, A. Wendler, A. K. Awasthi and R. Gilardi, *J. Am. Chem. Soc.*, **113**, 9374 (1991).
140. Y. Vannoorenberghe and G. Buono, *J. Am. Chem. Soc.*, **112**, 6142 (1990).
141. R. Burgada, *Ann. Chim.*, **8**, 347 (1963); L. Nesterov, R. A. Sabirova, N. E. Krepisheva and R. I. Mutalapova, *Dokl. Akad. Nauk SSSR*, **148**, 1085 (1963).
142. L. Anschutz and W. Broeker, *Chem. Ber.*, **61**, 1246 (1928); **76**, 218 (1943).
143. N. P. Grechkin, R. R. Sagidullin and L. N. Grishina, *Dokl. Akad. Nauk SSSR*, **161**, 115 (1965).
144. T. Reetz and J. F. Powers, *US Pat.*, 3 172 903 (1965).
145. R. Burgada, D. Houalla and R. Wolf, *C.R. Acad. Sci., Ser. C*, **264**, 356 (1967).
146. N. P. Grechkin, R. R. Shagidallin and G. S. Gubanova, *Izv. Akad. Nauk SSSR*, **8**, 1797 (1968).
147. H. Germa, M. Willson and R. Burgada *C.R. Acad. Sci., Ser. C.*, **270**, 1426 (1970); R. Burgada, H. Germa, M. Willson and F. Mathis, *Tetrahedron*, 5833 (1971).
148. D. Bernard, C. Laurenço and R. Burgada, *J. Organomet. Chem.*, **47**, 113 (1973).
149. R. Burgada and C. Laurenço, *J. Organomet. Chem.*, **66**, 255 (1974), and references cited therein.
150. A. Muñoz, *Bull. Soc. Chim. Fr.*, 728 (1977).
151. R. Mathis, R. Burgada and M. Sanchez, *Spectrochim. Acta, Part A*, **25**, 1201 (1969).
152. B. Garrigues, C. B. Cong, A. Muñoz and A. Klaebe, *J. Chem. Res. (S)*, 172 (1979).
153. J. J. C. Van Lier, R. J. M. Hermans and H. M. Buck, *Phosphorus Sulfur*, **19**, 173 (1984).
154. Y. Charbonnel and J. Barrans, *C.R. Acad. Sci., Ser. C*, **277**, 571 (1973); R. Mathis, M. Barthelat, Y. Charbonnel and J. Barrans, *C.R. Acad. Sci., Ser. C*, **280**, 809 (1975).
155. K. Karagiosoff, J. P. Majoral, A. Meriem, J. Navech and A. Schmidpeter, *Tetrahedron*, **24**, 2137 (1983).
156. J. E. Richman and T. J. Atkins, *Tetrahedron Lett.*, 4333 (1978); T. J. Atkins and J. E. Richman, *Tetrahedron Lett.*, 5149 (1978).

157. J. E. Richman and T. J. Atkins, *J. Am. Chem. Soc.*, **96**, 2268 (1974).
158. B. Garrigues, A. Muñoz, M. Koenig, M. Sanchez and R. Wolf, *Tetrahedron*, **33**, 635 (1977).
159. C. Malavaud and J. Barrans, *Tetrahedron Lett.*, 3077 (1975).
160. B. Tangour, C. Malavaud, M. T. Boisdon and J. Barrans, *Phosphorus, Sulfur Silicon*, **40**, 33 (1988); **45**, 189 (1989).
161. M. A. Pudovik, S. A. Terent'eva and A. N. Pudovik, *Dokl. Akad. Nauk SSSR*, **228**, 363 (1977).
162. R. Contreras, D. Houalla, A. Klaebe and R. Wolf, *Tetrahedron Lett.*, 3953 (1981), and references cited therein.
163. V. V. Ovchinnikov, M. A. Pudovik, V. I. Galkin, R. A. Cherkasov and A. N. Pudovik, *Izv. Akad. Nauk SSSR, Ser. Khim.*, 434 (1977).
164. S. D. Harper and A. J. Arduengo, *Tetrahedron Lett.*, 4331 (1980).
165. G. M. L. Cragg, B. Davidowitz, G. V. Fazakerley, L. Nassimbeni and R. J. Haines, *J. Chem. Soc., Chem. Commun*, 510 (1978).
166. S. D. Harper and A. J. Arduengo, *J. Am. Chem. Soc.*, **104**, 2497 (1982).
167. K. J. Schray and S. J. Benkovic, *J. Am. Chem. Soc.*, **93**, 2522 (1971); J. J. Steffens, I. J. Siewers and S. J. Benkovic, *Biochemistry*, **14**, 2431 (1975).
168. E. E. Nifant'ev and I. M. Matveeva, *Zh. Obshch. Khim.*, **39**, 1555 (Engl. Transl., 1525) (1969).
169. Y. Segal, I. Granoth, A. Kalir and E. D. Bergman, *J. Chem. Soc., Chem. Commun.*, 399 (1975).
170. B. Arbusov, A. O. Vizel, K. M. Ivanovskaya and E. I. Goldfarb, *J. Gen. Chem. USSR*, **45**, 1204 (1975).
171. F. Ramirez, M. Nowakowski and J. F. Marecek, *J. Am. Chem. Soc.*, **98**, 4330 (1976).
172. A. Muñoz, M. Gallagher, A. Klaebe and R. Wolf, *Tetrahedron Lett.*, 673 (1976).
173. G. Kemp and S. Tripett, *Tetrahedron Lett.*, 4381 (1976).
174. C. B. Cong, A. Muñoz, M. Sanchez and, A. Klaebe, *Tetrahedron Lett.*, 1587 (1977).
175. F. Ramirez, M. Nowakowski and J. F. Marecek, *J. Am. Chem. Soc.*, **99**, 4515 (1977).
176. I. Granoth and J. C. Martin, *J. Am. Chem. Soc.*, **100**, 5229 (1978).
177. I. Granoth and J. C. Martin, *J. Am. Chem. Soc.*, **101**, 4618, 4623 (1979).
178. G. W. Röschenthaler, R. Bohlen, W. Storzer, A. P. Sopchik and W. G. Bentrude, *Z. Anorg. Allg. Chem.*, **507**, 93 (1983).
179. A. Muñoz, B. Garrigues and M. Koenig, *Tetrahedron*, **36**, 2467 (1980).
180. A. Dubourg, R. Roques, G. Germain, J. P. Declercq, B. Garrigues, D. Boyer, A. Muñoz, A. Klaebe and M. Comtat, *J. Chem. Res. (S)*, 180 (1982).
181. D. Houalla, M. Sanchez, R. Wolf and F. H. Osman, *Tetrahedron Lett.*, 4675 (1978).
182. S. K. Das, P. Kumarz Amma, R. Validyanathaswamy, J. Weydent and J. G. Verkade, *Phosphorus Sulfur*, **30**, 301 (1987).
183. S. K. Tupchinenko, T. N. Dudchenko and A. D. Sinitsa, *Zh. Obshch. Khim.*, **59**, 1500 (1989).
184. A. Baceiredo, G. Bertrand, J. P. Majoral, U. Wermuth and R. Schmutzler, *J. Am. Chem. Soc.*, **106**, 7063 (1984).
185. H. Staudinger and J. Meyer, *Helv. Chim. Acta*, **2**, 635 (1919).
186. S. Patai (Ed.) *The Chemistry of the Azido Group*, Wiley, Chichester, 1970.
187. M. Berman, *Prog. Inorg. Chem. Radiochem.*, **14**, 1 (1972); M. Berman, *Top. Phosphorus Chem.*, **7**, 311 (1972).
188. J. I. G. Cadogan, I. Gosney, E. Henry, T. Naisby, B. Nay, N. J. Stewart and N. J. Tweddle, *J. Chem. Soc., Chem. Commun.*, 189 (1979); J. I. G. Cadogan, N. J. Stewart and N. J. Tweddle, *J. Chem. Soc., Chem. Commun.*, 191 (1979).
189. A. V. Kirsanov (Ed.), *Phosphazo Compounds*, Naukova Dumka, Kiev, 1965; I. N. Zhmurova and A. A. Kirsanov, *Zh. Obshch. Khim.*, **32**, 2576 (1962).
190. H. B. Stegmann, G. Bauer, E. Britmaier, E. Hermann and R. Scheffler, *Phosphorus*, **5**, 207 (1975).
191. M. Sanchez, J. F. Brazier, D. Houalla and R. Wolf, *Nouv. J. Chim.*, **3**, 775 (1979).
192. R. J. Dinger, W. Storzer and A. Schmutzler, *Z. Anorg. Allg. Chem.*, **555**, 154 (1987).
193. H. B. Stegmann, G. Wax, S. Peinelt and K. Scheffler, *Phosphorus Sulfur*, **16**, 277 (1983).
194. B. Köll, K. Totschning, J. Vögel, P. Peringer and E. P. Müller, *Phosphorus Sulfur Silicon*, **49/50**, 385 (1990).
195. E. Balogh-Hergovich and G. Speier, *Phosphorus Sulfur Silicon*, **48**, 223 (1990).
196. V. P. Kukhar and V. A. Giliarov, *Pure Appl. Chem.*, **52**, 891 (1980).
197. Yu. G. Gololobov, N. I. Gusar and M. Chaus, *Tetrahedron*, **41**, 793 (1985).
198. G. W. Roschenthaler, K. Sauerbrey and R. Schmutzler, *Chem. Ber.*, **111**, 3105 (1978).

199. J. A. Gibson, G. V. Röschenthaler, D. Schomburg and W. S. Sheldrick, *Chem. Ber.*, **111**, 1887 (1978).
200. V. P. Kukhar, E. V. Grishkun, V. P. Rudavskii and V. A. Giliarov, *Zh. Obshch. Khim.*, **50**, 1477 (1990).
201. V. P. Kukhar, T. M. Kashera, L. F. Kasukhin and P. N. Ponomarchuk, *J. Gen. Chem. USSR (Engl. Transl.)*, **54**, 1360 (1984).
202. R. Burgada, *Bull. Soc. Chim. Fr.*, **1**, 136 (1971).
203. S. A. Terent'eva, M. A. Pudovik and A. N. Pudovik, *J. Gen. Chem. USSR (Engl. Transl.)*, **56**, 632 (1986).
204. J. I. G. Cadogan, J. B. Husband and H. McNab, *J. Chem. Soc., Perkin Trans. 1*, 1449 (1984).
205. Yu. G. Gololobov, L. I. Nasterova, V. P. Khukhar and V. I. Luk'yanchuk, *Zh. Obshch. Khim.*, **51**, 477 (1981).
206. M. R. Marre, M. Sanchez, J. F. Brazier, R. Wolf and J. Bellan, *Can. J. Chem.*, **60**, 456 (1982).
207. R. A. Shaw, *Phosphorus Sulfur*, **4**, 101 (1978).
208. A. W. Johnson, *Ylide Chemistry*, Academic Press, New York, 1966.
209. H. J. Bestmann and R. Zimmermann, in *Organic Phosphorus Compounds* (Eds G. M. Kosolapoff and L. Maier), Vol. 3, Wiley–Interscience, New York, 1972, p. 1.
210. O. I. Kolodyazhnyi and V.P. Kukhar, *Russ. Chem. Rev.*, **52**, 1096 (1983).
211. H. Schmidbauer and H. Stuhler, *Angew. Chem., Int. Ed. Engl.*, **11**, 145 (1972).
212. R. Burgada, Y. O. El Khoshnieh and Y. Leroux, *Tetrahedron*, **41**, 1207 (1985).
213. R. Burgada, Y. O. El Khoshnieh and Y. Leroux, *Tetrahedron*, **41**, 1223 (1985).
214. R. Burgada, Y. U. El Khoshnieh and Y. Leroux, *Phosphorus Chemistry*, (*ACS Symposium Series*, Vol. 171) (Eds L. D. Quin and J. G. Verkade), American Chemical Society, Washington, DC, 1981, p. 607.
215. B. Ben Jaafar, D. El Manouni, R. Burgada and Y. Leroux, *Phosphorus Sulfur Silicon*, **47**, 67 (1990).
216. D. El Manouni, Y. Leroux and R. Burgada, *Tetrahedron*, **42**, 2435 (1986).
217. R. Burgada, Y. Leroux and Y. O. El Khoshnieh, *Phosphorus Sulfur*, **10**, 181 (1981).
218. R. Huisgen, K. Herbig, Z. Ziege and H. Heber, *Chem. Ber.*, **99**, 2526 (1966).
219. I. F. Wilson and J. C. Tebby, *J. Chem. Soc., Perkin Trans. 1*, 2830 (1972).
220. I. Granoth, R. Alkabets, E. Shirin, Y. Margalit and P. Bell, *Phosphorus Chemistry* (*ACS Symposium Series*, Vol. 171) (Eds L. D. Quin and J. G. Verkade), American Chemical Society, Washington, DC, 1981, p. 435.
221. R. W. Suter, H. C. Knachel, U. P. Petro, J. H. Howatson and S. G. Shore, *J. Am. Chem. Soc.*, **95**, 1474 (1973).
222. R. B. Dillon and T. P. Straw, *J. Chem. Soc., Chem. Commun.*, 234 (1991).
223. I. S. Sigal and F. H. Westheimer, *J. Am. Chem. Soc.*, **101**, 5329 (1979).
224. D. Camp and I. D. Jenkins, *J. Org. Chem.*, **54**, 3049 (1989).
225. R. A. McClelland, G. Patel and C. Cirinna, *J. Am. Chem. Soc.*, **103**, 6432 (1981); G. H. McGall and R. A. McClelland, *J. Am. Chem. Soc.*, **107**, 5198 (1985); G. H. McGall and R. A. McClelland, *J. Chem. Soc., Chem. Commun.*, 1222 (1982); G. H. McGall and R. A. McClelland, *Can. J. Chem.*, 2075 (1991).
226. J. W. Kelly and S. A. Evans, Jr, *J. Am. Chem. Soc.*, **108**, 7681 (1986).
227. A. Pautard–Cooper and S. A. Evans, Jr, *J. Org. Chem.*, **54**, 2485 (1989); A. Pautard–Cooper and S. A. Evans, Jr, *J. Org. Chem.*, **54**, 4974 (1989).
228. P. L. Robinson, J. W. Kelly and S. A. Evans, Jr, *Phosphorus Sulfur*, **31**, 59 (1987).
229. N. A. Eskew and S. A. Evans, Jr, *J. Chem. Soc., Chem. Commun.*, 706 (1990).
230. W. T. Murray, A. Pautard–Cooper, N. A. Eskew and S. A. Evans, Jr, *Phosphorus Sulfur Silicon*, **49/50**, 101 (1990).
231. R. Burgada, *Bull. Soc. Chim. Fr.*, 347 (1967); R. Burgada, *C.R. Acad. Sci., Ser. C*, **258**, 4789 (1964).
232. F. Ramirez, A. V. Patwardhan, H. J. Kugler and C. P. smith, *J. Am. Chem. soc.*, **89**, 6276 (1967); F. Ramirez, A. V. Patwandham, H. J. Kugler and C. P. Smith, *Tetrahedron*, **24**, 2275 (1968).
233. H. Fauduet and R. Burgada, *New J. Chem.*, **3**, 555 (1979); H. Fauduet and R. Burgada, *New J. Chem.*, **4**, 113 (1980); H. Fauduet and R. Burgada, *Synthesis*, 642 (1980).
234. D. B. Denney, D. Z. Denney, D. M. Gavrilovic, P. J. Hammond, C.-L. Huang and K.-S. Tseng, *J. Am. Chem. Soc.*, **102**, 7072 (1980).
235. D. B. Denney and S. D. Pastor, *Phosphorus Sulfur*, **16**, 239 (1983).

236. N. Lowther, P. D. Beer and C. D. Hall, *Phosphorus Sulfur*, **35**, 133 (1988).
237. A. V. Aganov, N. A. Polezhaeva, A. I. Khayarov and B. A. Arbusov, *Phosphorus Sulfur*, **22**, 303 (1985).
238. D. Bernard and R. Burgada, *Tetrahedron Lett.*, 3455 (1973).
239. D. V. Khasnis, M. Lattman and V. Siriwardane, *Inorg. Chem.*, **29**, 271 (1990).
240. D. V. Khasnis, M. Lattman and V. Siriwardane, *Inorg. Chem.*, **28**, 681 (1989).
241. A. Werner, *Chem. Ber.*, **44**, 1887 (1911).
242. F. Ramirez, K. Tasaka, N. B. Desai and C. P. Smith, *J. Am. Chem. Soc.*, **90**, 751 (1968).
243. F. Ramirez, G. Loweengart, E. A. Tsolis and K. Tasaka, *J. Am. Chem, Soc.*, **94**, 3531 (1972).
244. F. Ramirez, K. Tasaka and R. Herschberg, *Phosphorus*, **2**, 61 (1972).
245. W. C. Archie Jr, and F. H. Westheimer, *J. Am. Chem. Soc.*, **55**, 5955 (1973).
246. H. R. Allcock, *J. Am. Chem. Soc.*, **85**, 4050 (1963); **86**, 2591 (1964).
247. B. C. Chang, D. B. Denney, R. L. Powell and D. White, *J. Chem. Soc., Chem. Commun.*, 1070 (1971).
248. S. S. Chan and C. J. Willis, *Can. J. Chem.*, **46**, 1237 (1968).
249. D. B. Denney, D. Z. Denney and C. F. Ling, *J. Am. Chem. Soc.*, **98**, 6755 (1976).
250. E. Fluck and M. Vargas, *Z. Anorg. Allg. Chem.*, **437**, 53 (1977).
251. A. Kh. Vozhnesenskaya, N. A. Razumova, A. A. Petrov and O. Sheptienko, *Zh. Obshch. Khim.*, **47**, 1432 (1977).
252. D. K. Kemepohl and R. G. Cavell, *Phosphorus Sulfur Silicon*, **49/50**, 359 (1990); K. I. The, L. V. Griend, W. A. Whitla and R. G. Cavell, *J. Am. Chem. Soc.*, **103**, 1785 (1981); N. Burford, D. Kennepohl, M. Covie, R. G. Ball and R. G. Cavell, *Inorg. Chem.*, **26**, 650 (1987).
253. K. Galle and K. Utuary, *Monatsh. Chem.*, **119**, 53 (1988).
254. H. R. Allcock and E. C. Bissell, *J. Chem. Soc., Chem. Commun.*, 676 (1972).
255. L. Lopez, M. T. Boisdon and J. Barrans, *C.R. Acad. Sci., Ser. C*, **275**, 295 (1972).
256. R. Burgada and D. Bernard, *C.R. Acad. Sci., Ser. C*, **273**, 164 (1971).
257. M. Wieber and K. Foroughi, *Angew. Chem., Int. Ed. Engl.*, **12**, 419 (1973).
258. M. A. Pudovik, S. A. Terent'eva and A. N. Pudovik, *Dokl. Akad. Nauk SSSR*, **228**, 363 (1976).
259. R. Burgada, D. Bernard and C. Laurenço, *C.R. Acad. Sci., Ser. C*, **276**, 297 (1973).
260. D. Bernard, *Thèse de Doctorat ès Sciences*, Paris, 1974.
261. A. Muñoz, M. Koenig, G. Gence and R. Wolf, *C. R. Acad. Sci., Ser. C*, **278**, 1353 (1974); A. Muñoz, G. Gence, M. Koenig and R. Wolf, *C.R. Acad. Sci., Ser. C*, **280**, 395 (1975); A. Muñoz, M. Sanchez, M. Koenig and R. Wolf, *Bull. Soc. Chim. Fr.*, 2193 (1974).
262. A. Schmidtpeter, T. Griergern and K. Blank, *Z. Naturforsch. Teil B*, **33**, 583 (1976).
263. M. Koenig, *Thèse de Doctorat ès Sciences*, Toulouse, 1979.
264. F. Ramirez, V. A. V. Prasad and J. F. Maresek, *J. Am. Chem. Soc.*, **96**, 7269 (1974).
265. R. M. Minaev and V. I. Minkin, *Zh. Strukt. Khim.*, **20**, 842 (1979).
266. J. J. H. M. FontFreide and S. Tripett, *J. Chem. Res. (S)*, 218 (1981).
267. M. Koenig, A. Munož, D. Houalla and R. Wolf, *J. Chem. Soc. Chem. Commun.*, 182 (1974); C. B. Cong, G. Gence, B. Garrigues, M. Koenig and A. Munož, *Tetrahedron*, **35**, 1825 (1979).
268. D. Bernard and R. Burgada, *C.R. Acad. Sci., Ser. C*, **279**, 883 (1974); see also ref. 85.
269. V. I. Kal'chenko, V. V. Nebegretskii, R. Rudnyi, L. I. Atamas, M. I. Povolotskii and L. N. Markovskii, *Zh. Obshch. Khim.*, **53**, 932 (1983).
270. V. V. Nebegretskii, V. I. Kal'chenko, R. B. Runyi and L. N. Markovskii, *J. Gen. Chem. USSR (Engl. Transl.)*, **55**, 236, 1761 (1985).
271. L. H. Koole, W. J. M. van der Hofstad and H. M. Buck, *J. Org. Chem.*, **50**, 4381 (1985).
272. D. B. Denney, D. Z. Denney and M. N. Raab, *Phosphorus Sulfur*, **37**, 175 (1988).
273. F. Ramirez, S. Lee, P. Stern, I. Ugi and P. D. Gillespie, *Phosphorus*, **4**, 21 (1974).
274. D. V. Khasnis, M. Lattman and C. D. Gutsche, *J. Am. Chem. Soc.*, **112**, 9422 (1990).
275. D. V. Khasnis, J. M. Burton, M. Lattman and H. Zhang, *J. Chem. Soc., Chem. Commun.*, 562 (1991).
276. E. S. Kozlov and S. M. Gaidamaka, *Zh. Obshch. Khim.*, **39**, 933 (Engl. Transl. 902) (1969).
277. A. Schmidpeter, C. Weigand and E. Hafner–Roll, *Z. Naturforsch., Teil B*, **24**, 799 (1969); I. Yamashita and A. Masaki, *Kogyo Kagaku Zasshi*, **70**, 2031 (1967); *Chem. Abstr.*, **69**, 10322h (1968); G. I. Drozd, M. A. Sokalski, O. G. Strukov and S. Z. Ivin, *Zh. Obshch. Khim.*, **40**, 2396 (Engl. Transl. 2384) (1970); H. Nöth and H. J. Vetter, *Chem. Ber.*, **96**, 1816 (1963); D. H. Brown, K. D. Crosbie, J. I. Darragh, D. S. Ross and D. W. A. Sharp, *J. Chem. Soc. A*, 914 (1970).

278. D. G. Coe, H. H. Rydon and B. L. Tonge, *J. Chem. Soc. A*, 323 (1957); G. S. Harris and D. S. Payne, *J. Chem. Soc. A*, 3038 (1956).
279. D. C. Apperley, R. K. Harris, T. Kaukorat and R. Schmutzler, *Phosphorus Sulfur Silicon*, **54**, 227 (1990); T. Kaukorat, L. Ernst, R. Schmutzler and D. Schomburg, *Polyhedron*, **9**, 1463 (1990).
280. G. I. Drozd, S. Z. Ivin, V. V. Sheluchenko, B. I. Tetel'baum and A. D. Varshavski, *Zh. Obshch. Khim.*, **38**, 567 (Engl. Transl. 551) (1968); F. Seel, W. Gombler and K. H. Rudolph, *Z. Naturforsch., Teil B*, **23**, 387 (1968); E. R. Falardeau, K. W. Morse and J. G. Morse, *Inorg. Chem.*, **14**, 1239 (1975); R. Appel and A. Gilak, *Chem. Ber.*, **108**, 2693 (1975); R. K. Merat and A. F. Janzen, *Inorg. Chem.*, **19**, 798 (1980).
281. H. J. Emeleus and T. Onak, *J. Chem. Soc. A*, 1291 (1966).
282. N. A. Razumova, F. V. Bagrov and A. A. Petrov, *Zh. Obshch. Khim.*, **39**, 2369 (Engl. Transl. 2306) (1969); D. B. Denney, D. Z. Denney and Y. F. Hsu, *Phosphorus*, **4**, 217 (1974); B. A. Arbusov, A. O. Vizel, Yu. Yu. Samitov and K. M. Ivanovskaya, *Dokl. Akad. Nauk SSSR*, **159**, 582 (Engl. Transl. 1205) (1964) N. A. Razumova and A. A. Petrov, *Zh. Obshch. Khim.*, **33**, 3858 (Engl. Transl. 3796) (1963); Zh. L. Evtikhov, N. A. Razumova and A. A. Petrov, *Dokl. Akad. Nauk SSSR*, **177**, 108 (Engl. Transl. 972) (1967); N. A. Razumova, Zh. L. Evtikhov, A. Kh. Voznosenskaya and A. A. Petrov, *Zh. Obshch. Khim.*, **39**, 176 (Engl. Transl. 551) (1969).
283. L. D. Quinn, in *1–4 Cycloaddition Reactions* (Ed. J. Hamer), Academic Press, New York, 1967 p. 47; B. E. Ivanov and V. F. Zhetukhin, *Russ. Chem. Rev.*, **39**, 358 (1970).
284. N. A. Razumova, F. V. Bagrov and A. A. Petrov, *Zh. Obshch. Khim.*, **39**, 2369 (Engl. Transl. 2305) (1969); M. Wieber and W. R. Hoos, *Tetrahedron Lett.*, 4693 (1969).
285. A. A. Prischenko, M. V. Livantsov, S. A. Moshnikov and J. F. Lutsenko, *J. Gen. Chem. USSR* (Engl. Transl.), **57**, 1484 (1987).
286. N. A. Razumova, Z. L. Evtikhov and A. A. Petrov, *Zh. Obshch. Khim.*, **39**, 1419 (Engl. Transl. 1388) (1969); B. A. Arbusov, O. D. Zolova, V. S. Vinogradova and Y. Y. Samitov, *Dokl. Akad. Nauk SSSR*, **173**, 335 (Engl. Transl. 231) (1967); D. Gorenstein and F. H. Westheimer, *J. Am. Chem. Soc.*, **89**, 2762 (1967); A. K. Voznesenskaya, N. A. Razumova and A. A. Petrov, *Zh. Obshch, Khim.*, **39**, 1033 (Engl. Transl. 1004) (1969); B. A. Arbusov, N. A. Polezhaeva and V. S. Vinigradova, *Izv. Akad. Nauk SSSR, Ser. Khim.*, 2281 (Engl. Transl. 2185) (1967).
287. J. Albanbauer, K. Burger, E. Burgis, D. Marquarding, L. Schabl and I. Ugi, *Justus Liebigs Ann. Chem.*, 36 (1976).
288. H. Hund and G. V. Röschenthaler, *Phosphorus Sulfur Silicon*, **62**, 71 (1991).
289. E. Fluck, P. Kuhn and H. Riffel, *Z. Anorg. Allg. Chem.*, **567**, 39 (1988).
290. D. B. Denney, D. Z. Denney, P. J. Hammond, C.-L. Huang and K.-S. Tseng, *J. Am. Chem. Soc.*, **102**, 5073 (1980); D. B. Denney, D. Z. Denney, P. J. Hammond and K.-S. Tseng, *J. Am. Chem. Soc.*, **103**, 2054 (1981).
291. D. B. Denney, D. Z. Denney, P. J. Hammond and Y. P. Wang, *J. Am. Chem. Soc.*, **103**, 1785 (1981); D. B. Denney, D. Z. Denney, B. C. Chang and K. L. Marsi, *J. Am. Chem. Soc.*, **91**, 5243 (1969).
292. D. B. Denney, D. Z. Denney, L. T. Liu, *Phosphorus Sulfur*, **22**, 71 (1985); B. C. Burros, N. J. De'Ath, D. B. Denney, D. Z. Denney and I. J. Kipnis, *J. Am. Chem. Soc.*, **100**, 7300 (1978).
293. P. D. Bartlett, A. L. Baumstark and M. E. Landis, *J. Am. Chem. Soc.*, **95**, 6486 (1973); P. D. Bartlett, A. L. Baumstark, M. E. Landis and C. L. Lerman, *J. Am. Chem. Soc.*, **96**, 5266 (1974); A. L. Baumstark, C. J. McCloskey, T. E. Williams and D. R. Chrisope, *J. Org. Chem.*, **45**, 3593 (1980).
294. S. A. Bone and S. Trippett, *Tetrahedron Lett.*, **19**, 1583 (1975).
295. G. Baccolini, R. Dalpozzo and E. Mezzina, *Phosphorus Sulfur Silicon*, **45**, 255 (1989).
296. Y. Kimura, M. Miyamoto and T. Saegusa, *J. Org. Chem.*, **47**, 916 (1988).
297. R. N. Burangulova, Yu. G. Trishin, I. V. Konoralova, L. A. Burnaeva, V. N. Christokletov and A. N. Pudovik, *Zh. Obshch. Khim.*, **59**, 1979 (1989); I. V. Konovalova, I. S. Dokuchaeva, Yu. G. Trishin, L. Burnaeva, V. N. Christokletov and A. N. Pudovik, *Zh. Obshch. Khim.*, **59**, 1726 (1989); I. S. Dokuchaeva, I. V. Konovalova, Yu. G. Trishin, L. Burnaeva, T. Yu. Kazanina, V. N. Christokletov and A. N. Pudovik, *Zh. Obshch. Khim.*, **61**, 611 (1991); Yu. G. Trishin, I. V. Konovalova, R. N. Burangulova, V. N. Christokletov and A. N. Pudovik, *Tetrahedron Lett.*, 577 (1989).
298. F. Ramirez, *Bull. Soc. Chim. Fr.*, 3491 (1970); F. Ramirez, *Synthesis*, 90 (1974); F. Ramirez, A. V. Patvardhan and S. R. Heller, *J. Am. Chem. Soc.*, **86**, 514 (1964); F. Ramirez, J. F. Pilot, C. P.

Smith, S. B. Bhatia and A. S. Gulati, *J. Org. Chem.*, **34**, 3385 (1969); W. G. Bentrude, W. D. Johnson and W. A. Khan, *J. Am. Chem. Soc.*, **94**, 3058 (1972).
299. R. A. Goodrich and P. M. Treichel, *Inorg. Chem.*, **7** 694 (1968).
300. G. I. Drozd, S. Z. Ivin, V. L. Kulakova and V. V. Sheluchenko, *Zh. Obshch. Khim.*, **38**, 576 (Engl. Transl. 558) (1968).
301. K. N. Anisimov, N. E. Kolobova and A. N. Nesmeyanov, *Izv. Akad. Nauk SSSR, Ser. Khim.*, 799 (Engl. Transl. 689) (1954).
302. H. W. Roesky, H. Djarrah, D. Amizadeh-Asl and W. S. Sheldrick, *Chem. Ber.*, **114**, 1554 (1981).
303. S. C. Peake, M. J. C. Hewson and R. Schmutzler, *J. Chem. Soc. A*, 2364 (1970); M. J. C. Hewson and R. Schmutzler, *Z. Naturforsch., Teil B*, **27**, 879 (1972); R. Schmutzler, *J. Chem. Soc., Dalton Trans.*, 2687 (1973); S. C. Peake, M. J. C. Hewson, O. Schlak, R. Schmutzler, R. K. Harris and M. I. M. Wazeer, *Phosphorus Sulfur*, **4**, 67 (1978).
304. H. B. Stegman, M. V. Dumm and K. B. Ulmschneider, *Tetrahedron Lett.*, 2007 (1976).
305. R. Schmutzler, *J. Chem. Soc., Dalton Trans.*, 2687 (1973).
306. F. Ramirez, A. J. Bigler and C. P. Smith, *J. Am. Chem. Soc.*, **90**, 3507 (1968).
307. S. Z. Ivin and G. I. Drozd, USSR Patent 176 897 (1965); *Chem. Abstr.*, **64**, 12723 (1966).
308. G. V. Röschenthaler, R. Bohlen, W. Storzer, A. B. Sopchik and W. G. Bentrude, *Z. Anorg. Allg. Chem.*, 1256 (1983).
309. R. Burgada and H. Germa, *C.R. Acad. Sci., Ser. C*, **267**, 270 (1968); H. Germa and R. Burgada, *Bull. Soc. Chim. Fr.*, 2607 (1975); C. Laurenço and R. Burgada, *Tetrahedron*, **32**, 2089 (1976); C. Laurenço and R. Burgada, *Tetrahedron*, **32**, 2253 (1976); H. Willson and R. Burgada, *Phsophorus Sulfur*, **7**, 115 (1979).
310. R. Burgada, *C.R. Acad. Sci., Ser. C*, **282**, 849 (1976); R. Burgada and A. Mohri, *Phosphorus Sulfur*, **9**, 285 (1981); R. Burgada, *Phosphorus Sulfur*, **2**, 237 (1976); H. Fauduet and R. Burgada, *C.R. Acad. Sci., Ser. C*, **259**, 81 (1980).
311. L. Lamandé and R. Muñoz, *Tetrahedron Lett.*, **32**, 763 (1991).
312. J. Breker, P. G. Jones, R. Schmutzler and D. Schomburg, *Phosphorus Sulfur Silicon*, **62**, 139 (1991).
313. B. Tanjour, M. T. Boisdon, C. Malavaud and J. Barrans, *Tetrahedron*, **44**, 6087 (1988).
314. F. Bouvier, P. Vierling and J. M. Dupart, *Inorg. Chem.*, **27**, 1099 (1988).
315. P. Savignac, B. Richard, Y. Leroux and R. Burgada, *J. Organomet. Chem.*, **93**, 331 (1975).
316. M. Lattman, H. M. Olmstead, P. P. Power, D. W. Rankin and H. E. Robertson, *Inorg. Chem.*, **27**, 3012 (1988).
317. W. S. Sheldrick, F. Zwaschka and A. Schmidpeter, *Angew. Chem., Int. Ed. Engl.*, **18**, 935 (1979).
318. D. Schomburg, W. Storzer, R. Bohlen, W. Kuhn and G. W. Röschenthaler, *Chem. Ber.*, **116**, 3301 (1983).
319. J. M. Dupart, G. Le Borgne, S. Pace and J. G. Riess, *J. Am. Chem. Soc.*, **107**, 1202 (1985).
320. D. B. Denney, D. Z. Denney, C. D. Hall and K. L. Marsi, *J. Am. Chem. Soc.*, **94**, 245 (1972).
321. D. B. Denney, R. L. Powell, A. Taft and D. Twitchell, *Phsophorus*, **1**, 151 (1971).
322. D. B. Denney, D. Z. Denney and J. J. Gigantino, *J. Org. Chem.*, **49**, 2831 (1984).
323. P. L. Robinson, C. N. Barry, S. W. Bass and S. A. Evans, *J. Org. Chem.*, **48**, 5398 (1983).
324. D. Bernard and R. Burgada, *Phosphorus*, **5**, 285 (1975).
325. D. Hellwinkel, *Angew. Chem..*, **78**, 985 (1966); D. Hellwinkel, *Chem. Ber.*, **102**, 528 (1969).
326. T. J. Katz and E. W. Turnblom, *J. Am. Chem. Soc.*, **92**, 6701 (1970); E. W. Turnblom and T. J. Katz, *J. Am. Chem. Soc.*, **93**, 4065 (1971).
327. R. Schmutzler, *Inorg. Chem.*, **3**, 421 (1964); Y. Kobayashi, C. Akashi and K. Morinaga, *Chem. Pharm. Bull.*, **16**, 1784 (1968); W. C. Firth, S. Frank, M. Garber and V. P. Wystrach, *Inorg. Chem.*, **4**, 765 (1965); W. C. Smith, *J. Am. Chem. Soc.*, **82**, 6176 (1960); E. S. Kozlov, L. G. Dubenko and M. I. Povolotskii, *Zh. Obshch. Khim.*, **48**, 1900 (Engl. Transl. 1734); L. Horner, H. Hoffman and P. Beck, *Chem. Ber.*, **91**, 1583 (1958).
328. R. G. Cavell and R. D. Leary, *J. Chem. Soc., Chem. Commun.*, 1520 (1970).
329. H. Schmidbauer and P. Holl, *Chem. Ber.*, **112**, 501 (1979).
330. C. D. Reddy, S. S. Reddy and M. S. R. Naidu, *Synthesis*, 1004 (1980).
331. A. N. Bouin, A. N. Yarkevitch and E. N. Tsvetkov, *Heteroatom Chem.*, **1**, 485 (1990).
332. J. M. Dupart, *Phosphorus Sulfur*, **33**, 15 (1987).
333. R. Wolf, *Pure Appl. Chem.*, **52**, 1141 (1980).
334. C. W. Perkins, J. C. Martin, A. J. Arduengo, W. Lau, A. Alegria and J. K. Kuchi, *J. Am. Chem. Soc.*, **102**, 7753 (1980).

Reviews

335. M. Webster, *Chem. Rev.*, **66**, 87 (1966).
336. B. A. Arbusov and N. A. Polezhaeva, *Russ. Chem. Rev. (Usp. Khim.)*, **43**, 414 (1974).
337. R. A. Cherkasov, V. V. Ovchinnikov, M. A. Pudovik and A. N. Pudovik, *Russ. Chem. Rev. (Usp. Khim.)*, **51**, 746 (1982).
338. M. A. Pudovik, V. V. Ovchinnikov, R. A. Cherkasov and A. N. Pudovik, *Russ. Chem. Rev. (Usp. Khim.)*, **52**, 361 (1983).
339. O. I. Kolodyazhnyi, A. N. Kukhar, *Russ. Chem. Rev. (Usp. Khim.)*, **52**, 1096 (1983).
340. N. A. Polezhaeva and R. A. Cherkasov, *Russ. Chem. Rev. (Usp. Khim.)*, **54**, 1126 (1985).
341. R. A. Cherkasov and N. A. Polezhaeva, *Russ. Chem. Rev. (Usp. Khim.)*, **56**, 163 (1987).
342. L. N. Markovskii, N. P. Kolesnik and Yu. Shermolovich, *Russ. Chem. Rev. (Usp. Khim.)*, **56**, 894 (1987).
343. A. A. Kutyrev and V. V. Moskva, *Russ. Chem. Rev. (Usp. Khim.)*, **56**, 1028 (1987).
344. J. Gloede, *Z. Chem.*, **28**, 352 (1988).
345. O. G. Sinyashin, E. S. Batyeva and A. N. Pudovik, *Russ. Chem. Rev. (Usp. Khim.)*, **58**, 352 (1989).

Books

346. C. D. Hall, in *Organo Phosphorus Chemistry*, Specialist Periodical Reports, Royal Society of Chemistry, London, **16**, 51 (1986); **17**, 50 (1986); **18**, 52 (1987); **19**, 47 (1987); **20**, 512 (1988); **21**, 51 (1990); **22**, 48 (1991).
347. W. E. McEwen: 'Stereochemistry of Reactions of Organo Phosphorus Compounds', in *Topics in Phosphorus Chemistry*, Vol. 2 (Eds M. Grayson and E. J. Griffith), Wiley, New York, 1965, p. 1.
348. *Composés Organiques du Phosphore*, Colloque National No. 29, CNRS, Paris, 1966.
349. A. J. Kirby and S. G. Warren, *The Organic Chemistry of Phosphorus*, Elsevier, Amsterdam, 1967.
350. R. Schmutzler, 'Fluorophosphoranes', in *Halogen Chemistry*, Vol. 2, Academic Press, New York, 1967, p. 31.
351. K. D. Berlin and D. M. Hellwege, 'Carbon Phosphorus Heterocycles', in *Phosphorus Chemistry*, Vol. 6, Wiley, New York, 1969, pp. 1–186.
352. *Chimie Organique du Phosphore*, Colloque International No. 182, CNRS, Paris 1970.
353. F. G. Mann, *The Heterocyclic Derivatives of Phosphorus, Arsenic, Antimony, and Bismuth*, 2nd Ed., Wiley, New York, 1970.
354. D. Hellwinkel, in *Organic Phosphorus Compounds* (Eds G. M. Kosolapoff and L. Maier), Vol. 3, Wiley, New York, 1972, p. 185.
355. B. J. Walker, *Organo Phosphorus Chemistry*, Penguin, Harmondsworth, 1972.
356. J. F. Brazier, D. Houalla, M. Koenig and R. Wolf, 'NMR Parameters of the Proton Directly Bonded to Phosphorus', in *Topics in Phosphorus Chemistry* (Eds E. J. Griffith and M. Grayson), Vol. 8, Wiley, New York, 1976, p. 99.
357. D. E. C. Corbridge, *Phosphorus: an Outline of its Chemistry, Biochemistry, and Technology*, Elsevier, Amsterdam, 1978.
358. K. Burger, 'Pentacoordinated Phosphoranes in Synthesis', in *Organo Phosphorus Reagents in Organic synthesis* (Ed. J. I. G. Cadogan), Academic Press, London, 1979, p. 467.
359. D. J. H. Smith and R. S. Edmundson, 'Phosphorus Compounds', in *Comprehensive Organic Chemistry*, Vol. 2, Pergamon Press, Oxford, 1979.
360. L. D. Quin and J. G. Verkade (Eds), *Phosphorus Chemistry* (*ACS Symposium Series*, Vol. 171), American Chemical Society, Washington, DC, 1981.
361. R. Luckenbach, 'Organische Phosphorverbingungen II', in *Methoden der Organischen Chemie (Houben Weyl, Ergänzungswerk 2)*, Georg Thieme, Stuttgart, 1982.
362. J. G. Verkade and L. D. Quin (Eds), *Phosphorus ^{31}P NMR Spectroscopy in Stereochemical Analysis*, VCH Deerfield Beach, FL, 1987.
363. J. C. Tebby, *Handbook of Phosphorus 31 Nuclear Magnetic Resonance Data*, CRC Press, Boca Raton, FL, 1991.

See also references 9, 10, 30, 37, 38, 186, 189 and 208.

CHAPTER **4**

Structure, bonding and spectroscopic properties of phosphonium ylides

S. M. BACHRACH and C. I. NITSCHE

Department of Chemistry, Northern Illinois University, DeKalb, IL 60115, USA

The chemistry of organophosphorus compounds, Vol. 3
Edited by F. R. Hartley

I. INTRODUCTION

There has been a long-standing controversy regarding the nature of the bonding in phosphonium ylides (phosphoranes), centred about the relative contributions of the two resonance structures **A**, the *ylide* form, and **B**, the *ylene* form[1]. The ylene form includes a formal P=C double bond, which implies the participation of the *d* orbitals of phosphorus in the π bond with the carbon p orbital. The ylide form possesses a carbanion with an adjacent phosphonium centre. In this chapter, we shall address the bonding question by examining recent structural, spectroscopic and theoretical studies of phosphonium ylides. We shall limit our discussion to free ylides; Schmidbaur[2] has recently reviewed the structures and chemistry of metal-complexed phosphonium ylides.

$$\geq\overset{+}{P}-\overset{..}{C}\lessdot \longleftrightarrow \geq P=C\lessdot$$

(A) **(B)**

A number of structural questions will be addressed. What is the range of typical P=C bond lengths and how does this vary with substituent? Is the C planar or pyramidal, and does this depend on substituents? What is the geometry about P? The answers to these questions will determine the hybridization of the P and C atoms. We shall also examine the density distribution to determine the charges on the atoms and the degree of d orbital participation.

In Section II, we survey the structural analyses of ylides, summarizing the reported X-ray crystal structures, electron diffraction studies and calculational determinations. In Section III, we present spectroscopic data (primarily UV, NMR, ESCA and EPR) that pertain to characterizing the bonding in the ylides. Finally, we attempt in Section IV to bring together all the summarized evidence to present a cogent and coherent picture of the bonding in the ylides.

II. STRUCTURES OF PHOSPHONIUM YLIDES

A large effort has been expended in order to determine accurately the structures of the ylides using a variety of techniques. Single-crystal X-ray structures of over 40 ylides have been reported. Three small ylides have been characterized in the gas phase using electron diffraction techniques. While theoretical chemists had restricted their studies to methylenephosphorane, recent advances in computational techniques[3] and computer hardware have allowed for the geometry optimization of larger ylides.

A. X-ray Structures

Even though phosphonium ylides were first prepared in the 1890s[4,5] and had garnered widespread synthetic use with the discovery of the Wittig reaction in 1953[6], the first X-ray characterization of ylides did not appear until 1965, when Stephens reported the structures of **1**[7] and **2**[8] and Mak and Trotter[9] published the structure of **3**. These ylides are stabilized by conjugation with the carbonyl group. The first reported crystal structure of a non-stabilized ylide was of triphenyl(methylene)phosphorane (**4**) ($Ph_3P=CH_2$) by Bart[10] in 1969. Since then, over 40 other structures have been published.

$Ph_3P{=}C(COPh)(Cl)$ **(1)**

$Ph_3P{=}C(COPh)(I)$ **(2)**

$Ph_3P{=}C(CO_2Me)C(CO_2Me){=}N{-}C_6H_4{-}Br$ **(3)**

We have classified these structures into four categories based on their structural similarities: (1) non-stabilized ylides; (2) stabilized ylides possessing a substituent that conjugates with the P=C double bond; (3) ylides having the P=C=P functionality; and (4) ylides possessing the P=C=C functionality. The structural characteristics of each category will be discussed individually below, first by examining a few examples in detail and then summarizing the trends of the entire category.

1. Non-stabilized ylides

Non-stabilized ylides lack any substituent on carbon that can conjugate with the P=C double bond. In other terms, the substituents attached to the carbanion cannot assist in delocalizing the anion. Although these types of ylides are reactive, a large number of examples have been prepared and their single-crystal X-ray structures determined. We

TABLE 1. X-ray structural data for non-stabilized ylides (Class 1)[a]

Compound	P=C	R—P=C	P=C—R′	R′—C—R′	Sum[b]	δ^c	Ref.
$Ph_3P{=}CH_2$(**4**)	1.661	111.4 114.3 115.0	117.1 120.4	121.9	359.4	0	10
$Ph_3P{=}CH_2$(**4**)	1.692	109.9 112.8 117.8	116 116	119	351	28	11
$Fc_3P{=}CH_2^d$	1.629	113.5 114.9 117.5				~0	11
$Pr^i_3P{=}CMe_2$	1.731	107.7 114.9 117.5	119.8 120.4	114.0	354.2	23	12
$Ph_3P{=}C(CH_2)_2$(**6**)	1.696	109.7 110.2 117.0	117.1 117.8	58.8	293.7	58	13
$Ph_3P{=}C(CH_2)_3$(**7**)	1.668	108.6 108.7 123.5	130.0 132.5	91.4	353.9	19	14
$Ph_3P{=}C(CF_2)_3$	1.713	108.3 109.5 114.4	133.1 133.5	93.4	360.0	0	15

[a] All distances are in Å and all angles in degrees.
[b] Sum of the angles about C.
[c] The out-of-plane bending angle. See text for description.
[d] Fc indicates a ferrocenyl group.

TABLE 2. X-ray structural data for non-stabilized ylides (Class 2)[a]

Compound	P=C	R—P=C	P=C—R′	R′—C—R′	Sum[b]	δ[c]	Ref.
Ph_3P=$CHPMe_2$(**8**)	1.618	110.4 111.9 118.3	122.3				11
MeP=$CHPPh_2$(**9**)	1.684	117.7 114.8 114.8	122.2(P) 110(H)	128	360	0	11
Ph_3P=$C(PPh_2)_3$(**10**)	1.720	112.7 114.1 114.6	107.4 128.8	119.4	355.6	23	16
Ph_3P=$C(AsPh_2)_3$(**11**)	1.698	113.1 113.7 114.6	108.4 128.0	116.4	352.8	28	17
Ph_3P=$C(SbPh_2)_3$(**12**)	1.692	111.9 114.4 117.3	111.5 130.5	116.2	358.2	17	18
Me_3P=$C(NMe_2)(PFCF_3)$	1.727	111.1 112.7 113.0	116.6(P) 108.5(N)	131.2	356.3	22	19
Ph_3P=C(SPh)(SePh)	1.707	110.0 113.1 116.3	123.4(S) 113.5(Se)	122.2	359.1	14	20

[a] All distances are in Å and all angles in degrees.
[b] Sum of the angles about C.
[c] The out-of-plane bending angle. See text for description.

have split the non-stabilized ylides into two parts, the first class having H or saturated C substituents and the second having Group V or VI substituents on the carbanion.

A summary of the pertinent X-ray structural data for these classes are presented in Tables 1 and 2. For all X-ray structures, a few structural features are of primary interest. The P=C distance characterizes the strength of the interaction of either the π bond or the electrostatic attraction of the carbanion and the phosphonium (or some combination of the two). The R—P=C angle, defined by the atom directly connected to the P, indicates the hybridization of P. Defining the hybridization of C is complicated by defining the degree of pyramidalization. We approach this in two complementary ways. The sum of the angles about C will be 360° for a trigonal planar atom, and diminish as the atom becomes more pyramidal. Another method is to measure the deviation of the plane formed by C and its two substituents from the P=C bond. This angle is defined as δ and is shown in **5**. For a planar C, δ will equal zero, and δ increases as the atom becomes pyramidal.

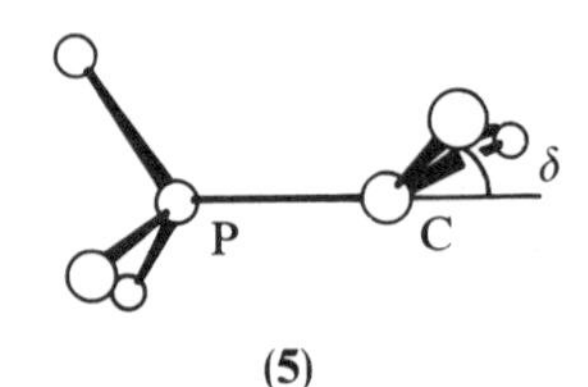

(5)

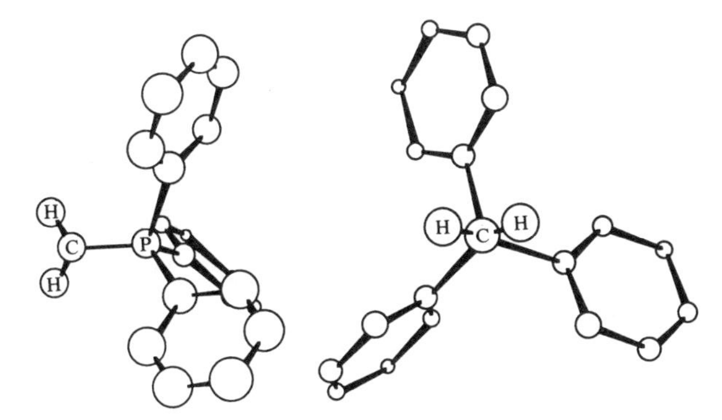

FIGURE 1. X-ray structure of **4**. The view on the right is a Newman projection down the P—C bond

*a. Triphenyl(methylene)phosphorane(***4***).* The X-ray crystal structure of triphenyl-(methylene)phosphorane was the first to be reported for a non-stabilized ylide[10]. Bart[10] was unable to locate the methylene hydrogens, prompting Schmidbaur *et al.*[11] to reinvestigate the structure using more advanced techniques and lower temperatures. Although two independent molecules are found in the crystal, they differ only by slight conformational changes. Therefore, we shall discuss only one of these molecules, and present in Figure 1 two views of the structure.

Schmidbaur *et al.*[11] reported the P═C distance as 1.692 Å (average of the two molecules), which is longer than Bart's value of 1.661 Å. Schmidbaur *et al.* suggested that the earlier study may suffer from, in addition to temperature effects, systematic errors.

Threefold symmetry is not present about the P atom; the P–phenyl distances are not identical. One phenyl group nearly bisects the methylene plane; its distance from the P atom is longer than the P—C bonds to the other two rings and it forms a wider Ph—P═C angle than the other two rings. This pattern is typical, as will be seen throughout this chapter. The Ph—P═C angles (109.9–117.8°) and Ph—P—Ph angles (104.0–107.0°) are suggestive of a tetrahedral environment, arguing against dsp^3 P hybridization.

Schmidbaur *et al.*[11] did locate the methylene hydrogens, and they distinctly form a pyramidal carbon. The methylene plane is bent 28° from planarity, *towards* the perpendicular phenyl ring, which can easily be seen in the Newman projection shown in Figure 1. The large hydrogen displacement factors, however, suggest a low barrier for inversion. As will be discussed in Section II.C.1, high-level theoretical calculations of related ylides support the notion of a non-planar carbon atom. Density deformation maps of **4** indicate a build-up of electron density on the ylidic carbon. Schmidbaur *et al.*[11] described this carbon as 'an easily pyramidalizable carbanion stabilized by the adjacent tetrahedral phosphonium centre'.

b. Triphenylphosphonium cyclopropylide (**6**). As shown in Tables 1 and 2, the value of δ is usually small, but non-zero, for the non-stabilized ylides. The announcement[13] of the X-ray structure of triphenylphosphonium cyclopropylide (**6**) was extremely exciting since the cyclopropyl ring deviates significantly from the P═C plane: $\delta = 58°$! The sum of the angles at carbon is very small, only 293.7°. This highly pyramidalized carbanion can be clearly seen in Figure 2.

Although the pyramidal carbon is the most unusual feature of this molecule, it is of

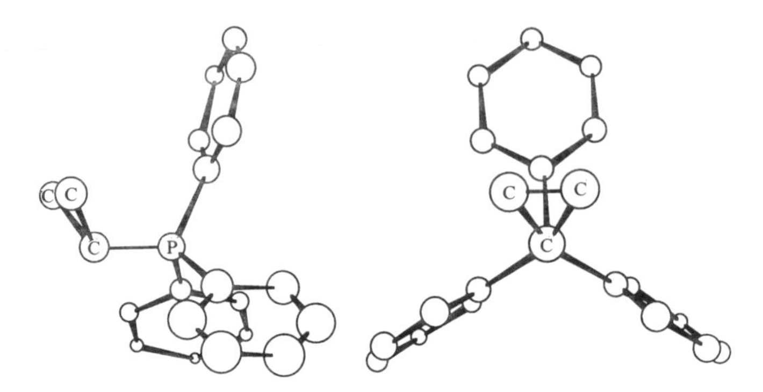

FIGURE 2. X-ray structure of **6**. The view on the right is a Newman projection down the P—C bond

considerable interest that the rest of the system is so ordinary. The P=C distance is 1.696 Å, within the normal range for ylides. The phosphorus atom has a distorted tetrahedral geometry, with the Ph—P=C angle largest for the perpendicular phenyl ring (117.0° vs 109.7° and 110.2°). The perpendicular phenyl ring virtually bisects the cyclopropylide group and the cyclopropylide group is bent towards this ring. Even though the carbanion is extremely pyramidal, the NMR spectrum indicates that rapid inversion of the carbanion occurs even at low temperature. These results are very similar to those of structure **4** discussed above and of other non-stabilized ylides.

c. Other Class 1 non-stabilized ylides. The non-stabilized ylides all possess very similar characteristics. The P=C distance ranges from 1.63 to 1.73 Å. The groups attached to P are arranged in a tetrahedral fashion. Even in the case where the substituents are bulky ferrocenyl (Fc) groups, the Fc—P=C angles are small (113–117.5°)[11]. Careful X-ray analysis indicates that in most cases, the ylidic carbon is pyramidal. The most extreme case is **6**, where $\delta = 58°$. When ring strain is reduced, as when the three-membered ring is replaced with a four-membered ring[14] [Ph_3 P=$C(CH_2)_3$] (**7**), the carbanion becomes more planar, but is still pyramidal, with $\delta = 19°$. The fluorinated analogue of **7** has been examined and the ylidic carbon is planar[15]. However, the fluorine atoms may be stabilizing the carbanion via electron withdrawal, and this ylide may be more akin to the stabilized ylides in Section II.B.

One of the P substituents always bisects the plane formed by the ylidic carbon and its substituents, and this plane is bent towards this perpendicular P substituent. Usually, the perpendicular substituent is further from P than the other two substituents and forms a larger angle to the ylidic carbon than the other substituents. The perpendicular substituent is clearly interacting with the carbanionic lone pair.

d. Triphenylphosphonium bis(diphenylphosphino)methylide and other Class 2 ylides. The non-stabilized ylides comprising Class 2 all have a Group V or VI atom bonded to the ylidic carbon. Of primary interest here is the interaction of the lone pair(s) of the heteroatom with the P=C π bond. Schmidbaur and coworkers have examined a related series of these compounds: Ph_3P=$CHPMe_2$(**8**)[11], Me_3P=$CHPPh_2$ (**9**)[11], Ph_3P=$C(PPh_2)_2$(**10**)[16], Ph_3P=C $(AsPh_2)_2$(**11**)[17] and Ph_3P=$C(SbPh_2)_2$(**12**)[18]. These compounds have very similar structures, typified by that of **10** shown in Figure 3.

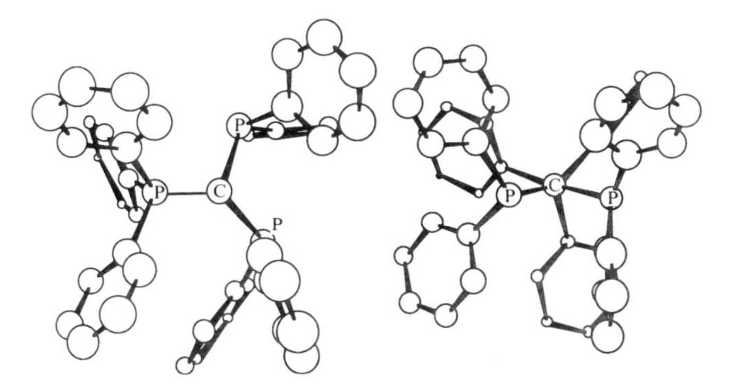

FIGURE 3. X-ray structure of **10**. The view on the right is a Newman projection down the P—C bond

The geometry about the P=C double bond for the Class 2 non-stabilized ylides is similar to the structure of the Class 1 ylides. The P=C distances range from 1.618 to 1.727 Å. The phosphorus atom is a slightly distorted tetrahedron and the carbon atom is slightly pyramidal. One of the phosphonium substituents bisects the plane formed by the ylidic carbon and its two substituents, and this plane is bent toward the perpendicular P substituent, as shown in Figure 3.

The lone pair on the P substituent in **8** and **9** lies in the plane of the P=C bond, in a *syn* relationship to the P=C bond. Thus, the lone pair is perpendicular to the P=C π bond. For the bis(diphenylpnictogen) species **10–12**, the lone pairs also lie perpendicular to the P—C π bond, with one arranged *syn* and the other *anti* to the P=C bond. This conformation of **10** is seen in Figure 3. The N and P lone pairs in $Me_3P{=}C(NMe_2)(PFCF_3)$ are both *syn* to the P=C bond[19].

Schmidbaur *et al.* argued that this *syn–anti* arrangement can be understood in simple electrostatic terms. If one assumes an ylide description, then the ylidic P carries a partial positive charge. One lone pair will be attracted to this centre, forming a *syn* conformer. Since the pnictogens are less electronegative than C, they will bear a partial positive charge, and the lone pair of the second pnictogen (P_n) will be attracted to the first pnictogen atom. This explanation is completely consistent with the observed angles about the ylidic carbon (see **13**–**15**): one small P=C—pN angle, a small pN—C—pN angle and a large P=C—pN angle.

$P^{\delta+}$ 128.8° 107.4° $C^{\delta-}$ $^{\delta+}P$ 119.4° $P_{\delta+}$ **(13)**

$P^{\delta+}$ 128.0° 108.4° $C^{\delta-}$ $^{\delta+}As$ 116.4° $As_{\delta+}$ **(14)**

$P^{\delta+}$ 130.5° 111.5° $^{\delta+}Sb$ 116.0° $Sb_{\delta+}$ **(15)**

The X-ray structure of a Group VI analogue, $Ph_3P{=}C(SPh)(SePh)$, has been reported[20]. The phenyl rings of the chalcogens are located on the opposite side of the P=C(S)(Se) plane, nearly perpendicular to this heavy atom plane. This orientation minimizes the electrostatic repulsion between the S and Se lone pairs.

2. Stabilized ylides

The stabilized ylides have at least one substituent on the ylidic carbon that can either conjugate with the P=C π bond (**16a**) or delocalize the carbanion charge (**16c**). Most of the crystal structures of the stabilized ylides have a carbonyl group attached to the ylidic carbon, although other groups are represented. The dominant resonance structures of the carbonyl stabilized ylides are shown in **16a**–**c**. A summary of the X-ray structural data for the stabilized ylides is given in Table 3.

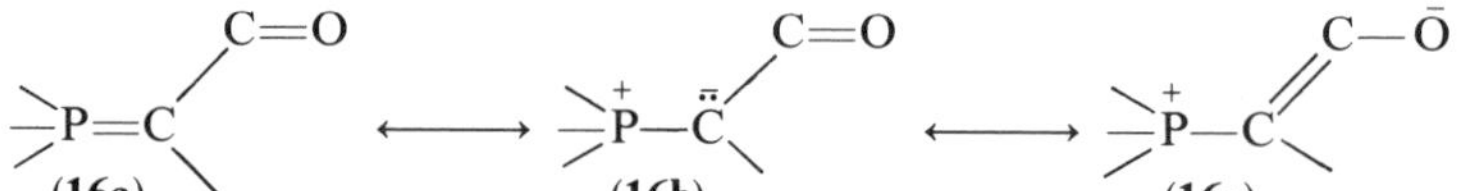

C=O / P=C (**16a**) ⟷ C=O / $\overset{+}{P}$—$\bar{C}$ (**16b**) ⟷ C—$\bar{O}$ / $\overset{+}{P}$—C (**16c**)

The important structural data for the stabilized ylides reflect the degree of conjugation between the P=C double bond (or carbanion) and the ylidic carbon substituent. The distance separating the ylidic carbon and the carbonyl carbon and the C=O distance directly reflect the interaction with the ylide. Typical C—C distances are about 1.5 Å and carbonyl C=O distances are about 1.2 Å. The participation of **16c** would be indicated by a longer C=O bond and a shorter C—C bond. Further, strong conjugation requires a planar ylidic carbanion (which can be judged by the sum of the angles about carbon) and the P—C—C—O dihedral angle γ (which will be 0° or 180° for a fully conjugated system and 90° for a non-conjugated system).

a. 2-Carboxy-1-methoxycarbonylethyltriphenylphosphorane (**17**). The X-ray structure[22] of **17** is shown in Figure 4. The P=C bond distance is 1.732 Å, slightly longer than in typical non-stabilized ylides. The geometry about the phosphorus atom is near tetrahedral, just as for the non-stabilized ylides. The sum of the angles about the ylidic carbon is 359.1°, indicating a very slight pyramidal nature, which can be seen in the Newman projection in Figure 4.

The ylidic carbon–carbonyl carbon distance is 1.392 Å, significantly shorter than the distance between the ylidic carbon and the other carbon connected to it (1.509 Å). The ester carbonyl bond length is 1.243 Å, longer than the acid carbonyl length of 1.197 Å. These distances clearly indicate the action of conjugation. The value of γ is 14.1°, which is close to the ideal arrangement for conjugation. The evidence for participation of resonance structure **16c** is overwhelming. The structures of the other ylides stabilized by a carbonyl group are very similar to the structure of **17**. The carbonyl group stabilizes the ylide functionality via strong delocalization of charge.

b. Triphenylphosphonium cyclopentadienylide (**18**). The X-ray structure[27] of triphenylphosphonium cyclopentadienylide (**18**) can be seen in Figure 5. The P=C distance is 1.718 Å, typical for the stabilized ylides. Once again, we note a tetrahedral phosphorus atom with a nearly planar ylidic carbon. The sum of the angles about the ylidic carbon is 358.0°. The cyclopentadienyl ring is nearly planar, and is bent upwards towards one of the phenyl rings. For this ylide we are interested in the participation of **19b** in the description of the electron distribution. The non-alternation of C—C distance within the five-membered ring would indicate that the delocalized resonance structure dominates.

TABLE 3. X-ray structural data for stabilized ylides[a]

Compound	P=C	C—C	C=O	R—P=C	P=C—R	R′—C—R′	γ^b	Ref.
$Ph_3P{=}C(COPh)(Cl)$ **(1)**	1.736	1.361	1.301	110.2 112.3 113.2	120.2(C) 118.3(Cl)	121.2	4.7	7, 21
$Ph_3P{=}C(COPh)(I)$ **(2)**	1.71	1.35	1.28	110 112 115	120(C) 116(I)	120	12	8, 21
$Ph_3P{=}C(CO_2Me)(CH_2CO_2H)$ **(17)**	1.732	1.392	1.243	108.7 111.1 115.4	117.7(CO) 123.8(C)	117.6	14.1	22
$Ph_3P{=}C(CO_2Me)(CH_2CO_2Bu^t)$	1.715	1.415	1.221	107.5 112.9 115.5	118.7(CO) 124.4(C)	116.7	2.5	22
$Ph_3P{=}C(CO_2Et)(CH_2CH_2C_2F_5)$	1.702	1.414	1.219	107.7 117.2 112.1	120.7(CO) 123.3(C)	115.8	8.2	23

(continued)

TABLE 3. *(continued)*

Compound	P=C	C—C	C=O	R—P=C	P=C—R	R′—C—R′	γ^b	Ref.
$Ph_3P{=}C(CO_2Me)C(CO_2Me){=}N{-}C_6H_4{-}Br$ (3)	1.698	1.473	1.201	108.4 111.9 113.7	128.3(CO) 116.8(CN)	114.9	2.3	9
$Ph_3P{=}CH{-}C(CH_3){-}OSnCl$	1.75	1.36	1.27		120.3(C)			24
$Ph_3P{=}C(CO_2Me){-}C(CO_2Me){=}CHPh$	1.752	1.42(CO) 1.45(CC)	1.35	108.3 111.4 117.5	113(CO) 121(CC)	126		25
$Ph_3P{=}C(CO_2Me){-}C(CO_2Me){=}CHPh$	1.718	1.42(CO) 1.48(CC)	1.219	110.7 113.3 114.0	115(CO) 122(CC)	121		25

(O(CH2CH2)2N)3P=CHCOPH	1.717	1,399	1.258	106.0 112.0 117.9	120.1			26
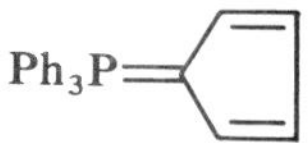 Ph_3P=C5H4 **(18)**	1.718	1.419 1.430		110.2 111.3 112.6	125.3 125.3	107.4		27
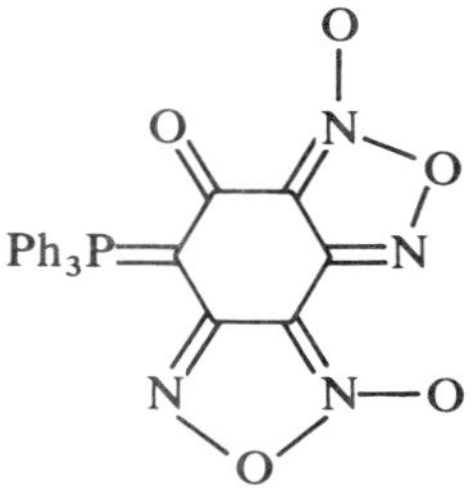 Ph_3P=C6(O)(bis-furoxano)	1.770	1.463	1.228	108 109 110	116.5(CO) 124(CN)	119	11.2	28
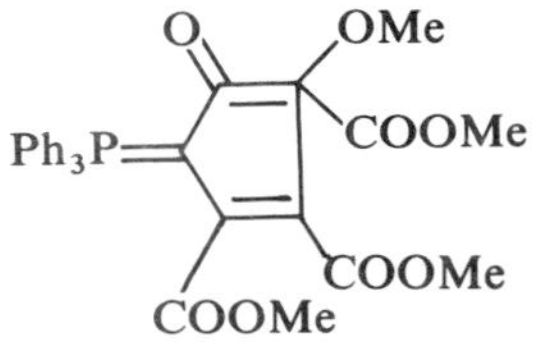 Ph_3P=C5(O)(OMe)(COOMe)3	1.728	1.48(CO) 1.41(CC)	1.22	109.2 112.3 112.9	118.9(CO) 135.1(CC)	105.8	0	29
Ph_3P=CHTs	1.709	1.686	1.444[c] 1.469[c]	105.2 112.6 118.2	123.9			21

(continued)

TABLE 3. (*continued*)

Compound	P=C	C—C	C=O	R—P=C	P=C—R	R′—C—R′	γ^b	Ref.
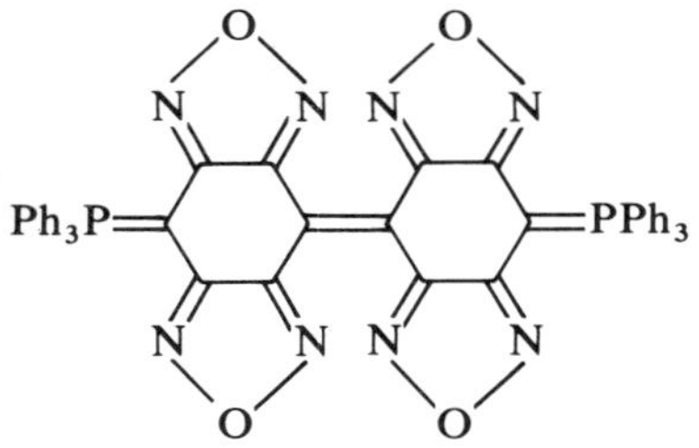 Ph_3P= (bis-furazano bicyclohexylidene) =PPh_3	1.708 1.767				125,125 123,123	110 113		3
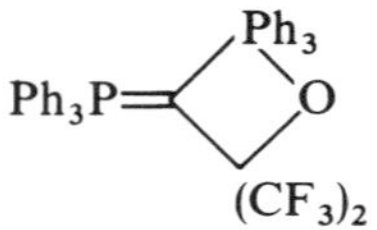 Ph_3P= (oxaphosphetane ring: Ph_3, O, $(CF_3)_2$)	1.745			112.5 113.4 116.3	133.4(P) 128.0	98.5		31
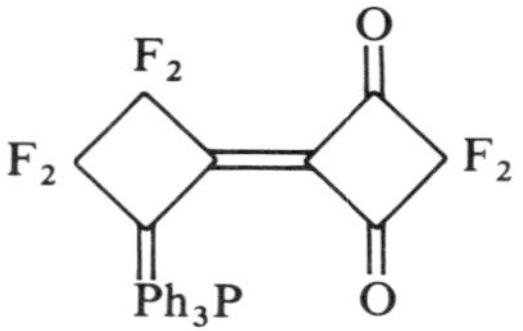 (F_2, F_2, Ph_3P, O, O, F_2 bicyclobutylidene)	1.773	1.388		106.4 109.5 110.2	137.4 128.6(F)	94.0		32
 $(Me_2N)_2P$ (ring) $Si(Bu^t)_2$	1.700			94.4(C) 115.5 121.0	91.4(Si)			33
$Me_3P{=}C(PMe_3)_2{}^{2+}\ 2I^-$	1.75			112.6 112.6 112.9	116.7 116.7	126.5		34, 35

[a] All distances are in Å and angles are in degrees.
[b] Defined as the P=C—C=O dihedral angle.
[c] The S=O distance.

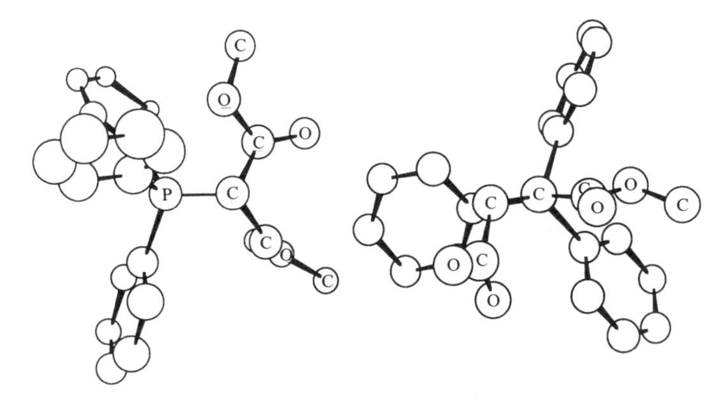

FIGURE 4. X-ray structure of **17**. The view on the right is a Newman projection down the P—C bond

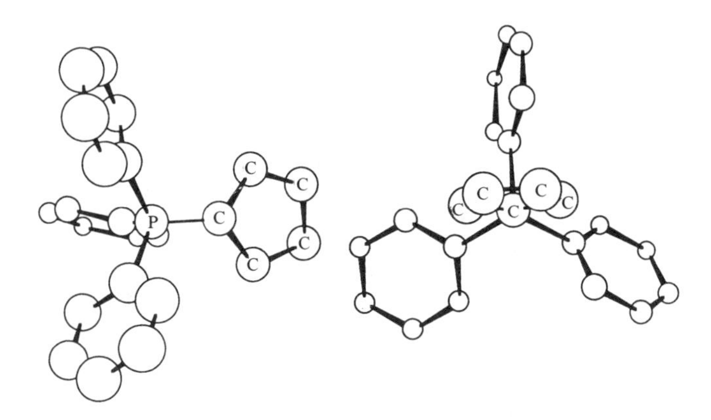

FIGURE 5. X-ray structure of **18**. The view on the right is a Newman projection down the P—C bond

The distances shown in the schematic drawing **20** display little bond alternation and support the notion of a delocalized carbanion.

PR_3 $\longleftrightarrow$ $\overset{+}{P}R_3$

Ph_3P 1.718 1.419 1.376 1.401 1.392 1.430

(19a) **(19b)** **(20)**

TABLE 4. X-ray structural data for P=C=P ylides[a]

Compound	P=C	P=C=P	X—P=C	Ref.
$Ph_3P{=}C{=}PPh_3$ (mol. 1) (**21**)	1.629,	143.8	109.2, 117.7, 117.8	37
As above (mol. 2)	1.633	130.1	109.9, 116.1, 117.4	37
Room temperature	1.610	134.4		39
− 160 °C	1.632, 1.638	131.7		39
$Ph_2MeP{=}C{=}PMePh_2$	1.648	121.8	110.9, 117.2, 117.2	40
$(Me_2N)_3P{=}C{=}P(NMe_2)_3$	1.584	180.0	120.5, 113.5, 113.4	44
$(Me_2N)_2FP{=}C{=}PF(NMe_2)_2$	1.598, 1.602	134.8	116.2(F), 113.9, 120.8 115.0(F), 112.7, 123.8	45
$Ph_2P{=}$ PPh_2 (cyclic)	1.645, 1.653	116.7	118.0(C), 113.2, 113.4 116.4(C), 114.1, 115.6	41, 42
$Ph_3PCH{=}PPh_3{}^+ Br^-$	1.695, 1.710	128.2	112.1, 112.5, 112.8 109.3, 112.4, 116.0	46
$W(CO)_5[O{=}P(Ph)_2CH{=}PPh_3]$	1.701, 1.681	127.9	118.0(O), 107.5, 111.9 111.8, 113.3, 112.2	47

[a] All distances are in Å and all angles are in degrees.

3. *P=C=P ylides*

Many examples of ylides possessing the P=C=P functionality have been synthesized and characterized. Symmetric, asymmetric, cyclic and acyclic examples are known[36–43]. The known crystal structures of these ylides are listed in Table 4. Whereas most small cumulenes are linear, cumulenes having more than two double bonds are frequently bent[48–50]. The P=C=P angle is the most characteristic feature of these types of ylides.

a. Hexaphenylcarbodiphosphorane (**21**). Hexaphenylcarbodiphosphorane (**21**) was first prepared in 1961[36]. The crystal structure of this species is dependent on the

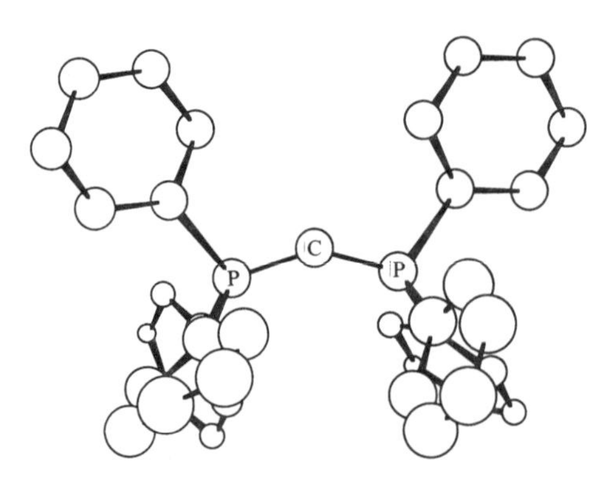

FIGURE 6. X-ray structure of **21**

crystallization method. Vincent and Wheatley[37] first reported the structure in 1972. They found two independent molecules in the unit cell, one of which is presented in Figure 6. The two molecules are very similar in most aspects: they have tetrahedral phosphorus atoms and similar P=C bond lengths (1.629 and 1.633 Å). They are both bent but to different extents; the P=C=P angles are 143.8° and 130.1°. Hardy *et al.*[39] soon thereafter reported another form of crystalline **21** which is triboluminescent. This low-temperature structure has non-equivalent P=C distances of 1.632 and 1.638 Å, and the P=C=P angle is 131.7°. Although the two studies indicate different geometries for **21**, they do agree that its backbone is distinctly bent.

All other structures of the P=C=P ylides have a bent backbone, except for $(Me_2N)_3P{=}C{=}P(NMe_2)_3$, which is linear[44]. The P=C=P backbone has been introduced into both a six- and a seven-membered ring[41]. The crystal structure of the smaller ring has been reported. As expected, in order to accommodate the ring angles, the P=C=P angle is small: 116.7°. Schmidbaur *et al.*[41] have argued that these ylides can be described by resonance structure **22**, which has a central sp^2 carbanion with an allyl cationic structure spread over the P—C—P framework.

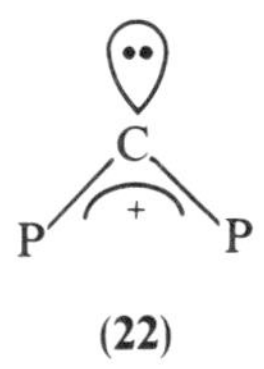

(22)

4. P=C=C ylides

Ylides containing the P=C=C functionality were first prepared in the early 1960s[51], when Matthews and Birum prepared $Ph_3P{=}C{=}C{=}O$ **23**[52], $Ph_3P{=}C{=}C{=}S$ **24**[52], $Ph_3P{=}C{=}C{=}NPh$ **25**[53] and $Ph_3P{=}C{=}C(CF_3)_2$[54]. A number of other examples have since been reported, and a few of these ylides have been characterized by X-ray crystallography. A summary of the X-ray structure is given in Table 5. As can be seen, the most striking feature of these ylides is the non-linear P=C=C angle.

a. (2,2-Diethoxyvinylidene)triphenylphosphorane (**26**). As an example of this type of ylide, Figure 7 shows the X-ray structure[58] of **26**. The P=C distance is 1.682 Å, typical for ylides. The phosphorus atom is tetrahedral. The acetal carbon is planar and the C=C distance is 1.314 Å. These distances are on the short end of the range of typical P=C and C=C bond lengths. Double bonds in cumulenes are usually shorter than normal, owing to the sp hybridization. The P=C=C angle is 125.6°, significantly distorted from the 180°

TABLE 5. X-ray structural data for P=C=C ylides[a]

Compound	P=C	C=C	C=X	R—P=C	P=C=C	Ref.
$Ph_3P{=}C{=}C{=}O$ (**23**)	1.648	1.210	1.185	110.6, 111.0, 113.0	145.5	55
$Ph_3P{=}C{=}C{=}S$ (**24**)	1.677	1.209	1.595	110.3, 110.6, 111.6	168.0	56
$Ph_3P{=}C{=}C{=}NPh$ (**25**)	1.677	1.248	1.252	109.7, 110.7, 114.7	134.0	57
$Ph_3P{=}C{=}C(OEt)_2$ (**26**)	1.682	1.314		113.9	125.6	58
$BeMn(CO)_4{=}C{=}C{=}PPh_3$	1.679	1.216	1.981	109.3, 110.5, 111.6	164.0	59

[a] All angles are in Å and all distances are in degrees.

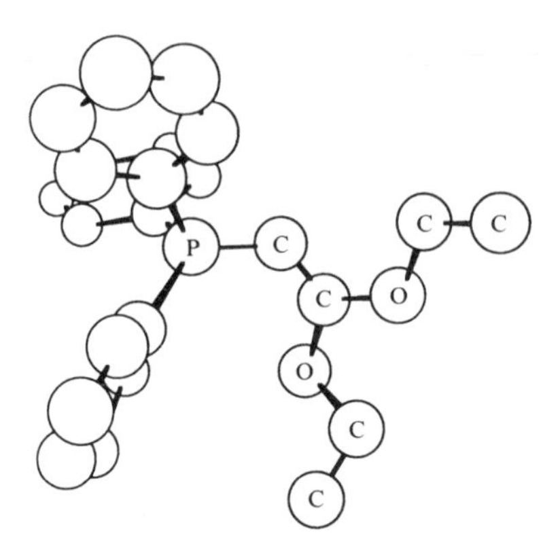

FIGURE 7. X-ray structure of **26**

expected for an allene-type system. This small angle suggests that the ylidic carbon is sp^2 and the bonding is best represented by **27b**.

$$\mathrm{P{=}C{=}C} \longleftrightarrow \overset{+}{\mathrm{P}}\mathrm{-}\overset{-}{\mathrm{C}}\mathrm{{=}C}$$

(27a) **(27b)**

Compounds **23**–**25** display a very bent P=C=C angle but the C=C=X angle is nearly linear. Further, the C=C distances are extremely short. These two features are attributed to the strong participation of **28b** and **28c** in the description of the electronic structure.

$\mathrm{Ph_3P{-}C{=}C{=}O}$ (P–C=C 145.5°; C=C=O 175.6°) **(23)**

$\mathrm{Ph_3P{-}C{=}C{=}S}$ (P–C=C 168.0°; C=C=S 178.3°) **(24)**

$\mathrm{Ph_3P{-}C{=}C{=}NPh}$ (P–C=C 134.0°; C=C=N 172.5°) **(25)**

$$\mathrm{P{=}C{=}C{=}X} \longleftrightarrow \overset{+}{\mathrm{P}}\mathrm{-}\overset{-}{\mathrm{C}}\mathrm{{=}C{=}X} \longleftrightarrow \mathrm{P^{+}{-}C{\equiv}C{-}X^{-}}$$

(28a) **(28b)** **(28c)**

TABLE 6. Electron diffraction data for **29–31**

Compound	P=C	C—P=C	δ	P=C=P	Ref.
$Me_3P{=}CH_2$ (**29**)	1.640	116.5	0	—	60
$Me_3P{=}CHSiH_3$ (**30**)	1.653	115.0	0	—	61
$Me_3P{=}C{=}PMe_3$ (**31**)	1.594	116.7	—	147.6	38

[a] All distances are in Å and angles are in degrees.

B. Electron Diffraction Structures

Ebsworth and coworkers have determined the structures of trimethyl(methylene)phosphorane (**29**)[60], trimethyl(silylmethylene)phosphorane (**30**)[61] and hexamethylcarbodiphosphorane (**31**)[38] in the gas phase using electron diffraction techniques. A summary of some of their results is given in Table 6 and schematic drawings of these structures are given in Figure 8.

Direct comparison of the electron diffraction results with the crystal structures is impossible, since no compound has been examined in both phases. Nevertheless, the P=C bond lengths for **29** and **30** are well within the range of the P=C distances found for the non-stabilized ylides. The P=C distance in **31** agrees well with the distances of similar P=C=P ylides discussed in Section II.A.3. These structures all display a tetrahedral phosphorus atom. Ebsworth and coworkers assumed a planar ylidic carbon in solving the structures of **29** and **30**, since they had insufficient data to determine whether the carbon is pyramidal. The P=C=P angle in **31** was found to be 147.6°, but a model involving free rotation of the PMe_3 groups fit the data effectively. It was argued that the molecule may in fact be linear with a very low P=C=P bending frequency.

C. Theoretical Structures

The earliest theoretical studies of phosphonium ylides were EHT studies of methylenephosphorane[62] and cyclopropylidenephosphorane[63] by Hoffmann and coworkers.

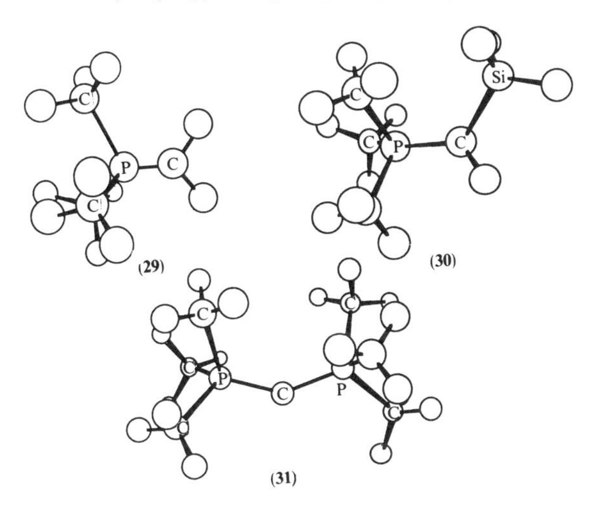

FIGURE 8. Electron diffraction structures of **29–31**

These studies suggested a planar ylidic carbon and definite participation of the P d orbitals in the description of the HOMO. The earliest *ab initio* study of methylenephosphorane found a very small rotational barrier [0.003 kcal mol^{-1} (1 kcal = 4.184 kJ)] about the P=C double bond[64]. These three points, the degree of pyramidalization at the ylidic carbon, the role of the P d orbitals and the P=C rotational barrier, remain the focal points of all theoretical studies of the ylides. A summary of the theoretical structures determined using *ab initio* techniques is given in Table 7.

1. *Methylenephosphorane* (**32**)

Since methylenephosphorane is the smallest phosphonium ylide, it is not surprising that it has attracted the most attention from theoreticians. The structure of **32** has been optimized at a variety of calculational levels with many different basis sets. The P=C bond distance is dependent on the quality of the basis set. Calculations using a non-polarized basis set, such as 3–21G, give a P=C bond length that is too long (1.728 Å[65]). Inclusion of polarization functions[66,69] (usually d orbitals on P at least) reduces the bond length to the range found in the crystal or electron diffraction studies. The addition of diffuse functions and/or optimization at correlated levels causes little change in the P=C distance.

Early calculations on **32** suggested that the ylidic carbon was planar or nearly so[62,63,68,69]. However, complete optimization using large basis sets has consistently supported a pyramidal ylidic carbon. The value of δ is about 22–25° at the HF level, with the inclusion of electron correlation increasing the value by a few degrees. The HF/6–31G* structure is shown in Figure 9. One should keep in mind that these calculations also show that the energetic consequence of pyramidalizing the ylidic carbon is small.

A number of studies of **32** have addressed the role of the phosphorus d orbitals. Early descriptions of the bonding in ylides suggested that the P=C double bond was formed by dπ–pπ interaction. The EHT calculations support this concept in that the d orbitals are used in describing the HOMO. However, the *ab initio* calculations argue against the presence of dπ–pπ bonding . Absar and Van Wazer[64] noted that the P—C π bond was present even when d orbitals were excluded from the basis set. Lishka's[70] study demonstrated that the HOMO is primarily carbanionic and that d orbitals do not affect the shape of the HOMO. Eades *et al.*[72] have shown that whereas the Mulliken population of the d orbitals is 0.37 in **32**, the population of the d orbitals in methylphosphine is 0.19, indicating only a small increase in d orbital participation in the ylide. Streitwieser *et al.*[65] examined the integrated charge distribution in **32** and how it changes with basis set modification. With the addition of d orbitals to first P and then C, the charge on carbon increases. The charge on the methylene group is − 1.165. Streitwieser *et al.* argued that the short P=C distance does not demand a double bond; rather, the electrostatic attraction of the adjacent carbanion and phosphonium is sufficient to cause a short bond. All of these studies suggest that phosphorus d orbitals serve only to polarize the charge distribution and do not participate in any dπ–pπ bonding. Nevertheless, the inclusion of polarization functions in the basis set is necessary in order to obtain reasonable structures.

The rotational barrier about the P=C bond is very small. Absar and Van Wazer[64] first estimated the barrier as 0.003 kcal mol^{-1}. The barrier is 0.13 kcal mol^{-1} at HF/DZ + P[70].

2. *Trimethyl(methylene)phosphorane* (**29**) *and trimethylphosphonium isopropylide* (**33**)

Using **32** as a model for phosphonium ylides is problematic since it is completely unsubstituted. Comparisons of the theoretical structure with the X-ray structures are not only complicated by the difference in phases, but also by the dramatic steric differences due to the substituents. Compounds **29** and **33** have methyl substituents instead of hydrogen

TABLE 7. Structural data from *ab initio* calculations[a]

Compound	Level[b]	P=C	R—P=C	X—C—X	Sum[c]	δ	Ref.
$H_3P=CH_2$ (**32**)	HF/3–21G	1.728	128.7	119.8		25.4	65
	HF/3–21G(*)	1.646				0	66
	HF/3–21G*	1.649	113.5, 113.5, 126.6	118.2	358.8	10.6	67
	HF/3–21 + G*	1.677	128.2	117.5		24.2	65
	HF/pseudopot	1.682		118.5	359.9	3.7	68
	HF/DZ + P(P)	1.64		118.5	359.9	1.5	69
	HF/DZ + O + d(C)	1.672	118	121	359	10	70
	HF/DZ + P + d(C)	1.675	112.1, 112.1, 127.4,	117.4	354.2	23.7	71
	HF/DZ + P + d(C)	1.668	117.7, 117.7, 118.6	120.4	358.8	10.8	72
	HF/6–31G*	1.667	112.2, 112.2, 128.4	117.3	353.6	25.1	73
	MP2/6–31G*	1.674	111.5, 111.5, 130.6	116.7	350.4	30.8	73
	HF/6–31 + G*	1.672	112.2, 112.2, 127.5	117.5	354.4	23.3	74
	MP2/6–31 + G*	1.680	111.5, 115.5, 129.6	116.8	351.0	29.6	74
	HF/6–31 + G**	1.672	112.3, 112.3, 127.2,	117.9	354.6	22.9	74
	MP2/6–31 + G**	1.679	111.5, 111.5, 129.4	117.0	350.6	30.3	74
$(CH_3)_3P=CH_2$ (**29**)	HF/3–21G(*)	1.651					66
	HF/6–31G*	1.671	111.1, 111.1, 122.8	116.8	353.7	24.6	74

(*continued*)

TABLE 7. (*continued*)

Compound	Level[b]	P=C	R—P=C	X—C—X	Sum[c]	δ	Ref.
$H_3P{=}C(CH_3)_2$	HF/6–31G*	1.669	111.2, 111.2, 130.0	116.5	356.5	18.2	74
$(CH_3)_3P{=}C(CH_3)_2$ (**33**)	HF/6–31G*	1.676	112.2, 112.2, 123.1	114.2	356.4	18.0	74
$H_3P{=}C(CH_2)_2$ (**34**)	HF/DZ + P	1.682	110.9, 110.9, 129.0	61.1	316.3	44.9	75
	HF/6–31G*	1.672	111.2, 111.2, 129.2	60.5	320.8	41.4	74
$(CH_3)_3P{=}C(CH_2)_2$	HF/6–31G*	1.683	110.0, 110.0, 123.0	60.3	315.5	45.1	74
$H_3P{=}CHF$	HF/DZ + P + d(C)	1.723	106.4, 109.6, 131.7	109.7	338.5		71
$H_3P{=}CF_2$	HF/DZ + P + d(C)	3.54	94.1, 94.1, 165.1	104.8	313.2		71
$H_3P{=}C(CF_3)_2$	HF/DZ + P + d(C)	1.707	112.4, 112.4, 118.9	118.2	359.8	0	71
$F_3P{=}CH_2$	HF/3–21G(*)	1.590					66
$H_3P{=}C{=}PH_3$	HF/STO–3G	1.63[d]	109.5[d]	113[e]			76
	HF/6–31G*	1.602	116.2, 120.3, 120.3	154.2[e]			

[a] All distances are in Å and all angles are in degrees.
[b] The geometry optimization was performed at the level designated as (level)/(basis set), so HF/6–31G* indicates a Hartree–Fock self-consistent field calculation using the 6–31G* basis set. A basis set designated with a P or d indicates the addition of polarization or diffuse functions, respectively.
[c] Sum of the angles about the ylidic carbon.
[d] Fixed.
[e] P=C=P angle.

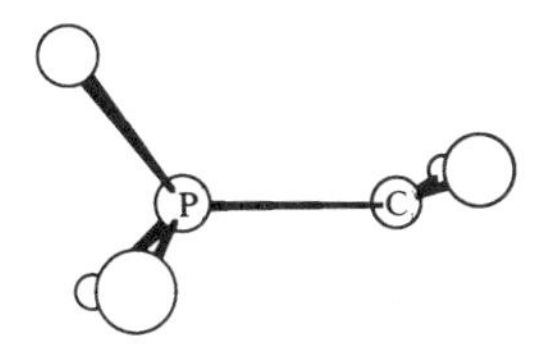

FIGURE 9. Calculated structure of **32**

and reflect some effects of increased steric demands. Comparing the HF/6–31G* geometry of **32** and **29**, one sees that methyl substitution on P causes very little change in the P=C bond length or the pyramidalization of the ylidic carbon. The angles at P do change to accommodate the larger methyl group, but on the whole one might expect that steric bulk on P will not severely affect the ylidic portion of the molecule. In moving from **29** to **33** by replacing the methylene hydrogens with methyl groups, again only small changes are seen. The largest change is that δ diminishes by 6.6°, but the carbon is still pyramidal.

Comparing the electron diffraction structure of **29** with the calculated structure reveals a few major differences. The experimental P=C distance is 0.031 Å shorter than the calculated structure. The model used in the experiment assumes all C—P=C angles are identical. This angle is found to be midway between the calculated angle. The experimental model also assumes a planar ylidic carbon whereas the calculation indicates that it is pyramidal. The difference in the bond lengths is probably due to the restrictive model used to solve the electron diffraction data.

3. *Phosphonium cyclopropylide* (**34**)

The geometry of phosphonium cyclopropylide (**34**) was examined by Vincent *et al.*[75] at the HF/DZ + P level in order to understand better the unusually bent structure of **6**[13]. The calculated structure is shown in Figure 10. The ylidic carbon is pyramidal with $\delta = 44.9°$. The cyclopropyl ring is bent toward one of the hydrogens on phosphorus, with this unique hydrogen bisecting the cyclopropyl ring. This hydrogen is further from P than the other two and also forms a larger H—P=C angle than the other hydrogen. The calculated structure is in excellent agreement with the experimental structure. Mulliken population analysis gives the atomic charges of P and C as +0.542 and −0.512, respectively. The HOMO of **34** is primarily formed of the C 2p orbital. For the carbon portion of the molecules, the optimized geometries[74] of $(CH_3)P{=}C(CH_2)_2$ and **34** are nearly identical,

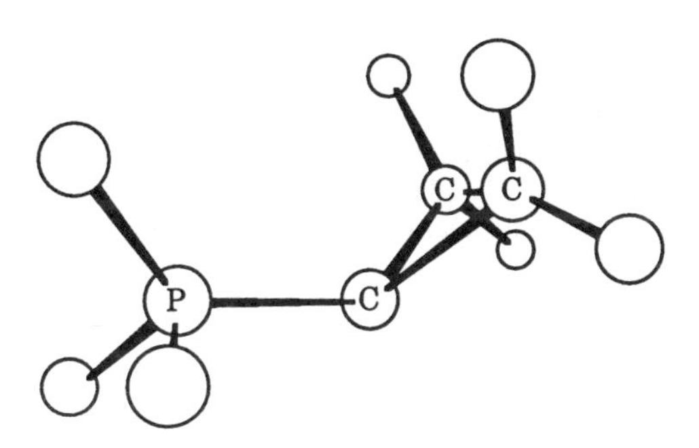

FIGURE 10. Calculated structure of **34**

indicating that large substituents on P are not inducing the non-planarity of the carbanion. Vincent *et al.*[75] also noted that although the ground-state structure is definitely bent, the barrier to inversion at carbon is only 6.3 kcal mol^{-1} and the rotational barrier about the P=C bond is only 5.8 kcal mol^{-1}. They concluded that the potential energy suface is very flat with respect to non-planar distortions at the ylidic carbon.

4. Fluorine-substituted phosphonium ylides

Dixon and Smart[71] examined a series of fluorine-substituted phosphonium ylides. Fluorine stabilizes carbanions via an inductive effect, favouring pyramidal carbanions. Dixon and Smart argued that the first substitution of F on the ylidic carbon ($H_3P{=}CHF$) causes the carbon to become very pyramidal (the sum of the angles at C is 338.5°), which reduces the potential overlap of the anionic orbital with any P orbitals. The leads to the long P—C distance of 1.723 Å. The second F substitution ($H_3P{=}CF_2$) actually breaks the P—C bond and the system is best described as a weak interaction of phosphine with $:CF_2$. Trifluoromethyl groups act to stabilize the anions via hyperconjugation. This leads to a planar ylidic carbon in $H_3P{=}C(CF_3)_2$.

III. SPECTROSCOPIC PROPERTIES OF PHOSPHONIUM YLIDES

A wealth of spectroscopic data for phosphonium ylides is available in the literature. We shall restrict our discussion here to results which directly pertain to the nature of the bonding in the ylides.

A. EPR Spectroscopy

In the mid-1960s, Lucken and Mazeline[77,78] reported two important EPR studies that shed a great deal of light on the nature of the P=C bond in ylides. While ylides themselves have no EPR signal, Lucken and Mazeline instead prepared two phosphorane radical cations by irradiating the appropriate crystalline precursor phosphonium salts, shown in Scheme 1.

$$Ph_3PCH_2COOH{\cdot}Cl \xrightarrow{h\nu} Ph_3\overset{+}{P}{-}\dot{C}HCOOH \quad \mathbf{(35)}$$

$$Ph_3PCH_3{\cdot}Cl \xrightarrow{h\nu_2} Ph_3\overset{+}{P}{-}\dot{C}H_2 \quad \mathbf{(36)}$$

SCHEME 1

The EPR spectra[77] of **35** give ^1H and ^{31}P hyperfine coupling constants and g values that indicate that the radical is delocalized into the carboxyl group, but not to the phosphorus atom. The ylidic carbon is nearly planar, with the radical in a 2p orbital. Since the phosphonium group is competing against the carboxyl group for delocalizing the radical, **35** may not fairly represent the ability of the P d orbital to participate. The EPR[78] of **36**, which has no conjugating groups on carbon, however, also indicates that little radical delocalization occurs to P.

B. UV and IR Spectroscopy

Table 8 presents the λ_{max} values for a series of stabilized[79] and non-stabilized[80] ylides. For the non-stabilized ylides, the reference compound is Ph_3PCH_2 (**4**), which has its λ_{max} at

TABLE 8. λ_{max} for phosphonium ylides

Compound	λ_{max} (nm)	Compound	λ_{max} (nm)
$Ph_3P{=}CH_2$(**11**)	341	BiphC=PPh_3^a (**37**)	382
$Ph_3P{=}CHMe$	374	BiphC=CBiph (**38**)	457
$Ph_3P{=}CHEt$	378	$BiphCH^-$ (**39**)	415–457
$Ph_3P{=}CH(CH_2)_4Me$	377	(BiphC=CH_2)C=PPh_3 (**40**)	574
$Ph_3P{=}CMe_2$	386	(BiphC=$CH)_2$C=CBiph (**41**)	443
$Ph_3P{=}(CH_2)_5$	391	(BiphC=$CH)_2CH^-$ (**42**)	635

[a]Biph =

341 nm. Introduction of the first alkyl group on to the ylidic carbon increases λ_{max} by 33 nm. The size of the alkyl chain has little effect. The second alkyl substituent further increases λ_{max} by 12 nm. Grim and Ambrus[80] interpreted this shift in terms of the electron-donating alkyl groups raising the π orbital, which is primarily the carbanion. The transition is either $\pi \rightarrow \pi^*$ or $\pi \rightarrow d$, with the higher energy orbital dominated by phosphorus atomic orbitals.

Of particular note for the stabilized ylides is that λ_{max} for the ylides is much closer to the value for the carbanion than for the analogous hydrocarbon system (compare **39** with **37** and **38** or **42** with **40** and **41**). The small shift brought on by the phosphonium group suggests that little $d\pi$–$p\pi$ bonding is present and that the ylidic nature dominates the description of the P—C interaction[79].

Iwata *et al.*[81] examined the UV spectra of Pr_3^n P=$C(CH)_4$, which is a stabilized ylide. They found three bands which are strongly dependent on the solvent. In hexane the bands appear at 4.63, 5–6 and 6.33 eV, whereas in methanol they appear at 4.88, 5–6 and 6.26 eV. The strong shift of the first band to higher energy on switching to a more polar solvent suggests that this band is associated with a transition from a more polar state to a less polar state. Iwata *et al.* argued, with the assistance of some theoretical work, that the ground state is predominantly P^+—C^- and the excited state corresponds to the ylene P=C.

A classical IR study of phosphonium ylides was carried out by Luttke and Wilhelm[82]. The P—C stretching frequencies in Ph_3PCH_2 and $Ph_3PCHCOPh$ are at 899 and 887 cm^{-1}, respectively. The P—C stretching frequencies is of their corresponding phosphonium salts are at 787 and 823 cm^{-1}, respectively. The ylides have only slightly larger P—C force constants than the salts. It was suggested that the bond order in the ylides is only 1.3. It is interesting to note that the P—C frequency of the stabilized ylide is slightly smaller than for the non-stabilized ylide.

C. ESCA and Photoelectron Spectroscopy

Ostoja Starzewski and Bock[83] reported the photoelectron spectra of an extensive series of phosphonium ylides, focusing on the substituent effects. Their lowest ionization potential (*IP*) is associated with the carbanion orbital. Their main results are given in Table 9. A number of important trends can be identified and interpreted. Replacement of the methyl groups on P with phenyl groups reduces the *IP* of the carbanion. The phenyl group is able to stabilize the P^+ charge, which reduces the ability of the phosphonium to stabilize the

TABLE 9. Vertical ionization potentials of phosphonium ylides

Compound	*IP* (eV)	Compound	*IP* (eV)
$Me_3P{=}CH_2$ (**29**)	6.81	$Me_3P{=}CHC(Me){=}CH_2$	6.20
$PhMe_2P{=}CH_2$	6.85	$Me_3P{=}CHCH{=}CH(Me)$	6.02
$Ph_2MeP{=}CH_2$	6.70	$Ph_3P{=}CHCH{=}CH(Me)$	5.95
$Ph_3P{=}CH_2$ (**4**)	6.62	$Me_3P{=}CHPh$	6.19
$Ph_3P{=}CHMe$	6.15	$Ph_3P{=}CHPh$	6.01
$Ph_3P{=}CMe_2$	6.04	$Ph_3P{=}C(CH)_4$ (**18**)	6.91
$Me_3P{=}CHCH{=}CH_2$	6.20		

carbanion. Methyl substitution at the ylidic carbon substantially lowers the *IP*, interpreted as hyperconjugative destabilization. Phenyl substitution at the ylidic carbon lowers the *IP*. The interaction of the carbanion with the phenyl ring lowers the energy of the phenyl group HOMO, while raising the energy of the carbanion[84]. Substitution of a vinyl group on to the ylidic carbanion induces a sharp decrease in the *IP*. The formation of an allyl anion system produces a low-energy allyl π orbital and a high-energy formally non-bonding orbital, which is dominated by the carbanion. These arguments are entirely consistent with a very polar $P^+{-}C^-$ bond.

Perry *et al.*[85] reported the core binding energies of some simple organophosphorus systems. The P core binding energies of phosphine and trimethylphosphine are 136.87 and 135.76 eV, respectively. The related values for **29** and $Me_3P{=}O$ are 137.03 and 137.63 eV, respectively. The phosphorus atom in the ylides must carry a large positive charge relative to the phosphines. The C binding energies of the methyl groups in these systems are slightly more than 290 eV. In contrast, the binding energy for the methylene group in **29** is only 287.83 eV, a substantial reduction due to the large negative charge borne by the ylidic carbon.

D. NMR Spectroscopy

The number of phosphonium ylides that have been characterized by NMR techniques is too large to list here or to discuss in detail. Therefore, we have restricted ourselves to some simple ylides that are directly related to the ylides discussed in the above sections. Also, we shall discuss only ^{13}C and ^{31}P NMR, since these techniques directly probe the nature of the P=C bond. A summary of the NMR data is given in Table 10.

Particularly striking in these data are the very small (upfield) ^{13}C chemical shifts of the ylidic carbon. For the non-stabilized ylides, the values range from about -14 to 9 ppm, whereas for the stabilized ylides the range is slightly more downfield, 28–80 ppm, with most clustered at the upfield end of this range. These values indicate a very shielded carbon, attributed to a large negative charge residing on the carbon[86–88]. The difference in the chemical shift for the ylidic carbon of **18** and **43** is 25 ppm, with the latter more shielded. Gray[86] has suggested that the anion is more localized in **43** than in **18**. The chemical shifts for the carbon in P=C=P ylides are also far upfield.

The phosphorus–carbon coupling constant (J_{P-C}) has been used to determine the hybridization at carbon. The very large values of this coupling constant for most ylides indicates that the carbon is sp^2 hybridized. The stabilized ylides have J_{P-C} ranging from 112 to 131 Hz. Since the ylidic carbon in this species is nearly (if not exactly) planar, these values can be used as a benchmark for comparing the pyramidal nature of other ylides. Schlosser *et al.*[89] argued that the coupling constant of **4** (about 101 Hz) is too low for a planar carbanion. They suggested that the carbanion is a very flat pyramid, which is in excellent agreement with the recent X-ray structure[11].

TABLE 10. NMR data for selected phosphonium ylides

Compound	$\delta(^{13}C)$ (ppm)	$\delta(^{31}P)$ (ppm)	J_{P-C} (Hz)	Ref.
$Me_3P{=}CH_2$ (**29**)			90	90
	−2.3		90.5	91
$Et_3P{=}CH_2$	−14.2		86.8	91
$Et_3P{=}CHMe$	−7.9		113.2	91
$Ph_3P{=}CH_2$ (**4**)	−6.4		100.3	92
	−5.4	18.4	101	89
$Ph_3P{=}CHMe$ (**50**)	3.2	15.0	110.7	88
	1.7		118.8	92
$Ph_3P{=}CMe_2$ (**51**)	9.0	9.8	121.5	88
	7.6		124.7	92
$Ph_3P{=}C(CH_2)_2$ (**6**)	4.3	16.7	132.8	88
	0.13	15.6	3.9	13
$(c\text{-}C_3H_5)_3P{=}C(CH_2)_2$ (**44**)	−19.44	20.9	18.3	93
$Ph_3P{=}C(CH_2)_3$ (**7**)	14.6	16.5	77.3	88
	14.72		83.0	14
$Ph_3P{=}CHPh$ (**49**)	28.0	7.0	128.0	88
$Ph_3P{=}CHC{=}CH_2$ (**45**)	28.7	10.7	131.4	88
	28.2		125.2	89
$Ph_3P{=}CHCOMe$ (**46**)	51.3	14.6	108.0	88
$Ph_3P{=}CHCOPh$ (**47**)	50.4	16.8	111.7	88
$Ph_3P{=}C(Me)COOEt$ (**48**)	31.7		120.7	86
$Ph_3P{=}C_5H_4$ (cyclopentadienylide) (**18**)	78.3		113.1	86
$Ph_3P{=}$fluorenylidene (**43**)	53.3		128.7	86
$Ph_3P{=}C{=}PPh_3$ (**21**)		4.3		94
$Me_3P{=}C{=}PMe_3$	10.8	−29.6	78.1	95
$Me_3P{=}C{=}PPh_3$		−10.1, −21.31		43
$Ph_2P{=}$ (cyclic, ring) PPh_2	−3.4	−1.7	75.7	41

The ylides having the ylidic carbon in a small ring (**6** and **44**) have very small values of J_{P-C}. (Schmidbaur *et al.*[13] discounted the measurements of Albright *et al.*[88] as erroneous.) These values indicate that the carbanion is highly pyramidal and decidedly sp^3-like. Even the less strained ylide **7** has a small value for J_{P-C}, which Schmidbaur *et al.* contended indicates a non-planar geometry about the ylidic carbanion. The J_{P-C} values corroborate well the X-ray structures of **6** and **7**; the former is much more pyramidal than the latter (see Table 1).

Schlosser *et al.*[89] and Albright *et al.*[88] also attempted to determine the degree of delocalization of the anion in stabilized ylides. Schlosser *et al.* examined **45**, attempting to determine the participation of **45a** vs **45b**. If **45a** dominates, one would expect $\delta_{C(\alpha)}$ would be about -5 ppm and $\delta_{C(\gamma)}$ at a normal olefinic position of ca 120 ppm. If **45b** dominates, $\delta_{C(\alpha)}$ would be at ca 120 ppm and $\delta_{C(\gamma)}$ would be appreciably below this value. The experimental values are $\delta_{C(\alpha)}$ 28.2 ppm and $\delta_{C(\gamma)}$ 90.1 ppm, i.e. 3:1 ratio for the participation of **45a**:**45b**. In a similar vein, for carbonyl stabilized ylides, the chemical shift of the ylidic carbon should be about -5 ppm if the ylide resonance form dominates or about 100–120 ppm if the 'betaine' form dominates. The values for **46**–**48** are approximately midway between the extremes; both resonance structures play a strong role in describing these ylides.

$$\mathrm{Ph_3\overset{+}{P}\overset{-}{C}HCH{=}CH_2} \longleftrightarrow \mathrm{Ph_3\overset{+}{P}CH{=}CH\overset{-}{C}H_2}$$

(**45a**) (**45b**)

Albright's[88] approach was to compare the chemical shifts of the ylides with those of their analogous phosphonium salts. In comparing the salt with **45**, $\delta_{C(\alpha)}$ moves downfield by only 1.1 ppm whereas $\delta_{C(\gamma)}$ moves upfield by 33.6 ppm. The *ortho* and *para* carbons of **49** appear 10.6 and 14.5 ppm upfield from their respective positions in the salt. Clearly, negative charge on the ylidic carbon is delicalized into the π system. The ^{31}P chemical shifts are downfield of phosphonium salts. The ^{31}P chemical shifts of the salts of **50** and **51** are 25.5 and 31.3 ppm, respectively. Schmidbaur *et al.*[87] have shown that δ_P increases in the series Me_3PO, Me_4P^+ and Me_3PCH_2. Phosphorus in the ylides carries a fractional positive charge.

IV. BONDING IN THE PHOSPHONIUM YLIDES

The structural analyses in this chapter present a fairly unified description of the bonding in phosphonium ylides. The bonding in these compounds can best be described by the 'ylide' resonance form **A** with a small contribution from the 'ylene' form **B**. The evidence for this conclusion is summarized below.

1. The X-ray crystal structures of the ylides all indicate a slightly distorted tetrahedral geometry about phosphorus, precluding the dsp^3 hybridization necessary for the ylene. The geometry about carbon is dependent on the substituents. For non-stabilized ylides, the carbon is pyramidal and its hybridization lies between sp^2 and sp^3. Especially important is the detailed analysis of **4**, showing that this prototypical ylide is pyramidal. Ring strain leads to a very pyramidal carbon (see **6**) that is nearly pure sp^3. Stabilized ylides generally have a planar, or nearly planar, ylidic carbon. The bond distances in the conjugating group support the strong participation of resonance structures **16b** and **16c** (and not the ylenic **16a**) in describing the electronic structure of these ylides.
2. All theoretical studies that include polarization function in the basis set of small unsubstituted ylides find a pyramidal ylidic carbon and a distorted tetrahedral phosphorus atom. Population analyses conclude that the charge distribution is $P^+—C^-$.
3. EPR studies of two radical cations of ylides show little delocalization of the radical on to phosphorus, discounting the ylene structure.
4. UV and ESCA studies support the notion that the HOMO of the ylides is primarily a C *p* orbital, i.e. a nearly pure carbanion.
5. The P—C stretching frequency suggests a P—C bond order only slightly above one.
6. The very upfield ^{13}C chemical shifts for the ylidic carbon are indicative of carbanion character. The J_{P-C} coupling constants are large and imply that carbon is sp^2 hybridized, a particularly for the stabilized ylides. The values of J_{P-C} for the

non-stabilized ylides are smaller and suggest that the carbon is a shallow pyramid. The ^{13}C chemical shifts of the stabilized ylides support the dominant participation of **16b** and **16c** in describing these systems.

While these systems are dominated by the ylidic form, the phosphorus atom does aid in stabilizing the carbanion. It is well known that the trialkylphosphonium ylides are more reactive than the triarylphosphonium ylides. The aryl substituents remove some of the positive charge from phosphorus (as seen in the ESCA studies listed in Table 9), overall stabilizing the molecule.

The mechanism by which the phosphorus onium centre stabilizes the carbanion is not universally accepted. The theoretical studies suggest that phosphorus simply acts to polarize the carbanion, the extreme position[65] being that the adjacent positive and negative centres are sufficient to produce a short P—C bond. This view discounts the participation of phosphorus d orbitals in covalent bonding with the carbon p orbitals, i.e. no $d\pi-p\pi$ bonding. Johnson[1] succinctly noted a major problem in attempting to determine the role of the d orbitals: '$\cdots$ it appears virtually impossible to *prove* the d orbitals are involved in π bonding' (his italics). Nevertheless, the theoretical studies show that P d orbitals do not induce the carbanion to become planar and that the d orbital population is small.

Schier and Schmidbaur[93] performed a clever experiment that addressed part of this question: does the orientation of the carbanion relative to the phosphorus atom play any role? Scheme 2 shows two syntheses of ylides involving cyclopropyl substituents. In the first reaction, since the pK_a of cyclopropane is considerably below that of propane, the expected product is the cyclopropylide. However, the isopropylide is the only recovered product. The second reaction also demonstrates the avoidance of the cyclopropylide product. The cyclopropylide possesses a very pyramidal carbanion that is directed away from phosphorus, allowing for minimal orbital overlap. The isopropylide is much less pyramidal and phosphorus can better assist in stabilizing the carbanion. While this stabilization does not require explicit orbital overlap (the electrostatic interaction of the carbanion with the onium is expected to be smaller in the cyclopropylide since it is directed away from P), it does suggest that some orbital interactions are involved. Hence, although the ylene contribution is small, it is unlikely that the ylene contribution is nil.

In conclusion, the dominant resonance structures for the ylides are as follows: for non-stabilized ylides, **A**; for stabilized ylides, **16b** and **16c**; for P=C=P ylides, **22** (although further study on this class is advised to determine if the gas-phase structure may in fact be linear); and for P=C=C ylides, **28b** and **28c**.

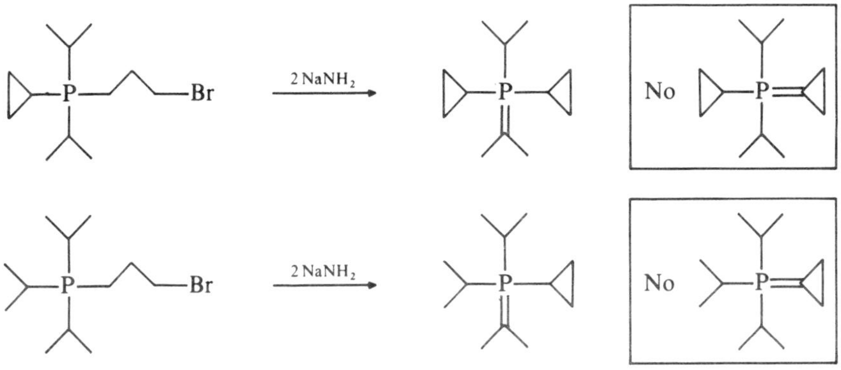

P Br 2 NaNH$_2$ P No P

P Br 2 NaNH$_2$ P No P

SCHEME 2

V. ACKNOWLEDGEMENTS

The authors thank Joseph Bachrach and Aaron Goldfarb for assistance in preparing this chapter.

VI. REFERENCES

1. A. J. Johnson, *Ylid Chemistry*, Academic Press, New York, 1966.
2. H. Schmidbaur, *Angew. Chem., Int. Ed. Engl.*, **22**, 907 (1983).
3. W. J. Hehre, L. Radom, R. v. R. Schleyer and J. A. Pople, *Ab Initio Molecular Orbital Theory*, Wiley, New York, 1986.
4. A. Michaelis and H. V. Gimborn, *Ber. Dtsch. Chem. Ges.*, **27**, 272 (1884).
5. A. Michaelis and E. Kohler, *Ber. Dtsch. Chem. Ges.*, **32**, 1566 (1899).
6. G. Wittig, and G. Geissler, *Justus Leibigs Ann. Chem.* **580**, 44 (1953).
7. F. S. Stephens, *J. Chem. Soc.*, 5658 (1965).
8. F. S. Stephens, *J. Chem. Soc.*, 5640 (1965).
9. T. C. W. Mak and J. Trotter, *Acta Crystallogr.*, **18**, 81 (1965).
10. J. C. J. Bart, *J. Chem. Soc. B*, 350 (1969).
11. H. Schmidbaur, J. Jeong, A. Schier, W. Graf, D. L. Wilkinson and G. Muller, *New J. Chem.*, **13**, 341 (1989).
12. H. Schmidbaur, A. Schier, C. M. F. Frazao and G. Muller, *J. Am. Chem. Soc.*, **108**, 976 (1986).
13. H. Schmidbaur, A. Schier, B. Milewski-Mahrla and U. Schubert, *Chem. Ber.*, **115**, 722 (1982).
14. H. Schmidbaur, A. Schier and D. Neugebauer, *Chem. Ber.*, **116**, 2173 (1983).
15. M. A. Howells, R. D. Howells, N. C. Baenziger and D. J. Burton, *J. Am. Chem. Soc.*, **95**, 5366 (1973).
16. H. Schmidbaur, U. Deschler and B. Milewski-Mahrla, *Chem. Ber.*, **116** 1393 (1983).
17. H. Schmidbaur, P. Nusstein and G. Muller, *Z. Naturforsch., Teil B*, **39**, 1456 (1984).
18. H. Schmidbaur, B. Milewski-Mahrla, G. Muller and C. Kruger, *Organometallics*, **3**, 38 (1984).
19. J. Grobe, D. Le Van and J. Nientiedt, *New J. Chem.*, **13**, 363 (1989).
20. H. Schmidbaur, C. Zybill, C. Kruger and H. Jurgen-Kraus, *Chem. Ber.*, **116**, 1955 (1983).
21. A. J. Speziale and K. W. Ratts, *J. Am. Chem. Soc.*, **87**, 5603 (1965).
22. A. F. Cameron, F. D. Duncanson, A. A. Freer, V. W. Armstrong and R. Ramage, *J. Chem. Soc., Perkin Trans. 2*, 1030 (1975).
23. H. Trabelsi, E. Rouvier, A. Cambon and S. Jaulmes, *J. Fluorine Chem.*, **39**, 1 (1988).
24. J. Buckle, P. G. Harrison, T. J. King and J. A. Richards, *J. Chem. Soc., Chem. Commun.*, 1104 (1972).
25. U. Lingner and H. Burzlaff, *Acta Crystallogr., Sect. B*, **30**, 1715 (1974).
26. G. O. Nevstad, K. Maartmann-Moe, C. Romming and J. Songstad, *Acta Chem-Scand., Ser. A*, **39**, 523 (1985).
27. H. L. Ammon, G. L. Wheeler and P. H. Watts, Jr, *J. Am. Chem. Soc.*, **95**, 6158 (1973).
28. A. S. Bailey, J. M. Peach, T. S. Cameron and C. K. Prout, *J. Chem. Soc. C*, 2295 (1969).
29. O. Kennard, W. D. S. Motherwell and J. C. Coppola, *J. Chem. Soc. C*, 2461 (1971).
30. T. S. Cameron and C. K. Prout, *J. Chem. Soc. C*, 2292 (1969).
31. G. Chioccola and J. J. Daly, *J. Chem. Soc. A*, 568 (1968).
32. F. W. B. Einstein and T. Jones, *Can. J. Chem.*, **60**, 2065 (1982).
33. H. Schmidbaur, R. Pichl and G. Muller, *Chem. Ber.*, **120**, 789 (1987).
34. H. H. Karsch, B. Zimmer-Gasser, D. Neugebauer and U. Schubert, *Angew. Chem., Int. Ed. Engl.*, **18**, 484 (1979).
35. B. Zimmer-Gasser, D. Neugebauer, U. Schubert and H. H. Karsch, *Z. Naturforsch., Teil B*, **34**, 1267 (1979).
36. F. Ramirez, N. B. Desai, B. Hansen and N. McKelvie, *J. Am. Chem. Soc.*, **83**, 3539 (1961).
37. A. T. Vincent and P. J. Wheatley, *J. Chem. Soc., Dalton Trans.* 617 (1972).
38. E. A. V. Ebsworth, T. E. Fraser, D. W. H. Rankin, O. Gasser and H. Schmidbaur, *Chem. Ber.*, **110**, 3508 (1977).
39. G. E. Hardy, J. I. Zink, W. C. Kaska and J. C. Baldwin, *J. Am. Chem. Soc.*, **100**, 8001 (1978).
40. H. Schmidbaur, G. Hasslberger, U. Deschler, U. Schubert, C. Kappenstein and A. Frank, *Angew. Chem., Int. Ed. Engl.*, **18**, 408 (1979).

41. H. Schmidbaur, T. Costa, B. Milewski-Mahrla and U. Schubert, *Angew. Chem., Int. Ed. Engl.*, **19**, 555 (1980).
42. U. Schubert, C. Kappenstein, B. Milewski-Mahrla and H. Schmidbaur, *Chem. Ber.*, **114**, 3070 (1981).
43. H. Schmidbaur, R. Herr and C. E. Zybill, *Chem. Ber.*, **117**, 3374 (1984).
44. R. Appel, U. Baumeister and F. Knoch, *Chem. Ber.*, **116**, 2275 (1983).
45. E. Fluck, B. Neumuller, R. Braun, G. Heckmann, A. Simon and H. Borrmann, *Z. Anorg. Allg. Chem.*, **567**, 23 (1988).
46. P. J. Corroll and D. D. Titus, *J. Chem. Soc., Dalton Trans.* 824 (1977).
47. S. Z. GHoldberg and K. N. Raymond, *Inorg. Chem.*, **12**, 2923 (1973).
48. L. Farnell and L. Radom, *J. Am. Chem. Soc.*, **106**, 25 (1984).
49. L. Farnell and L. Radom, *Chem. Phys. Lett.*, **91**, 373 (1982).
50. G. Markl, H. Sejpka, S. Dietl, B. Nuber and M. L. Zeigler, *Angew. Chem., Int. Ed. Engl.*, **25**, 1003 (1986).
51. C. N. Matthews and G. H. Birum, *Acc. Chem. Res.*, **2**, 373 (1969).
52. C. N. Matthews and G. H. Birum, *Tetrahedron Lett.*, 5707 (1969).
53. G. H. Birum and C. N. Matthews, *Chem. Ind. (London)*, **20**, 653 (1968).
54. G. H. Birum and C. N. Matthews, *J. Org. Chem.*, **32**, 3554 (1967).
55. J. J. Daly and P. J. Wheatley, *J. Chem. Soc., A*, 1703 (1966).
56. J. J. Daly, *J. Chem. Soc., A*, 1913 (1967).
57. H. Burzlaff, E. Wilhelm and H.-J. Bestmann, *Chem. Ber.*, **110**, 3168 (1977).
58. H. Burzlaff, U. Voll and H.-J. Bestmann, *Chem. Ber.*, **107**, 1949 (1974).
59. S. Z. Goldberg, E. N. Duesler and K. N. Raymond, *Inorg. Chem.*, **11**, 1397 (1972).
60. E. A. V. Ebsworth, T. E. Fraser and D. W. H. Rankin, *Chem. Ber.*, **110**, 3494 (1977).
61. E. A. V. Ebsworth, D. W. H. Rankin, B. Zimmer-Gasser and H. Schmidbaur, *Chem. Ber.*, **113**, 1637 (1980).
62. R. Hoffmann, D. B. Boyd and S. Z. Goldberg, *J. Am. Chem. Soc.*, **92**, 3929 (1970).
63. D. B. Boyd and R. Hoffmann, *J. Am. Chem. Soc.*, **93**, 1064 (1971).
64. I. Absar and J. R. Van Wazer, *J. Am. Chem. Soc.*, **94**, 2382 (1972).
65. A. Streitwieser Jr., A. Rajca, R. S. McDowell and R. Glaser, *J. Am. Chem. Soc.*, **109**, 4184 (1987).
66. M. M. Francl, R. C. Pellow and L. C. Allen, *J. Am. Chem. Soc.*, **110**, 3723 (1988).
67. M. T. Nguyen and A. F. Hegarty, *J. Chem. Soc., Perkin, Trans. 2*, 47 (1987).
68. G. Trinquier and J.-P. Malrieu, *J. Am. Chem. Soc.*, **101**, 7169 (1979).
69. A. Strich, *Nouv. J. Chim.*, **3**, 105 (1979).
70. H. Lishka, *J. Am. Chem. Soc.*, **99**, 353 (1977).
71. D. A. Dixon and B. E. Smart, *J. Am. Chem. Soc.*, **108**, 7172 (1986).
72. R. A. Eades, P. G. Gassman and D. A. Dixon, *J. Am. Chem. Soc.*, **103**, 1066 (1981).
73. B. F. Yates, W. J. Bouma and L. Radom, *J. Am. Chem. Soc.*, **109**, 2250 (1987).
74. S. M. Bachrach, *J. Org. Chem.*, **57**, 4367 (1992).
75. M. A. Vincent, H. F. Schaefer, III, A. Schier and H. Schmidbaur, *J. Am. Chem. Soc.*, **105**, 3806 (1983).
76. T. A. Albright, P. Hofmann and A. R. Rossi, *Z. Naturforsch., Teil B*, **35**, 343 (1980).
77. E. A. C. Lucken and C. Mazeline, *J. Chem. Soc., A*, 1074 (1966).
78. E. A. C. Lucken and C. Mazeline, *J. Chem. Soc., A*, 439 (1966).
79. H. Fischer and H. Fischer, *Chem. Ber.*, **99**, 658 (1966)
80. S. O. Grim and J. H. Ambrus, *J. Org. Chem.*, **33**, 2993 (1968).
81. K. Iwata, S. Yoneda and Z. Yoshida, *J. Am. Chem. Soc.*, **93**, 6745 (1971).
82. W. Luttke and K. Wilhelm, *Angew. Chem., Int. Ed. Engl.*, **4**, 875 (1965).
83. K. A. Ostoja Starzewski and H. Bock, *J. Am. Chem. Soc.*, **98**, 8486 (1976).
84. K.-H. A. Ostoja Starzewski, H. Bock and H. tom Dieck, *Angew. Chem., Int. Ed. Engl.*, **14**, 173 (1975).
85. W. B. Perry, T. F. Schaaf and W. L. Jolly, *J. Am. Chem. Soc.*, **97**, 4899 (1975).
86. G. A. Gray, *J. Am. Chem. Soc.*, **95**, 5092 (1973).
87. H. Schmidbaur, W. Buchner and D. Scheutzow, *Chem. Ber.*, **106**, 1251 (1973).
88. T. A. Albright, M. D. Gordon, W. J. Freeman and E. E. Schweizer, *J. Am. Chem. Soc.*, **98**, 6249 (1976).
89. M. Schlosser, T. Jenny and B. Schaub, *Heteratom Chem.*, **1**, 151 (1990).

90. K. Hildenbrand and H. Dreeskamp, *Z. Naturforsch., Teil B*, **28**, 226 (1973).
91. H. Schmidbaur, W. Richter, W. Wolf and F. H. Kohler, *Chem. Ber.*, **108**, 2649 (1975).
92. Z. Liu and M. Schlosser, *Tetrahedron Lett.*, **31**, 5753 (1990).
93. A. Schier and H. Schmidbaur, *Chem. Ber.*, **117**, 2314 (1984).
94. G. H. Birum and C. N. Mathews, *J. Am. Chem. Soc.*, **88**, 4198 (1966).
95. H. Schmidbaur, O. Gasser and M. S. Hussain, *Chem. Ber.*, **110**, 3501 (1977).

CHAPTER **5**

Electrochemistry of ylides, phosphoranes and phosphonium salts

K. S. V. SANTHANAM

Chemical Physics Group, Tata Institute of Fundamental Research, Colaba, Bombay 400 005, India

I. INTRODUCTION

The synthesis of alkenes through the Wittig reaction has generated an impressive understanding of the chemistry of organophosphorus compounds. The generated carbanions stabilized by a phosphoryl moiety can be considered as ylide anions, and the

The chemistry of organophosphorus compounds, Vol. 3
Edited by F. R. Hartley

increase or decrease in their reactivity towards ketones and aldehydes due to this stabilization has invoked a great deal of interest in stereochemical studies[1,2]. Electrochemical investigations on organophosphorus compounds have produced results giving a detailed picture of the reactivity of ylides, phosphoranes and phosphonium salts. Among these three classes of compounds, the last has been investigated in detail in electrochemistry, providing positive support for the route by which ylides are formed and their reactivity[3].

II. PHOSPHONIUM YLIDES

A. Synthesis

The synthesis of ylides with an open–chain isopropyl group has been accomplished through deprotonation of phosphonium cation $[Me_2(CH)_nR_{4-n}P]^+$, where R = cyclopropyl. The pyramidal carbanion geometry of cyclopropylides is less favoured than the planar carbanion geometry of isopropylides[4]. The pioneering reviews by Walker[5–9] on the chemistry of ylides and related compounds cover the general areas of this topic.

Utley and coworkers[10,11] studied the stereochemical course of Wittig-type reactions at a cathode through variations of the electrolyte cations. The reaction sequence is represented as

$$B + 2e \rightleftharpoons B^{2-} \tag{1}$$

$$B^{2-} + Ph_3\overset{+}{P}{-}CH_2{-}R \longrightarrow Ph_3P{=}CH{-}R + BH^- \tag{2}$$

where B is a base which should undergo reduction at a low cathodic potential to give a strong base. The probases of the type fluoren-9-ylidenemethane are useful reagents for this purpose (see Table 1):

R=H, X=CN; R=Br, X=CN; R=H, X=COOEt

The formation of ylides is observed in cases where the probase reduction at the electrode is reversible in cyclic voltammetric recordings, as this provides a control to monitor the

R R NC X

TABLE 1. Cyclic voltammetric peak potentials of phosphorus ylides[a]. Reproduced by permission of the Royal Society of Chemistry from Reference 10

Ylide	$-E_p$ (V)[b]
$PhCH{=}PPh_3$	2.02
$PhCH{=}CHCH{=}PPh_3$	1.89
$Me_2C{:}CHCH_2C(Me){:}CHCH{:}C(Me)CH_2CH{=}PPh_3$	2.02

[a] Working electrode: vitreous carbon cathode. Medium: *N*, *N*-dimethylformamide (dmf) containing 0.1 $M Bu_4NI$. Sweep rate: 300 $mV s^{-1}$.
[b] vs Ag/AgI.

TABLE 2. Cyclic voltammetric features of probases

Probase	E_p^{I} (V)	E_p^{II} (V)
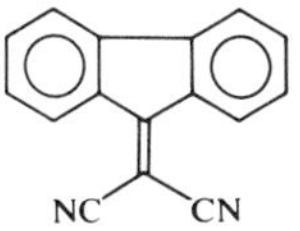 NC, CN	−0.11	−0.88
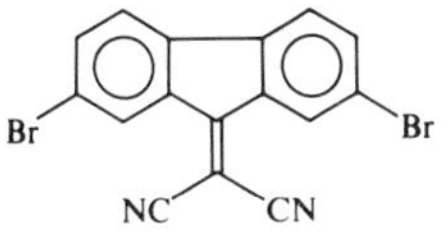 Br, Br, NC, CN	0.13	−0.59
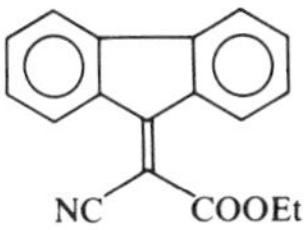 NC, COOEt	−0.21	−0.75

reactions of the base. The regeneration of probase in the reaction would be a route in the continuous conversion of the salt to an ylide.

In situ Wittig synthesis through an electrogenerated base has been carried out in a divided cell at a mercury pool cathode in an inert atmosphere; the potential of the cathode is controlled at the peak value of the corresponding probase as given in Table 2.

The potentials in Table 2 are referred to Ag/AgI in a medium of dmf + Bu_4NI. While both the peaks are reversible or quasi-reversible in the cyclic voltammetric scans, the electrochemical reduction is carried out at the first peak, where an anion is generated. Dianion formation occurs at a more negative potential but is at a lower potential than the phosphonium salt. On controlling the potential of the electrode at the first reduction of the probase and carrying it to 2 F mol^{-1}, the ylide is produced.

The synthesis of phosphonium ylides by reaction of (*E*)-2-benzylideneoxazolidine-4, 5-dione with methylenetriphenylphosphoranes has been reported[12]; these ylide phosphoranes have the structure $PhCH_2CONHCOCOCR{=}PPh_3$ with R = COOMe, COOEt and COPh.

B. Wittig Reaction: Alkene Synthesis

The electrochemical formation of ylides described in reactions 1 and 2 has been utilized for producing alkenes by reaction with aldehydes through reaction 3. The potential of the cathode is controlled at or near the cathodic reduction peak potential of the probase. The yields of the alkenes are given in Table 3.

$$Ph_3P{=}CHR + R'CHO \longrightarrow RCH{=}CHR' + Ph_3PO \qquad (3)$$

The reactive ylides give rise to *cis*-rich mixtures; the moderately reactive ylides give rise to mixtures of *cis* and *trans* isomers. Using this approach, vitamin A acetate has been synthesized by the route shown in reaction 4.

As the phosphonium salt has $E_p = -1.39$ V and the aldehyde component in the reaction has $E_p = -1.36$ V, it was necessary to select probases of the type shown in Table 2 to carry

TABLE 3. Yields of alkenes through electrochemical Wittig reaction[a]. Reproduced by permission of the Royal Society of Chemistry from Reference 10

Probase	Electrolyte	Stilbene[b]		Yield (%)	1,4-Diphenylbutadiene[c]		Yield (%)
		cis (%)	*trans* (%)		*cis* (%)	*trans* (%)	
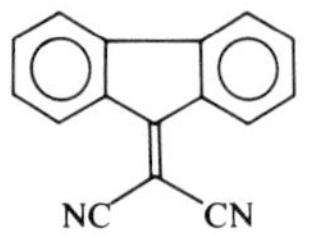 NC CN	Bu_4N^+	39	61	46	7	93	7
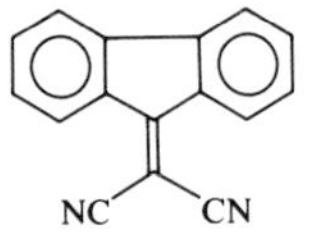 NC CN	Li^+	59	41	74	52	48	54
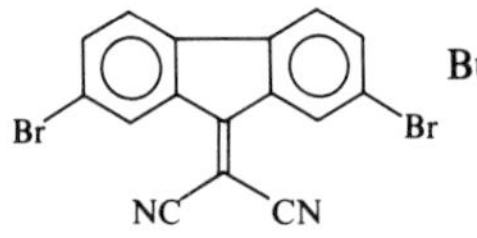 Br Br NC CN	Bu_4N^+	52	48	38	4	96	67
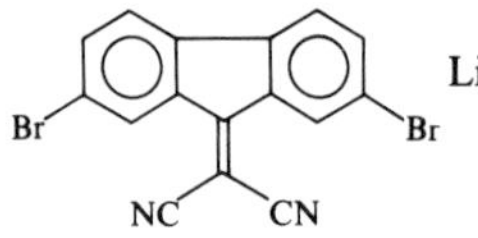 Br Br NC CN	Li^+	50	50	21	15	85	35
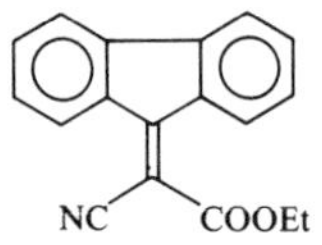 NC COOEt	Bu_4N^+	50	50	68	—	—	—
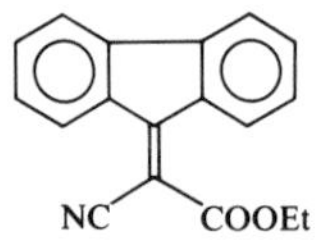 NC COOEt	Li^+	46	54	94	—	—	—

[a] Electrolysis is carried out using a divided cell in the potential range −0.60 to −0.90 V. Medium: *N, N*-dimethylformamide containing the electrolyte at a concentration of 0.1 M.
[b] Using phosphonium salt $PhCH_2P^+ Ph_3NO_3^-$ and benzaldehyde.
[c] Benzaldehyde and cinnamyltriphenylphosphonium nitrate.

out the electrolysis at −0.80 or −0.90 V. The yield of vitamin A acetate is about 30–40% when a Li^+ salt is used as the electrolyte; with a Bu_4N^+ salt as the electrolyte the yield is zero[10]. The stereoselectivity and the effect of cation are discussed with reference to the stabilization of ylide and the pseudo-rotation that is required for the breaking the C—P bond of the phosphetane in an apical position.

The reactivity of phosphorus ylides towards carbonyl compounds such as ketones, esters, carbonates and amides has also been explored[13]. The mechanism shown in

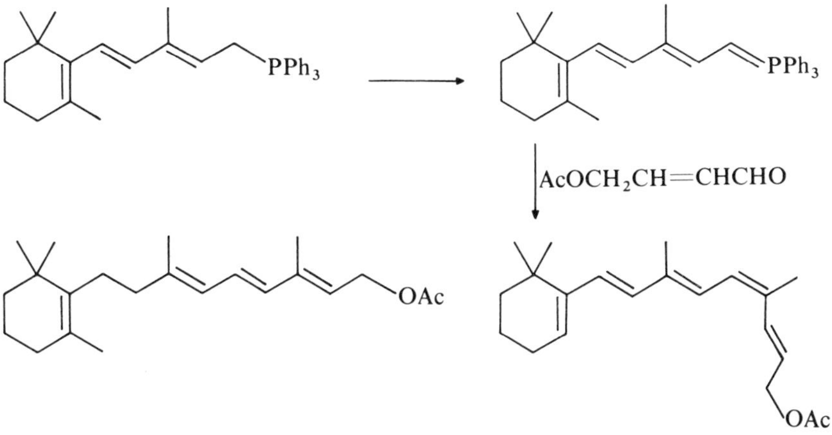

PPh_3 ⟶ PPh_3

$AcOCH_2CH{=}CHCHO$

OAc OAc

(4)

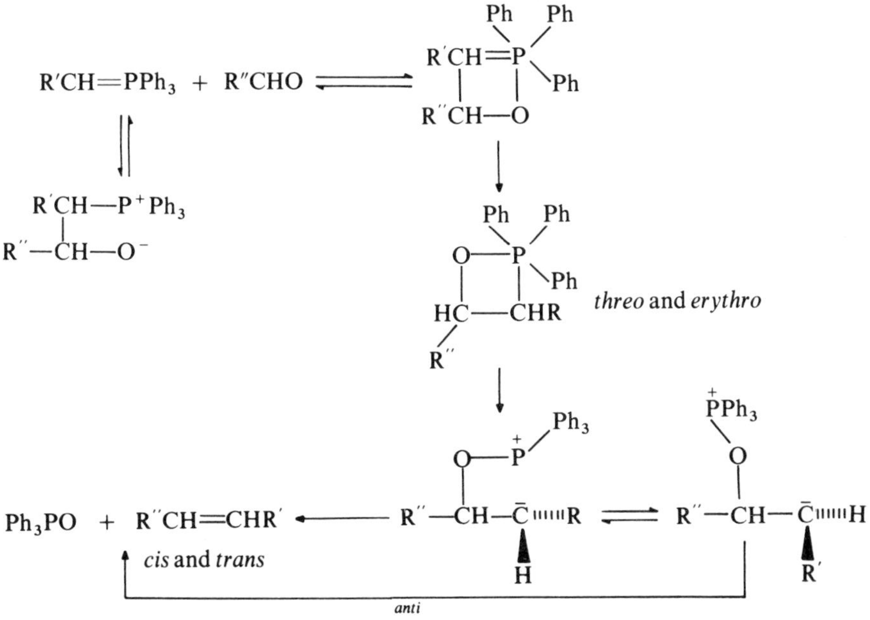

$R'CH{=}PPh_3 + R''CHO \rightleftharpoons$ $R'CH{=}P(Ph)_2Ph$ / $R''CH{-}O$

$R'CH{-}P^+Ph_3$ / $R''{-}CH{-}O^-$

$O{-}P(Ph)_2Ph$ / $HC{-}CHR$ / R'' *threo* and *erythro*

$Ph_3PO + R''CH{=}CHR'$ ⟵ $R''{-}CH(O{-}\overset{+}{P}Ph_3){-}\bar{C}(R)H \rightleftharpoons R''{-}CH(O{-}\overset{+}{P}Ph_3){-}\bar{C}(H)R'$

cis and *trans*

anti

(5)

equation 5 has been postulated. Here the rotation about the C—C bond of the carbanion is predicted to be faster than the inversion at the carbon centre. The lifetime of the carbanion is greater for the *trans* than the *cis* configuration owing to decreased steric interference with delocalization of the negative charge[10,14].

The conversion of benzyl, allyl, cinnamyl and phenylphosphonium salts through a one-electron reduction to generate ylides in 50% yield has been demonstrated[14]. Figure 1

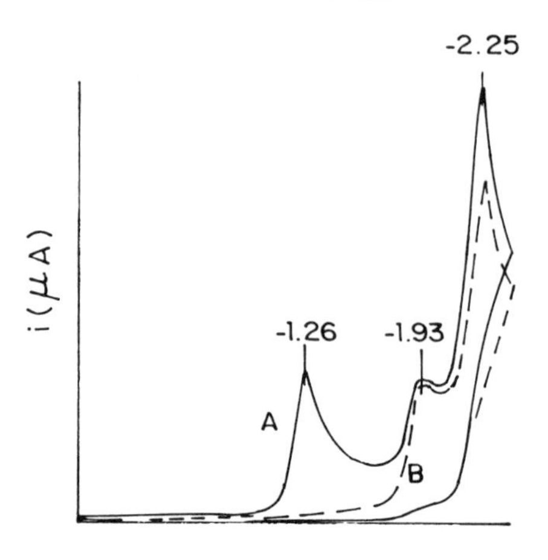

FIGURE 1. Cyclic voltammetric curves of $PhCH{=}CHCH_2\overset{+}{P}Ph_3NO_3^-$ (cinnamyltriphenylphosphonium nitrate) in dmf containing Bu_4NClO_4. (A) First scan; (B) Second scan. Reproduced by permission of the Royal Society of Chemistry from Reference 11

shows the cyclic voltammetric curve for $PhCH{=}CHCH_2\overset{+}{P}Ph_3NO_3^-$. The ylide formation follows the reaction scheme

$$RCH_2\overset{+}{P}Ph_3 + e \longrightarrow R\dot{C}H_2 + PPh_3 \quad (6)$$

$$R\dot{C}H_2 + e \longrightarrow RCH_2^- \quad (7)$$

$$PPh_3 + e \rightleftharpoons PPh_3^- \quad (8)$$

$$RCH_2^- + PPh_3^{-\cdot} + RCH_2\overset{+}{P}Ph_3 \longrightarrow RCH{=}PPh_3 + RCH_3^- \quad (9)$$

The ylide is formed on electrochemical reduction of PPh_3 at -2.77 V[15]; this is conclusively demonstrated by controlling the potential sweeps in cyclic voltammetry. With benzyltriphenylphosphonium nitrate, the above reaction scheme proceeds smoothly up to 2 F mol^{-1} when benzaldehyde and dicyano(fluoren-9-ylidene)methane are used in the reaction.

C. Structural and Redox Properties

The structural data for phosphorus ylides obtained through NMR spectroscopy and X-ray diffraction were summarized and discussed by Schmidbaur *et al.*[16]. A strong *gauche* effect due to the lone pair of electrons at the ylidic carbon atom was indicated in those studies. The synthesis and properties of phosphorus–halogen ylides have been discussed[17] with special reference to their chemical properties. The electronic transitions of phosphorus ylides are markedly characterized by longer wavelength transitions; the ylides of the type $R_2R^1P^+CR^2R^3$ (R = MeN, Et, EtO, Me_2CHO; R^1 = Me_2N, alkylamino, NH_2,

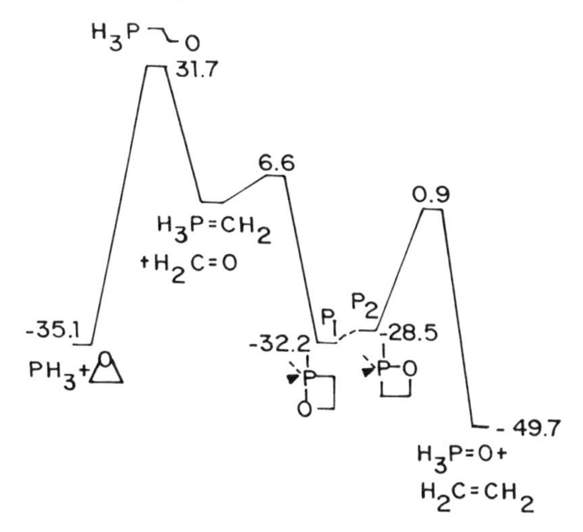

FIGURE 2. Energy profile for the Wittig-type reaction. Reprinted with permission from Reference 19. Copyright 1984, American Chemical Society

alkyl, EtO, EtS; R^2 = COOMe, COOEt; R^3 = H, COOEt, COOMe) show an increase in the dipole moment in the excited state[18]. The charge transfer from the ylide carbon to phosphorus and a linear correlation of λ_{max} with substituent have been discussed and established.

The activation energy for Wittig-type reactions has been discussed and compared with the sulphur ylide[19]. An energy profile for this reaction is shown in Figure 2. The activation energy to form the first intermediate P_1 (oxaphosphetane) is 6.6 kcal mol^{-1} (1 kcal = 4.184 kJ); here an oxygen atom occupies an axial position of the trigonal bipyramidal phosphorus. Interchange of the positions of the oxygen and the carbon atoms is produced by the pseudo-rotation at the phosphorus centre. This results in weakening of the P—C bond. The activation energy required for this pseudo-rotation is very small[20] and leads to a second intermediate oxaphosphetane, P_2, with an axial carbon; the more electronegative atom prefers the axial position in a trigonal bipyramid[21]. The decomposition of P_2 to give the alkene is markedly lower than the activation energy for the decomposition reaction to give the ylide: 29.4 vs 38.8 kcal mol^{-1}. The decomposition of oxaphosphetane is concerted in a geometrical sense with four atoms being coplanar.

The reactivities of phosphonium ylides and ylidones have been predicted by *ab initio* molecular orbital calculations[22] and these results have been correlated with neutralization–reionization mass spectrometry.

III. PHOSPHORANES

A. Anion Radicals

The electrode reactions of organic compounds containing two atoms of P per molecule were investigated in *N*, *N*-dimethylformamide containing 0.1 M Et_4NClO_4 by simultaneous electrochemical reduction and observation of the ESR signal[23]. The one-electron reduction is reversible and forms an anion radical; further reduction of the anion is irreversible. Phosphoranyl radicals having three electron bonds have also been generated by X-irradiation of single crystals of methylthiotriethylphosphonium iodide; these radicals show a large ^{31}P hyperfine coupling[24].

Redox systems involving R_2P and R_2PX substituted benzenes and biphenyls have been studied by cyclic voltammetry and ESR spectrometry. The successive addition of one electron has been postulated[25]. The formation of a free radical on one-electron reduction has been confirmed by the observation of an ESR signal showing splittings from ^{31}P. The synthesis of phosphonium salts in nearly quantitative yields from phosphane has also been established.

The formation of anion radicals in unsaturated clusters, $[Fe_4(CO)_{11}(PR)_2]$ (R = CMe_3, Ph), has been postulated in the electrochemical reduction; it is characterized by two reversible one-electron transfers[26]. The saturated analogue $[Fe_4(CO)_{12}(PPh)_2]$ can also be oxidized reversibly. The ESR spectra of the cluster ions show strong spin–spin coupling with μ_4-bridging of PR units and the terminal ligands. The double and triple clusters can also be electrochemically reduced to stable mono- and di-anions. The cyclic voltammetric and ESR spectral data provide some evidence for an electronic interaction between the two redox centres[26].

The redox behaviour of polynuclear clusters has been examined for electrocatalytic properties[27]. Clusters of the type $[Co_4(CO)_{10-x}(\mu_4\text{-}PPh)_2L_x]$, where L is a mono-, or di-phosphine having the structures of the type shown:

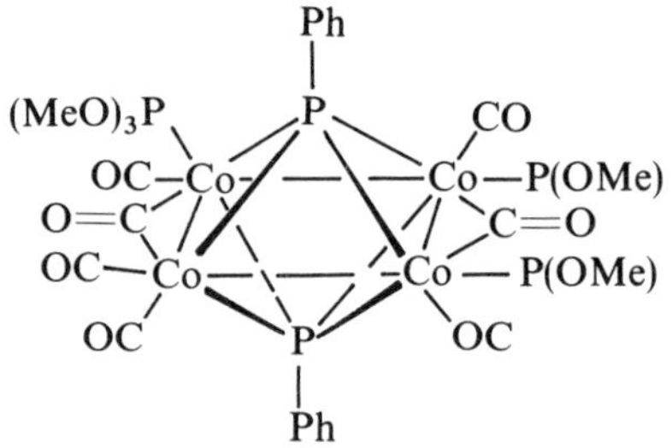

Ph
P
$(MeO)_3P$
CO
OC—Co
Co—P(OMe)
O=C
C=O
OC—Co
Co — P(OMe)
OC
P
OC
Ph

Oxidation of the tetracobalt clusters is made easier by the successive phosphine substitution. The anion radical derived from the 0/−1 redox couple of the cluster is substitutionally labile. The cyclic voltammetric pattern of tetracobalt carbonyl cluster is shown in Figure 3.

The one-electron redox couple represents reaction 10.

$$[Co_4(CO)_{10}(PPh)_2] \underset{-e}{\overset{+e}{\rightleftharpoons}} [Co_4(CO)_{10}(PPh)_2]^{-\cdot} \qquad (10)$$

B. Conductivity

The electrical conductivities of Me_3PF_2, Me_2PF_3 and $MePF_4$ were determined in acetonitrile; the fluoromethylphosphoranes behave as weak electrolytes[28]. The 1:1 and 2:1 adducts (R_3PX_2 and R_3PX_4) were isolated in the conductometric titrations of tributyl-, trioctyl- and tricyclohexyl-phosphine with Br_2, I_2 and iodine bromide and their electrical conductivities were measured in acetonitrile[29]. Table 4 lists the conductances of several of these adducts.

The phosphoranes have been used to increase electrical conductivity[30]; the phosphorane structure also helps in providing improved performance of non-aqueous batteries[31–33]. The formation of excess of SO_2 during high rates of discharge is prevented by providing PCl_5 at the cathode or the electrolyte, and this prevents the breakage of the battery[32].

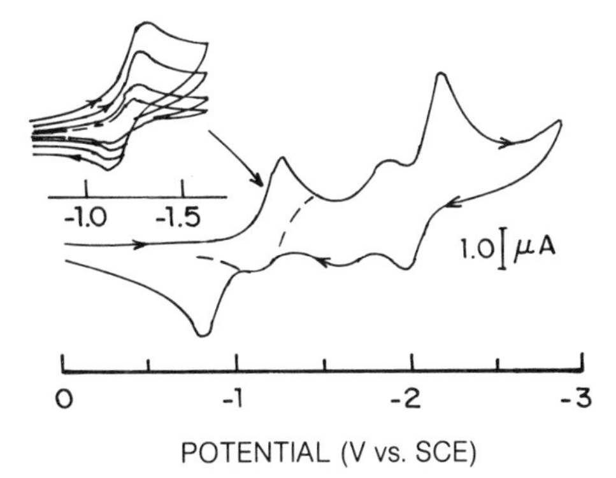

FIGURE 3. Cyclic voltammetric curves for the tetracobaltcarbonyl cluster in thf containing 0.1 M tetrabutylammonium perchlorate. Sweep rate, 200 mVs^{-1}. The inset shows the scans of the first (0/−1) redox couple at 1000, 500, 200 and 100 mVs^{-1}. Reprinted with permission from Reference 27. Copyright 1986, American Chemical Society

TABLE 4. Conductances of phosphoranes. Reproduced with permission from Reference 29. Copyright 1985, Pergamon Press PLC

Adduct	Colour	Λ ($S\,cm^2\,mol^{-1}$)[a]
Bu_3PBr_2	White	93.3
Bu_3PI_2	White	90.0
Bu_3PIBr	Pale yellow	103.7
Bu_3PBr_4	Orange	137.4
Bu_3PI_4	Red–brown	150.8
$Bu_3PI_2Br_2$	Dark red	136.4
Oc_3PI_2	White	77.0
Oc_3PBr_2	White	87.4
Oc_3PIBr	Pale yellow	83.0
Oc_3PBr_4	Orange	124.0
Oc_3PI_3	Dark red	109.5
Oc_3PI_4	Dark red	123.9
$Oc_3PI_2Br_2$	Dark red	120.8
$Hex_3PBr_2 \cdot CH_3CN$	Pale yellow	115.6
$Hex_3PI_2 \cdot CH_3CN$	Pale yellow	117.8
$Hex_3PIBr \cdot CH_3CN$	White	98.0

[a] Measurements made in acetonitrile.

C. MIS Devices

Metal–semiconductor–metal or metal–insulator–semiconductor (MIS) devices have been constructed using polymerized phthalocyanine derivatives with PF_5 as the dopant (phosphorane structure)[34].

IV. PHOSPHONIUM SALTS

A. Reductive Cleavage

The one-electron electrochemical reduction of 1,2-vinylene and buta-1,4-dienylene bisphosphonium salts at a mercury cathode produces an ylide character by the reaction pathway depicted in reactions 11–13. The mechanism is altered when OH^- is generated in the unbuffered aqueous–organic medium; this reaction is depicted as reaction 14. The electrochemical reduction of phosphonium salt in the presence of tri-*p*-anisylphosphine produces a mixture of the saturated or semi-saturated bisphosphonium salts through either reaction scheme 15 or alternatively 16.

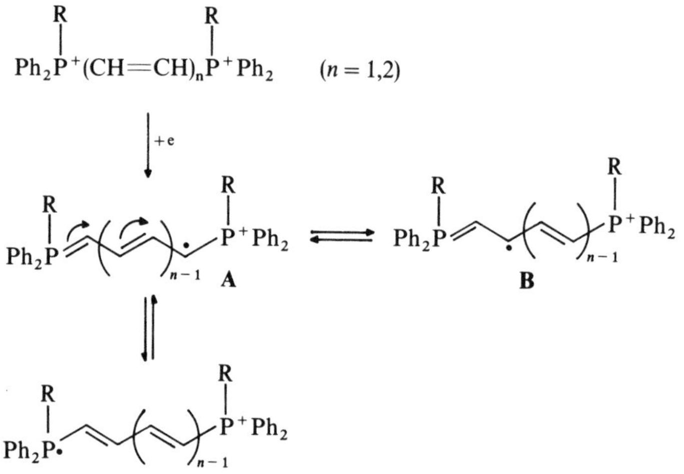

$Ph_2P^+(R)(CH{=}CH)_nP^+(R)Ph_2 \quad (n = 1,2)$

+e

A ⇌ B (11)

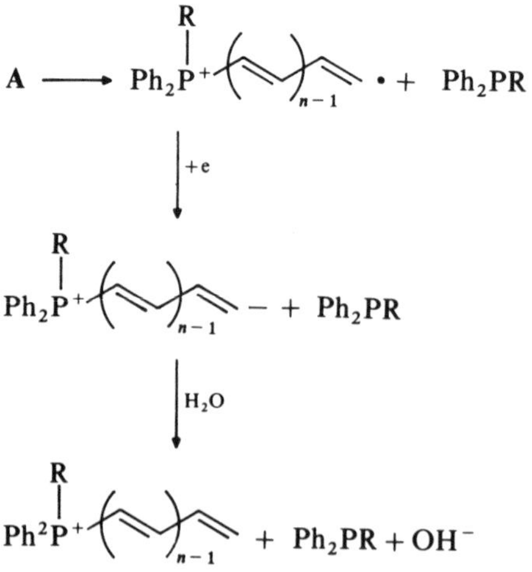

$Ph_2P^{\bullet}(R)$–$(CH{=}CH)_{n-1}$–$CH{=}CH$–$P^+(R)Ph_2$

$\mathbf{A} \longrightarrow Ph_2P^+(R)$–$(CH{=}CH)_{n-1}$–$CH{=}CH^{\bullet} + Ph_2PR$

+e

$Ph_2P^+(R)$–$(CH{=}CH)_{n-1}$–$CH{=}CH^- + Ph_2PR$ (12)

H_2O

$Ph^2P^+(R)$–$(CH{=}CH)_{n-1}$–$CH{=}CH_2 + Ph_2PR + OH^-$

$$\mathbf{B} \xrightarrow{H_2O} Ph_2\overset{R}{\overset{|}{P}}{}^+CH(H)\dot{C}(CH{=}CH)_{n-1}\overset{R}{\overset{|}{P}}{}^+Ph_2$$

$$\downarrow +e$$

$$Ph_2\overset{R}{\overset{|}{P}}{}^+CH(H)\bar{C}(CH{=}CH)_{n-1}\overset{R}{\overset{|}{P}}{}^+Ph_2$$

$$\downarrow H_2O \qquad (13)$$

$$Ph_2\overset{R}{\overset{|}{P}}{}^+CH_2(CH{=}CH)_{n-1}CH_2\overset{R}{\overset{|}{P}}{}^+Ph_2$$

$$Ph_2\overset{R}{\overset{|}{P}}{}^+CH{=}CH\overset{R}{\overset{|}{P}}{}^+Ph_2$$

$$\downarrow OH^-$$

$$Ph_2\overset{R}{\overset{|}{P}}{}^+CH{=}CH\overset{OH}{\overset{|}{P}}Ph_2$$

$$\downarrow H_2O \mid OH^- \qquad\qquad (14)$$

$$Ph_2\overset{R}{\overset{|}{P}}{}^+CH{=}CH^- + Ph_2\overset{R}{\overset{|}{P}}(O) \qquad Ph_2\overset{R}{\overset{|}{P}}{}^+CH\underset{Ph_2}{\underset{|}{C}}H{-}\overset{O}{\overset{\|}{\underset{R}{\underset{|}{P}}}}{-}Ph_2$$

$$\downarrow \qquad\qquad \downarrow OH^- \mid H_2O$$

$$Ph_2CH{=}CH_2 + Ph_2\overset{R}{\overset{|}{P}}(O) \qquad Ph_2\overset{R}{\overset{|}{P}}{}^+CH_2CH\overset{O}{\overset{\|}{P}}Ph_2$$

$$\begin{array}{c} Ph_2\underset{\displaystyle Ph}{P^+}\,CH{=}CH\underset{\displaystyle Ph}{P^+}\,Ph_2 \quad 2Br^- + (p\text{-MeOPh})_3P \\ \Big\downarrow {\scriptstyle +2e}\ \big|\ {\scriptstyle H_2O\ (4\ equil.\ MeCOOH)} \\ Ph_2\underset{\displaystyle Ph}{P^+}\,(CH_2)_2\underset{\displaystyle Ph}{P^+}\,Ph_2 \quad 2Br^- + (p\text{-MeOPh})_3P^+(CH_2)_2\underset{\displaystyle Ph_3}{P^+} \end{array} \tag{15}$$

$$\begin{array}{c} Ph_3P^+(CH{=}CH)_2P^+Ph_3 \quad 2Br^- + (p\text{-MeOPh})_3P \\ \Big\downarrow {\scriptstyle H_2O}\ \big|\ {\scriptstyle 2e,\ 4\ equil.\ MeCOOH} \\ Ph_3P^+CH_2CH{=}CHCH_2P^+Ph_3 + \\ (p\text{-MeOPh})_3P^+CH_2CH{=}CHCH_2P^+Ph_2 2Br^- + Ph_3P(O) + (p\text{-MeOPh})_3PO \end{array} \tag{16}$$

The kinetic parameters of the cleavage reaction of 4*H*-flavylium and Ph_3P have been evaluated from polarographic measurements of the limiting currents[35]. Figure 4 shows typical cyclic voltammetric curves for the 1,2-vinylene bisphosphonium salt in acetonitrile–water in the presence of 0.1 M Et_4NBr. The locations of the peak potentials for the different substituents are given in Table 5.

The three cyclic voltammetric peaks are irreversible owing to the follow-up chemical reactions proposed in reactions 11–16. In comparison with the monophosphonium salts reported earlier[37–39], the potentials in Table 5 are lower owing to $d\pi$–$p\pi$ conjugative interaction in the radical between phosphorus and the unsaturation.

The cyclic voltammetry of methyltriarylphosphonium tosylates exhibits irreversible electrochemical reduction and oxidations at glassy carbon and platinum electrodes[40].

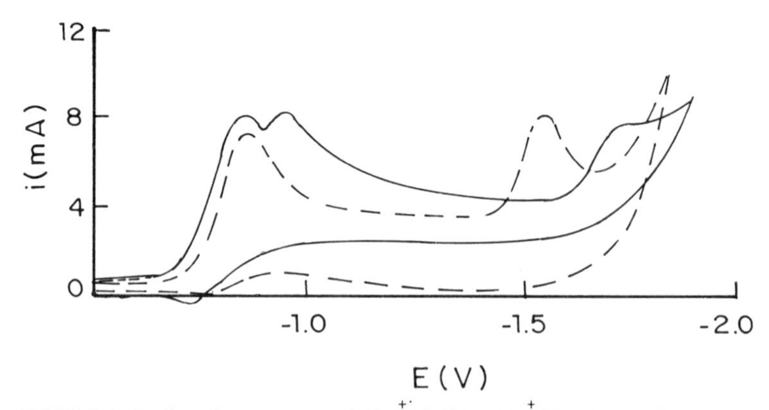

FIGURE 4. Cyclic voltammogram of $Ph_2\overset{+\cdot}{P}PhCH{=}CH\overset{+}{P}Ph_3\ 2Br^-$ (solid line and $Ph_3\overset{+}{P}(CH{=}CH)_2\overset{+}{P}Ph_3\ 2Br^-$ (dashed line)) in acetonitrile–water containing 0.1 M tetraethylammonium bromide. Reproduced by permission of Pergamon Press from Reference 36

TABLE 5. Cyclic voltammetric peak potentials for 1,2-vinylene and buta-1,4-dienyliene bisphosphonium salts[a]. Reproduced with permission from Reference 36, Copyright 1984, Pergamon Press PLC

Structure	E_p^{I} (V)	E_p^{II} (V)	E_p^{III} (V)
$Ph_2\overset{+}{P}CH{=}CH\overset{+}{P}Ph_2$	−0.83	−0.95	−1.71
$Ph_2\overset{+}{P}(CH_2CH_2CN)CH{=}CH\overset{+}{P}Ph_2(CH_2CH_2CN)$	−0.80	−0.89	−1.80
$Ph_2\overset{+}{P}(CH_2{-}CH{=}CH_2)CH{=}CH\overset{+}{P}Ph_2(CH_2CH{=}CH_2)$	−0.91	−1.00	−1.79
$Ph_3\overset{+}{P}(CH{=}CH)_2\overset{+}{P}Ph_3$	−0.83		−1.45
$(PhCH_2)_3\overset{+}{P}(CH{=}CH)_2\overset{+}{P}(CH_2Ph)_3$	−0.82		−1.48

[a] Medium: acetonitrile–water containing 0.1 M LiBr. Potentials are referred to SCE. Working electrode: Hg. Sweep rate: 0.10 Vs^{-1}.

These potentials have been correlated with the highest occupied molecular orbital (E_{HOMO}) and the lowest unoccupied molecular orbital (E_{LUMO}). The potentials and the energies of E_{HOMO} and E_{LUMO} are given in Tables 6 and 7. The peak potentials have been correlated with HOMO parameters as

$$E_{pa} = 1.05(-E_{HOMO}) - 4.46 \tag{17}$$

TABLE 6. Cyclic voltammetric cathodic peak potentials for methyltriarylphosphonium tosylates. Reproduced by permission of VCH from Reference 40

Structure	E_{pc} (V)[a]	E_{HOMO} (eV)[a]
$(Me_2N(C_6H_4))_3\overset{+}{P}CH_3\bar{O}SO_2(C_6H_4Me)$	−2.45	−11.45
$(OMeC_6H_4)_3\overset{+}{P}CH_3\bar{O}SO_2(C_6H_4Me)$	−2.12	−12.57
$(MeC_6H_4)_3\overset{+}{P}CH_3\bar{O}SO_2(C_6H_4Me)$	−2.08	−12.81
$(C_6H_5)_3\overset{+}{P}CH_3\bar{O}SO_2(C_6H_4Me)$	−1.97	−12.89
$(ClC_6H_4)_3\overset{+}{P}CH_3\bar{O}SO_2(C_6H_4Me)$	−1.71	−13.19
$(MeOOCC_6H_4)_3\overset{+}{P}CH_3\bar{O}SO_2(C_6H_4Me)$	−1.41	−11.02
$(F_3CC_6H_4)_3\overset{+}{P}CH_3\bar{O}SO_2(C_6H_4Me)$	−1.60	−13.82

[a] Potentials measured with reference to ferrocene ($E_{1/2} = 0.423$ V).

TABLE 7. Cyclic voltammetric anodic peak potentials for methyltriarylphosphonium tosylates. Reproduced by permission of VCH from Reference 40

Structure	E_{pa} (V)[a]	E_{LUMO} (eV)[a]
$Ph_3\overset{+}{P}CH_3\overset{-}{O}SO_2C_6H_4NH_2$	0.93	3.40
$Ph_3\overset{+}{P}CH_3\overset{-}{O}SO_2C_6H_4Me$	2.04	3.08
$Ph_3\overset{+}{P}CH_3\overset{-}{O}SO_2C_6H_4OMe$	1.79	2.94
$Ph_3\overset{+}{P}CH_3\overset{-}{O}SO_2Ph$	2.23	3.23
$Ph_3\overset{+}{P}CH_3\overset{-}{O}SO_2C_6H_4Cl$	2.33	2.71
$Ph_3\overset{+}{P}CH_3\overset{-}{O}SO_2C_6H_4NO_2{}^{-}p$	2.48	1.70
$Ph_3\overset{+}{P}CH_3\overset{-}{O}SO_2C_6H_4NO_2{}^{-}$-$m$	2.52	1.43
$Ph_3\overset{+}{P}CH_3\overset{-}{O}SO_2C_6H_4(COOEt)_2$	2.49	1.99

[a] Potentials measured with reference to ferrocene ($E_{1/2} = 0.429$ V).

and

$$E_{pc} = 0.67(-E_{LUMO}) - 5.13 \tag{18}$$

These correlations have to be regarded as approximate as the true E° values are likely to be shifted by the subsequent chemical reactions.

Cristau *et al.*[41] reported an electrochemical deprotection of γ-thioacetalated phosphonium salts which involves the elimination of the base of the phosphorus group. The reaction scheme is represented by reaction 19. This conversion proceeds through an

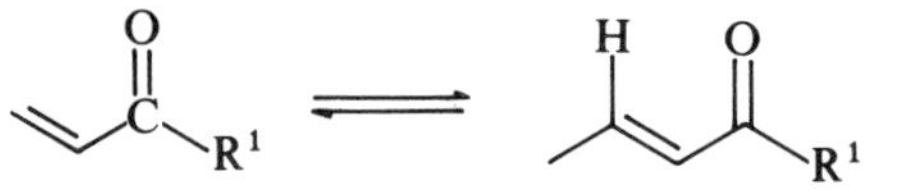

(19)

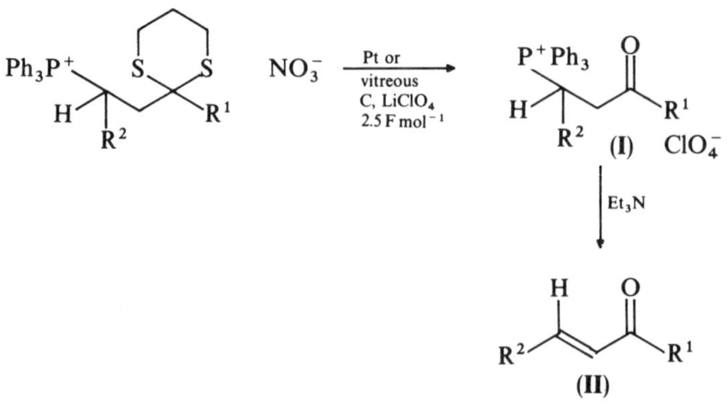

Pt or vitreous C, $LiClO_4$ 2.5 F mol^{-1}

(I) ClO_4^-

Et_3N

(II)

(20)

TABLE 8. Yields of α, β-ethylenic ketones. Reprinted with permission from Reference 41. Copyright (1983) American Chemical Society

R^1	R^2	Yield (%)	
		I	II
Me	H	96	65
Me	Me	95	80
Me	Bu^n	100	82
Me	Ph	100	93
H	Me	75	85
H	Ph	80	74

electrochemical oxidation step (reaction 20). The percentage yields for the different substituents R^1 and R^2 of **I** and **II** are given in Table 8.

B. Effect of Substituents on Reductive Cleavage

The phosphonium salts undergo reductive cleavage by reaction 21[42]:

$$R^{\pi}-\overset{+}{P}R_3 \xrightarrow{2e} (R^{\pi})^{-\bullet} + PR_3 \tag{21}$$

where R^{π} represents an aromatic π-system such as Ph, PhPh, $C_{10}H_8$, C_4SH_5, styrene, phenylacetylene and ferrocene with R = Me or Ph. The effect of substituents on the half-wave potentials of 23 organophosphorus compounds have been reported. Table 9 gives a list of half-wave potentials of a few selected phosphonium salts.

The sequence in which the reduction occurs is represented as

$$PMe_2 \ll PPh_2 < P(OMe)_2 < PSMe_2 = P(NR)Me_2 = PSeMe_2 < {}^+PMe_3.$$

Statistical analysis within a π-perturbation model and comparison with 1,4-disubstituted benzene derivatives supports the above sequence of increasing acceptor effect of substituents. The substituent effect is observed for the other series such as $R^{\pi}(PR_2)_n$, $R^{\pi}(P^+R_3)_n$ and $R^{\pi}(PyR_2)_n$, where Y = O, S, Se, NR and $n = 1, 2$.

TABLE 9. Half-wave potentials of phosphonium salts

Salt	$E^{I}_{1/2}$ (V)[a]	$E^{II}_{1/2}$ (eV)[a]
p-$Me_2\overset{+}{P}C_6H_4PMe_2$	−2.07	−2.65
m-$Me_2P^+C_6H_4\overset{+}{P}Me_3$	−1.68	−2.65
$Me_3P^+C_6H_4\overset{+}{P}Me_3$	−1.57	−1.83
$Me_3P^+C_6H_4\overset{+}{P}Me_3$	−1.52	−1.98
$Ph_3P^+C_6H_4\overset{+}{P}Ph_3$	−1.14	−1.50

[a] Potentials measured in dmf–0.1 M tetrabutylammonium perchlorate (tbap) with reference to SCE.

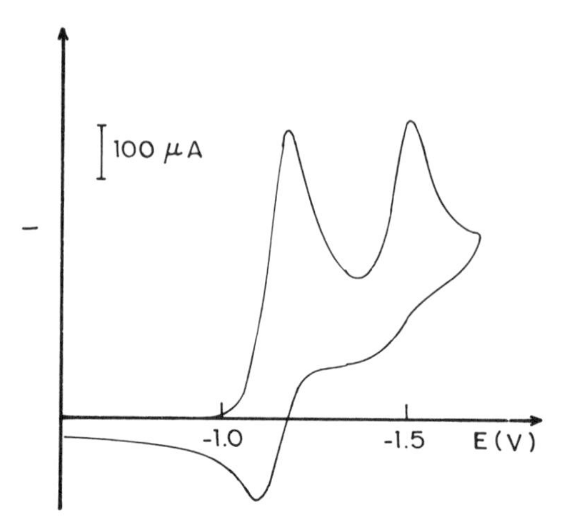

FIGURE 5. Cyclic voltammogram of 1,4-bis(triphenylphosphonio)benzodiiodide in dmf containing 0.1 M tetrabutylammonium perchlorate. The abscissa is with reference to SCE. Reproduced by permission of VCH from Reference 42

Figure 5 shows a typical cyclic voltammogram of 1,4-bis(triphenylphosphonio)benzodiiodide in dmf; it is characterized by a reversible couple followed by an irreversible reduction. The shift in half-wave potential with the substituent has been correlated with the energy of the Hückel molecular orbital[42].

C. Radical Anions

The mechanisms of the electrochemical reduction of phosphonium salts have been investigated by monitoring the products by electron spin resonance spectroscopy[42,43]. The electrolysis of tetraphenylphosphonium tetrafluroborate (tpp) at -2.15 V vs AgI/Ag produces an ESR spectrum in the presence of a nitroxide spin marker (*N-tert*-butyl-α-phenyl nitrone) by consuming 1.3 electrons per molecule of tpp[43]. The ESR spectrum is shown in Figure 6. The spectrum can be fitted with the coupling constants $a_N = 14.1$ G and $a_H = 2.5$ G. The same ESR features are obtained on adding PhLi to the nitrone through air oxidation. The nitroxide produced is identified to have the following structure:

```
      H   O·
      |   |
Ph—C—N—Bu^t
      |
      Ph
```

A few other phosphonium salts have been shown to produce free radicals during electrolytic reduction[42].

D. Probes for Membrane Studies

The phosphonium cations of the type $Ph_3P^+(CH_2)_nMe$ ($n = 0$–5) and tpp^+ have been used in determining the transport rate through membranes of *Halobacterium halobium*[44].

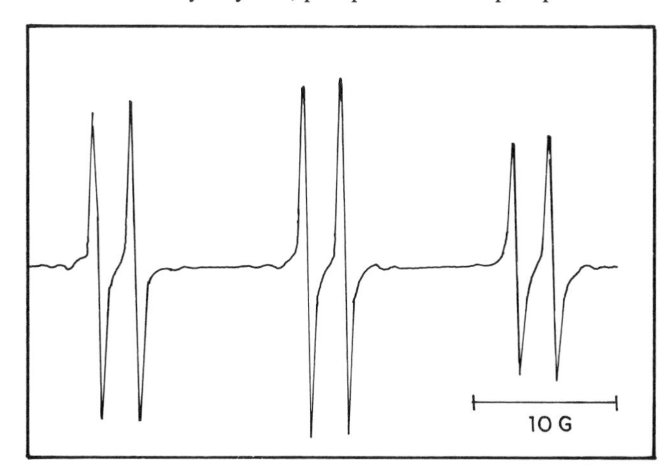

FIGURE 6. ESR spectrum of 2×10^{-4} M tpp tetrafluroborate in dmso containing 0.1 M tetrabutylammonium tetrafluroborate electrolysed at −2.15 V after passing a number of coulombs corresponding to 1.3 electrons per molecule of the substrate. Reproduced by permission of Elsevier Science Publishers from Reference 43

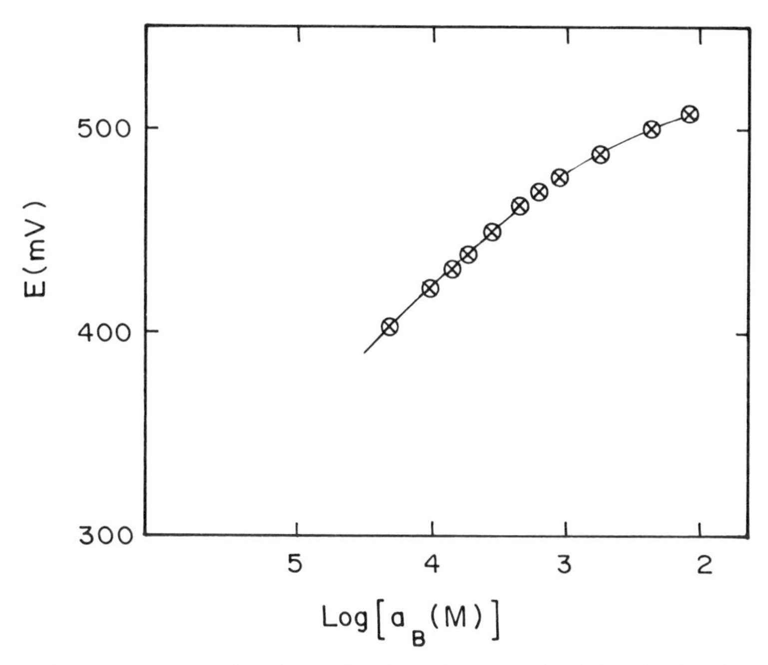

FIGURE 7. Response of tba^{+}-sensitive electrode to tetraphenylphosphonium ion concentration. Membrane: 2×10^{-4} M tbatpb in nitrobenzene. Reproduced by permission of Elsevier Science Publishers, from Reference 45

The membrane potential increased with increasing hydrocarbon chain length. Figure 7 shows the transport rate constants of various phosphonium ions as a function of the membrane potential. The selectivity coefficient of tpp^+ has also been determined using a nitrobenzene—based membrane electrode[45]; a value of $\log K^{\text{Pot}} = 8.75$ was reported, which is lower than that obtained with the arsenic analogue.

E. Chiral and Optical Salts

Horner and Dickerhof[46] demonstrated the reductive cleavage of chiral and optical quaternary phosphonium salts. The reduction of (*S*) − (+)-$PhCH_2\overset{+}{P}MePh$ Br^- gave (*S*) − (+)-PMePhBr; with onium salts containing both benzyl and *tert*-butyl groups; the reductive elimination of the latter groups predominates. Although reductive cleavages are also observed using alkali metal amalgams at the cathode, the cleavage is less efficient at the cathode.

F. Inhibitory Effect of Phosphonium Salts

The adsorption of phosphonium salts on mercury and their effect on interfacial tension have been discussed in Volume 2[3]. The adsorption of tetraphenylphosphonium chloride produces selective inhibition[47] of electrode reactions of chromium(III), europium(III), vanadium(III) and zinc(II). While the electrode reactions are completely inhibited, the reductions of copper(II), O_2, H_2O_2 and indium are partially inhibited. The effect of Ph_4PCl on the polarogram of $CuCl_2$ is shown in Figure 8. A similar trend is observed in the reduction of cadmium(II) and zinc(II) at a dropping mercury electrode[48]. The formal and standard rate constants are determined in the presence of benzyltriphenylphosphonium cations. A logarithmic relationship between rate constants and the degree of surface coverage of the phosphonium cation has been established. The interaction between surfactant cation and the activated complex particles has been suggested to be mainly electrostatic[48].

G. Formation of Ylides

The formation of an ylide by the reaction of phosphonium salt and electrogenerated base has been confirmed by cyclic voltammetry[10,14]. The essential conditions required for

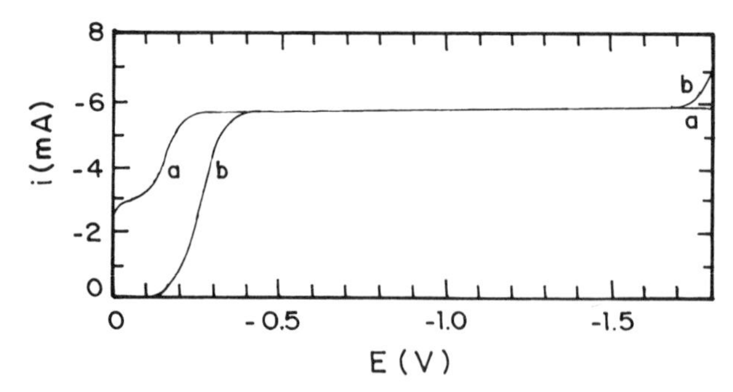

FIGURE 8. Polarograms of 0.6 mM $CuCl_2$ in (a) 0.1 M NaCl and (b) 10 mM Ph_4PCl. The abscissa is with reference to Hg/Hg_2Cl_2. Reproduced by permission of Elsevier Science Publishers from Reference 48

TABLE 10. Cyclic voltammetric features of metal-bonded phosphonium salts

Salt[a]	E_{pc} (V)	E_{pa} (V)	i_{pa}/i_{pc}	Ref.
$[Ph_4P]_2Fe_4S_4(SPh)_2(Et_2dtc)_2$	1.28[b]			50
	−1.12	−1.01	0.45	
$[Ph_4P]_2Fe_4S_4Cl_3(Et_2dtc)$	−1.24			50
	−1.16			
$[Ph_4P]_2Fe_4S_4Cl_3(Et_2dtc)$	−1.10	−1.00	<0.2	50
	−0.90	−0.81	<0.3	
$[Fe_4S_4(LS_3)]_2[C_6H_2Me_2S_2][Ph_4P]_4$	−1.20			49
$[Fe_4S_4(LS_3)]_2[C_6H_4S_2][Ph_4P]_4$	−1.15			49
$[Fe_4S_4(LS_3)]_2[SCH_2CH_2S][Ph_4P]_4$	−1.28			49
$[Ph_4P][Co(S\text{-}2,4,6\text{-}^iPr_3C_6H_4)]$	−1.60			57
$[Ph_4P]_2[Br_2FeS_2MoS_2FeBr_2]$	−0.56			58
$[Ph_4P]_2[Cl_2FeS_2WS_2FeCl_2]$	−0.81			58
$[PPh_4]Na[VO(edt)_2]\cdot 2EtOH$	−0.19[b]			56
$[PPh_4]Na[VS(edt)_2]\cdot xEt_2O$	−0.46[b]			56

[a] dtc = Dithiocarbamate; edt = Ethane-1, 2-dithiolate.
[b] Refers to E_{pa} (initial anodic peak).

this method are as follows: (a) the electrochemical reduction of the probase should precede the reduction of the phosphonium salt; (b) cyclic voltammetric reversibility of the probase such that the stronger base that is generated at the electrode is stable for reaction with the phosphonium salt; (c) the probase peak potential can be shifted by substitution such that a wide range of phosphonium salts can be reacted; and (d) the formation of ylide and regeneration of probase would be preferable for carrying out the reaction in a cyclic process. The ylides prepared by this method are listed in Table 1; they have been characterized by cyclic voltammetric peak potentials[10,14].

H. Miscellaneous Phosphonium Salts

A large number metal-bridged phosphonium salts have been synthesized and characterized by cyclic voltammetry[49–58]. With these coordinated compounds, the electrochemical features generally arise from the ionized species. For example, in the phosphonium salts shown in Table 10, the cyclic voltammetric reduction occurs from the anions.

V. ACKNOWLEDGEMENTS

The author is grateful to Dr M. Sharon and Dr R. Sundaresan for supplying photocopies of various publications from their libraries and to Mr S. K. Haram for help in drawing the structures.

VI. REFERENCES

1. H. Schmidbaur, *Angew. Chem. Int. Ed. Engl.*, **22**, 907 (1983).
2. H. J. Cristau, Y. Ribeill, F. Plenat and L. Chiche, *Phosphorus Sulfur*, **30**, 135 (1987).
3. K. S. V. Santhanam, in *Chemistry of Organophosphorus Compounds* (Ed. F. R. Hartley), Vol. 2, Wiley, Chichester, 1992, p. 77.
4. A. Schier and H. Schmidbaur, *Chem. Ber.*, **117**, 2314 (1984).
5. B. J. Walker, *Organophosphorus Chem.*, **14**, 231 (1983).
6. B. J. Walker, *Organophosphorus Chem.*, **11**, 192 (1980).

7. B. J. Walker, *Organophosphorus Chem.*, **17**, 316 (1986).
8. B. J. Walker, *Organophosphorus Chem.*, **16**, 282 (1986).
9. B. J. Walker, *Organophosphorus Chem.*, **15**, 218 (1985).
10. R. R. Mehta, V. L. Pardini and J. H. P. Utley, *J. Chem. Soc., Perkin Trans. 1*, 2921 (1982).
11. V. L. Pardini, L. Roullier and J. H. P. Utley, *J. Chem. Soc., Perkin Trans. 2*, 1520 (1981).
12. M. M. Sidky, M. R. Mahrn, A. A. El–Kateb, I. T. Hennawy and M. A. Abdel– Malek, *Phosphorus Sulfur*, **29**, 227 (1987).
13. H. J. Cristau, Y. Riberll, F. Plenat and L. Chiche, *Phosphorus Sulfur*, **30**, 125 (1987).
14. S. Wawzonek and J. H. Wagenknecht, *Polarography 1964*, Macmillan, London, 1966, p. 1035.
15. (a) K. S. V. Santhanam and A. J. Bard, *J. Am. Chem. Soc.*, **90**, 1118 (1968); (b) K. S. V. Santhanam, in *Chemistry of Organophosphorus Compounds* (Ed. F. R. Hartley), Vol. 1, Wiley, Chichester, 1990, p. 109.
16. H. Schmidbaur, A. G. Bowmaker, U. Deschler, C. Doerzbach, R. Hern, B. Milewski-Mahila, A. Schier and C. E. Zybill, *Phosphorus Sulfur*, **18**, 167 (1983).
17. O. I. Kolodiazhnyi and V. P. Kukhar, *Phosphorus Sulfur*, **18**, 191 (1983).
18. R. Loktionova, I. E. Boldeskul and V. P. Lysenke, *Teor. Eksp. Khim.*, **19**, 30 (1983); *Chem. Abstr.*, **98**, 178479g (1983).
19. F. Volatron and O. Eisenstein, *J. Am. Chem. Soc.*, **106**, 6117 (1984).
20. R. Holler and H. Lischka, *J. Am. Chem. Soc.*, **102**, 4632 (1980).
21. R. Hoffmann, J. M. Howell and E. L. Muetterties, *J. Am. Chem. Soc.*, **94**, 567 (1972).
22. B. F. Yates, W. J. Boama and L. Radom, *J. Am. Chem. Soc.*, **106**, 5805 (1984).
23. A. V. Il'yasov, M. K. Kadirov and Yu. M. Kargin, *Dokl. Akad. Nauk SSSR*, **294**, 1155 (1987); *Chem. Abstr.*, **107**, 143500b (1987).
24. R. A. J. Janssen, O. M. Aagaard, H. J. Van der Woerd and H. M. Buck, *Chem. Phys. Lett.*, **171**, 127 (1990).
25. W. Kaim, U. L. Knoblauch, P. Hanel and H. Bock, *J. Org. Chem.*, **48**, 4206 (1983).
26. J. T. Jaeger, J. S. Field, D. Collison, G. P. Speck, M. B. Peake, J. Hachnle and H. Vahrenkamp, *Organometallics*, **7**, 1753 (1988).
27. M. G. Richmond and J. K. Kochi, *Inorg. Chem.*, **25**, 656 (1986).
28. S. G. Harris, F. Inglis and R. Schmuzler, *J. Fluorine Chem.*, **16**, 293 (1980).
29. G. S. Harris and J. S. McKechnie, *Polyhedron*, **4**, 115 (1985).
30. T. Hagiwara, T. Demura and K. Iwata, *Nippon Kagaku Kaishi*, 356 (1986); *Chem. Abstr.*, **104**, 2127211v (1986).
31. Toshiba Corporation, Toshiba Battery Co. Ltd, *Chem. Abstr.*, **102**, 116678f (1985).
32. G. R. Ramsay and D. J. Salmon, *Chem. Abstr.*, **102**, 98475f (1985).
33. Citizen Watch Co., *Chem. Abstr.*, **95**, 105356k (1981).
34. A. Yamada and Y. Osada, *Chem. Abstr.*, **106**, 167292r (1987).
35. L. I. Kisarova, V. T. Abaev, E. A. Arzumaryants and A. A. Bumber, *Elektrokhimiya*, **24**, 1562 (1988).
36. H. J. Cristau, B. Chabaud, L. Labaudiniere and H. Christol, *Electrochim. Acta.*, **29**, 381 (1984).
37. K. S. V. Santhanam, in *Chemistry of Organophosphorus Compounds* (Ed F. R. Hartley), Vol. 2, Wiley, Chichester, 1992, p. 127.
38. R. D. Rieke, R. A. Copenhafer, C. K. White, A. Aguion, C. Williams, Jr, and M. S. Chattha, *J. Am. Chem. Soc.*, **99**, 6656 (1977).
39. C. K. White and R. D. Rieke, *J. Org. Chem.*, **43**, 4639 (1978).
40. D. E. Bugner, *J. Org. Chem.*, **54**, 2580 (1989).
41. H. J. Cristau, B. Chabaud and C. Niangoran, *J. Org. Chem.*, **48**, 1527 (1983).
42. H. Bock, U. L. Knoblauch and P. Hamel, *Chem. Ber.*, **119**, 3749 (1986).
43. J. Simonet, M. C. Badre and G. Mousset, *J. Electroanal. Chem.*, **286**, 163 (1990).
44. M. Demura, N. Kamo and Y. Kobatake, *Bioelectrochem. Bioenerg.*, **14**, 439 (1985).
45. V. Sustacek and J. Senkyr, *J. Electroanal. Chem.*, **279**, 31 (1990).
46. L. Horner and K. Dickerhof, *Phosphorus–Sulfur*, **15**, 213 (1983).
47. R. Srinivasan and R. DeLevie, *J. Electroanal. Chem.*, **205**, 299 (1986).
48. A. Anastopoulos, A. Christodoulu and I. Moumtzis, *Chem. Chron.*, **17**, 74 (1988).
49. T. D. P. Stack, M. J. Carney and R. H. Holm, *J. Am. Chem. Soc.*, **111**, 1670 (1989).
50. M. G. Kanqtzidis, D. Coucovaris, A. Simopoulos, A. Kostikas and V. Papefthymino, *J. Am. Chem. Soc.*, **107**, 4925 (1985).

51. C. A. Gryygon, W. C. Fultz, A. L. Rheingold and J. L. Bumeister, *Inorg. Chim. Acta*, **144**, 31 (1988).
52. M. G. Kanatzidis, *Inorg. Chim. Acta*, **168**, 101 (1990).
53. J. R. Dilworth, S. K. Ibrahim, S. R. Khan, M. B. Hursthouse and A. A. Karaulov, *Polyhedron*, **9**, 1323 (1990).
54. M. G. Kanatzidis and S. Dhingra, *Inorg. Chem.*, **28**, 2024 (1989).
55. A. Muller, R. Jostes, W. Eltzner, C. S. Nie, E. Diemann, H. Bogge, M. Dartmann, U. R. Vogell, S. Che, S. J. Cyvin and B. N. Cyvin, *Inorg. Chem.*, **24**, 2872 (1985).
56. J. K. Money, J. C. Huffman and G. Christou, *Inorg. Chem.*, **24**, 3297 (1985).
57. R. L. Fikar, S. A. Koch and M. M. Millar, *Inorg. Chem.*, **24**, 3311 (1985).
58. D. Coucouvanis, E. D. Simhon, P. Stremple, M. Rayan, D. Swenson, N. C. Baenziger, A. Simpoulos, V. Papaefthymiou, A. Kostikas and V. Petrouleas, *Inorg. Chem.*, **23**, 741 (1984).

CHAPTER **6**

Photochemistry of phosphonium salts, phosphoranes and ylides

M. DANKOWSKI

Degussa AG, ZN Wolfgang, AC-PT-PS, Postfach 13 45, 63403 Hanau, Germany

I. INTRODUCTION

During the last 40 years, the number of reports in the area of preparative photochemistry has increased considerably. Schönberg wrote one of the first systematic reviews on the topic in 1958, updated in 1968[1]. A number of important papers are now currently surveyed annually[2] and some selected results on the photochemistry of organophosphorus compounds[3] are presented in an annual publication[4]. Recent reviews dealing with the irradiation of organophosphorus(III) compounds and the photochemistry of phosphine chalcogenides by Dankowski appeared in Volumes 1[5] and 2[6] of this series. In this chapter,

The chemistry of organophosphorus compounds, Vol. 3
Edited by F. R. Hartley

the photochemical reactions of phosphonium salts, phosphoranes and ylides are covered. The emphasis is on the preparative aspects. Reactions due to stabilizers, polymers or complexes are not included.

II. PHOSPHONIUM SALTS

A. Photolysis of Phosphonium Halides

During the photolysis of triarylphosphines, fission of the carbon–phosphorus bond leads to aryl- and diaryl-phosphinyl radicals[7], whereas in the presence of aryl halides tetraarylphosphonium salts are obtained as stable products[8]. The irradiation of triphenylphosphine and lithium chloride leads to biphenyl, diphenylphosphine, methyl diphenylphosphinate, methyldiphenylphosphine, triphenylphosphine and triphenylphosphine oxide as soon as tetraphenylphosphonium iodide is treated with potassium iodide in aqueous solution[7]. The photoreactions of phosphines with aryl halides and phosphonium salts in various mixtures, and also in the presence of alkali metal salts, have been reported with reference to homolytic C—P bond cleavage[9]. The photolytic decomposition of tetraarylphosphonium halides has been investigated and described in great detail[7,10]. Thus, the photolysis of benzyltriphenylphosphonium chloride in a solution of benzene and ethanol produces as detectable products biphenyl, diphenylmethane, bibenzyl, diphenylphosphine and its oxidation products, in addition to triphenylphosphine. Mixed phosphines (benzylphenyl, benzyldiphenyl and benzylethylphenyl) and the chlorine-containing compounds were not detected[11].

The products obtained suggest the following reaction sequence: (a) one-electron transfer with generation of benzyltriphenylphosphoranyl radical[12-14] and a chlorine atom; (b) radical decomposition to yield benzyl radicals; (c) radical decomposition to yield phenyl radicals; (d) coupling reactions; (e) reaction of radicals with the solvent; and (f–h) subsequent reactions of the initial photoproducts[11] (equation 1).

$$
\begin{array}{lll}
\text{(a)} & [PhCH_2Ph_3]^+ Cl^- & \rightleftharpoons [PhCH_2PPh_3]^{\cdot} + Cl^{\cdot} \\
\text{(b)} & [PhCH_2PPh_3]^{\cdot} & \rightleftharpoons PhCH_2{\cdot} + Ph_3P \\
\text{(c)} & [PhCH_2PPh_3]^{\cdot} & \rightleftharpoons Ph^{\cdot} + PhCH_2PPh_2 \\
\text{(d)} & 2Ph^{\cdot} & \longrightarrow Ph{-}Ph \\
 & 2PhCH_2^{\cdot} & \longrightarrow PhCH_2CH_2Ph \\
\text{(e)} & Ph^{\cdot} + C_6H_6 & \longrightarrow Ph{-}Ph \\
 & PhCH_2^{\cdot} + C_6H_6 & \longrightarrow PhCH_2Ph \\
\text{(f)} & PhCH_2PPh_2 & \longrightarrow PhCH_2^{\cdot} + Ph_2P^{\cdot} \\
\text{(g)} & Ph_3P & \longrightarrow Ph^{\cdot} + Ph_2P^{\cdot} \\
\text{(h)} & PhCH_2PPh_2 & \longrightarrow Ph^{\cdot} + [PhCH_2PPh]^{\cdot}
\end{array}
\tag{1}
$$

If the assumption of this reaction sequence is correct, the photolysis of tetraphenylphosphonium chloride must then only lead to biphenyl, diphenylphosphine, ethyl diphenylphosphinate and triphenylphosphine and its oxidation products. After 2 h of irradiation, biphenyl, diphenylphosphine and its oxidation products, triphenylphosphine and triphenylphosphine oxide, in a ratio of 3:1:5, along with raw material, are obtained. Ethyl diphenylphosphinate was detected in trace amounts[7]. These results support the postulate of the reversibility of phosphoranyl radical formation in such systems and indicate one-electron transfer processes[15] in the formation and decomposition of the tetraarylphosphonium cation. This reaction is comparable to the observation of an electron transfer from halide ions to hydroxyl radicals or hydrogen atoms in aqueous solutions[16], or the initiation step of tetrakis(*p*-dimethylaminophenyl)ethylene diiodide in ethylene chloride[17]. Mechanistic investigations have been performed on the photolysis of carbethoxymethyltriphenylphosphonium salts (**1**)[18,19]. After irradiation in acetonitrile

$$[Ph_3PCH_2CO_2Et]^+ X^-$$
$$X = Cl, Br, I, NO_3, BF_4, ClO_4$$

(**1**)

with a high-pressure mercury lamp, the following products were identified and determined: triphenylphosphine (TPP), carbethoxymethyldiphenylphosphine (DCP), halobenzene (PhX), diethyl succinate, ethyl phenylacetate, diphenyl, benzene, ethyl acetate and a Brønsted acid (HX). From the results, especially the quantum yields of product formation, one can recognize the relative reactivities of the salts, which depend strongly on the anion in the photochemical reactivity of these species:

$$ClO_4^-, NO_3^- < Cl^- < Br^- < BF_4^- < I^-$$

The yields of TPP and DCP also depend on the photolysis of **1** in various solvents (Table 1)[18,19].

After investigating the photolytic P–Ph bond fission of ylides[20] and the decomposition mode of phosphonium salts, some mechanistic differences in the bond fission in the photolysis of ammonium[21–23] and sulphonium salts[24,25] were observed[19].

Pyrenylmethyltriphenylphosphonium salts, prepared by quaternization of triphenylphosphine with 1-bromomethylpyrene (the corresponding chloride and perchlorate salts are generated by an exchange reaction with silver chloride and silver perchlorate, respectively, in methanol), undergo photochemical solvolysis in alcoholic solutions and produce the corresponding alkoxymethylpyrene and triphenylphosphine compounds. Whereas the analogous naphthylmethyl species undergo similar reactions, the corresponding tributylphosphonium salts are photostable under the same conditions[26].

Whereas the reaction of benzyltriphenylphosphonium[11] and carbethoxymethyltriphenylphosphonium[19] salts produces homolysis of the C—P bond, 2-(pyrenyl)methyltriphenylphosphonium salts form products through heterolysis[26], and 1-(naphthyl)methyltriphenylphosphonium salts show both homolytic and heterolytic fission of this bond[27]. After these controversial results, a new investigation on the photoreactivity of 9-(anthryl)methyltriphenylphosphonium chloride[28] was started. No photoproducts of homolysis were detected, but the reaction gives, apart from triphenylphosphine, the heterolysis product isopropyl 9-(anthryl)methyl ether. The mechanistic explanation of the reaction leads to a radical cation (equation 2). The nature of the aromatic ring has a strong influence on the mechanism of this photoprocess[28].

TABLE 1. Photolysis of carbethoxymethyltriphenylphosphonium bromide in various solvents (1.6 mmol, 45 min, high pressure Hg lamp)

Solvent	TPP[a]	DCP[a]
MeOH	6	0
EtOH	98	6
Pr^{i}OH	101	6
MeCN	79	6
Dmf	165	15
PhH–EtOH	124	20

[a] 10^{-3} mmol.

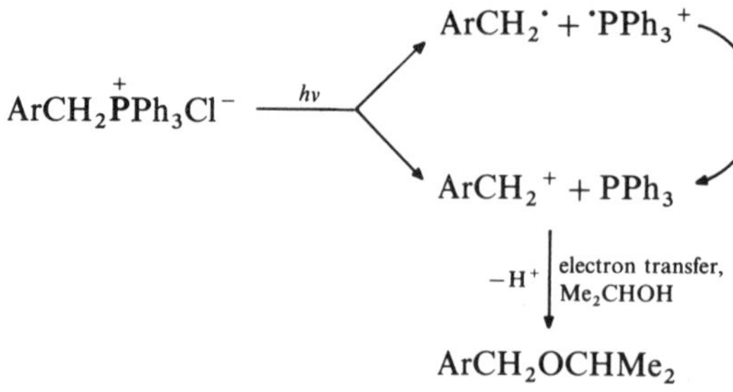

$$\mathrm{ArCH_2\overset{+}{P}Ph_3Cl^-} \xrightarrow{h\nu} \begin{cases} \mathrm{ArCH_2{}^{\bullet} + {}^{\bullet}PPh_3{}^{+}} \\ \mathrm{ArCH_2{}^{+} + PPh_3} \xrightarrow[-\mathrm{H^+}]{\text{electron transfer, } \mathrm{Me_2CHOH}} \mathrm{ArCH_2OCHMe_2} \end{cases} \quad (2)$$

$$(2) + \mathrm{Ph_3\overset{+}{P}CH{=}CH_2\ Br^-} \xrightarrow[-\mathrm{PPh_3},\ -\mathrm{HBr}]{h\nu} (4) \quad (3)$$

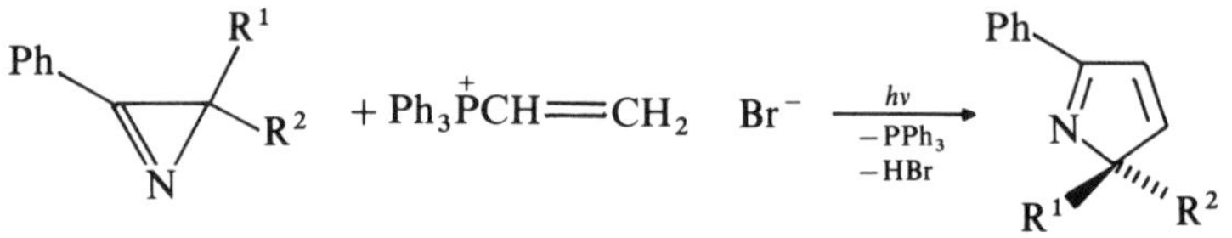

(2) (3) (4)

$R^1 = \mathrm{Me, Ph}$
$R^2 = \mathrm{Me, H}$

Irradiation of 3-phenyl-2*H*-azirines (**2**) with triphenylvinylphosphonium bromide (**3**) in acetonitrile forms 2*H*-indoles (**4**) (equation 3)[29]. The probable mechanism of formation of **4** leads, via benzonitrile ylide and 1,3-dipolar cycloaddition, to the C=C double bond of the vinylphosphonium bromide via the assumed intermediate (**5**).

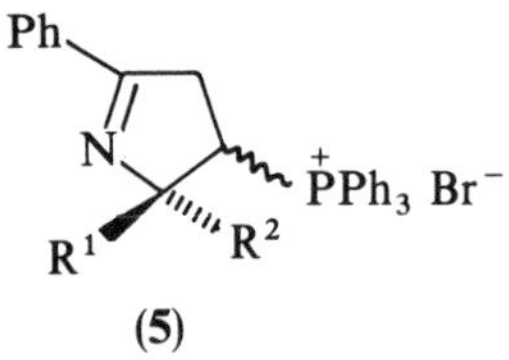

$\mathrm{\overset{+}{P}Ph_3\ Br^-}$

(5)

Triphenylvinylphosphonium salts are usually used as dipolarophiles against azides and diazoalkanes[30]. 2,3-Diphenyl-2*H*-azirine and 1-azido-1-phenylpropene as precursors of 2-methyl-3-phenyl-2*H*-azirine produce, after irradiation, 2,5-diphenylpyrrole and 2-methyl-5-phenylpyrrole, respectively[29].

B. Photolysis of Diazophosphonium Compounds

During the photolysis of α-diazophosphonium salts, the skeletons of the reactants are preserved. Excluding a plausable Wolff rearrangement, only the O/H insertion product of the carbene with the solvents can be obtained. Diazophosphonium tetrafluoroborate (**6**), formed by UV irradiation in methanol under a nitrogen atmosphere and subsequent anion exchange with sodium tetraphenylborate, yields the stable (2-methoxy-2-oxopropyl)-triphenylphosphonium tetraphenylborate (**7**) (equation 4)[31].

The ^{1}H NMR spectrum shows the nature of **7**[31]. The nature of the starting material was determined from analytical and spectroscopic data. The diazo valence resonance lies within the 2140 cm^{-1} short-wavelength range and the carbonyl frequency within the

$$\underset{\mathbf{(6)}}{\mathrm{Ph_3\overset{+}{P}C(=N_2)CMe(=O)\ BF_4^-}} \xrightarrow[-N_2]{h\nu} \mathrm{Ph_3\overset{+}{P}\overset{-}{C}C(=O)Me\ BF_4^-} \xrightarrow{\text{(i) MeOH; (ii) NaBPh}_4} \underset{\mathbf{(7)}}{\mathrm{Ph_3\overset{+}{P}CH(OMe)C(=O)Me\ BF_4^-}} \quad (4)$$

$1675\,cm^{-1}$ long-wavelength range for α-diazocarbonyl compounds[32]. This behaviour is consistent with the diazonium enolate structure of the major species in comparison with the other possible electron isomers (equation 5).

$$\mathrm{Ph_3\overset{+}{P}C(=\overset{+}{N}=\ddot{N}^{:-})C(=O)Me} \longleftrightarrow \mathrm{Ph_3PC(\overset{+}{N}\equiv N:)C(=O)Me} \longleftrightarrow \mathrm{Ph_3\overset{+}{P}C(\overset{+}{N}\equiv N:)C(O^-)Me} \quad (5)$$

C. Photolysis of Other Phosphonium Compounds

The photolysis of zwitterion **8**, a stable 1:1 adduct formed by the reaction between phosphorus trisdimethylamide and tetracyclone in benzene solution, affords 1,2,3,4,5-pentaphenylcyclopenta-1,3-diene (**9**) (equation 6)[33].

$$\underset{\mathbf{(8)}}{\text{tetraphenylcyclopentadienide–O}\overset{+}{P}(NMe_2)_3} \xrightarrow[PhH]{h\nu} \underset{\mathbf{(9)}}{\text{pentaphenylcyclopentadiene}} \quad (6)$$

A small amount of anisomeric material is also obtained. The result of this reaction is comparable to the photochemical behaviour of 1-diazo-2,3,4,5-tetraphenylcyclopenta-2,4-diene. In contrast, the adduct **8** generates, if pyrolysed in the presence of an excess of triphenyl- or tributyl-phosphine or phosphorus trisdimethylamide, hydrocarbon **10** as the major product; this is not observed in the photochemical reaction[33]. The adducts of diazo compounds and phosphines are normally known as stable ylides[34].

(**10**)

Another resonance-stabilized intramolecular phosphonium salt (**11**), the addition product of ethyl 2-cyanocinnamate with triethyl phosphonate, yields after irradiation in cyclohexane solution ethyl 2-cyano-3-diethoxyphosphinyl-3-phenylpropanoate (**12**) and ethylene (**13**) (equation 7)[35]. The structure of **12** was confirmed by IR spectra and an independent synthesis from corresponding reactants.

$$\text{Ph}\diagdown\text{CHCCO}_2\text{Et},\ (\text{EtO})_3\overset{+}{\text{P}}-\text{C}=\text{C}=\text{N}^- \xrightarrow[C_6H_{12}]{h\nu} \text{Ph}\diagdown\text{CHCHCO}_2\text{Et}\ [(\text{EtO})_2\text{P}{=}\text{O},\ \text{CN}] + \text{CH}_2{=}\text{CH}_2 \quad (7)$$

(**11**) (**12**) (**13**)

Photolysis of tris(dimethylamino)azidophosphonium hexafluorophosphate (**14**) leads, via release of nitrogen, to the iminophosphenium salt **15** (synthesis of the first stable iminophosphenium salt[36]) (equation 8)[37]. The photoreaction of cation **14** shows a high dependence on the anion. The hexafluorophosphate achieves the results shown in equation 8, but the bromide yields the iminophosphorane **16**. When the anion is a good

$$(\text{Me}_2\text{N})_3\overset{+}{\text{P}}\text{N}_3\ \text{PF}_6^- \xrightarrow[-\text{N}_2]{h\nu} (\text{Me}_2\text{N})_2\overset{+}{\text{P}}{=}\text{NNMe}_2\ \text{PF}_6^- \longleftrightarrow (\text{Me}_2\text{N})_2\text{PN}{=}\overset{+}{\text{N}}\text{Me}_2\ \text{PF}_6^- \quad (8)$$

(**14**) (**15**)

nucleophile such as Br^-, the formation of an iminophosphorane is certain. On the other hand, the poor nucleophilicity of PF_6^- stabilizes the reaction intermediate formed by migration of one phosphorus substituent on to the nitrogen atom, which generates the iminophosphenium salt **15**[37].

$$(\text{Me}_2\text{N})_3\text{P}{=}\text{NH}$$

(**16**)

The first example of an iminophosphenium salt (**18**) with a P—C bond was obtained by irradiation of the azidophosphonium salt **17** during which migration of a dimethylamino group occurs; migration of the phenyl group leads to **19** (equation 9)[37].

$$\text{Ph}(\text{Me}_2\text{N})_2\overset{+}{\text{P}}\text{N}_3\ \text{PF}_6^- \xrightarrow[-\text{N}_2]{h\nu,\ \text{MeCN}} \text{Ph}(\text{Me}_2\text{N})\overset{+}{\text{P}}{=}\text{NNMe}_2\ \text{PF}_6^- + (\text{Me}_2\text{N})_2\overset{+}{\text{P}}{=}\text{NPh}\ \text{PF}_6^- \quad (9)$$

(**17**) (**18**) (**19**)

D. Radiolysis

The results of the pulse radiolysis in aqueous solution of substituted phosphonium ions have been investigated[38] in addition to those of the electrode reduction of benzyltriphenylphosphonium ion[39] and quaternary phosphonium salts with hydrated electrons[40–42]. The radiolysis of methyltriphenylphosphonium, dimethyldiphosphonium and trimethylphenylphosphonium ions in neutral aqueous solutions leads to phosphoranyl radicals which show transient absorption spectra. This result, along with the kinetic data and information about electrode reduction, suggests that the bond strength of P—Ph is stronger than that of P—Me in their respective phosphoranyl radicals. This trend agrees with the fact that the P—C bond strength is 77 kcal mol^{-1} for Ph_3P and 67 kcal mol^{-1} (1 kcal = 4.184 kJ) for Me_3P[43].

E. Miscellaneous

ESR data obtained on γ-irradiation of tetraalkylphosphonium salts in solution (96% sulphuric acid) at 77 K differ markedly from those for the trialkylphosphonium ions[44]. The latter irradiation produces the expected phosphabetaine radical cation and, for example, tetrabutylphosphonium iodide has features assigned previously to I_2^- [45]. On the other hand, PH_3^+ was not detected[44,46]. Trialkylphosphonium salts, such as trimethyl and triethyl compounds, produce radicals whose ESR spectra are characteristic of the radical cations of R_3P^+. The central region of the ESR spectra is dominated by features from $R_2\dot{C}\overset{+}{P}HR_2$[44,47]. These results are comparable to those for the irradiated trialkylarsonium salts[48].

Experiments on $R_2\dot{C}XR_2$ radicals, which have generally been prepared by the γ-irradiation of the corresponding alkyl derivatives such as R_4Sn, R_4Pb, R_4P^+ hal^- and R_4As^+ hal^-, have been studied by ESR[49]. Further compounds involving Group IV or V anions in placed X have also been discussed[49–55]. All these observations lead to considerable evidence against significant dπ–pπ bonding[49].

III. PHOSPHORANES

A. Photolysis of Phosphoranes

The reaction of tetraarylphosphonium salts with phenyllithium results in pentaarylphosphoranes[56,57]. On the other hand, phosphonium salts which have hydrogen atoms attached to one of the carbons bonded to phosphorus lead to ylides[58–61]. One such system that does not lead to the formation of an ylide is the pentaphosphorane **20**[62]. Homocubyltriphenylphosphorane (**20**) yields, after irradiation through a Pyrex filter at room temperature in a solution of $CHCl_3$ or C_6D_6, tricyclooctadiene (**21**) (equation 10). Cyclooctatetraene, which is normally found in thermal fragmentation, is not obtained[63,64].

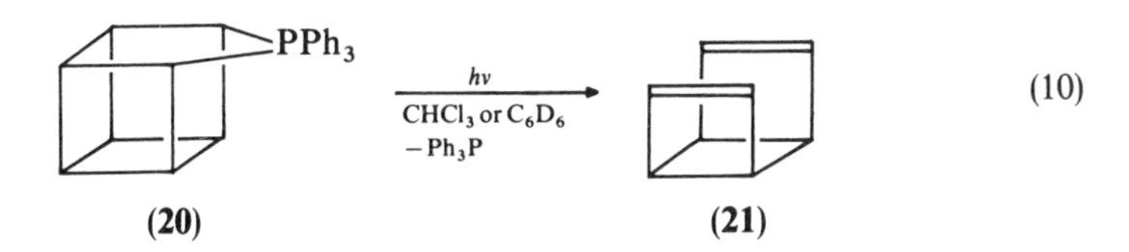

(10)

The photolysis of the first stable pentaalkylphosphorane, homocubyltrimethylphosphorane, in cyclohexene or benzene yields analogous results[63,64].

B. Photolysis of Heterophosphoranes

Through the photolytic extrusion of dimethyl phenylphosphonate from oxazaphosphoranes, the corresponding carbazoles are formed by a phosphorus analogue of the Graebe–Ullmann reaction. The irradiation of 2,2-dimethoxy-3-(4-methoxyphenyl)-2-phenyl-2,3-dihydrobenz-1,3,2-oxazaphospholine (**22**) in benzene gives 3-methoxycarbazole (**23**) and a dimethylphosphonate (**24**) (equation 11)[65]. The unsubstituted species, such as the methoxymethyl derivative, show similar results.

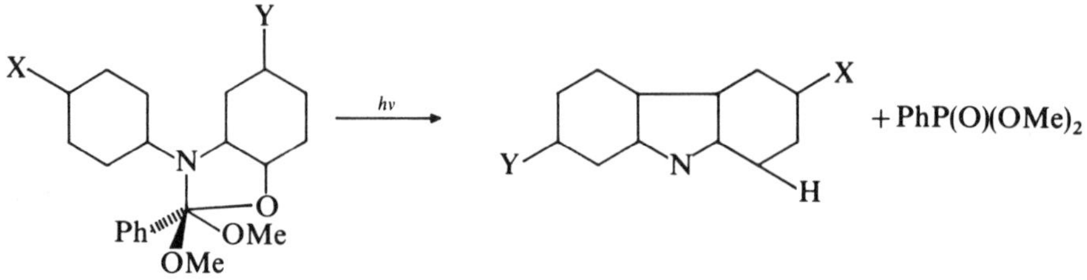

(**22**) (**23**) (**24**)

$+ PhP(O)(OMe)_2$

X = OMe, Y = H
X = Y = H
X = OMe, Y = Me

(11)

Photolysis of pentacoordinate pentavalent phosphorus azides leads to a Curtius-type reaction[36,66]. Further investigations into the photolytic behaviour of λ^5-phosphorus azide derivatives showed that three reaction types can be observed, depending on the nature of the substituents: (a) Curtius-type rearrangement; (b) tautomeric equilibrium between cyclic and open azides; and (c) hydrogen abstraction reaction[66].

Irradiation of the azidophosphorane **25** in acetonitrile solution yields the two diastereoisomers **26** and **27** (equation 12)[66] via to loss of nitrogen. Their structures (*cis* and *trans* species) were confirmed by ^{31}P NMR spectra.

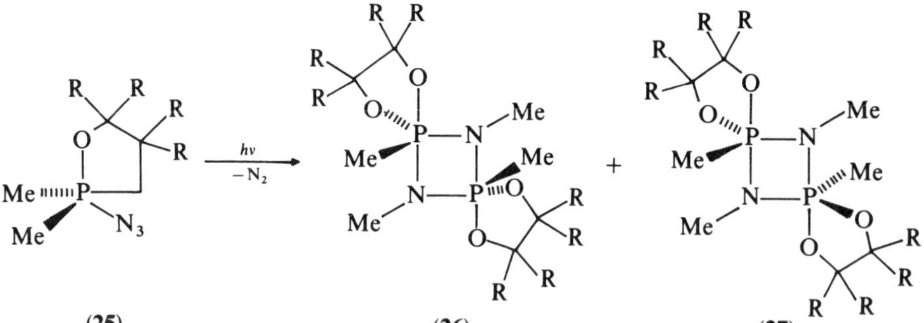

(**25**) (**26**) (**27**)

$R = CF_3$

(12)

Under UV radiation, the phosphorane azide **28**, which is one of the first thermally stable derivatives in which λ^5-phosphorus is bonded to five nitrogen atoms, is quantitatively converted into the new spirocycle phosphorane **29** (equation 13)[66].

Photolysis of the phosphorane azide **30** in benzene at 254 nm yields after 48 h 2-amino-2,2′-spirobi(1,3,2-benzodioxa-λ^5-phosphole) (**31**) (equation 14). Its structure was confirmed by comparing its ^{31}P NMR and IR spectra with those of the authentic material[66].

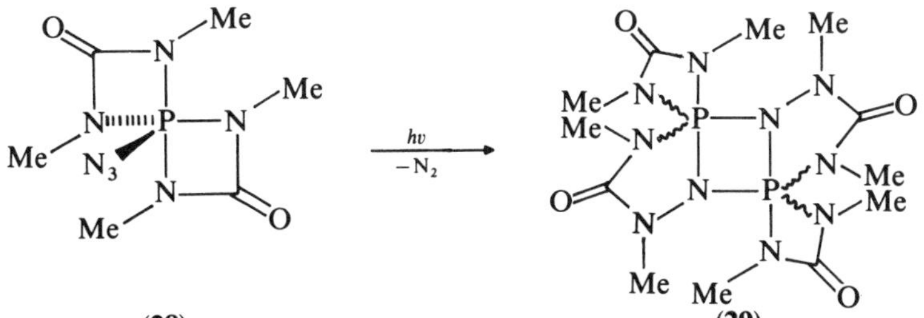

(28) $\xrightarrow[-N_2]{h\nu}$ (29) (13)

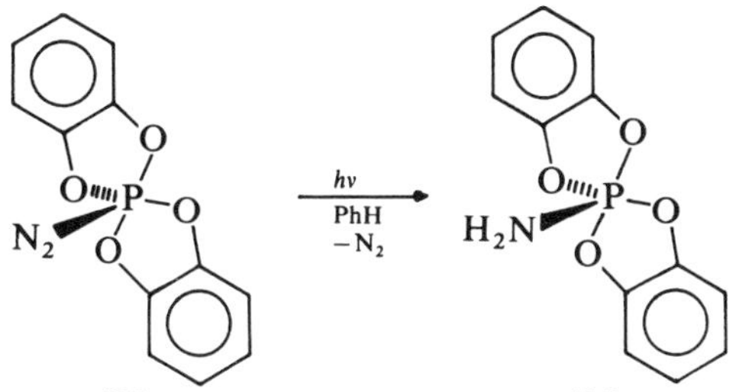

(30) $\xrightarrow[\text{PhH}, -N_2]{h\nu}$ (31) (14)

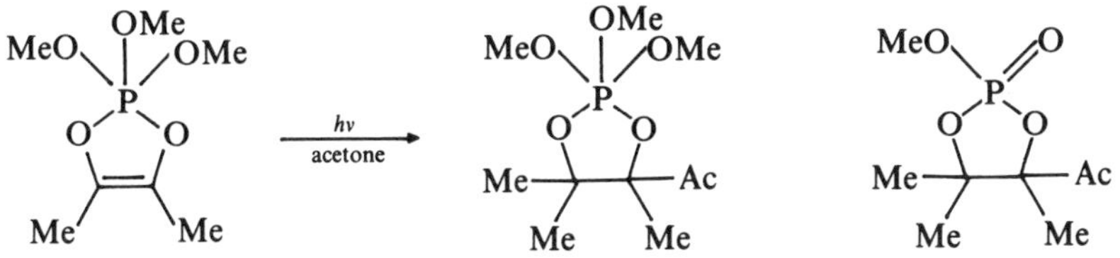

(32) $\xrightarrow[\text{acetone}]{h\nu}$ (33) (34) (15)

Irradiations of 1,3,2-dioxaphosph(V)oles in various solvents with and without sensitizers have also been described[67]. 2,2,2-Trimethoxy-4,5-dimethyl-1,3,2-dioxaphospholene (**32**) added acetone under photo-irradiation to yield the oxaphospholane **33** (equation 15)[68]. Using slightly damp acetone, the cyclic phosphate **34** is obtained. However it could be shown that the oxetane **35** is formed as an intermediate in the photocondensation

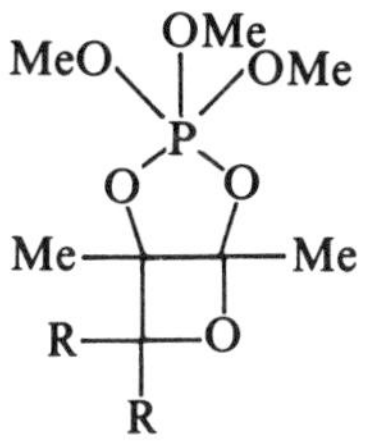

(35)

$R = CD_3$

of acetone-d_6 with the phosphole **32**[69]. The photochemical reaction of phosphole **32** with biphenyl produces a different product ratio of the possible isomers of **34** from those obtained from the thermal reaction[69].

The two-step deoxygenation of benzoyl cyanide to phenylcyanocarbene, which added quickly to alkenes via photolysis of the appropriate 2,2-dihydro-1,3,2,-dioxaphospholanes, has been performed[70]. A homolytic fragmentation mechanism was assumed. When the 1,3,2-dioxaphospholane **36** in *trans*-butene was irradiated, the range of observed substances included **37** and **38** as the major products, in addition to fluorenone and bifluorenylidene (equation 16)[70].

hν, *trans*-MeCH = CHMe

(36) **(37)** **(38)**

R =

(16)

The UV irradiation of the 4,5-dihydro-1,3,5-oxazaphosph(V)olene **39** leads to a cyclic elimination. Analogously to the thermal [5 + 3 + 2] cyclo elimination, the photochemical reaction generated nitrile ylides reacted with alkynes and alkenes to yield the 2*H*-pyrrole **40** and the pyrrol-1-ine **41**, respectively (equation 17)[71].

hν, $-PO(OMe)_3$; $MeO_2CC{\equiv}CCO_2Me$; $MeO_2CCH{=}CHCO_2Me$

(39) **(40)** **(41)**

X = H, *p*-Me

(17)

Using unsymmetric substituted alkenes as dipolarophiles, a mixture of 2- and 3-substituted 1-pyrrolines were obtained[71]. The thermolyses and the photolyses of several oxazaphosph(V)oles (**42**) produce, after $PO(OR^1)_3$ elimination, the corresponding nitrile ylides which are trapped by addition of the intercepting dipolarophiles[72,73].

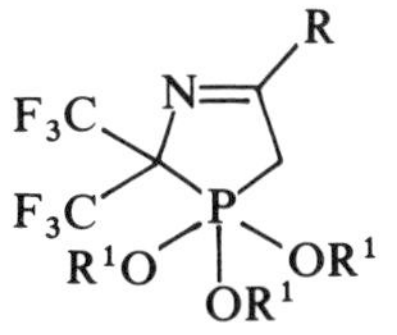

(**42**)

$R = CMe_3$, Ph, $p\text{-}MeC_6H_4$, $CHMe_2$
$R^1 = Et$, Me

C. Miscellaneous

Investigations into the stereoisomerization of the tricyclotrioxaazoniaphosphaundecane tetrafluoroborate phosphoranyl radical[74] and phosphorus hybridization in the equatorial and apical directions of trigonal bypyramids[75] in the latter system (**43**) by ESR studies have been reported.

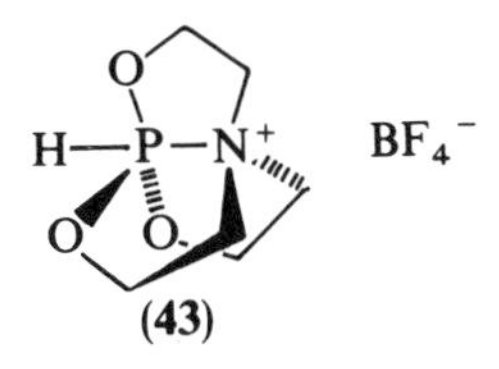

(**43**)

The synthesis of thiols by the photochemical reaction of an olefinic compound with hydrogen sulphide is possible in the presence of an initiator which contains the phosphorane group[76].

IV. YLIDES

A. Photolysis of *P*-Ylides

Phosphorus ylides are useful synthons[77–79] and there is great interest in the bond between phosphorus and carbon[80–82]. Today, the zwitterionic character of this bond is undisputed[82]. After an early indication of a thermal fission of alkoxy-substituted alkylidenephosphoranes[83], there have been few subsequent studies that have suggested decomposition of the phosphorus ylides into phosphanes and carbenes[84–88].

Diphenylmethylenetriphenylphosphine (**44**), synthesized by heating benzophenone triphenylphosphazine under a reduced nitrogen atmosphere[89,90], produces, after irradiation in cyclohexene solution, diphenylmethane (**45**), 1,1,2,2-tetraphenylethane (**46**), 1,1′-bicyclohex-2-ene (**47**) and triphenylphosphine (**48**) (equation 18)[84].

Photolysis of carbethoxymethylenetriphenylphosphine in cyclohexene yields benzene, ethyl acetate, ethyl cyclohexylacetate, ethyl cyclohex-2-ene-1-acetate, phenylcyclohexane 1,1′-bicyclohex-2-ene (Quantum yield measured by use of a low pressure mercury lamp as a light source; no yield in material) and diphenyl phosphinic acid. In this case, no triphenylphosphine is produced. On the other hand, pyrolysis of this carbethoxymethylene compound shows that only P=C bond fission occurs[91]. Using acetylmethylenetriphenylphosphine, the observed products are analogous[20]. However, the irradiation of

$$Ph_3P{=}CPh_2 \xrightarrow[\text{cyclohexene}]{h\nu} CH_2Ph_2 + Ph_2CHCHPh_2 + \text{(3,3'-bicyclohexenyl)} + Ph_3P \qquad (18)$$

(44) (45) (46) (47) (48)

diphenylmethylenetriphenylphosphine generates triphenylphosphine whose formation depends on the wavelength of the exciting radiation. No triphenylphosphine is produced at wavelengths below 300 nm. Table 2 shows the product distribution formed in quartz and Pyrex tubes[92].

The quantum yields of the products[92] also depend on the wavelength. The radiation intensity was determined by ferrioxalate actinometry[93,94]. During the irradiation of $Ph_3P{=}CPh_2$ using radiation at wavelengths above 320 nm, the P=C bond is broken first via an intramolecular charge transfer or by light absorption at a benzhydryl moiety, and Ph_3P is produced. The radiation energy at 253.7 nm is absorbed mainly at a triphenylphosphinyl moiety, and then the P—Ph bond is broken[92]. Another example of this photolysis pathway is the irradiation of dimethylmethylenetriphenylphosphine. Short-wavelength irradiation produces benzene and a resinous compound. Long-wavelength irradiation results in Ph_3P and $Me_2C{=}CMe_2$ with small amounts of $Me_2CHCHMe_2$ and $MeCH{=}CH_2$ and traces of products formed by reaction with the solvent[95].

Unsensitized and sensitized photolyses of benzoylmethylene(triphenyl)phosphorane (**49**) in cyclohexene produce acetophenone (**50**) and 7-norcaryl phenyl ketone (**51**) in different ratios (equation 19)[87]. Direct irradiation also leads to triphenylphosphine (**52**) and other products[87]. By analogy with the reaction of diazoacetophenone[96] and dimethyl-

TABLE 2. Photolysis of $Ph_3P{=}CPh_2$ in cyclohexene solution

Product	Yield (%)	
	Quartz tube	Pyrex tube
PhH	78	5
Ph_2CH_2	6	—
Ph_2CH–(cyclohexenyl)	57	56
$Ph_2CHCHPh_2$	10	—
Ph_3P	0[a]	90
(bicyclohexenyl)	—	—
Ph–(cyclohexyl)	11	—

[a] $Ph_2P(O)OH$ is obtained in 18% yield on oxidation of the photolysed solution.

$$Ph_3P{=}CHC(O)Ph \xrightarrow{h\nu,\ \text{cyclohexene}} PhC(O)Me + C_6H_{10}\text{-}C(O)Ph + Ph_3P \quad (19)$$

(49) → hν, cyclohexene: (50) : (51) = 1 : 1.2; sensitizer: (50) : (51) = 6.5 : 1; + Ph_3P (52)

sulphonium phenacylide[97] in cyclohexene, which show an identical distribution of photoproducts, a triplet benzoylcarbene intermediate can be assumed[87].

In the course of investigations on α-sulphanyl carbenes, a new cleavage reaction was found. After irradiation the phosphorus ylide **53** yields nearly quantitatively the thiaazaphosphetane **54**, in addition to triphenylphosphine (**55**) (equation 20)[98]. An insertion of the corresponding carbene intermediate in to the methine—CH bond is plausible. The structure of **54** was confirmed by single-crystal X-ray structure analysis[98].

$$Ph_3\overset{+}{P}\overset{-}{C}(SR)P(S)(NPr^i_2)_2 \xrightarrow{h\nu} \mathbf{54} + Ph_3P \quad (20)$$

(53) $R = 4$-MePh; (54); (55)

Bis(diisopropylamino)thiophosphoranyl(triphenylphosphoranylidene)ethane (**56**) and 1-[bis(diisopropylamino)thiophosphoranyl]-1-(triphenylphosphoranylidene)methane (**57**) show a similar photolytic behaviour. However, the failure of attempted cyclopropanation reactions of alkenes and the absence of the expected 1,2-H shift product leave doubts as to the presence of free carbenes[99].

$$Ph_3\overset{+}{P}{-}\overset{-}{C}(R)P(S)(NPr^i_2)_2$$

(**56**) R = Me
(**57**) R = H

The photo-oxygenation of phosphorus ylides with singlet oxygen is a simple reaction. A typical example is methoxycarbonylbenzylidenetriphenylphosphorane (**58**), which yields methylphenylglyoxylate (**59**) quantitatively after irradiation (equation 21)[100].

$$Ph_3P{=}C(Ph)(CO_2Me) \xrightarrow{^1O_2} PhCOCO_2Me + \text{unidentified side products} \quad (21)$$

(**58**) (**59**)

The behaviour of isobutylidenephosphorane and ethylidenephosphorane is analogous and the photoreactions produce α-ketoisobutyrate and methyl pyruvate, respectively[100]. All these reactions were carried out in the presence of oxygen and *meso*-tetraphenylporphin or Rose Bengal. The following reaction mechanism shown in equation 22 can be assumed[100].

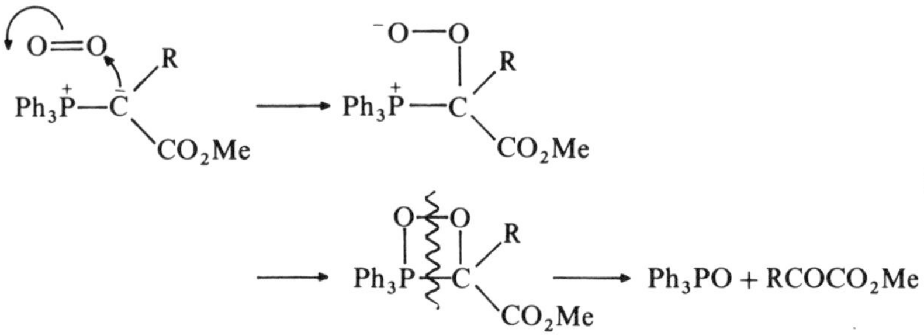

$$O{=}O + Ph_3\overset{+}{P}{-}\overset{-}{C}(R)(CO_2Me) \longrightarrow Ph_3\overset{+}{P}{-}C(R)(CO_2Me){-}O{-}O^- \longrightarrow [Ph_3P{-}C(R)(CO_2Me){-}O{-}O \text{ (cyclic)}] \longrightarrow Ph_3PO + RCOCO_2Me \quad (22)$$

B. Photolysis of Phosphazines

Diarylmethylenetriphenylphosphazines can be regarded as nitrogen analogues of acyclic conjugated dienes with a phosphorus atom in one part of the system and, according to this investigation[101], can also be considered to be homogolous ylides as the two possible resonance structures show (equation 23). Irradiation of diphenylmethylenetriphenylphosphazine (**60**) in benzene or decalin yields benzophenone, triphenylphosphine, tetraphenylethylene, triphenylphosphine oxide, tetraphenylethane, diphenylmethyl azine and bisdiphenylmethyl ether. A comparison of the products obtained by

$$Ar_2C{=}N{-}N{=}PPh_3 \longleftrightarrow Ar_2\bar{C}{-}N{=}N{-}\overset{+}{P}Ph_3 \quad (23)$$

(**60**)

R = Ph

both the thermal and photolytic decompositions of diarylmethylenetriphenylphosphazines indicate the same major fragmentation pathway which produces carbene, nitrogen and triphenylphosphine. The product distributions can be obtained on the basis of carbene stability, concentration, hydrogen atom availability and oxygen availability. The product analysis suggests a more radical-like behaviour of the carbene produced in the photolytic reaction; more triplet carbene production and/or availability might be inferred[101].

Photolysis of benzylidenehydrazono-, diphenylmethylenehydrazono-, 9-fluorenylidenehydrazono-, 9-xanthenylidenehydrazono- and 10-methyl-9,10-dihydroacridin-9-ylidenehydrazono-triphenylphosphorane (**61a**–**e**) in CH_2Cl_2 with methylene blue at −78 °C with a tungsten–halogen lamp through a yellow optical filter in the presence of oxygen gives the corresponding carbonyl compounds (**62a**–**e**) and triphenylphosphine oxide (**62**) with chemiluminescent light emission (equation 24)[102,103].

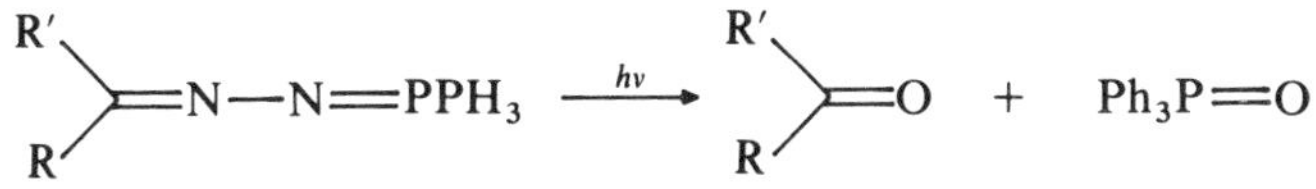

$$\mathrm{RR'C{=}N{-}N{=}PPh_3} \xrightarrow{h\nu} \mathrm{RR'C{=}O} + \mathrm{Ph_3P{=}O}$$

(61a–e) **(62a–e)** **(63)**

(**a**) R = H, R′ = Ph
(**b**) R = R′ = Ph
(**c**) R + R′ =
(**d**) R + R′ =
(**e**) R + R′ =

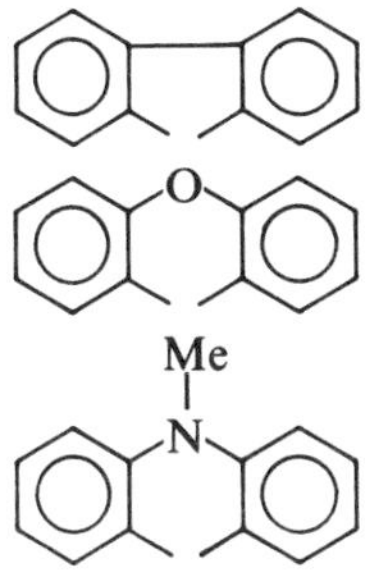

(24)

It can be shown by ^{13}C Fourier transform NMR spectroscopy at −78 °C that a six-membered peroxide is an intermediate which disappears completely on warming, with formation of the observed reaction products. The extrusion of nitrogen from the six-membered peroxide yielded the four-membered peroxide (equation 25) accompanied by light emission[103].

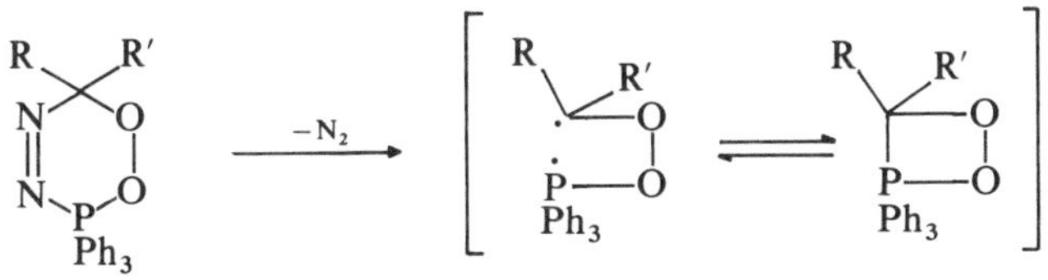

$\xrightarrow{-N_2}$ (25)

Comparative investigations have been performed on the photo-oxygenation of azines[104–108], nitrogen analogues of acyclic conjugated dienes and phosphazines[102,103] in connection with chemiluminescent systems[109,110]. In addition to the pathway discussed above, a free-radical pathway initiated by singlet oxygen to give a linear peroxide polymer has been proposed[107]. Photoreaction of 1,1,3,3-tetramethylindan-2-onetriphenylphosphazine with oxygen at 15 °C leads to the corresponding ketone, triphenylphosphine oxide and 2,2,5,5-tetramethyl-3,4-benzopent-3-en-5-olide. When the reaction is carried out at −78 °C followed by heating to room temperature, strong light emission is observed[111]. Further experiments established the existence of the phosphadiazine and the phosphadioxetane[112].

C. Miscellaneous

(Trimethylsilylimino) {(trimethylsilyl)[tris(dimethylamino)phosphoranylidene]methyl} phosphine (**64**), produced by the reaction of [(dichlorophosphanyl)methyl]tris (dimethylamino)phosphonium tetraphenylborate with sodium bis(trimethylsilyl)amide,

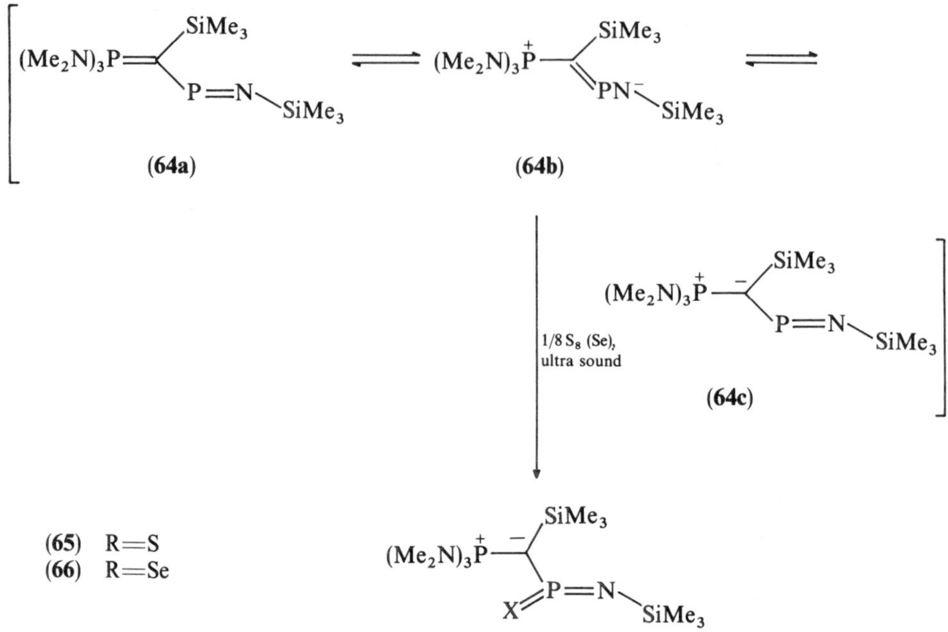

$$\left[(Me_2N)_3P{=}C(SiMe_3)P{=}N{-}SiMe_3 \;(\mathbf{64a}) \rightleftharpoons (Me_2N)_3\overset{+}{P}{-}C(SiMe_3){=}PN^{-}{-}SiMe_3 \;(\mathbf{64b}) \rightleftharpoons (Me_2N)_3\overset{+}{P}{-}\overset{-}{C}(SiMe_3)P{=}N{-}SiMe_3 \;(\mathbf{64c}) \right]$$

1/8 S_8 (Se), ultra sound

$$(Me_2N)_3\overset{+}{P}{-}\overset{-}{C}(SiMe_3){-}P(=X){=}N{-}SiMe_3$$

(**65**) R=S
(**66**) R=Se

(26)

reacts with sulphur or selenium under ultrasonic conditions to form the corresponding thioxo- and selenoxo-(trimethylsilylimino){(trimethylsilyl)[tris(dimethylamino)phosphoranylidene]methyl}phosphoranes (**64** and **66**, respectively) (equation 26)[113].

Because the $C_{(1)}$—$P_{(1)}$ bond in **64** is shortened, the $(Me_2N)PC(SiMe_3)$ group is a strong π-donor in this compound and the mesomeric structures **64b** and **64c** give a good description of the heteroallyl anion[113].

X-irradiation of monocrystalline triphenylphosphinetrichloroborane and triphenylphosphinetrifluoroborane results in Ph_3P^+. The δ tensor and ^{31}P coupling tensors depend on the matrix. ESR findings and investigations into isotropic and anisotropic compounds of the ^{31}P coupling constants have been discussed in terms of electronegativity differences between the A and B groups of AB_3. Comparison of these results with those of CNDO calculations have been reported[114]. ESR data on radicals formed by the irradiation of methylenephosphoranes[54,55], carbomethylenephosphoranes[49], carboethyoxymethylenephosphoranes[54,55,115] and diarylmethylenephosphazines[101,116] in a single-crystal matrix indicates that $Ph_3P^+—^{13}\dot{C}H_2$ radicals are formed[117] in addition to an allylic species involving the participation of a phosphoranyl moiety[80,118,119].

The photoelectron spectra of 17 salt-free phosphorus ylides have been recorded and the results compared with those of CNDO calculations; good agreement was formed[80,120,121].

V. REFERENCES

1. (a) A. Schönberg, *Präparative Organische Photochemie*, Springer, Berlin, Göttingen, Heidelberg, 1958; (b) A. Schönberg, G. O. Schenk and O. A. Neumüller, preparative Organic photochemistry, 2nd ed., Springer,Berlin, Heidelberg, New York, 1968.

2. D. Bryce-Smith, *Photochemistry*, Vols 1–, Chemical Society, London, 1969–.
3. M. Dankowski, *Chem. Ztg.*, **108**, 303 (1984).
4. R. S. Davidson, *Organophosphorus Chemistry*, Vols 1–, Chemical Society, London, 1970–.
5. M. Dankowski, in *The Photochemistry of Organophosphorus Compounds* (Ed. F. R. Hartley), Vol.1, Wiley, Chichester, 1990, p. 489.
6. M. Dankowski, in *The photochemistry of organophosphorus Compounds* (Ed. F. R. Hartley), Vol. 2,Wiley, Chichester, 1992, p137.
7. M. L. Kaufman, *Diss. Abstr. B*, **27**, 2295 (1967); University Microfilms, Ann Arbor, MI, Order No. 66–13874.
8. J. B. Plumb and C. E. Griffin, *J. Org. Chem.*, **27**, 4711 (1957).
9. M. L. Kaufman, *Tetrahedron Lett.*, 769 (1965).
10. C. E. Griffin, *U. S. Dep. Commerce. off. Tech. Serv., AD Rep.*, AD 62 72 69, 1965.
11. C. E. Griffin and M. L. Kaufman, *Tetrahedron Lett.*, 773 (1965).
12. G. Kamai and **Z.** Kbarrasova, Zh. Obshch. Khim., **27**, 953 (1957).
13. C. Walling and Rabinowitz, *J. Am. Chem. Soc.*, **75**, 5326 (1957).
14. F. Ramirez and N. McKelvie, *J. Am. Chem. Soc.*, **79**, 5829 (1957).
15. R. Stewart, *Oxidation Mechanisms*, Benjamin New York, 1964.
16. A. O. Allen, *The Radiation Chemistry of Water and Aqueous Solutions*, Van Nostrand, princeton, NJ, 1961.
17. D. H. Anderson, R. M. Elofson, H. S. Gutowsky, S. Levine and R. B. Sandin, *J. Am. Chem. Soc.*, **83**, 3157 (1961).
18. Y. Nagao, K. Shima and H. Sakurai, *Tetrahedron Lett.*, 1101 (1971).
19. Y. Nagao, K. Shima and H. Sakurai, *Bull. Chem. Soc. Jpn.*, **45**, 3122 (1972).
20. Y. Nagao, K. Shima and H. Sakurai, *Kogyo Kagaku Zasshi*, **72**, 236 (1969).
21. J. W. Knapczyk and W. E. MacEven, *J. Org. Chem.*, **35**, 2539 (1970).
22. A. L. Maucock and G. A. Berchtold, *J. Org. Chem.*, **35**, 2533 (1970).
23. T. Laird and H. Williams, *Chem. Commun.*, 561 (1961).
24. T. D. Walsh and R. C. Long, *J. Am. Chem. Soc.*, **89**,3943 (1967).
25. C. Pac and H. Sakurai, *Chem. Commun.*, 20 (1969).
26. M. E. C. Dias Real Oliveira and L. C. Pereira, *J. Photochem.*, **31**, 373 (1985).
27. E. O. Alonso, M. E. R. Marcondes and V.G. Toscano, *Phosphorus Sulfur*, **30**, 737 (1987).
28. T. M. S. Peranovich, M. E. R. Marcondes and V. G Toscano, *Phosphorus Sulfur Silicon Relat. Elem.*, **51/52**, 314 (1990).
29. N. Gakis, H. Heimgartner and H. Schmidt, *Helv. Chim. Acta*, **57**, 1403 (1974).
30. D. Redmore, *Chem. Rev.*, **71**, 315 (1971).
31. M. Regitz, M. Abd El-Razak and H. Heydt, *Liebigs Ann. Chem.*, 1865 (1981).
32. M. Regitz, *Diazoalkane*, 1st ed., Georg Thieme, Stuttgart, 1977, p.13
33. M. J. Gallagher and I. D. Jenkins, *J. Chem. Soc., C*, 2605 (1969).
34. D. Lloyd, M. I. C. Singer, M. Regitz and A. Liedhegener, *Chem Ind. (London)* 324 (1967).
35. C. -G Shin, Y. Yonezawa, Y. Sekine and J. Yoshimura, *Bull. Soc. Chem. Jpn.*, **48**, 1321 (1975).
36. M. Sanchez, M. R. Marre, J. F. Brazier, J. Bellan and P. Wolf, *Phosphorus Sulfur*, **14**, 331 (1983).
37. J. -P. Majoral, G. Bertrand, A. Baceiredo, M. Mulliez and R. Schmutzler, *Phosphorus Sulfur*, **18**, 221 (1983).
38. H. Horii, S. Fujita, T. Mori and S. Taniguchi, *Radiat. Phys. Chem.*, **19**, 231 (1982).
39. H.Hori, S. Fujita, T. Mori and S. Taniguchi, *Bull. Chem. Soc. Jpn.*, **52**, 3099 (1979).
40. H. Hori, S. Fujita, T. Mori and S. Taniguchi, *Radiat. Phys. Chem.*, **8**, 521 (1976).
41. L. Horner and A. Mentrup, *Justus* Liebgs *Ann. Chem.*, **646**, 65 (1961).
42. L. Horner and J. Haufe, *Chem, Ber.*, **101**, 2913 (1963).
43. W. G. Bentrude, in Free Radicals (Ed. J. K. Kochi), Vol. 2, Wiley–Interscience, New York, 1973, Chapter 22.
44. A. Begun, A. R. Luons, M. C. R. Symons, *J. Chem. Soc. A*, 2388 (1971).
45. C. L. Marquardt, *J. Chem. Phys.*, **48**, 994 (1968).
46. A. Begum, A. R. Lyons and M. C. R. Symons, *J. Chem. Soc. A*, 2290 (1971).
47. A. R. Lyons, G. W. Neilson and M. C. R. Symons, *J. Chem. Soc., Faraday Trans. 2*, **68**, 1063 (1972).
48. A. R. Lyons and M. C. R. Symons, *J. Am. Chem. Soc.*, **95**, 3483 (1973).
49. A. R. Symons, G. W. Neilson and M. C. R. Symons, *J. Chem. Soc., Faraday Trans. 2*, **68**, 807 (1972).

50. P. J. Krusic and J. K. Kochi, *J.Am. Chem. Soc.*, **91**, 6161 (1969).
51. R. W. Fessenden, *J. Phys. Chem .*, **71**, 74 (1967).
52. J. H. Mackey and D. E. Wood, *Mol. Phys.*, **18**, 783 (1970).
53. A. Hudson and H. A. Hussain, *J. Chem. Soc. B*, 793 (1969).
54. E. A. C. Lucken and C. Mazeline, *J. Chem. Soc. A*, 1074 (1966).
55. E. A. C. Lucken and C. Mazeline, *J, Chem. Soc. A*, 439 (1967).
56. G. Wittig and M. Rieber, *Justus Liebigs Ann. Chem.*, **562**, 187 (1949).
57. G. Wittig and M. Rieber, *Justus Liebigs Ann. Chem.*, **562**, 177 (1949).
58. G. Wittig and G. Geissler, *Justus Liebigs Ann. Chem.*, **580**, 44 (1953).
59. D. Seyferth, W. B. Hughes and J. K. Heeren, *J. Am. Chem. Soc.*, **87**, 2847 (1965).
60. A. Maercker, *Org. React.*, **14**, 270 (1965).
61. D. Seyferth, W. B. Hughes and J. K. Heeren, *J. Am. Chem. Soc.*, **87**, 3467 (1965).
62. T. J. Katz and E. W. Turnbloom, *J. Am. Chem. Soc.*, **92**, 6701 (1970).
63. E. W. Turnbloom and T. J. Katz, *j. Am. Chem. Soc.*, **95**, 4292 (1973).
64. T. J. Katz and E. W. Turnbloom, *J. Am. Chem. Soc.*, **93**, 4065 (1971).
65. J. I. G. Cadogan, B. S. Tait and N. J. Tweddle, *J. Chem. Soc., Chem. Commun.*, 847 (1975).
66. A. Baceiredo, G. Bertrand, J. -P. Majoral, U. Wermuth and R. Schmutzler, *J. Am. Chem. Soc.*, **106**, 7065 (1984).
67. W. G. Bentrude, *Chem. Commun.*, 174 (1967).
68. W. G. Bentrude and K. R. Darnall, *Tetrahedron Lett.*, 2511 (1967).
69. W. G. Bentrude and K. R. Darnall, *J. Chem. Soc. D*, **15**, 862 (1969).
70. P. Petrellis and G. W. Griffin, *Chem. Commun.*, 1099 (1968).
71. K. Burger and J. Fehn, *Tetrahedron Lett.*, 1263 (1972).
72. K. Burger and J. Fehn, *Chem, Ber.*, **105**, 3814 (1972).
73. K. Burger, J. Albanbauer and F. Manz, *Chem. Ber.*, **107**, 1823 (1974).
74. J. H. H. Hamerlinck, P. Schipper and H. M. Buck, *J. Chem. Soc., Chem. Commun.*, 1148 (1981).
75. J. H. H. Hamerlinck, P. Schipper and H. M. Buck, *j. Org. Chem.*, **48**, 306 (1983).
76. E. Arretz, C. Landoussy, A. Mirassou and J. Ollivier (Soc. Nat. Elf Aquitaine), Eur. Pat., 60 754 (1982).
77. H. J. Bestmann and R. Zimmermann, in *Houben*–Weyl–Müller: Methoden der Organischen Chemie (Ed. M. Regitz), Vol. El, Thieme, Stuttgart, 1982, p. 616.
78. H. Schmidbauer, *Angew., Chem.*, **95**, 980 (1983): *Angew. Chem., Int. Ed. Engl.* **22**, 907 (1983).
79. W. C. Kaska, *Coord. Chem. Rev.*, **48**, 1 (1983).
80. K. A. Starzewski and H. Bock, *J. Am. Chem. Soc.*, **98**, 8486 (1976).
81. H. J. Bestmann, A. J. Kos, K. Witzgall and P. von R. Schleyer, *Chem. Ber.*, **119**, 1331 (1986).
82. M. M. Francl, R. C. Pellow and L. C. Allen, *J. Am. Chem. Soc.*, **110**, 3723 (1983).
83. G. Wittig and W. Böll, *Chem. Ber.*, **98**, 2527 (1962).
84. H. Tschesche, *Chem. Ber.*, **98**, 3318 (1965).
85. H. Bestmann and T. Denzel, *Tetrahedron Lett.*, 3591 (1966).
86. A. Ritter and B. Kim, *Tetrahedron Lett.*, 3449 (1968).
87. R. R. Da Silva, V. G. Toscano and R. G. Weiss, *J. Chem. Soc., Chem. Commun.*, 567, (1973).
88. E. Anders, T. Clark and T. Gassner, *Chem. Ber.* **119**, 1350 (1986).
89. H. Staudinger and J. Meyer, *Helv. Chim. Acta*, **2**, 633 (1919).
90. H. Staudinger and J. Meyer, *Helv. Chim. Acta*, **2**, 641 (1919).
91. Y. Nagao, K. Shima and H. Sakurai, *Bull. Chem. Soc. Jpn.*, **43**, 1885 (1970).
92. Y. Nagao, K. Shima and H. Sakurai, *Tetrahedron Lett.*, 2221 (1970).
93. C. A. Parker, *Proc. R. Soc. London, Ser. A*, **220**, 104 (1953).
94. C. G. Hatchard and C. A. Parker, *Proc. R. Soc. London, Ser. A*, **235**, 518 (1956).
95. H. Dürr, D. Barth and M. Schlosser, *Tetrahedron Lett.*, 3045 (1974).
96. D. O. Cowan, M. M. Couch, K. R. Kopecky and G. -S. Hammond, *J. Org. Chem.*, **29**, 1922 (1964).
97. B. M. Trost, *J. Am. Chem. Soc.*, **89**, 138 (1967).
98. G. Sicard, H. Grützmacher, A. Baceiredo, J. Fischer and G. Bertrand, *J. Org. Chem.*, **54**, 4426 (1989).
99. H. Grützmacher, *Z. Naturforsch., Teil B*, **45**, 170 (1990).
100. C. W. Jefford and G. Barchietto, *Tetrahedron Lett.*, 4531 (1977).
101. D. R. Dalton and S. Liebman, *Tetrahedron*, **25**, 3321 (1969).
102. N. Suzuki, S. Wakatsuki and Y. Izawa, *Tetrahedron Lett.*, **21**, 2319 (1980).

103. N. Suzuki, K. Sano, S. Wakatsuki and Y. Izawa, *Bull. Chem. Soc. Jpn.*, **55**, 842 (1982).
104. G. Rio and J. Berthelot, *Bull. Soc. Chim. FR.*, 1509 (1970).
105. P. Leuchtken, *Z. Naturforsh., Teil B*, **31**, 1436 (1976).
106. S. S. Talwar, *Indian J. Chem.*, **16**, 980 (1978).
107. M. E. Landis and D. C. Madoux, *J. Am. Chem. Soc.*, **101**, 5106 (1979).
108. W. Ando, R. Sato, H. Sonobe and T. Akasaka, *Tetrahedron Lett.*, **25**, 853 (1984).
109. J. Lind, G. Merenyi and T. E. Eriksen, *J. Am. Chem. Soc.*, **105**, 7655 (1983).
110. S. Ljunggren, G. Mereny and J. Lind, *J. Am. Chem. Soc.*, **105**, 7666 (1983).
111. T. Akasaka, R. Sato, Y. Miyama and W. Ando, *Tetrahedron Lett.*, **26**, 843 (1985).
112. T. Akasaka, R. Sato and W. Ando, *J. Am. Chem. Soc.*, **107**, 5539 (1985).
113. U. Krüger, H. Pritzkow and H. Grützmacher, *Chem. Ber.*, **124**, 329 (1991).
114. T. Berclaz and M. Geoffroy, *Mol. Phys.*, **30**, 549 (1975).
115. M. Zanger and R. Poupko, *Spectrosc. Lett.*, **10**, 739 (1977).
116. D. R. Dalton, S. A. Liebman, H. Waldman and R. S. Sheinson, *Tetrahedron Lett.*, 145 (1968).
117. M. Geoffroy, L. Ginet and E. A. C. Lucken, *Mol. Phys.*, **34**, 1175 (1977).
118. M. Geoffroy, G. Rao, Z. Tancic and G. Bernardinelli, *J, Am. Chem. Soc.*, **112**, 2826 (1990).
119. J. A. Baban, C. J. Cooksey and B. P. Roberts, *J. Chem. Soc., Perkin Trans. 2*, 781 (1979).
120. H. Schmidbauer and W. Tronich, *Chem. Ber.*, **101**, 595 (1968).
121. J. C. J. Bart, *J. Chem. Soc. B*, 350 (1969).

CHAPTER **7**

Chemical analysis of organophosphorus compounds

H. FEILCHENFELD

Department of Organic Chemistry, The Hebrew University of Jerusalem, Jerusalem 91904, Israel

The chemistry of organophosphorus compounds, Vol. 3
Edited by F. R. Hartley

I. INTRODUCTION

As organophosphorus compounds have been playing an increasingly important role in our lives, from plastic stabilizers and fire retardants to agricultural pesticides and nerve gases, their analysis has been the subject of intense development. Thanks to the advent of ever more sensitive instruments, the detection and determination methods have undergone a steady process of miniaturization, from large-scale analysis to the assessment of micro-, nano- and even pico-gram amounts. At the same time, analytical procedures have evolved from the simple determination of total phosphorus content to the more sophisticated study of individual compounds in multi-component mixtures.

II. ANALYSIS OF PHOSPHORUS AND PHOSPHATES

Most chemical methods for the analysis of organophosphorus compounds involve initial decomposition of the substance and conversion of the phosphorus into orthophosphate or orthophosphoric acid, which can then be determined by a number of widely differing procedures. Breakdown and oxidation are accomplished by dry fusion with alkali, by wet combustion or by oxygen flask combustion. The final determination, essentially an inorganic problem, may be carried out by gravimetry, titrimetry, colorimetry, amperometry and many other techniques. The analysis of elemental phosphorus in organic materials was discussed some years ago[1], and reports on new advances are published biennially[2,3]. Reviews on the analysis of inorganic phosphorus in aqueous media, mainly in the form of phosphate ions, also appear biennially[4]. In addition to the chemical methods, several instrumental techniques allow the direct determination of the total phosphorus content of compounds, without previous oxidation and decomposition; the most frequently used are atomic absorption spectrometry, flame photometry and inductively coupled plasma atomic emission spectrometry.

A. Decomposition and Oxidation Methods

1. *Alkali fusion*

Organophosphorus substances can easily be mineralized by heating them with alkali metal hydroxides, sodium carbonate, sodium nitrate or zinc oxide[1]. Ashing with magnesium nitrate[5-7] and fusion with sodium peroxide[8,9] or with sodium fluoride and potassium pyrosulphate[10] have also been used. Decomposition with metallic potassium in a micro-bomb yields potassium phosphide, which is oxidized with permanganate solution to phosphate[11]. In these methods stable inorganic phosphates are formed, which can be dissolved in acid and determined by one of the many analytical methods available. The procedures are convenient and efficient, but volatile phosphorus compounds are liable to be lost in the early stages of the heating, unless the reaction is carried out in a closed vessel.

2. *Wet combustion procedures*

Wet combustion was originally carried out by the Carius method[12], i.e. by heating for several hours a 0.1–0.2 g sample with nitric acid in a sealed tube at 200–300 °C. The reaction is time consuming and potentially hazardous and has progressively been replaced by more convenient procedures, although recent publications show that slightly modified Carius combustion is still routinely used in some laboratories[13-15]. A recommended mineralization procedure[16] involves, for instance, introduction of 2 mg of the organophosphorus compound into a Pyrex tube, addition of 0.250 ml of 70% sulphuric acid, sealing, heating at 450–470 °C for at least 30 min, cooling and diluting with water.

Kjeldahl's method[17], which consists of oxidation by a mixture of sulphuric acid and potassium permanganate, was originally developed for the determination of nitrogen in organic compounds, and was later found to be suitable for the analysis of a number of elements, including phosphorus. Alternative versions include reaction of the sample with various mixtures of sulphuric, nitric and perchloric acids with hydrogen peroxide or ammonium nitrate[1]. Potassium peroxodisulphate has recently been used for wet oxidation in flow-injection determinations of phosphorus in aqueous samples[18,19] (see Section II.D). A typical Kjeldahl digestion[20] involves mixing a 5–15 mg sample of organophosphorus compound with 1–2 ml of the oxidizing reagent, heating slowly to the boiling point and keeping at boiling temperature for 1.5 h; during that period an additional 1.0 ml of oxidizing solution is added dropwise to the reaction mixture. A large number of slightly modified procedures have been published and recommended[16]. A complication that sometimes arises during the Kjeldahl combustion is the possible loss of volatile phosphorus-containing components; evaporation of phosphoric acid is unlikely as long as an excess of water is present and only becomes a problem in the later stages of the wet digestion, when the samples approach dryness and the temperature rises; it may be avoided by the use of condensers and reflux techniques[1].

Specific wet oxidation procedures should be chosen for each type of compound as a function of the ease or difficulty of complete combustion. The method applied after oxidation for the determination of the inorganic phosphorus should also be taken into account.

3. *Oxygen flask combustion*

Direct combustion of organic samples in an oxygen atmosphere was first used by Hempel[21] to determine their sulphur content. It was re-examined about 40 years ago by Mikl and Pech[22,23] and adapted by Schoniger[24–26] to the microanalysis of sulphur and halogens in organic substances. The procedure was found to work also for the analysis of phosphorus[5,27–35]. Schoniger's method uses a flask provided with a ground-glass stopper, to which a platinum wire basket or spiral is attached. The sample is wrapped in an ashless filter-paper and inserted into the platinum holder. Before the combustion a small volume of an acidic (sulphuric or nitric acid) or basic (sodium hypobromite) solution is introduced into the flask to absorb the combustion products later. The flask is filled with oxygen, the filter-paper is ignited and the flask is closed immediately[24]. Alternatively, the platinum wire may be fused through the glass stopper and connected to an electrical ignition device to be activated by remote control after the sample has been introduced into the flask and the flask has been closed; this allows the combustion to take place in a totally enclosed system and avoids the potential hazards of manual firing[36]. The flask is shaken and allowed to stand for complete absorption of the combustion products to take place, then the stopper and neck of the flask are washed into the bottom and the solution is allowed to boil for at least 10 min in order to convert all phosphorus from possibly present metaphosphates or polyphosphates into orthophosphoric acid.

Total combustion of the samples is not always achieved by Schoniger's technique. Small amounts of soot or carbon deposits are sometimes detected after the reaction. These deposits may retain some P in the form of 'phosphorus carbon' of undetermined composition and lead to low results[34]. The phosphorus-containing substances which are difficult to oxidize completely include compounds with several isolated or condensed aromatic rings[34], phospholipids[37], heavy metal-containing compounds[37], nucleic acids and proteins[38] and molecules in which the phosphorus atom is directly linked to a carbon, such as phosphines and phosphoranes[39]. A number of successful improvements have been suggested in order to attain complete combustion of the sample. Increases in the amount of oxygen available for the firing have been proposed, either by use of a larger flask or by addition of small amounts of an oxygen-rich substance such as ammonium peroxodisulphate to the sample[34]. Higher temperatures also yield better mineralization and several attempts have been made to modify the design of the sample holder in order to reduce its heat content or its heat conduction and thus to minimize the heat losses during the combustion[38–42].

The choice between oxidation by the Kjeldahl or the Schoniger technique is not clear-cut. Both methods will usually yield quantitative reactions, both will occasionally give incomplete combustion and low results, depending on the family of substances analysed and on the slight modifications introduced into the procedures in different laboratories.

B. Determination of Orthophosphates

1. Precipitation and titration techniques

The oldest method for the determination of phosphate was its precipitation as magnesium ammonium phosphate, followed by filtration, drying and weighing of the precipitate as magnesium pyrophosphate. A number of other gravimetric methods may be found, such as precipitation as ammonium molybdophosphate or silver phosphate[1], but simple gravimetry is now only rarely used, and has widely been replaced by titration procedures involving precipitation of insoluble salts of the phosphate. Two main techniques are in use: titration by precipitation of the phosphate with some heavy cation until the appearance of excess cation can be detected or quantitative precipitation of the phosphate by a known amount of cation and titration of the excess cation remaining after all the phosphate has been removed from the solution.

a. Precipitation as magnesium or zinc ammonium phosphate. This method was originally proposed by Flaschka and Holasek[43] for the microdetermination of phosphorus in amounts smaller than 100 μg, and was later adapted to semimicro analysis of samples containing about 1 mg of phosphorus[28]. The basic procedure is precipitation of the phosphate as the magnesium ammonium double salt, filtration and washing of the precipitate, dissolution with dilute acid, neutralization of the solution with ammonia or NaOH to a pH of 8–10 and titration of the magnesium with disodium edta in presence of a pH 8–10 buffer. Bennewitz and Tanzer[31] applied a simplified technique in which the phosphate is precipitated as the magnesium ammonium double salt by a known excess amount of $MgCl_2$ and the surplus magnesium is titrated with edta. Zinc ammonium phosphate may be precipitated in a similar fashion, filtered, washed and dissolved in HCl; the solution is then brought to a pH of 10 and the amount of zinc determined by titration[44,45].

b. Titration with cerium. Puschel and coworkers[33,46,47] determined semimicro and micro amounts of orthophosphates by titration with cerium(III), using Eriochrome Black T as indicator. In this method the solution is buffered to pH 7 with hexamethylenetetramine and the phosphate is precipitated at boiling point by cerium(III), the indicator changing colour from blue to red when the first free cerium appears. The procedure is suitable in the presence of a large excess of sulphate but a number of other anions (arsenate, antimonate, vanadate, tungstate, fluoride, silicate or citrate) affect the results. Interfering cations may be removed by ion exchange.

c. Titration with lanthanum. Horacek[48] titrated the phosphates at boiling temperature with lanthanum nitrate in an hexamethylenetetramine buffer solution. The phosphate precipitates as insoluble lanthanum salt according to reaction 1:

$$H_2PO_4^- + La^{3+} \longrightarrow LaPO_4 + 2H^+ \qquad (1)$$

and the Chrome Azurol S indicator changes colour from yellow to violet when an excess of lanthanum is added. Barium, if present, must be precipitated with sulphuric acid before the phosphate determination; calcium does not interfere, as the indicator is less affected by the presence of heavy cations than Eriochrome T. In a modified procedure[37] the reaction is carried out without buffer, and the titration is monitored by pH measurements, the pH remaining constant when the end-point is attained. Barium must again be removed beforehand; silver, mercury or cadmium ions are precipitated with potassium iodide. If selenium is present it may be precipitated with thioacetamide[20].

d. Titration with zirconium. Zirconium was used in an indirect complexometric determination of phosphates[49]. A known excess of zirconium sulphate is added to the phosphate solution, the mixture is heated until all the zirconium phosphate has precipitated and the remaining zirconium is titrated with edta until the Xylenol Orange indicator turns from pink to yellow.

e. Titration with silver. Silver is often used in preference to other metals in the determination of phosphates, because the low solubility of silver phosphate (about 1×10^{-5} M) allows quantitative precipitation of even micro amounts of phosphate. A simple gravimetric method[50] is based on the formation of a complex between silver and ammonia and on the resulting solubility of silver phosphate in ammoniacal solutions[51]: silver nitrate, in ammonia solution, is added to the phosphate, the mixture is slowly heated in order to evaporate the ammonia and the silver phosphate precipitates in large, easily filtered crystals.

In some procedures silver phosphate is formed by addition of silver nitrate to a slightly alkaline[52] or acidic[53] phosphate solution, then the precipitate is dissolved in ammoniacal potassium cyanonickelate, according to reaction 2. The displaced nickel is titrated with

$$2Ag_3PO_4 + 3K_2[Ni(CN)_4] \longrightarrow 3K_2[Ag_2(CN)_4] + 3Ni^{2+} + 2PO_4^{3-} \qquad (2)$$

disodium edta until the Murexide indicator turns from yellow to purple. The presence of a number of anions affects this determination: halides and sulphates, in particular, react with the silver ion and precipitate quantitatively with the phosphate. This interference can be avoided if the silver nitrate is not added in the conventional manner, but is gradually generated in the phosphate solution by slow decomposition of an ammoniacal silver nitrate complex. In this technique the different precipitates are formed sequentially in order of increasing solubility and the final product is a mixture of precipitates instead of a mixed precipitate[54]. The silver phosphate is then easily dissolved in dilute nitric acid, while silver halides or sulphate remain undissolved; the silver ions are determined with potassium thiocyanate by Volhard's method.

The low solubility of silver phosphate, which is comparable to that of silver chloride, has led to repeated investigations of the potentiometric titration of phosphates by silver nitrate. The acid salts AgH_2PO_4 and Ag_2HPO_4 are very soluble, and an alkaline environment, leading to Ag_3PO_4 precipitation, is required for the titration; however, the pH should not exceed 9.5, in order to avoid silver oxide formation. The earliest titration was carried out in a borate buffer at pH 9.0 with a silver indicating electrode, but the end-point was not as clear, or the titration curve as symmetric, as in the case of silver chloride[50]. The use of aqueous ethanol or 2-propanol as solvent further reduced the solubility of silver phosphate and led to easier estimates of the end-point[55,56]. A silver sulphide ion-selective electrode was also used for the potentiometric titration of phosphate with silver nitrate in borate-buffered aqueous methanol[57].

In a different potentiometric procedure, a known excess of silver nitrate was added to the sample in a pH 7–8 buffer, the precipitated silver phosphate was removed and the remaining silver ions were determined by KBr titration[50]. Heavy cations which form insoluble phosphate salts, such as Ca^{2+}, Al^{3+} or Fe^{3+}, often interfere with these potentiometric titrations[50] and should be removed beforehand, e.g. by ion exchange[58]. The presence of anions does not, in most cases, affect the results; silver halides precipitate before the phosphate, they do not interfere and may actually be determined simultaneously in the same titration[58]. Amperometric titration was also attempted[55].

f. Titration with lead. Lassner *et al.*[46] were able to determine phosphates, in the absence of sulphates and other anions, by precipitation of $Pb_3(PO_4)_2$ with 4-(2-pyridylazo)resorcinol as indicator; the procedure yields excellent results for small amounts of phosphate.

The extremely low solubility of lead phosphate in water (about 6×10^{-15} M) again suggests potentiometric analysis. Selig[57,59] determined micro amounts of phosphate by precipitation with lead perchlorate in aqueous medium. The sample was buffered at pH 8.25–8.75 and a lead-selective electrode was used to establish the end-point. The detection limit is about 10 μg of phosphorus. Anions which form insoluble lead salts, such as molybdate, tungstate or chromate, interfere with the procedure. Similar direct potentiometric titrations of phosphate by precipitation as insoluble salts of lanthanum(III), copper(II) or cadmium(II) are suggested, the corresponding ion-selective electrodes being used to detect the end-point.

g. Determination with molybdenum. The use of molybdate complexes, either by precipitation or more often by colorimetry, is the most common method for the analysis of phosphates. In consequence, a large number of variations have been reported, in attempts to improve the stability of the complexes, to increase the sensitivity of the procedures and to eliminate as many of the interferences as possible.

The reaction of an acidic molybdate solution with phosphate yields heteropoly anions, in which a central PO_4 ion is surrounded by MoO_3 oxomolybdate units; in the dodecamolybdophosphate ion, for instance, twelve MoO_6 octahedra are found around the central phosphorus oxoanion. Ammonium dodecamolybdophosphate precipitates as a lemon yellow solid when an acidic solution of ammonium molybdate is added to phosphate[1,60]. Quinoline molybdate may be used instead of ammonium molybdate[27,30,61]; quinoline molybdophosphate also appears as a yellow precipitate, less soluble in water than ammonium molybdophosphate. Both the ammonium and the quinoline complexes seem to retain undefined amounts of water or other contaminants in their crystal lattices and hence are not suited for gravimetric work. The phosphates are determined volumetrically by filtering the precipitates, dissolving them in a known amount of sodium hydroxide and titrating the excess hydroxide with hydrochloric acid. In a related procedure, the formation of solid cetylpyridinium molybdophosphate was used for the determination of phosphates; cetylpyridinium chloride was added as titrant in the direct constant-current potentiometric titration of orthophosphate, after conversion of the phosphate into dodecamolybdophosphate[62,63].

Instead of being determined volumetrically, however, the ammonium or quinoline molybdophosphates are usually reduced to another molybdenum complex, 'molybdenum blue' (heteropoly blue), which is then analysed spectrophotometrically (see Section II.B.2.c).

h. Simultaneous analysis of phosphorus and sulphur. In a large number of organophosphorus compounds, one or more oxygen atoms are replaced by sulphur. The presence of phosphorus often interferes with the microanalysis of sulphur, and most methods eliminate the phosphorus by precipitation as silver or magnesium phosphate before the sulphur determination. Some procedures, however, allow the simultaneous analysis of phosphate and sulphate. Titration with barium chloride, and tetrahydroxyquinone as indicator, yields good results if the phosphorus to sulphur ratio is known[14]. A more generally applicable polarographic method[64] involves precipitation of the sulphate by barium chloride and determination of the excess barium, followed by precipitation of the phosphate as quinoline molybdophosphate and determination of the excess molybdenum.

2. Colorimetric and fluorimetric analysis

a. Complexation with aluminium. Aluminium is known to form a fluorescent complex with morin (2′, 3′, 4′, 5, 7-pentahydroxyflavone); the wavelength of maximum emission of the chelate is observed in the vicinity of 500 nm and the fluorescence is most intense at $pH > 5.5$. If phosphate is present, it will compete with the morin to react with the aluminium, leading to a reduction in the aluminium–morin concentration and quenching of the fluorescence intensity. Phosphate concentrations down to $50\,ng\,ml^{-1}$ have been determined by measuring the decrease in fluorescence of a standard solution of aluminium–morin on addition of phosphate samples[65]. A large number of strongly complexing, coloured or metallic ions interfere with the procedure and should be removed before the fluorimetric analysis; a calibration graph must be prepared under conditions identical with those of the measurement.

b. Determination with iron. The complex formed between iron and phosphate ions exhibits an absorption maximum at 310 nm. A suggested procedure for the analysis of phosphates is the addition of iron perchlorate to the unknown sample; the complexing ability of perchlorate is much weaker than that of phosphate and the perchlorate is displaced. Absorption measurements, carried out at 340 nm in order to minimize background interference, allow phosphate detection at levels down to 12 nmol. The complexes of many other anions with iron exhibit absorption maxima in the same wavelength range and these anions should be separated from the phosphate before the measurement. If high-performance liquid chromatography is used for the separation, the simultaneous determination of a number of different anions is possible[66].

c. Molybdenum complex methods. As already mentioned, molybdenum derivatives are those most frequently used in the colorimetric determination of phosphates. Molybdophosphate complexes are easily obtained by addition of molybdate to a phosphate solution. A number of analytical techniques are available, including direct determination of the molybdophosphoric acid, reduction of the molybdophosphate to molybdenum blue, conversion to molybdivanadophosphoric acid or formation of various ion pairs, for instance with Malachite Green. Some of the methods have been reviewed and discussed previously[1,16].

i. Direct determination of molybdophosphoric acid. The yellow ammonium or quinolinium molybdophosphate formed on addition of ammonium or quinoline molybdate to a phosphate solution may be extracted with a solvent such as isobutanol[1], amyl acetate[16] or a mixture of butanol and chloroform[60]. The optical absorption of the organic solution is determined at 310 nm or in the 400–430 nm region; in the latter case care must be taken to thermostat both the solution and the instrument. Alternatively, the molybdophosphate is re-extracted into an aqueous solution with a basic agent and the absorption measured in the near-ultraviolet region[1]. These methods are not often applied now, as more sensitive procedures have become available.

ii. Molybdenum blue methods. Partial reduction of molybdophosphoric acid yields molybdenum blue, a strongly colored blue heteropoly complex in which the molybdenum is thought to be present in hydroxo form, with oxidation number V or VI. This complex exhibits an intense absorption maximum in the red region of the visible spectrum; for colorimetric or spectrophotometric phosphate determination the absorbance is measured at wavelengths between 600 and 900 nm, often at 820 nm in aqueous solutions or at 780 nm in organic solvents. A similar complex may be obtained by precipitation of the phosphate as magnesium ammonium phosphate, dissolution of the precipitate in sulphuric acid, subsequent addition of ammonium molybdate and reduction to molybdenum blue[67,68].

The reduction is mostly carried out directly in the aqueous phase[32,69,70], but the results may be affected by the presence of anions such as arsenate and silicate. Although the analysis becomes more cumbersome, a number of workers advocate extraction of the molybdophosphate complex by an organic solvent before reduction, in order to eliminate the interfering species. An added advantage of the extraction is the removal of molybdic anions remaining in excess after the reaction and of the resulting 'non-specific' blue colour sometimes observed. Amyl acetate[16], butanol[71,72], octyl alcohol[73], mixtures of amyl alcohol with benzene[5] or with diethyl ether[74], of isobutanol with benzene[29,75] or of chloroform with butanol[76] have been adopted for the extraction; the reduction is then carried out in the organic phase. A large number of reducing compounds have been proposed, e.g. aminonaphtholsulphonic acid[67,77,78], methylamidophenol sulphate[32], hydroquinone[72,79], hydrazine sulphate[60,80] and tin(II) chloride[5,16,29,60,71,73,75,81]. Ascorbic acid, originally suggested by Lowry *et al.*[82], is now a very frequently used reducing agent[6,28,35,69,70,74,83–86], as is tin(II) chloride[85,87,88]. Reduction with ascorbic acid does not affect labile organophosphorus compounds such as phosphate esters and allows the separate determination of the inorganic and organic phosphates[83]. Microgram amounts of phosphorus are routinely measured. Trace concentrations of phosphate as low as 20 ng l^{-1} can be determined by applying a preconcentration technique based on collecting the complex on a nitrocellulose or acetylcellulose membrane in the presence of a cationic surfactant, then dissolving the complex with the membrane in dimethyl sulphoxide and measuring the absorbance at 710 nm[89].

The fact that twelve molybdate ions are bound to each phosphate unit in molybdophosphate has been exploited as an amplifying factor. The molybdophosphoric acid is extracted from the original solution into a mixture of diethyl ether or chloroform with butanol, in order to eliminate the excess of molybdate reagent; it is then re-extracted into an aqueous phase and broken down with alkali. The molybdate ions are reacted with ammonium rhodanide[90] or aminochlorobenzenethiol[76] to give coloured products with absorption maxima at 470 and 710 nm, respectively. One phosphate ion thus yields twelve molybdate units, the optical absorbance is amplified with respect to that of the heteropoly phosphate complex and its measurement provides a sensitive determination procedure.

Succinyldihydrazide[91] or malonyldihydrazide[92,93] have also been proposed as reducing agents in the determination of phosphorus-containing residues in either aqueous or organic solution. In this method the sample is directly mixed with ammonium molybdate in concentrated sulphuric acid and with the reducing agent and is heated for a few minutes; the procedure mineralizes the residues and transforms the phosphates into molybdenum blue in one operation. The blue complex is extracted with butanol and the optical absorption is measured at 780 nm. The experimental conditions applied in each case, the choice of the reducing compound, the pH and the temperature of the mixture while the colour develops, the time elapsed after addition of the reducing agent and the nature of the solvent (aqueous or organic) were all shown to influence the intensity and the wavelength of the molybdenum blue absorption band and the stability of the complex[1,16,60,69,72,94]. Standardization of the analytical protocol is imperative; after constant conditions have been chosen, however, the procedures yield very precise and reproducible results.

iii. Molybdivanadophosphate methods. Addition of ammonium molybdate to an acidic mixture of phosphate and ammonium vanadate results in the formation of a yellow heteropoly complex known as molybdivanadophosphate. Mission[95] first described the use of this compound for the colorimetric determination of phosphate. The method was later elaborated for semimicro- and micro-determination of phosphates in organic substances[34,96–98]. The procedure involves preparing an acidified mixture of the phosphate and of ammonium vanadate, then adding an excess of ammonium molybdate and waiting a few minutes for the colour to develop. The exact order of mixing of the reagents is important; if it is changed several other complexes may be formed simultaneously and interfere with the analysis. The absorption maximum of molybdivanadophosphate appears at about 350 nm, but the absorbance in the violet region of the visible spectrum is intense enough to allow the colorimetric measurements to be carried out between 400 and 480 nm.

iv. Ion-associates formation. Molybdophosphate reacts with organic bases or cationic dyes to form intensely coloured or fluorescent ion pairs. These associates can be isolated by precipitation from the aqueous phase or by flotation at the boundary between aqueous and organic solutions, and are then dissolved in an organic solvent. Measurement of their optical properties provides sensitive methods for phosphate determination. Alternatively, the solubility of the complexes in water may be increased by addition of a surface-active agent and the determination can be carried out directly in the aqueous solution; care must be taken to eliminate any excess of free dyes by protonation or oxidation[99].

Safranine[100], Methylene Blue[101,102], Crystal Violet[99,103,104], Methyl Green[104,105], Iodine Green[103] and a number of other dyes have been used for the formation of ion-associates[104,106]. As a typical example, Ethyl Violet was found to yield one of the highest molar absorbances; extraction of the Ethyl Violet–ion complex from aqueous solution by a mixture of cyclohexane and methylpentanone, followed by absorbance measurement at 602 nm, allows the determination of 2 parts per billion (ppb) of phosphorus in water[106]. Molybdophosphate may also be complexed with Resazurine or *p*-rosolic acid to give stable, intensely coloured solutions; no extraction is required for the measurement of the absorbance at 495 or 470 nm, respectively[107]. A cobalt complex, bis[2-(5-chloro-2-pyridylazo)-5-diethylaminophenolato]cobalt(III), proposed previously for the analysis of sulphonated or sulphated surfactants[108], forms an associate with molybdophosphate. This compound may be collected on an organic membrane filter, then dissolved with the filter in *N*, *N*-dimethylformamide; the absorbance of the solution is determined at 560 nm[109]. Most of these methods allow determinations down to less than 1 ppb.

The complex formed by association of three Malachite Green molecules and one molybdophosphate unit shows an even larger absorbance, about three times as intense at its absorption maximum of 650 nm as that of the Ethyl Violet associate[110–112]. The complex has the added advantage of being relatively soluble in water, so that the optical absorbance can be measured directly in aqueous solution. Some workers nonetheless extract the molybdophosphate–Malachite Green complex by addition of diethyl ether to the aqueous mixture; the complex floats at the phase interface, the aqueous solution is discarded, methanol is added to dissolve the ion pair and the absorbance is determined at 620 nm (diethyl ether was chosen because it floats the ion pair, but neither floats nor extracts the excess Malachite Green dye)[113]. In a different procedure a mixture of toluene and methylpentanone was used for the extraction[112]. As in the molybdenum blue method, the limit of detection may be lowered by collecting the molybdophosphate–Malachite Green aggregates on a nitrocellulose membrane filter, then dissolving the membrane, together with the complex, in an organic solvent[114]; this last technique, incidentally, allows the simultaneous determination of phosphates and arsenates at concentrations down to 0.3 ppb. A method for determining phosphates outside the laboratory under 'field conditions' is based on this ion-pair formation between molybdophosphate and Malachite Green, followed by visual comparison of the colour obtained with a series of standards; the authors claim to be able to determine 10 ppb of phosphate in this way[115].

Addition of quinine to an acidic molybdophosphate solution results in precipitation of the quinine molybdophosphate complex, which after filtration and washing may be dissolved in a mixture of acetone and sulphuric acid, yielding a stable and intensely fluorescent solution[116]. Rhodamine B was found to combine in 3:1 ratio with the molybdophosphate ion in acidic solution[117], the excess Rhodamine was extracted with chloroform and the complex was dissolved in a mixture of chloroform and butanol. Measurement, at 575 nm, of the intense fluorescence of the solution yielded a sensitive and highly selective phosphorus determination. In more recent procedures, the associates formed between Rhodamine B or Rhodamine 6G and molybdophosphate in acidic solution are floated at the boundary between the aqueous phase and a diethyl ether layer and are then dissolved in the organic phase by addition of methanol. The fluorescence intensities of the organic solutions are proportional to the phosphate concentration and ppb amounts of phosphorus can be determined by fluorimetry at 573 nm[118] or 548 nm[119], respectively.

The presence of low concentrations of phosphate has long been known to interfere with other fluorimetric analytical procedures; this ability of trace amounts of phosphate to quench the fluorescence of chelates has been applied to the determination of the phosphates themselves. Several chelates were investigated; measurement of the decrease of fluorescence of the aluminium–morin complex after phosphate addition proved to give the most satisfactory results[65] (see Section II.B.2.a).

d. Iodometric determination. Determination of the phosphoric acid obtained by combustion of organophosphorus compounds in the presence of hydrogen peroxide has been carried out by iodometry[120]. Reaction with KI and KIO_3 according to reaction 3

$$6H_3PO_4 + KIO_3 + 5KI \longrightarrow 3I_2 + 3H_2O + 6KH_2PO_4 \qquad (3)$$

liberates iodine, which is titrated by thiosulphate in a 1:1 mixture of acetone and water, the iodine itself serving as indicator.

In an improved procedure[121], the combustion products were absorbed in water, saturated bromine water was added for quantitative conversion of all phosphorus products into orthophosphoric acid and the excess bromine removed by boiling. Orthophosphoric acid normally reacts as a monobasic acid with KI and KIO_3. However, if zinc and ammonium ions are added, it functions as a tribasic acid through precipitation of insoluble zinc ammonium phosphate and liberation of three hydrogen ions, instead of one, per molecule of orthophosphoric acid. A threefold amount of iodine is obtained on addition of KI and KIO_3, considerably increasing the sensitivity and accuracy of the method. The iodine was titrated with thiosulphate, using starch as indicator[121].

e. Simultaneous analysis of phosphorus and sulphur. A colorimetric procedure has recently been developed to solve the frequently arising problem of the simultaneous determination of phosphorus and sulphur in organophosphorus compounds[15]. Sulphate is determined at 530 nm after reaction with barium chloranilate[122] and phosphate at 430 nm after the previously described conversion into molybdivanadophosphate.

3. Other methods

a. Electrochemical methods. A number of methods for the determination of phosphates rely on electrochemical techniques. Molybdivanadophosphate, formed from phosphate by a procedure similar to that applied in colorimetric analyses, may be reduced at a glassy carbon electrode, at a potential of 0.5 V versus the saturated calomel electrode (SCE), and determined with excellent reproducibility down to 10^{-6} or 10^{-7} M concentrations by differential-pulse voltammetry[123]. In a related technique, phosphates are converted into molybdophosphates, which are then determined by using a glassy carbon electrode as a voltammetric detector[124,125]; the method, which yields accurate results down to 10^{-6} M phosphate concentrations, has been applied to the analysis of phosphates in turbid waters, where spectrophotometric methods are difficult to use[126].

b. Atomic absorption spectrometry. Atomic absorption spectrometry is based on the electronic excitation of atoms by light absorption. A sample in solution is sprayed into a flame, where it is vapourized and atomized. An external light beam is passed through the flame, and the spectral lines corresponding to transitions of the atoms from ground to excited state are absorbed; in many instruments the light source is a hollow-cathode discharge lamp emitting the same spectral lines which are absorbed by the element under investigation. The transmitted radiation is analysed with a monochromator, and the intensities at the exciting wavelengths in the presence and absence of the test substance are compared. The amount of light absorbed depends on the population of the ground state, and is proportional to the concentration of the sample solution. Although the overall sensitivity of atomic absorption spectrometry is lower than that of spectrophotometric procedures, the ease and rapidity of the determination, together with its high selectivity and the absence of interference by foreign elements, often compensate for the lower precision. A large number of analytical procedures based on atomic absorption spectrometry have therefore been proposed.

Phosphorus itself is not directly accessible to atomic absorption spectrometry, as its most characteristic atomic lines appear in the vicinity of 200 nm. Indirect methods are necessary if the technique is to be applied, and the procedures often require previous conversion of the phosphorus into phosphate. One possible approach is the formation of a well defined stoichiometric complex between phosphate and another ion, which may then be isolated and determined by atomic absorption spectrometry.

The molybdophosphate complex again seems ideally suited to this purpose, because of its constant chemical composition and because of the amplification provided by the twelve molybdenum atoms associated with each phosphate ion. Before the measurements can be carried out, however, all traces of unused molybdate reagent must be eliminated; this is usually done by extraction of the heteropoly compound into an organic solvent. The organic solution is then injected directly into the air–acetylene or nitrous oxide–acetylene burner of the spectrometer[127,128]. Of the different wavelengths available for the determination of molybdenum, the 313.26 nm line is the most intense[127] and is usually employed. Phosphorus could be determined down to concentrations of 0.08 ppm[128]. The ion associate formed in acidic aqueous solution between molybdophosphate and Malachite Green was used in a similar way[113]; it was extracted with methylpentanone (the complex floated at the boundary between the aqueous and the organic phase and was dissolved in methylpentanone) and the atomic absorption of the molybdenum was measured at 313.26 nm.

In addition to molybdenum, other ions, e.g. copper and bismuth, have been used for the complexation of some organophosphates, extraction of the complexes and determination of the metals by atomic absorption spectrometry[129]. It is interesting that the ability of organophosphorus compounds to form complexes with a large number of metals has been applied to the determination of those metals in dilute solutions such as waste waters; the organophosphorus reagent, in organic solution, is shaken with the sample in order to form the complex, which is then extracted into the organic phase, back-extracted with an acid and finally determined by atomic absorption spectrometry[130,131].

C. Total Phosphorus Determination

A number of instrumental analytical techniques can be used to measure the total phosphorus content of organophosphorus compounds, regardless of the chemical bonding of phosphorus within the molecules, as opposed to the determination of phosphate in mineralized samples. If the substances are soluble, there is no need for their destruction and for the conversion of phosphorus into phosphate, a considerable advantage over chemical procedures. The most important methods are flame photometry and inductively coupled plasma atomic emission spectrometry; the previously described atomic absorption spectrometry is sometimes useful.

1. Flame photometry

This method is based on the emission of light by atoms returning from an electronically excited to the ground state. As in atomic absorption spectrometry, the technique involves introduction of the sample into a hot flame, where at least part of the molecules or atoms are thermally stimulated. The radiation emitted when the excited species returns to the ground state is passed through a monochromator. The emission lines characteristic of the element to be determined can be isolated and their intensities quantitatively correlated with the concentration of the solution.

An early, indirect, flame photometric technique[132] was based on the depressing effect that phosphorus compounds have on the absorption and the emission signals of some elements such as calcium, strontium or barium; phosphates, phosphines and a number of organophosphorus substances were shown to influence the spectra in various ways [133–138]. The quenching phenomenon is probably due to formation in the burner of stable compounds of phosphorus and the alkaline earth metals, which have only a limited tendency to decompose into atoms and thus reduce the concentration of non-bonded calcium, strontium or barium in the flame[134]. If the metals themselves are to be determined, this depressive effect may be eliminated by adding large amounts of a releasing agent such as lanthanum or edta[134,137], or by using a hotter air–acetylene flame[136,137]. In the proposed procedure, however, the emission quenching was used for phosphate analysis. The solution to be analysed was first passed through a cation-exchange column to remove all interfering cations present in the sample. A known amount of calcium was then added and the intensity of the calcium flame emission was measured. Comparison with a calibration graph prepared with the same amount of calcium and different amounts of phosphate yielded accurate results for 0.005–0.012 M concentrations[132].

More recent flame photometric methods rely on direct measurement of the phosphorus emission. If organophosphorus compounds are injected into a hydrogen flame, a continuous emission is obtained in the 490–650 nm region. A broad band system, with an intensity maximum at 526 nm, is superimposed on this background[139]; it is attributed to the HPO species formed in the flame. An early determination of phosphorus at 0.01–0.04 M concentrations was based on examination of the continuous emission; standard and sample solutions were injected into the burner and the intensities were measured at 540 nm; the calibration graph was linear down to the detection limit of 10^{-4} M phosphorus; sodium or calcium, if present in the sample, interfered with the results[140].

Better precision is obtained when the 526 nm HPO line is isolated with the help of filters; the emission intensity is proportional to the phosphorus concentration and phosphorus can be determined down to 0.1 ppm concentrations[139,141,142]. Most cations produce a depressive effect and should be eliminated by passage of the sample through an ion exchanger. The HPO emission has been used for the development of a sensitive, phosphorus-specific, flame photometric detector for gas chromatography, in which the 526 nm band is isolated by a narrow bandpass interference filter and its intensity is measured with a photomultiplier; this detector is able to sense ppb concentrations of organophosphorus compounds[143] (see Section III.B.3.b).

Organic solvents have been reported to quench the HPO emission, and in early determinations the samples were introduced into the burner as aqueous solutions, sometimes after mineralization by ashing[144]. Interferences due to solvents can be overcome by use of a molecular emission cavity (an open, flat-bottomed, cylindrical container drilled into a metallic holder) to introduce the sample into the flame; with this device the solvent is evaporated before the sample is inserted into the burner[145]. The time elapsed between sample ignition and maximum emission (t_m) is characteristic of the compound investigated under a given set of experimental conditions. Time-dependent recording of the flame emission can resolve peaks from compounds with sufficiently different t_m values. Water cooling of the cavity enhances the sensitivity and improves the resolution of mixtures. Measurements of the maximum emission intensity at time t_m allows the determination of nanogram amounts of phosphorus[145].

2. *Inductively coupled plasma atomic emission spectrometry*

Inductively coupled plasma atomic emission spectrometry is similar to flame photometry, but uses a more powerful source of excitation. An inert gas such as argon flows continuously through a quartz tube and is excited to plasma, the energy required to support the plasma being supplied by a radiofrequency coil looped around the tube. The sample is injected into the plasma torch where its atoms are electronically excited. The emitted light is analysed with a grating spectrometer and the intensities of the lines characteristic of the element under investigation are measured, to yield the concentration of this element in the injected sample. A plasma source presents several advantages with regard to a flame: its high temperature provides enough energy to evaporate the solvent and eliminate the solvent interferences often found troublesome in flame photometry; it has the potential to generate short-wavelength emission lines not observable in flame spectrometry and can therefore be used to investigate non-metallic elements such as phosphorus which exhibit their most intense lines in the vicinity of 200 nm; it also yields much better detection limits than can be obtained with a flame[146].

Samples are usually in aqueous solutions[146–150]; if the organophosphorus compounds are not soluble in water or acid they may first be decomposed into phosphates by oxygen flask combustion or Kjeldahl digestion[151]. Organic solvents have a tendency to form carbon deposits, to clog the nebulizer and to increase the background emission. However, changes in the operating conditions and in the torch design are able to overcome these difficulties and allow continuous injection of organic solutions[152]. Mixtures of water and ethanol have also been used to improve the solubility and the nebulization efficiency[151]. Gaseous samples can be introduced into the plasma torch with the argon carrier stream, and an inductively coupled plasma detector for gas chromatography has been built on this principle[153]. The most intense and sensitive lines for the determination of phosphorus are at 177.49 and 178.29 nm and require the use of a vacuum instrument[149,151]. Above 200 nm the best lines appear at 213.62 and 214.91 nm, but interferences may occur at these wavelengths; in particular molybdenum, when present, exhibits an intense emission at 213.61 nm which produces considerable background at 213.62 nm[150]; copper also affects the 213.62 and 214.91 nm lines[147,149,151]. Calibration graphs of emission intensity versus the known phosphorus concentrations of reference samples are linear over several orders of magnitude[146,147,150]. The detection limit, defined as three times the background noise, is 0.07 $\mu g\ ml^{-1}$ for direct injection of the sample into the plasma[149,150].

D. Recent Developments, Flow-injection Analysis

In recent years, the methods applied to the analysis of elemental phosphorus or phosphate have hardly changed. Most of the current research is aimed at improving the accuracy, selectivity and speed of the determinations. Both batch analysers and continuous flow-injection systems have been developed, allowing very fast monitoring of the phosphorus content of a large number of samples. Originally proposed by Ruzicka and Hansen[154], continuous flow analysis involves injection of the sample into a rapidly moving carrier stream. This stream often contains the reagents necessary for the analysis, or alternatively the reagents are injected either before or after the sample. Several of the analytical techniques discussed above for phosphate determination have been adapted to flow-injection methods.

Sulphuric acid, ammonium molybdate and ascorbic acid may, for instance, be introduced into the continuous stream; when a phosphate-containing sample is injected, the characteristic molybdenum blue colour is formed and the optical absorbance may be measured at 830 or 670 nm in a flow-through cell[86,155–157]. In a different method, the phosphate sample is injected first into a water stream, ammonium molybdate is added to form the yellow molybdophosphoric acid, then ascorbic acid or tin(II) chloride is injected for the reduction to heteropoly blue and the absorbance is determined in the red region of the visible spectrum[85,87,88]. An acidic solution of ammonium heptamolybdate and Malachite Green may be used as carrier, the samples being injected into this mixture and the absorbance measured in a flow-through cell at 650 nm; addition of ethanol to the reagent stream prevents coagulation and adsorption of the complex on the internal surface of the tubing[158]. A mixture of ammonium molybdate and ammonium vanadate in nitric acid has also been injected into the carrier stream in order to determine phosphates by absorbance measurement at 470 nm, as described earlier for the molybdivanadophosphate procedure[150]; the presence of nitric acid in the flow system inhibits blockage of the tubing by precipitation of the reagents.

Better precision is obtained when the flow-injection system is coupled with a phase extractor[159]. In such a procedure dilute sulphuric acid is used as the carrier, the sample is injected first, followed by the reagent mixture of molybdate and Malachite Green in sulphuric acid; after the stream has passed through a mixing coil, an organic solvent (benzene and methylpentanone) is injected at right-angles to the main flow direction and the phases are separated with a polytetrafluoroethylene membrane permeable to organic solvents but not to aqueous solutions; the optical absorbance is measured at 630 nm, allowing phosphate determination at amounts of 0.1 ng ml^{-1} at a rate of 40 samples per hour.

In addition to spectrophotometric methods, other detection techniques can be applied to flow-injection analysis. Some of the electrochemical procedures described above may be used; for instance, molybdivanadophosphate or molybdophosphate formed in the eluent stream has been reduced at a glassy carbon electrode and determined voltammetrically, allowing the analysis of phosphates at 10^{-6} or even 10^{-7} M concentrations[19,123,124]. Flame photometry, relying on measurement of the 526 nm emission of HPO, is able to measure nanogram amounts of phosphates or organophosphorus compounds[160]. Inductively coupled plasma atomic emission at 177.50 nm, in a vacuum instrument, has been applied to the determination of total phosphorus in water samples at a rate of 80 measurements per hour[150].

In more complex flow-injection systems the organophosphorus samples are injected without previous mineralization, and all decomposition reactions are carried out in the instrument itself: oxidation of organic compounds by digestion with, for instance, peroxodisulphate in a heated capillary tube[18,19] or hydrolysis of di- and tri-phosphates with acid at 140 °C[155]. Ultraviolet photo-oxidation in the presence of alkaline peroxodisulphate has been applied to the destruction of the organic compounds, before injection of the ammonium molybdate and reduction to molybdenum blue by tin(II) chloride[161].

III. ANALYSIS OF ORGANOPHOSPHORUS COMPOUNDS

Organophosphoric substances are known to play a fundamental role in a large number of natural biochemical processes; progress in the understanding of these biological effects depends on the identification and determination of the individual compounds present at the reaction sites. On the other hand, organophosphorus derivatives are increasingly used in industry, e.g., as plastic stabilizers, flame retardants, lubricant additives and hydraulic fluids, requiring analytical control methods. One of their most profitable applications is in agriculture, for the treatment of crops with insecticides, pesticides and fungicides; the compounds, however, are then absorbed by the livestock which feed on the treated plants and are later found in meat and milk; they migrate into the soil and the water through leaching and into the air with the help of winds; fish, both in fresh water and in the sea, may be contaminated. These organophosphorus compounds do not remain unchanged, they usually react and are transformed into toxic products different from the substances originally used. As a consequence, traces of poisonous organophosphorus material are present in the air, water and food, leading to serious environmental hazards in many aspects of our daily lives. A related problem may result from the disposal or dumping of unserviceable nerve gases into the oceans, where with time they could leak into the water. All these factors create a pressing need for the detection and identification, in picogram or even smaller amounts, of the compounds involved and also of their metabolites and degradation products, which often are just as dangerous to living organisms.

The analytical task implied differs completely from the previously discussed determination of phosphates or total phosphorus, which is based on the destruction and loss of identity of the organophosphorus compounds. The substances must first be extracted, with as little alteration and loss as possible, from their inorganic or biological surroundings (water, soil, animal or vegetal tissues); they have then to be cleaned from interfering, non-organophosphorus co-extracts. Identification and determination are usually carried out at a later stage by instrumental techniques, the most frequently applied methods relying either on chromatography or mass spectrometry[162]. The problem is complicated by the steadily increasing number of relevant compounds and by the fact that they appear under a variety of chemical designations or commercial trade names; a dictionary of organophosphorus compounds, listing over 5000 entries, has been published, and should help with nomenclature uncertainties and identification of the substances[163].

A. Extraction and Clean-up

The extraction and clean-up methods used for the recovery of organophosphorus derivatives do not differ basically from those applied to other groups of substances. However, because of the high toxicity of the compounds involved, the techniques must be extremely sensitive and able to detect sub-nanogram amounts. A number of reviews have been published on the subject[164–171]. As previously mentioned, the compounds to be analysed may be found in living matter as well as in soil, sediments, water and air; the experimental methods vary accordingly.

1. Extraction

An early method for the isolation of an organophosphorus compound from animal or plant tissues was based on extraction by hexane and selective partitioning with acetonitrile[172,173]. More general multi-residue procedures for direct extraction by acetonitrile or ethyl acetate were developed by Getz[174] and Storherr and coworkers[175–178]; acetone was later substituted for acetonitrile[164,179–181] because acetone is more volatile and easier to remove, reduces the amount of clean-up work required and decreases the losses of soluble polar residues. The methods were further improved by use of the whole extract instead of an aliquot, by concentration of the solutions and by better clean-up processes[182].

a. Extraction from solids. Vegetal and animal matter may simply be immersed in an organic solvent and left to stand for several hours; usually, however, it is mechanically chopped, ground, blended and homogenized with the extracting solvent; the liquid is then separated by decantation, filtration or centrifugation; Soxhlet reflux is often applied. The solvents vary from the highly polar acetone[166,183–188], acetonitrile[165,182,189–192], methanol[168,193], ethyl acetate[169,194,195] or dichloromethane[166,195–197] to non-polar hexane[168,172,173,195], light petroleum[198,199] or benzene[199]; mixtures such as chloroform with methanol[200–202], or more complex mixtures, are also used[203,204]. The water-miscible solvents are often preferred for biological samples, because of their greater ability to penetrate into the cells[164]. After the extraction, the aqueous phase is saturated with sodium chloride and partitioned with dichloromethane[166,185,193], ethyl acetate[175], mixtures of dichloromethane and hexane[188] or other solvents; the final organic solution is dried with sodium sulphate. In some procedures the sample is thoroughly ground with sodium sulphate until a dry power is obtained from which the organophosphorus substances can be extracted, avoiding the cumbersome separation of the aqueous phase required in other techniques[164,172,173,197,205,206].

Samples of soil and sediments are refluxed with mixtures of methanol and water[168], of dichloromethane and methanol[168] or of hexane and acetone[168,207]. The solvents may then be evaporated, the solid residues dissolved in water and extracted with dichloromethane; the organic solution is finally dried with sodium sulphate[168,208]. Alternatively, the solution obtained after extraction can be partitioned directly with dichloromethane and dried[207]. Another method proposed for soil extraction is shaking with an aqueous mixture of acetone and ammonium acetate[166].

b. Extraction from liquids. Liquid samples such as oil, beer or milk are extracted by shaking with a mixture of hexane and acetonitrile[209,210], with ethyl acetate[211,212] or with acetone[200,201]. Aqueous solutions may be directly partitioned with dichloromethane[164,166,168,213–215], hexane[164,215], heptane or chloroform[216–218]; ethyl acetate is probably the best extractant[215]; adding sodium chloride to the aqueous phase improves the separation[215]. However, the recovery of minute amounts of organophosphorus compounds from water by adsorption or reversed-phase adsorption techniques has become increasingly popular[164,170]. For instance, the isolation of ionic dimethyl or diethyl phosphates and thiophosphates from large volumes of aqueous media was carried out by passing the solutions through an acidified XAD (macroreticular cross-linked polystyrene) resin bed, then eluting with acetone; interfering inorganic phosphates were eliminated by addition of ammonium molybdate to form the yellow heteropoly molybdophosphate which was then precipitated as its safranine complex[219]. Passage through XAD-type resins and elution with ethyl acetate[215,218,220,221] dichloromethane[221] or a mixture of acetone and hexane[222] are now routine procedures for the extraction of large numbers of organophosphorus derivatives and of their oxidation and decomposition products from dilute aqueous solutions. Enrichment may be effected by reversed-phase adsorption on a diphenylsilane- or an octadecyltrichlorosilane-bonded column[223], e.g. a Sep-Pak C_{18} cartridge pretreated with methanol, followed by elution with acetonitrile[224] or other solvents[195,213,225]. This procedure requires application of pressure to the column and is usually carried out with the help of a high-performance liquid chromatograph (see

Section III.B.5)). A recently reported technique for the extraction of phosphates and organophosphorus compounds from large volumes of sea water relies on the selective adsorptions either of inorganic phosphorus molybdenum blue complexes onto cation-exchange acrylic fibres or of organophosphorus substances onto iron hydroxide-coated acrylic fibres[226]. Natural water samples are often mixed with solid matter, organic or inorganic, which adsorbs some of the organophosphorus compounds and makes the extraction problematic; the best method in these cases is to separate the solid particles from the liquid and deal with the two phases separately[164,213]. Adsorption on carbon has occasionally been used[227,228] to remove organophosphorus compounds from aqueous solutions.

c. Extraction efficiency. The effectiveness of an extraction method is not easy to gauge. It is usually determined by adding known amounts of some organophosphorus derivative, sometimes radioactively labelled, to the samples; the extraction procedure is applied and the recovered fraction measured. It is difficult, however, to compare results obtained from freshly spiked samples with those of naturally aged ones, as the organophosphorus compounds may have migrated with time in the organic matter and become more difficult to extract, or may have reacted and been degraded to different substances. The efficiency varies according to the specific sample treated and the substances to be isolated. Methanol extraction, for instance, seems to yield the highest recoveries from grain[167], whereas ethanol[229] and acetone are the solvents most commonly used[168,183,184,188] in general procedures.

2. Clean-up

The previously described extractions normally yield mixtures of organophosphorus compounds, water, organic solvents and co-extracted extraneous matter which interfere with the identification and determination procedures ; the presence of lipids is particularly troublesome. Clean-up of the crude product is therefore carried out before additional analysis. The first step is usually the selective extraction of the organophosphorus substances into an organic phase by liquid–liquid partitioning. This has already been described in the discussion of the extraction methods; recovery with dichloromethane, followed by drying with sodium sulphate, is the most commonly applied procedure[179,193,207]. Several techniques are availble for further treatment : adsorption chromatography, sweep co-distillation and size-exclusion (or gel-permeation) chromatography are the most commonly applied, others include low-temperature precipitation of lipids[165,181,230–232], blending of the extraction mixture with deactivated alumina[233] or ion-exchange chromatography[234,235]. These methods, however, always involve some loss of the compounds of interest, in particular of water-soluble polar derivatives, and attempts are being made to reduce or eliminate completely the need for clean-up by developing analytical procedures directly applicable to the extracted mixture[179,180,216,229]. The advances, in recent years, of miniaturization and automation now permit integration of the clean-up process with separation and determination methods such as gas–liquid or high-perfomance liquid chromatographies. Samples recovered from water by one of the new reversed-phase adsorption techniques usually require no other purification.

a. Adsorption chromatography. Clean-up on columns containing alumina or magnesium oxide[166,167,181.203,217], charcoal[174,178,217,236,237], magnesium silicate (Florisil)[188,204,205,238,239], mixtures of charcoal and Florisil[198], of charcoal, magnesium oxide and silica[166,175,182,207] or of related adsorbents[165,181,191] was first described many years ago and is still widely applied. Passage through a silica gel column and elution with different solvents not only purifies, but also roughly separates, the organophosphorus derivatives to be analysed later[166,167,200,201,240]; prepacked disposable Kieselguhr columns for liquid–liquid extraction of lipophilic substances from aqueous samples are often used[241,242]. Clean-up of organophosphorus substances from grain[167] or from fish, beef and butter fats[206] can be carried out efficiently on a Florisil column, using acetonitrile as the stationary phase and eluting with hexane. Adsorption on an alumina column, followed by elution with a mixture of hexane and diethyl ether, was also used to clean-up fish and sediment extracts[168]. Commercially available columns, such as the Sep-Pak C_{18} or Extrelut cartridges, are now in increasing service[192,193,205,195]. A multi-cartridge apparatus has recently been described for the single-step separation of organophosphorus residues from vegetable oil extracts[210].

b. Sweep co-distillation. A sweep co-distillation instrument was developed specifically to separate organophosphorus compounds from co-extracted matter[176,177]. Its main component is a tube packed with glass-wool and heated to about 170 °C; the crude mixture, dissolved in ethyl acetate, is injected at the entrance of the tube and carried through by a nitrogen current; fractions are collected at the exit with a condensing coil; most of the extraneous matter remains on the glass-wool. Slightly modified commercial extractors, with various packing materials and adjustable temperatures, allow time-saving treatment of a large number of samples[170,212,243].

c. Gel-permeation chromatography. Gel-permeation chromatography, originally used to isolate chlorinated pesticides from lipids[244,245], is now widely applied to the clean-up of organophosphorus compounds. Separation is based primarily on differences in molecular sizes: the molecular masses of most pesticides are between 200 and 400, whereas those of the lipids vary between 500 and 1500; elution with cyclohexane from various cross-linked polystyrene gels (in particular from Bio-Beads S-X2 and S-X3) can separate microgram amounts of pesticides from as much as 0.5 g of lipids. Labour-saving automated instruments yield results superior to those obtained with liquid–liquid partitioning followed by column adsorption chromatography[170,184,185,188,197,245–249]. For the elution, the originally used cyclohexane may be replaced by ethanol[185,217] or by various mixtures of hexane, cyclohexane, toluene, dichloromethane, diethyl ether, ethyl acetate or acetone[168,169,184,186–188,197,246–248]. Data on the gel-permeation chromatography of over 400 compounds have been reported, together with information on further clean-up, should it be needed, by chromatography on a silica column[186,187].

B. Separation, Detection, Identification and Determination

Organophosphorus substances are often found in mixtures containing several compounds and their decomposition products. The analysis of such a multi-component sample requires the separation of the individual derivatives before identification and determination are possible. A number of techniques of general applicability are available to this end, mostly based on chromatography and mass spectrometry. In addition, methods have been developed for the analysis of individual compounds requiring no previous separation.

1. Paper chromatography

An early attempt to analyse organophosphorus compounds, specifically phosphoric esters, was made with paper chromatography[250]. The esters were spotted on filter-paper and developed with a variety of solvent mixtures in order to isolate the different components of the samples. The paper was dried, sprayed with an acidic molybdate solution, heated to about 85 °C in order to hydrolyse the compounds and finally immersed in gaseous hydrogen sulphide for reduction. The position of the esters on the chromatograms was detected as intense molybdenum blue spots on a buff background. The amounts of the components were quantitatively measured by cutting out small paper areas around the spots, ashing them with concentrated sulphuric and perchloric acids, then determining the resulting phosphates by the molybdenum blue method. In further developments both one- and two-dimensional chromatography were applied to the screening of unknown samples of various origin[174,181,251]. As an example, paper chromatography has been used for the separation of inositol mono- and tri-phosphates originating from biological systems[252]. The necessity of working with clean extracts soon became evident, particularly in samples of fatty origin, as fatty or waxy co-extracts interfered with the procedure. Paper chromatographic separation became routinely coupled with some prior clean-up, the fats and waxes being either frozen out or removed by solvent partitioning and column elution[181]. A variety of developing mixtures and chromogenic reagents were investigated for the separation and detection of both polar and non-polar compounds[174,175,181,251,253–255]. Paper chromatography has now generally been replaced by thin-layer chromatography for the analysis of organophosphorus mixtures, although in some instances it is used in conjunction with thin-layer chromatography[256].

2. *Thin-layer chromatography*

As a rule, thin-layer chromatography (TLC) is faster than paper chromatography, it usually requires no previous washing of the chromatographic layer and both the development and the detection are speedier; it is more sensitive, the spots are often more compact and the resolution is sharper[196,257]. Gas chromatography and high-performance liquid chromatography are more sensitive and provide better separations than TLC, and therefore are often preferred for analytical methods. Nevertheless, the speed and versatility of TLC and the fact that no expensive instrumentation is required have led to its extensive use, even sometimes for further identification of compounds separated by more sophisticated chromatographic techniques. An ever increasing number of organophosphorus compounds are being separated and characterized by TLC[188]. Practical advances such as the commercial availability of automatic scanners and of a variety of ready-made chromatographic plates (e.g. fluorescent silica gel plates and reversed-phase plates) have greatly increased the convenience of the method, which is now widely applied from preliminary screening and detection to identification and semi-quantitative determination of small amounts of substances.

a. Adsorbents, stationary and mobile phases. Several adsorbents may be employed: a cellulose support was chosen when the similarity with paper chromatography was to be maintained[258], magnesium silicate has been investigated[259] and thin layers of polyamide powder have been used in attempts to exploit the hydrogen bonding forces between the —CONH— groups of the polyamide and the oxygen- or sulphur-containing adsorbates[260,261]. The most common supports, however, are silica and alumina[162,262], which are applied in layers 0.10–0.25 mm thick for detection work and up to 0.5–1.0 mm thick for preparative chromatography. Silica and alumina have intrinsic clean-up properties of their own and thus largely reduce the need for preliminary treatment. TLC itself may serve as a clean-up method because it allows the handling of as much as 100 mg of sample on a single plate; the solvent chosen in this case should isolate the organophosphorus compounds from extraneous matter rather than separate them from each other[196,263,264]. Nevertheless, as in other forms of chromatography, fatty or waxy impurities must be removed beforehand, usually by some solvent partitioning stage[253].

Whereas paper chromatography essentially serves as a liquid–liquid partitioning technique, TLC may be used either as a liquid–liquid or as a liquid–solid process. Liquid–liquid procedures require the adsorbent layer to be deactivated and impregnated with a stationary phase solvent; organophosphorus as well as organothiophosphorus compounds and their oxygen analogues may be analysed by dipping the chromatographic plates into dimethylformamide after the samples have been spotted and dried, then developing them with methylcyclohexane, trimethylpentane or a mixture of trimethylpentane and benzene[257,258]; alternatively the plates can be pretreated with formamide[256] or cresol[265], then developed with dichloroethylene or a more complex solvent mixture. Reversed-phase liquid–liquid procedures have also been described, with mineral oil as the stationary solvent and a combination of acetonitrile and water for the development[258]. Pretreated plates for reversed-phase TLC are now commercially available[266].

In liquid–solid procedures only one, mobile, solvent is required, as the adsorbent layer constitutes the stationary phase. Under low-humidity conditions the plates may be used without further treatment; however, they are often activated by heating to 100–130 °C before the spotting stage[190,196,259,263,267–273]. The choice of the developing system is dictated by differences in the polarities of the organophosphorus compounds to be separated. Pure liquids such as hexane, benzene, toluene, xylene, diethyl ether, ethyl acetate, acetone, chloroform, carbon tetrachloride and many others are sometimes used[190,214,270–272], but more often mixed systems are chosen, so that the distinction between liquid–liquid and liquid–solid chromatography becomes blurred. In an example of such a procedure, part liquid–liquid, part liquid–solid partition, over 60 pesticides were separated and characterized by using mixtures of a stationary 'background' solvent (hexane, benzene or chloroform) with various amounts of mobile 'eluting' solvents of different types (diethyl ether, ethyl acetate, acetone, methanol or acetic acid); hexane was found to be the best for low-polarity compounds and benzene and chloroform for more polar compounds; the displacement of the spots could be controlled by the concentrations of the eluting solvents[263]. The most commonly used system for the separation of organophosphorus compounds is a 4:1 mixture of hexane and acetone[189,264,268,269,274,275], although other ratios are found to be useful[267,276]; inorganic phosphorus compounds are not eluted if the mixture contains less than 70% of acetone[196]. An almost endless list of binary, ternary or more complex solvent combinations have been investigated, each of them tailored to the specific compounds under scrutiny[190,214,217,259,260,270–272,277–280]. Usually no single solvent system clearly separates all components and several plates and eluting mixtures must be used for complete resolution[261,275,280]. It is helpful to apply two-dimensional chromatography, in which the spotted plate is first developed in the usual way, then dried, rotated through an angle of 90° and treated with a different solvent mixture. This method allows the separation of compounds which, either fortuitously or because of their polarities, moved together in the first direction; it is sometimes coupled with chemical treatment such as exposure to bromine vapour between the development stages[165,256,260,270,278,281].

b. Detection methods. Detection of the spots is carried out by several basic procedures, which have been studied in many variations and combinations. Some compounds, e.g. those containing aromatic rings, may be directly viewed by fluorescence[258,282] or as dark spots in transmitted ultraviolet light[260,271]. Most plates, however, are treated, after the development, with chromogenic agents which react with the organophosphorus substances to yield coloured or fluorescent products; these can then be detected visually either in ordinary or in ultraviolet illumination. Enzymatic inhibition techniques are often applied.

i. Colorimetric and fluorimetric detection. A large number of indicators have been sprayed onto chromatograms for the detection of spots[190,261,262,271,272,277] and new ones are still being probed[275]; it is almost impossible to list all the substances investigated and only a few of them will be mentioned here. Some of the reagents are of general applicability and will help to locate many different organophosphorus derivatives, and sometimes non-phosphorus compounds too. The phosphorus compounds can, for instance, be hydrolysed by wetting the chromatographic plate with a solution of hydriodic and acetic acids and heating to 180 °C, then treating it with ammonium molybdate and reducing the molybdophosphoric acid with tin(II) chloride; this procedure yields characteristic molybdenum blue spots on a buff background[217]. Spraying with silver nitrate, then illuminating with ultraviolet light, can be used for general detection, producing grey spots on a light-brown background[190,258,271,280]; the plates may first be treated with bromine and fluorescein for better spot visibility[251,263]. Exposure to iodine vapour is an excellent technique[263,265] and spraying with 4-(*p*-nitrobenzyl)pyridine, sometimes followed by tetraethylenepentamine, is in widespread use[190,255,256,259,261]. Fluorogenic agents, e.g. dansyl chloride[214,283], may also be sprayed onto the TLC plates for labelling of the compounds and sensitive detection of the spots.

As a considerable proportion of the organophosphorus pesticides contain sulphur, chromogenic reagents specific to thiophosphorus compounds have been investigated[277]. Treating the plates with tetrabromophenolphthalein ethyl ester, silver nitrate and citric acid yields blue or purple spots on a yellow background, characteristic of thiophosphate compounds[165,190,257]; spraying with trichlorobenzoquinoneimine results in an orange colour on a white background[266,275]. Reaction of sulphur-containing compounds with bromine vapour yields hydrobromic acid, which may colour a pH-sensitive indicator or release a fluorescent ligand from a non-fluorescent complex[284]. Exposure to bromine followed by spraying with iron(III) chloride and hydroxyphenylbenzoxazole thus produces fluorescent spots for thiophosphorus compounds and some of their breakdown products[190,277]. Treating with bromine, then spraying dimethylphenylazoaniline[267], Congo Red or fluorescein onto the plate has been suggested[190,260]. Application of a silver nitrate and Bromophenol Blue or a silver nitrate and Bromocresol Green solution, followed by immersion in citric acid also identifies thiophosphorus derivatives[258,259]. Palladium chloride has been used to distinguish between sulphur-containing and other organophosphorus compounds[261]. The exchange reaction between a palladium chloride–calcein complex and thiophosphorus compounds, in which the fluorescent calcein is released and the palladium chloride becomes selectively bound to the sulphur, is similar[264,276,284]. Other fluorogenic agents have also been applied, such as Rhodamine B[165] and dichlorodicyanobenzoquinone[273].

ii. Enzymatic detection. Many of the organophosphorus compounds to be analysed possess enzyme–inhibiting properties, which may be put to efficient use for the detection of TLC spots. In this method the chromatographic plate obtained from development of the sample is sprayed with an enzyme solution and a sufficient incubation time is allowed for the organophosphorus substances to interact with the enzyme. A substrate is then sprayed onto the chromatogram, and reacts with the enzyme over the entire surface, except at the spots where the active sites of the enzyme molecule were previously blocked by organophosphorus compounds. When the plate is later exposed to a chromogenic agent, and colour due to the product of the enzyme–substrate reaction is formed, the areas where the enzyme action was inhibited appear as white spots on a coloured background; other indicators may show the inhibited spots as coloured against a colourless background[285]. An often used enzyme is cholinesterase (extracted from animal livers), which reacts with acetylcholine[258,285], acetyl- and butyryl-thiocholine[279] or naphthyl acetate[253,268,269,285] to yield hydrolysis products detectable with the help of Methylene Blue, dithiobisnitrobenzoic acid, nitrobenzenediazonium fluoroborate and other reagents. The reaction products of cholinesterase with other substrates, e.g. indophenyl, indoxyl or bromoindoxyl acetate, are fluorescent and become coloured in the presence of oxygen, so that no further step is required for detection of the spots[165,188,189,274,278,285]. In some instances the plates are exposed to bromine vapour before the enzyme treatment, in order to convert the thion compounds into their more strongly inhibiting, and therefore more clearly detectable, oxon analogues[189,258,269,274,286].

iii. Radiometric detection. Derivatization of dolichyl phosphate to [^{14}C]phenyl chloroformate, followed by TLC and detection of the radioactivity on the thin layer with a radiochromatogram scanner has been reported[202]. This technique allows the assay of dolichyl phosphate in samples of animal origin at sub-nanomolar concentrations.

c. Identification. Identification of the organophosphorus compounds isolated by TLC can be carried out by eluting a spot from the developed plate or by scraping if off and dissolving it; the solution is then analysed by infrared and nuclear magnetic resonance spectroscopy, different chromatographic techniques or other instrumental methods.

Most TLC screening procedures rely on measurement of R_f values. The R_f factor is defined as the ratio of the distance moved by a certain spot to the distance moved by the developing solvent during the same period. Its value depends on the physical circumstances at the time of the development, such as the purity of the sample, the nature of the thin layer, the ambient temperature and mainly the composition of the solvent. Under given, well specified, experimental conditions, the R_f value is characteristic of a substance, although several compounds may show the same R_f factor. Papers reporting TLC work usually give the observed values, and by now considerable knowledge is available on the R_f data of a large number of organophosphorus derivatives in different solvents and solvent mixtures. It is difficult however, to duplicate the exact conditions under which these R_f factors were determined; in order to make certain of the identity of the compounds it is recommended to add to the chromatographic plate spots containing pure standards of the compounds suspected to be present, and to compare their migrations with those of the sample spots[270]. If impurities are likely to interfere and affect the observed R_f values, the standards should be mixed with the sample to be studied, so as to confirm their behaviour under circumstances identical with those of the unknown[252,262].

d. Determination. TLC is basically a highly sensitive, but qualitative, technique. Detection limits vary with the specific procedure applied. The colorimetric or fluorimetric detection limit is usually of the order of 0.1 μg per spot[258,264,266,273,275–277,284]; however, detection at 0.01 ppm concentrations has been reported for two-dimensional TLC[270], and dansyl chloride labelling allows fluorimetric detection of 10–25 ng amounts[214]. Enzyme inhibition techniques exhibit a much higher sensitivity, in particular if the thio compounds have first been converted into their more potent oxo analogues; 0.1 ng amounts are easily detected[268,269]. Some attempts have been made to apply TLC to the semi-quantitative determination of organophosphorus compounds; the spots are eluted or scraped from the plate and the solutions analysed chemically[196], colorimetrically or fluorimetrically[277]; as an example, pyrazophos, isolated and identified by TLC, can be quantitatively determined down to 0.2 ppm concentrations by ultraviolet absorption measurements at 245 nm[282]. In a different approach, the size and intensity of spots may be estimated and compared with calibration graphs prepared from known standards[217,257,277]. Amounts in the 10–100 μg range can be determined with chromogenic or fluorogenic techniques[264,267,269,276]. Automatic scanning improves the sensitivity and gives straight calibration graphs in the 0.2–1.25 μg range[266,275]. Fluorimetric labelling with dansyl chloride and spot intensity measurements show a linear relationship between fluorescence intensity and concentration in the 50–500 ng per spot range[214], allowing more precise determinations. The best results were again obtained with enzyme inhibition methods, which allowed quantitative estimates of 5–50 ng amounts[268,269].

3. Gas chromatography

Gas–liquid chromatography (GLC) is, with high-performance liquid chromatography, the most frequently used method for the separation and identification of organophosphorus substances, in particular in the analysis of pesticide residues. The number of papers reporting on some aspect or other of this subject is so large that it is practically impossible to cite them all. Nevertheless, as the samples must first be gasified, the applicability of the technique is limited to mixtures volatile enough to be easily vaporized and care must be taken not to heat the analytes above their decomposition temperatures (about 200 °C).

The main component of a gas–liquid chromatograph is a column filled with the stationary phase, a non-volatile liquid saturating a solid support. The mobile phase, an inert gas such as oxygen-free nitrogen, argon or helium, flows at a steady rate through the chromatograph. The sample is introduced into the instrument, evaporated, then carried through the column by the inert gas. The compounds in the injected mixture are temporarily adsorbed by the liquid substrate, then later released into the continuing flow of carrier gas. The time needed for the desorption depends on the polarity and on the molecular mass of the compounds and varies from substance to substance. Each component of the mixture thus passes through the column at a different rate and produces a separate signal (or chromatographic band) in the detector which is situated at the exit of the column. The detector output is recorded as function of the time or of the amount of gas flowing through the instrument, yielding a band chromatogram. The retention time, t_r, is defined as the length of time elapsed between the injection of the sample and the emergence of each compound at the other end of the column and the retention volume, V_r, is the volume of carrier gas which flows through the instrument during the retention time. Usually a small bubble of air is injected with the sample into the instrument; the air passes through the column without adsorption, at the same rate as the carrier gas, and results in a signal on the chromatogram. The adjusted retention time, t'_r, is obtained by subtracting the 'dead time' needed by the air to pass through the column from the previously defined retention time and the adjusted retention volume, V'_r, is the volume of carrier gas which flowed through the chromatograph during t'_r. Under completely defined experimental conditions the retention times and volumes are characteristic of the different substances; they may be obtained by injecting pure compounds into the column and measuring their t'_r and V'_r. The relative retention time of two compounds is the ratio of their t'_r values; the greater this ratio, the better is their separation on a given column. Gas chromatography is normally carried out under isothermal conditions, although temperature programming may be used for the separation of compounds with different volatilities.

a. Supports and liquid phases. The columns used for the study of organophosphorus compounds may basically be of two forms, as follows.

The first type consist of relatively short tubes, with an inner diameter of several millimetres, packed with an inert support on which the stationary liquid phase is held. The preferred sorbents are variously calcined, acid- or base-washed and silanized forms of diatomaceous earth (Chromosorb, Gas Chrom Q, Supelcoport)[165,168,178,180–184,191,184,195,197,200,201,204–206,209,219,221,236,237,239–242,247,249,287–295].

The second type consist of open capillary tubes, in which a thin film of the liquid phase is bonded to the inner wall[192]. They are extremely long, usually between 10 and 30 m, although much greater lengths are sometimes used; their inner diameter varies from 0.1 to 0.3 mm. These columns, which require the samples to be cleaner[167], are far superior to the packed columns for organophosphorus residue analysis[167,296,297] and their use is rapidly increasing[207,243,298]. The so-called 'megabore' columns apply the same principle to 0.5–1.0 mm diameter tubes, achieving better capacity than the narrow capillaries[299]; conventional packed-column chromatographs may easily be adapted for use with this type of tube[299].

The stationary phases coated on the inert supports are similar in the two classes of columns. Liquids of different polarities are chosen for the separation of non-polar or polar compounds; a relatively small number of solvents, chemically inert and heat resistant, are sufficient for most residue analyses. The most commonly used for organophosphorus compounds vary from low-polarity hydrocarbons, methylsilicones, phenylsilicones, phenylmethylsilicones or carboranesiloxanes[165,168,175,178,180,182–184,192,195,197,198,200,201,204,205,207,209,225,227,237,239,240,243,288,290–292,294,295,298–302] to more polar liquids such as fluorosilicones and cyanopropylphenylsilicone[183,210,241,242,288,292,293,298,301]. These solvents provide a variety of polarities from which a column suitable for almost every purpose may be derived. Mixtures of liquids with different polarities are frequently used; they allow the preparation of columns customized to the separation of virtually any organophosphorus sample[180,182,194,206,219,221,227,237,239,247,249,287,288,291,294,295]. Some other liquids often employed for the separation of polar organophosphorus compounds are polyethylene glycol[291,292], polypropylene glycol adipate[287,289,2291], polyethylene glycol ter-octylphenyl ether[227], diethylene glycol succinate (DEGS)[165,178,183,184,236,237,295] and neopentyl glycol succinate (NPGS)[294]. These very polar solvents are mainly used when the less polar silicones are unable to effect the required separations, e.g. with mixtures of oxons and their parent thions[165], or in the separation between an organophosphorus compound and its oxidation and hydrolysis products[289].

b. Detectors. Only a small number of sensors are able to monitor organophosphorus compounds when they emerge from from the chromatographic column. Some of these were developed specifically to this end, and research is still continuing in attempts to find others[303,304]. At present, electron-capture or phosphorus-specific thermionic and flame photometric detectors are the most sensitive to phosphorus-containing molecules.

i. Electron-capture detectors. Electron-capture detectors consist of a chamber in which electrons emitted by a radioactive source (e.g. tritium or ^{63}Ni) are collected on an anode, yielding a steady background electric current. When an electrophilic substance passes from the chromatographic column into the chamber, some of the electrons are captured and a decrease in the current is noted; the instrument is usually set up to give linear responses. This detector is useful for the study of molecules containing oxygen, nitrogen, sulphur or phosphorus atoms; it is extremely sensitive, the limits of detection reaching picogram amounts in the case of organophosphorus derivatives[165,183,184,191,197,198,201,203,205,243,247,288,289,298,302,305]. An examination of the correlation between structure and electron affinity in phosphorus-containing compounds has shown that replacement of oxygen by sulphur in homologue substances substantially increases the intensity of the signals[305].

ii. Flame ionization detectors. Flame ionization detectors work on a different principle. The gas stream emerging from the chromatographic column is mixed with hydrogen and introduced, with an excess of oxygen, into a burner. The hydrogen flame ionizes the organic molecules moving with the stream and the gas surrounding the flame becomes an electric conductor. The conductivity is measured by applying a potential between the burner and an electrode placed around the flame; signals proportional to the amounts of ionized organic substance present in the flame are obtained. Phosphorus itself does not markedly affect the response of an ordinary flame ionization detector and organophosphorus compounds cannot be differentiated and selectively studied in this way. However, special sodium- or potassium-modified detectors, in which the burners are coated with sodium sulphate, sodium hydroxide or potassium chloride, enhance by several orders of magnitude the ionization signals obtained from organophosphorus and organothiophosphorus compounds, while only slightly increasing the bands due to halogenated substances and leaving unchanged the responses to other molecules[300]. They constitute extremely sensitive and selective thermionic detectors which allow the determination of phosphorus-containing derivatives with minimal interference from other compounds[300]. The original burners have now been replaced by electrically conducting beads of alkali metal silicate or alkali metal-coated ceramic[306,307]. These devices are heated by electric current, atomic alkali metal is vaporized and ionization of the analyte takes place in the vicinity of the beads. At the end of the process the positively charged alkali metal ions are recaptured by the negatively charged beads, which prevents both depletion of the alkali metal supply and contamination of the detector housing. Thermionic detectors are now often incorporated in commercial gas chromatographs[165,178,179,182,184,186,195,197,206,217,219,222,225,236,237,239,294]. Similar caesium- or rubidium-modified burners[287,290,227,294] are also in use.

iii. Flame photometry. The HPO emission previously described (see Section II.C.1) has been applied to the development of a sensitive, phosphorus-specific, flame photometric detector in which the 526 nm band is isolated by a narrow pass interference filter and its intensity measured with a photomultiplier. The intensity of the signals is proportional to the concentration, and parts per billion amounts of organophosphorus compounds can be determined[143]. Responses to different groups of substances, however, cannot be directly compared: at equal molar concentrations of phosphorus the intensities of the signals vary from compound to compound, both the structure of the derivative and the nature of the solvent used for the sample injection affecting the relative responses[308]. In this respect the substituent groups on the phosphorus atom play a considerable role; halogen substitution, for instance, enhances the signal whereas dialkylaminoalkyl groups substantially reduce it[309]. The flame photometric detector is now in wide use for the quantitative analysis of insecticides, pesticides and their decomposition products[165,179,180,182,183,184,194,201,204,206,210,218,220,241,242,247,249,291,293,298,301,302]. Many organophosphorus compounds, such as fenthion, VC-506 (a phenyl phosphonothioate derivative), ronnel and fenitrothion, can be determined with high precision in the presence of their oxygen analogues and their metabolites[200,201,218,221,240]. Phosphorus compound are sometimes reduced to phosphines before determination by flame photometry[165,310]; the procedure, although cumbersome, increases the sensitivity by eliminating interferences from co-existing materials.

The responses of these three detectors to a variety of organophosphorus molecules have often been compared; the results seem to vary from compound to compound[165,179,217,238,289]. In many instruments the stream is split and passes through several detectors for parallel measurements[184,187,298]. The flame photometric detectors are the most selective for phosphorus, yield less extraneous peaks and make the identification work easier[188].

GLC instruments are now often coupled with mass spectrometers for more detailed studies of the column effluents. This technique will be discussed in connection with mass spectrometry (see Section III.B.6).

c. Separation and identification. A considerable amount of information has been accumulated on the retention data of organophosphorus compounds; their metabolites, their hydrolysis and oxidation products, often as toxic as the original compound, have also been studied[167,180,200,236,290,291,294,295,298,299,301,302]. Recent years have witnessed a dramatic increase in the number of compounds found in residue analyses and a single column can no longer separate them all; it is often necessary to chromatograph the same samples on several columns of different polarities, sometimes equipped with two or three detectors, and to characterize substances by more than one t'_r value[183,291,294,295,298,302,311]. Even then, identification of the components of a mixture is not always easy. The gas chromatographic determination of other elements present in the organophosphorus compounds, nitrogen, sulphur or halogens, is useful for confirmation of the results[164]. Combination with other chromatographic techniques such as gel chromatography, thin-layer chromatography, high-performance liquid chromatography or fractionation on a silica gel column is often used for better identification[167,183,188,201,217,288,292,294].

Preliminary chemical treatment sometimes improves the ability to separate the components of a mixture. Oxidation of fenthion[209] or other compounds[290], methylation of ethephon[194] or dimethoate[188], trifluoroacetylation of monocrotophos[293], basic hydrolysis followed by methylation or by treatment with pentafluorobenzyl bromide of parathion[188] and other organophosphorus compounds[287,288,292] have been attempted. The derivatives obtained differ from the original compounds in both retention times and intensities of detector responses, leading to easier separation of the residues and confirmation of their identity[288]. Compounds which tend to decompose during passage through the chromatographic column may gain stability and be more easily analysed after derivatization[287].

If no previous information is available on the compounds to be expected in a given sample, its complete residue analysis may require a very lengthy and complicated procedure[183]. The amount of work can be reduced by automation, e.g. by splitting of the carrier stream, or by pneumatic switching between different columns and detectors, or by computerized data processing[298,301,302]. As in the case of thin-layer chromatography, the samples must be cleaned from extraneous matter, in particular from fatty substances, before injection into the instrument. If a sample has not been well enough cleaned the retention times may be modified[167]; in those cases it is recommended that small reference amounts of the compounds expected to be found are introduced into the sample and their retention times compared with those of the unknowns under the actual conditions of the measurement[302].

d. Quantitative determination. Quantitative analyses are easily carried out by gas chromatography by measuring the area under each peak of the chromatogram. The method is based on the assumption that this area is proportional to the amount of compound which was eluted from the column, i.e. that the detector response itself is linear with the concentration or the amount of analyte. As different compounds generate different responses in the detectors, the correlation factor must be determined for each component of a sample by studying the behaviour of pure compounds or of well known mixtures under the same experimental conditions and by preparing calibration graphs[198,205,227,302]. The phosphorus-specific detectors are extremely sensitive and determinations to better than 1 ppb or less than nanogram amounts per millilitre are sometimes possible[191,198,209]. Detection limits lower than 0.1 ppb have been reported[195,200,227,240,288] and derivatization often improves the results[209,288,293]. When mixtures contain a large number of compounds the reported sensitivities are not as good[183], although the megabore capillary columns previously mentioned allow detection down to 0.1 ng and determination down to 1.0 ng in such cases[299].

The sensitivity of the gas chromatographic equipment is usually not the limiting factor in the detection and determination of the organophosphorus compounds found in natural environments. The extraction and clean-up procedures discussed earlier are the most important steps and the percentage recoveries probably play the primary role in this regard. In particular the use of reversed-phase adsorption cartridges allows enrichment and concentration of the samples and reduces the lower limits of analysis[195].

4. Liquid chromatography

Attempts to apply conventional liquid–chromatography to the separation of organophosphorus compounds were reported early. Fenthion and VCS-506, for instance, were isolated from their metabolites on alumina or silica gel columns and characterized by their retention times or by subsequent GLC examination[200,201]. Identification of the effluents if often difficult, in particular for those substances which are transparent in the ultraviolet spectral range. Electrochemical methods have sometimes been used; specifically, polarographic detection and determination of *p*-nitrophenyl-containing thiophosphorus derivatives is made possible by the reduction of the nitro group on a dropping mercury electrode[312–314]. This procedure has been applied to the determination of parathion and some of its derivatives, with a detection limit of 30 ppb[229,315].

Ion-exchange and the related ion-exclusion chromatographies are modifications of liquid chromatography; they rely on differences in the degree of dissociation of sample molecules under various eluent pH conditions, and on the distinctive affinities of the sample ions and counterions towards specially treated chromatographic supports. These types of chromatography may be applied successfully to the analysis of polar organophosphorus compounds[316], although detection of the non-light-absorbing, non-fluorescent effluents again presents a frustrating problem which requires indirect approaches. One possible technique is to add to the mobile phase light-absorbing ions having the same charge as the analyte ions; the light-absorbing ions selectively displace the adsorbed sample ions from the column and the appearance of 'transparent' sample ions, replacing some of the light-absorbing ions in the effluent stream, is signalled by 'dips' or 'troughs' in the baseline optical absorbance[317]. In a different procedure, phosphorus oxo acids were separated on a quaternary ammonium anion-exchange column, a post-column reagent solution of aluminium–morin complex was mixed with the effluent and the quenching of the aluminium–morin fluorescence was measured as described previously[318] (see Section II.B.2.a). Conductivity measurements have been applied to the study of phosphonic acids[319] and inductively coupled plasma atomic emission spectrometry has been used for the determination of orthophosphates in water[320], after passage of the samples through ion-exchange columns. Nitrosation of the effluent, followed by dropping mercury polarography, as detailed above, has been applied to the determination of glyphosate residues[321]. A colorimetric method based on hydrolysis of condensed phosphates and formation of molybdenum blue in the presence of ascorbic acid has also been reported[84,322].

Liquid chromatography, however, relies on gravitational flow through wide-bore columns, its efficiency is low and it is time consuming. It is more commonly used for the clean-up of samples than for their complete analysis. In fact, it only became popular with the tremendous increase in speed and resolving power resulting from the development of high-performance liquid chromatography described below.

5. High-performance liquid chromatography

In high-performance (or high-pressure) liquid chromatography (HPLC), the size of the adsorbent support particles is considerably smaller than in conventional liquid chromatography in order to obtain more efficient separation, while the mobile phase is driven by high pressure through the column in order to shorten the flow-through time. HPLC is now increasingly complementing or altogether replacing GLC in the analysis of organophosphorus compounds. It is particularly important in the cases of organophosphorus derivatives which have low vapour pressures or are thermally unstable, and may therefore be in danger of decomposition by heating during GLC analyses[216]. HPLC is applied in a number of variations, including, in addition to liquid–liquid chromatography, techniques such as liquid–solid adsorption, reversed-phase liquid partition, gel-permeation and ion-exchange chromatography[170,182,322,323]. These forms of chromatography exhibit excellent resolution of pesticide residues analyses. The introduction of fully automated instruments and sample miniaturization, and the development of highly sensitive detectors, have greatly increased the potential of the method[84,171,322].

a. Columns and liquid phases. For the separation of organophosphorus compounds, the support generally used in HPLC columns consists of 5–10 μm diameter particles of silica or alumina bonded to stationary phases of various polarities. Ion-echange resins are sometimes chosen[84,292,322,324]. The small grain dimensions result in dense packing, requiring application of pressure for the mobile solvent to pass through the column within a reasonable amount of time.

Regular columns have inner diameters of a few millimetres. They may be packed with silica for simple adsorption chromatography[214,224], or with silica impregnated with a stationary phase such as oxydipropionitrile[216,325] or compounds bearing aminopropyl groups[326] for liquid–liquid separations. The eluents vary from heptane[216] to mixtures of chloroform and hexane[214], aqueous acetonitrile[224] and buffered methanol solutions[326]. Reversed-phase adsorption is becoming increasingly popular and is now the technique most often applied; the procedures often combine extraction, separation and analysis in a single step. For instance, Abate was extracted from dilute aqueous solutions with a diphenylsilane-bonded bead column, then eluted with acetonitrile and quantitatively determined[223]. Accidentally spilled organophosphorus pesticides were recovered from aqueous samples by using a Sep-Pak C_{18} cartridge[213]. A μBondapak C_{18} column was used for the separation and determination of azinphos-methyl and its oxon derivative[193]. Insecticides and fungicides extracted from agricultural crops[183,199], as well as various commercial samples[224,327,328], could be analysed on LiChrosorb RP-18, Spherisorb ODs C_{18} or Zorbax RP-8 columns. The mobile phases used in these cases usually are mixtures of water with acetonitrile[193,199,327,328] or methanol[183,193,224,327,328] in various ratios.

The general trend towards instrument miniaturization has led to the introduction of columns with sub-millimetre dimensions. Microcolumns include capillary columns (with inner diameters of the order of 0.1–0.5 mm) of both the open-tabular and the packed types, and microbore columns (with inner diameters of 0.03–0.07 mm), obtained by packing a capillary tube with silica or alumina, drawing it out to the appropriate diameter, then impregnating it with solutions of bonding agents such as octyl- or octadecyltriethoxysilane and methyltrimethoxysilane[329–331]. These columns are convenient for the high-resolution analysis of small amounts of residues and permit direct coupling with mass spectrometers (see Section III.B.6.a) or with specially designed miniaturized detectors[331–333]. Intermediate narrow-bore columns with inner diameters of 0.5–1.0 mm are sometimes used for increased capacity. Reversed-phase chromatography is routinely applied and the columns are packed with C_{18}-bonded silica, e.g. LiChrosorb RP-18[307,334,335], Spherisorb ODS-2[208,336,337], μBondapak C_{18}[338], Hypersil ODS[339] or Nucleosil C_{18}[336]. As in wider columns, the eluents are aqueous methanol[334,336,337] or acetonitrile[337,338], although the use of more complex systems has been reported, e.g. mixtures of water, methanol and dichloromethane[340] or of water, acetonitrile and chloroacetonitrile[339].

b. Detectors. The early detection methods applied in HPLC wee basically similar to those applied in TLC or in simple LC. Miniaturization of the columns, however, now allows direct introduction of the HPLC effluent into more sensitive phosphorus-specific detectors similar to those developed for GLC.

i. Colorimetric and fluorimetric methods. Most organophosphorus compounds do not absorb visible or ultraviolet light and are not amenable to photometric detection. Photometry can be used only for compounds bearing chromophoric groups; the Abate molecule, for instance, contains a phenylene residue and may be determined by absorption measurements at 254 nm[216,223]. Optical flow-through cells, often micro cells, can be connected with the outlet of the chromatographic column for continuous monitoring of such derivatives[328,331,339,341]. Detection of azinphos-methyl and its oxon was carried out at 224 nm[193], whereas the best wavelength for fenamiphos and its metabolites is 200 nm[224]. Some pesticides extracted from agricultural produce may be observed at 220 nm[199,336]. Dolichyl phosphates extracted from animal tissues were detected directly by on-line measurements at 210 nm[342].

Chromogenic or fluorogenic labelling of organophosphorus compounds which are devoid of ultraviolet absorption or of fluorescence has been found useful. Alkyl methylphosphonic acids, for instance, were esterified with *p*-bromophenacyl bromide, allowing subsequent UV detection[338]; alkyl phosphates, thiophosphates and dithiophosphates were similarly treated with pentafluorobenzyl bromide for photometric determination at 260 nm[292]. Derivatization of dolichyl phosphates to phenyl chloroformate allows spectrophotometric detection at 257 nm[202]. Organophosphorus amines, phenols and imidazoles can be dansylated or treated with chloronitrobenzoxadiazole to yield fluorescent derivatives[214,283,325]. Labelling with 9-fluorenyl methylchloroformate allows, in a similar way, the determination of glyphosate and its metabolite aminomethylphosphonic acid[326]. Conversion of dolichyl phosphates into their methyl 3-(9-anthryl)prop-2-enyl derivatives, followed by passage through a reversed-phase HPLC column and fluorimetric monitoring, has been reported[343]. Fluorimetric detection yields better sensitivity than was observed with TLC[214]; it has the advantage of eliminating most of the preliminary clean-up steps, as the interfering co-extracts in a residue mixture usually do not undergo the same derivatization[325].

ii. Other detectors. Electrochemical methods have sometimes been applied to the determination of organophosphorus compounds, in spite of the exceptional stability exhibited by many of these substances with regard to electrochemical oxidation or reduction[344]. A review on the subject[345] mentions several analytical procedures based on polarography at a dropping mercury electrode; substances containing a $>$P=S, a $\equiv$P—S or an NO_2 group, and nitrosated derivatives of organophosphorus compounds, are accessible to this technique. The polarographic reduction of *p*-nitrophenyl drivatives such as parathion or methylparathion can be applied to the HPLC microanalysis of these substances[323]. Other electrochemical methods have also been modified for monitoring HPLC effluents, e.g. a small-volume electrochemical cell, with a pyrolytic graphite electrode, has been successfully used for the analysis of submicrolitre samples eluted from a small-diameter capillary column[331].

The enzyme-inhibiting properties of many organophosphorus derivatives, dicussed in connection with TLC (see Section III.B.2.b), may also be put to efficient use for HPLC detection. An autoanalyser sensitive to cholinesterase-inhibiting compounds, coupled with an HPLC system, was shown to exhibit higher sensitivity than the spectrophotometric methods, with detection limits as low as 1 pg[346]. A different enzymatic procedure uses a post-column reactor loaded with alkaline phosphatase to hydrolyse phosphate esters, such as inositol bis- and tris-phosphate or other organic phosphates, after their separation by ion-exchange HPLC; a molybdate solution is then injected into the stream and the inorganic phosphates formed can be detected down to the nanomoles level[324]. Dolichyl phosphates may similarly be dephosphorylated to dolichols by using phosphatase, the dolichols then being analysed by a suitable method[347].

The recently introduced microcolumns allow the use of specially adapted, more sensitive, devices. A flame photometric detector, similar to those used in GLC for measuring the HPO emission (see Section III.B.3.b), has been modified for this purpose[329,335]. The total column effluent can be nebulized into the flame without noticeable quenching by the organic solvents. Detection limits of nanograms are attained and linear determination is possible in the 2–100 ng range[329].The performance of such a detector, directly connected to the chromatographic column, was investigated as a function of the mode of introduction of the effluent into the flame and of the composition and flow rate of the flame gases[335].

A phosphorus-specific thermionic detector was also adapted from GLC (See Section III.3.b) for use with small-bore HPLC columns[208,307,330,334]. Based on an electrically heated rubidium salt bead, it permits detection limits of 0.2–0.5 ng of phosphorus and its response is linear with the amount of phosphorus over several orders of magnitude. This detector yields good results with phosphates which cannot be detected by UV spectrophotometry or by fluorescence measurements.

Mass spectrometry is perhaps the most powerful method for studying HPLC effluents. As in GLC, HPLC columns are increasingly being coupled with mass spectrometers in order to identify the eluted compounds. This method is important enough to warrant separate treatment (see Section III.B.6).

c. Separation and quantitative determination. Separation between components of a mixture by HPLC is extremely efficient, once the optimal experimental conditions have been found. For instance, fenamiphos, which is not volatile and is oxidized in the soil to its sulphoxide and sulphone derivatives, could be separated from these metabolites both on silica and on a C_{18}-bonded column; the sulphone was resolved from the sulphoxide on the silica column only, while showing a similar retention time to that of the sulphoxide on the bonded column[224]. A μBondapak C_{18} column yields complete resolution between azinphos-methyl and its oxon, the retention time for the compound being double that of its oxon[193]. The literature on the HPLC analysis of organophosphorus residues (including many of the papers already cited) abounds with examples of excellent separations of complex samples. For instance, a mixture of seven compounds extracted from grapes was totally resolved in about 15 min[199]. The separation of highly polar compounds such as methylphosphonic acids may be improved by esterification, which reduces the polarity and allows better resolution by reversed-phase adsorption[338]. Identification of the HPLC effluents usually relies on comparison of the flow-through times with those of known reference samples passing through the same instrument under identical experimental conditions.

The extremely high sensitivity of HPLC detectors permits the accurate measurement of nanogram and sub-nanogram amounts of organophosphorus compounds. Ultraviolet spectrophotometry, when suitable, gives good results[199,338]; Abate, for instance, can be detected down to 1 ng and determined in the 1–150 ng ml^{-1} range by absorption measurements at 254 nm, after reversed-phase adsorption chromatography[216,223]; in a similar procedure, use of a calibration graph for azinphos methyl and its oxon[193] yields a detection limit of about 0.1 ppm and a linear peak height vs concentration relationship in the 0.1–10.0 ppm range; fenamiphos and its sulphoxide and sulphone can be determined down to 10 pg ml^{-1} after preconcentration of the aqueous samples on a Sep-Pak C_{18} reversed-phase cartridge[224]. Fluorogenic labelling allows the determination of glyphosate and its major metabolite over a wide concentration range; detection is possible down to 0.3 ng[326]. The use of microbore columns in conjunction with a flame photometric detector leads to a detection limit of nanogram amounts of phosphorus and linear determination in the 2–100 ng range[329]. The phosphorus-specific thermionic detectors described previously, measuring the vaporized effluents of small-bore HPLC columns, permit detection limits of 0.2–0.5 ng of phosphorus, with a linear response to the amount of phosphorus over several orders of magnitude[208,307,334].

As with other techniques, a single instrument is usually not sufficient for full analysis of organophosphorus residues. HPLC is often combined with other instrumental methods, mostly GLC, TLC or mass spectrometry. A recently suggested procedure for the analysis of residues found in potatoes requires the use of three different HPLC and three different GLC instruments for the study of one sample[183].

d. Flow-injection techniques. The expanding application of HPLC to a variety of analyses has led to the development of labour- and time-saving flow-injection methods (see Section II.D), in which the sample is introduced into a carrier stream passing into the chromatographic column while the effluents are continuously monitored by a suitable detector.

With a molecular emission cavity flame detector, relying on measurement of the 526 nm emission of HPO (see Section II.C.1 and III.B.3.b), nanogram amounts of phosphates or organophosphorus compounds can be assessed in automated systems[160,348]. Determination of t_m, the time elapsed between sample ignition and maximum emission, allows the resolution of ternary or more complex mixtures of insecticides. Dicrotophos, dimethoate, malathion and parathion mixed in aqueous solution were separated and identified in nanograms per millilitre concentrations[348].

The inhibition by paraoxon of acetylcholinesterase immobilized on glass beads was monitored by visible spectrophotometry at 500 nm, after hydrolysis of α-naphthyl acetate by the remaining non-inhibited enzyme and reaction of the resulting α-naphthol with *p*-nitrobenzenediazonium fluoroborate (see Section III.B.5.b); whereas the manual method required about 30 min for each determination, the automated instrument was able to carry out 30 analyses per hour[349]. A related detection method, based on immobilization of the enzyme on magnetic support particles, was also applied to the flow-injection analysis of pesticides[350]. In a recent investigation, the inhibition of immobilized acetylcholinesterase by a series of organophosphorus insecticides (azinphos-methyl, azinphos-ethyl, bromophos-methyl, dichlorvos, fenitrothion, malathion, paroxon, methylparathion and ethylparathion) was monitored by pH measurements[351], the detection limits varying between 0.5 and 275 ppb.

6. Mass spectrometry

The analysis of organophosphorus compounds and other pollutants is complicated by the fact that they often constitute only a small part of the very complex mixtures extracted from natural environments, and that they usually appear in extremely low concentrations. Mass spectrometry (MS) is one of the most powerful techniques developed for the identification of very small amounts of substances. In recent years it has increasingly been coupled with GLC or HPLC for the examination of the effluents obtained after separation on the chromatographic columns[352]. A complete mass spectrum can be recorded extremely fast compared with the time scale of chromatographic processes; this allows repeated scanning of each chromatographic band and thus analysis of each compound separated by the column at the maximum of the chromatographic peak. The interfacing of chromatographs with mass spectrometers was originally hampered by a number of technical problems, but several procedures are now available for both the introduction of the chromatographic samples into the mass spectrometers and their subsequent ionization. GLC–MS and HPLC–MS are now well established as analytical methods[222,311,327,336,337,339,340,341,353–355].

a. Introduction of samples into the mass spectrometer. Samples should normally be injected into the ionization chamber of a mass spectrometer in gaseous form. This is not always easy to accomplish, in particular with organophosphorus compounds which are difficult to volatilize.

The coupling of a GLC column with the sample inlet system of a mass spectrometer is relatively easy, as the effluents are already in gaseous form. The main problem is the relatively high pressure at which these effluents reach the spectrometer and the excess of carrier gas in the stream. Several experimental devices now allow separation of the sample from the carrier gas, either by an effusion process or with the help of a thin, semi-permeable membrane[222,353]. The use of capillary columns permits direct insertion of the GLC effluent into the ion source without overtaxing the pumping capacity of the mass spectrometer[311,355,356].

HPLC–MS offers major advantages over GLC–MS in the analysis of thermally labile or highly polar compounds, but the interfacing of HPLC with a mass spectrometer poses more serious problems, as the pumping systems of many mass spectrometers are unable to handle the large volumes of liquid solvents eluted from the HPLC column at the same time as the compounds of interest. The different LC–MS interfacing techniques have recently been reviewed[352,357–359]; the most promising, now commercially available, include transport interfaces, direct liquid introduction and thermospray interfacing.

In the first of these methods, the HPLC effluent is sprayed or otherwise deposited on a moving belt, most of the solvent is removed by heating and pumping, then the sample is carried directly into the ionization compartment[352]. Ionization may be of the electron impact or chemical type (see Section III.B.6.b).

In direct liquid introduction, the HPLC effluent stream is split and a fraction of about 1–5% is allowed into the spectrometer[342]. The solvent vapour is then removed by pumping out the ion source. Injecting only a small part of the available sample restricts the applicability of the MS analysis to the major components of a mixture and reduces its usefulness for trace detection. The new microbore columns, however, allow the total effluent to be injected, leading to much improved sensitivity. With direct liquid introduction of the sample, ionization is solely of the chemical type, the carrier solvent itself often serving as ionization reagent (see Section III.B.6.b). The fact that electron impact ionization cannot be used is a marked drawback of the method, as it makes useless the numerous data accumulated for the identification of the electron impact mass spectra of organophosphorus compounds. In addition, the procedure has a tendency to plug the entrance orifice; a capillary silica interface has recently been proposed to overcome the problem[360].

In the thermospray technique, the total HPLC effluent passes through a heated capillary tube and is introduced as a jet of fine droplets into a specially designed ion source. An auxiliary pump is coupled with the ionization chamber to remove the excess solvent vapour[361,362]. A buffer such as ammonium acetate or formic acid is often added to the chromatographic stream in order to ionize the sample by chemical reaction (see Section III.B.6.b), although electron impact ionization may also be applied. Thermospray suffers from fewer clogging problems than direct liquid introduction, the instrument design is simpler and the electrochemical ionization facilitates the study of high-polarity, low-volatility compounds. The method is therefore being increasingly applied to organophosphorus compounds analyses.

b. Ionization techniques. The main ionization methods for organophosphorus compounds are electron impact and chemical ionization, the latter technique appearing in numerous variations. Thermospray, already mentioned above, simultaneously volatilizes and ionizes the sample. Other less frequently applied techniques include field desorption, fast atom bombardment and laser desorption[352]. Vacuum-sensitive samples can be laser-ionized under atmospheric pressure, although the resulting spectra may be significantly different from those obtained under low-pressure conditions[363].

i. Electron impact ionization. Classical electron impact (EI) ionization, still the most frequently applied technique, relies on bombarding the volatilized sample with a medium-energy (70–100 eV) electron beam to knock out one or several electrons from the molecules, generating positively charged ions which are then analysed. A large number of organophosphorus compounds have been studied in this way; the ionization usually leads to considerable fragmentation of the molecules and yields characteristic ions[341,353,355,364–366].

ii. Chemical ionization. Chemical ionization (CI), both positive and negative, results from the collision of the sample molecules with ions rather than with electrons.

In positive chemical ionization (PCI), a set of ions is generated in the mass spectrometer by bombarding gas molecules with high-energy electrons. Sample molecules, introduced into the instrument, react with these positive ions and become ionized themselves[367–371]. The main advantage of PCI in the identification of organophosphorus compounds is that it produces 'quasimolecular' ions such as $[M + 1]^+$ or $[M - 1]^+$, with little fragmentation, and yields simple spectra and easy determination of molecular mass. The EI mode, in contrast, was shown to cause fragmentation and give more information of a structural nature. Methane is one of the most often employed CI reagents; its ionization by a 70–200 eV electron beam yields as the main ions $[CH_5]^+$, $[C_2H_5]^+$ and $[C_3H_5]^+$, which function primarily as proton donors or abstractors[353,356,366,367,372]. Other gases may be used, e.g. ammonia and deuteriated ammonia[368,372], the ionization agents, obtained at 500 eV, being $[NH_4]^+$ and $[ND_4]^+$; an argon–water mixture has been employed[370], the reagent ions in this case being $[Ar]^+$, $[Ar_2]^+$ and $[H_3O]^+$; nitric oxide, generating $[NO]^+$, has been studied in this respect[371]; ethylene and isobutane both yield proton-donating ions[353].

In negative chemical ionization (NCI), low-energy electrons (0–10 eV) are captured by the reagent gas molecules, producing negative ions which react with the sample molecules. An alternative ionization mechanism could be a mode in which the gas is not a reagent but a mediator present in excess, which reduces the energy of the electrons sufficiently for them to be directly captured by the analyte molecules[366]. Application of this technique requires appropriate field reversals in the ionization chambers. As in PCI, the spectra remain simple and the amount of fragmentation is low; the relative abundance of molecular anions is often greater than that observed with PCI, and the sensitivity is better. Peaks at masses higher than the molecular ion can be detected, representing associations of the parent compound with ions derivied from the reagent gas[352,369]. Methane, isobutane, dichloromethane and ammonia have been introduced as reagents[355,356,369]; methane, oxygen and mixtures of methane and oxygen have also been used[365]. In more recent applications, the HPLC mobile solvent present in excess, e.g. acetonitrile mixed with water, serves directly as the CI reagent[339,342].

iii. Thermospray ionization. The sample introduced into the mass spectrometer by thermospray (TSP) (see Section III.B.6.a) may be ionized by EI, or by CI using the residual solvent molecules as reagent (this is the filament-on mode, in which either sample or solvent molecules are initially bombarded with hot filament-generated electrons). It has been found, however, that if ionic species are present in the HPLC effluent, ionization occurs by reaction between these ions and the sample, eliminating the need for additional processes (this is the filament-off mode, in which no electron source is required)[361,362]. To carry out this electrochemical ionization, a volatile buffer is now often added to the HPLC mobile phase, either before or after passage through the column[373]. The most common is

0.1 M ammonium acetate, which yields $[NH_4]^+$, clustered $[NH_4]^+$ and $[CH_3COO]^-$ ions; protonation to $[M + H]^+$ or ammonium addition to $[M + NH_4]^+$ is observed and the spectra are similar to ammonia CI or field desorption spectra[327,374]; possible reactions between gas-phase ions and analyte molecules have been investigated[374]. The ionization mechanism is still not entirely understood, nor is the influence of a number of experimental parameters such as the nature and concentration of the electrolyte, the flow rate and the thermospray temperature[375]. Thermospray ionization exhibits better sensitivity than conventional CI and is widely used in pesticide residue analyses[327,336,337,340].

c. Identification of organophosphorus compounds. All the previously described techniques, and some other less common modes, have been employed to obtain the mass spectra of organophosphorus compounds and extensive data libraries have been built[376]. As repeatedly noted, the spectra were found to vary widely with the ionization method applied. Modern instruments now allow easy switching from one ionization procedure to another and the identification of unknown substances is often based on the examination of several different mass spectra of the same sample[366].

Fragmentation and rearrangement of tri- and penta-valent phosphorus esters were studied early[377,378]. A large number of compounds representative of four classes of pentavalent organophosphorus derivatives, phosphates, phosphorothiolates, phosphorothionates and phosphorodithioates, and some of their metabolites, were studied with EI ionization[364,365] after direct introduction into the mass spectrometer, in order to later allow identification of the compounds in natural environments. Almost all yield weak molecular peaks[364]. Characteristic ions at m/z 93 $\{[(MeO)_2P]^+$ or $[HO(EtO)P]^+\}$, m/z 97 $\{[(HS)_2P]^+$ or $[(HO)_2P{=}S]^+\}$, m/z 109 $\{[(MeO)_2P{=}O]^+$ or $[HO(EtO)P{=}O]^+\}$, m/z 121 $\{[(EtO)_2P]^+\}$ and m/z 125 $\{[(MeO)_2P{=}S]^+$ or $[HO(EtO)P{=}S]^+\}$ allow classification into groups; additional peaks make individual identification possible[365]. PCI spectra from some common organophosphorus pesticides related to the previous ones and from their decomposition products were obtained with methane or ammonia as ionization gas[372]; NCI spectra of the same and similar pesticides were determined with methane, oxygen or mixtures of methane and oxygen as reagents (and were labelled MENI, OENI and MOENI, respectively)[366]. Comparison of the spectra resulting from the different ionization methods for a number of these organophosphorus compounds showed OENI to yield low-resolution outputs, whereas MENI and MOENI generate fewer and better resolved, but essentially similar, peaks. The sensitivity of the NCI techniques was found to be far greater than that of PCI; for some halogenated aromatic derivatives the detection limit was as low as 25 pg[366]. These results were applied to the HPCL–NCI—MS trace analysis of a series of phosphates, phosphorothioates, phosphorodithioates and phenylphosphonothioates[342].

Organophosphates of the phosphorothioate, phosphorodithioate and dimethyl phosphate groups were screened by tandem mass spectrometry (MS–MS), first applying PCI with methane as reagent gas or NCI with methane or ammonia, then recording the daughter-ion spectra of the $[M + 1]^+$ or $[M - R]^-$ ions obtained (R = [Me] or [Et])[356]. The results were similar to those reported by other workers[364,365]: in the daughter-ion spectra of the $[M + 1]^+$ ions of phosphorothioates and phosphorodithioates the previously mentioned m/z 125 fragments appear 16 out of 22 times; the $m/z = 109$ ion is present in 5 out of 6 spectra for dimethyl phosphorothioates; in the daughter-ion spectra of all dimethyl phosphates an m/z 127 fragment may be found {probably due to $[(MeO)_2P(OH)_2]^+$}; all dimethyl phosphorothioates show a neutral loss of 32 mass units (assigned to MeOH), all diethyl phosphorothioates show neutral losses of 28 units (assigned to C_2H_4); these observations are useful for differentiation between classes of compounds. In electron-capture NCI the $[M{-}R]^-$ fragment was chosen for further examination in preference to $[M + 1]^-$ because of its higher abundance; all dimethyl phosphorothioates and phosphorodithioates exhibit a characteristic neutral loss of 110 mass units [attributed to MeO(O)PS]; diethyl phosphorothioates and phosphorodithioates show a neutral loss of 124 units [presumably due to EtO(O)PS]; in the dimethyl phosphate spectra the most common ion is $[MeO(HO)PO]^-$ with m/z 111.

The mass spectra obtained by both EI and PCI, with methane, ethylene and isobutane as reagents, from a series of nerve gases and related compounds[353] again showed the EI ionization to break up the molecules, generating a large number of ions and much structural information, whereas PCI more often led to the formation of protonated molecular ions $[M + H]^+$. Methane PCI exhibited the highest detection sensitivity, reaching sub-nanogram levels.

GLC–MS and HPLC–MS (with EI, CI or TSP ionization) have been applied to the detection and identification of a large number of additional organophosphorus compounds: Guthoxon and benzazimide, two degradation products of Guthion (or azinphos-methyl) could be studied down to 10 and 5 ng, respectively[341]; nerve gases such as sarin, soman and related compounds, in blood or in contaminated soil, were studied in this way[379,380]. Soils and sediments[339,340], biological samples[354], agricultural crops and foods[311,327,355] have all been analysed for organophosphorus compounds and for their degradation products by these techniques.

7. *Analysis of individual compounds*

In some instances a single organophosphorus substance or a mixture of very few such compounds are known to be present in a sample. In such cases separation and identification are not necessary and it may be expedient to use a simple method of limited applicability for their determination, rather than a sophisticated instrumental technique. Many procedures designed for the analysis of specific organophosphorus compounds of special interest have been published, of which only a few examples will be discussed here.

Malathion is one of the pesticides for which distinctive analytical methods have been developed. An early procedure[381] involved decomposition of the malathion to sodium dimethyl dithiophosphate and conversion of the dithiophosphate into a copper complex which was extracted into carbon tetrachloride, yielding an intensely yellow solution; colorimetric determination of the malathion was carried out, with the help of a calibration graph, by measuring the absorbance of the complex at 420 nm. The copper adduct, however, turned out to be unstable; modifications of the method, based on complexation of the hydrolysis product of malathion with palladium[382] or bismuth[383–385], were shown to yield more stable complexes, leading to the determination of malathion at ppb concentrations. An alternative approach is to use a known amount of copper, extract the complex from the solution with chloroform, determine iodometrically the remaining excess copper and indirectly estimate the initial amount of malathion[386]. Another suggestion is to extract the copper complex, mineralize the extract with concentrated nitric acid, then use atomic absorption spectrometry to measure the amount of copper[387]; a similar procedure applies atomic absorption spectrometry to copper or bismuth complexes of malathion and related compounds without requiring mineralization of the complex[129]. Other organic thiophosphates may be determined by a simple colorimetric procedure involving reaction of the P=S group with benzophenone, to yield the blue thiobenzophenone which is determined at its absorbance maximum, 599 nm[388]; substances containing sulphur but no phosphorus do not interfere.

Some general methods can also be adapted to the determination of individual organophosphorus compounds. For instance, the cholinesterase-inhibiting properties of methylparathion, dimethoate and their oxygen analogues were used for their determination: a known amount of enzyme is incubated with the sample, then addition of a chromogenic agent such as *p*-nitrobenzenediazonium fluoroborate yields a coloured product with the excess cholinesterase, which can be determined spectrophotometrically at 500 nm[389–391].

Nerve gases, because of their special applications and the secrecy associated with them, are usually studied separately from other organophosphorus substances. Their high toxicity calls for extreme caution and the low concentrations at which they must be detected require special extraction procedures[379,392–395]. A review of the known detection methods available for these compounds has been published[396]; they include colorimetric, fluorimetric, electrochemical and other approaches. The anticholinesterase activity of nerve gases may also be applied to their determination[397,398]; parts per trillion (ppt) concentrations in sea water and picogram amounts in air are easily detected[399,400]. Reversed-phase ion chromatography combined with conductivity detection has been reported[319]. However, gas chromatography seems to be the most sensitive technique[395]. The specific retention times of a number of different alkyl methylphosphonofluoridates have been reported for both polar and apolar supports; the changes caused in these retention times by various substitutions, such as oxygen in the place of sulphur, or fluorine instead of hydrogen, have been investigated[401]. Isopropyl methylphosphonofluoridate could be determined in nano- to pico-gram amounts with a GLC instrument equipped with a flame photometric detector, even in the presence of much larger amounts of other organophosphorus compounds[392]. A point of particular interest is the gas chromatographic separation of stereoisomers, which is sometimes useful in the study of nerve gases. Because of the presence of two chiral centres in its molecule, soman (1, 2, 2-trimethylpropylphosphonofluoridate), for instance, has four stereoisomers widely different in their biological properties. These can easily be separated by GLC, by the use of an optically active stationary phase such as copolymeric organosiloxane bound to L-valine-*tert*-butylamide; four clearly distinct signals are observed[394,402,403]. Similar observations were reported by other workers[393,401]. Mass spectrometry may be combined with GLC for the identification of the isomers[379].

IV. REFERENCES

1. E. Q. Laws, in *Treatise on Analytical Chemistry*, Part II (Eds I. M. Kolthoff and P. J. Elving), Vol. 11, Wiley, New York, 1965, pp. 499–555.
2. T. S. Ma and M. Gutterson, Organic Elemental Analysis, Analytical Reviews, *Anal. Chem.*, **36**, 155R (1964); **40**, 149R (1968); **42**, 109R (1970); **44**, 445R (1972); **46**, 437R (1974).
3. The last two reviews are T. S. Ma, Organic Elemental Analysis, Analytical Reviews, *Anal. Chem.*, **60**, 177R (1988); **62**, 79R (1990).
4. The two last reviews are P. McCarthy, R. W. Klusman, S. W. Cowling and J. A. Rice, Water Analysis, Analytical Reviews, *Anal. Chem.*, **63**, 301 R (1991); **65**, 244R (1993).
5. H. Gubser, *Chimia*, **13**, 245 (1959).
6. B. N. Ames, *Methods Enzymol.*, **8**, 115 (1966).
7. A. D. Cembella, N. J. Antia and F. J. R. Taylor, *Water Res.*, **20**, 1197 (1986).
8. H. Buss, H. W. Kohlschutter and M. Preiss, *Fresenius' Z. Anal. Chem.*, **214**, 106 (1965).
9. F. S. Malyukova and A. D. Zaitseva, *Plast. Massy*, 57 (1966).
10. E. A. Terent'eva and N. N. Smirnova, *Zavod. Lab.*, **32**, 924 (1966).
11. M. N. Chumachenko and V. P. Burlaka, *Izv. Akad. Nauk SSR*, 560 (1962).
12. G. L. Carius, *Fresenius' Z. Anal. Chem.*, **4**, 451 (1865); *Ann. Chem. Pharm.*, 1361 (1865).
13. C. DiPietro, R. E. Kramer and W. A. Sassaman, *Anal. Chem.*, **34**, 586 (1962).
14. V. G. Shah, S. S. Ramdasi, R. B. Malvankar, S. Y. Kulkarni and V. S. Pansare, *Indian J. Chem.*, **12**, 419 (1974).
15. R. B. Malvankar and V. S. Pansare, *Microchem. J.*, **33**, 359 (1986).
16. W. J. Kirsten and M. E. Carlsson, *Microchem. J.*, **4**, 3 (1960).
17. M. J. Kjeldahl, *Fresenius' Z. Anal. Chem.*, **22**, 357 (1883).
18. M. Aoyagi, Y. Yasumasa and A. Nishida, *Anal. Chim. Acta*, **214**, 229 (1988).
19. S. Hinkamp and G. Schwedt, *Anal. Chim. Acta*, **236**, 345 (1990).

20. J. Binkowski and A. Rudnicki, *Mikrochim. Acta*, **I**, 371 (1977).
21. W. Hempel, *Z. Angew. Chem.*, **5**, 393 (1892).
22. O. Mikl and J. Pech, *Chem. Listy*, **46**, 382 (1952).
23. O. Mikl and J. Pech, *Chem. Listy*, **48**, 1039 (1953).
24. W. Schoniger, *Mikrochim. Acta*, 123 (1955).
25. W. Schoniger, *Mikrochim. Acta*, 869 (1956).
26. W. Schoniger, *Fresenius' Z. Anal. Chem.*, **181**, 28 (1961).
27. R. Belcher and A. M. G. Macdonald, *Talanta*, **1**, 185 (1958).
28. K. D. Fleischer, B. C. Southworth, J. H. Hodecker and M. M. Tuckerman, *Anal. Chem.*, **30**, 152 (1958).
29. L. E. Cohen and F. W. Czech, *Chemist-Analyst*, **47**, 86 (1958).
30. M. Corner, *Analyst*, **84**, 41 (1959).
31. R. Bennewitz and I. Tanzer, *Mikrochim. Acta*, 835 (1959).
32. W. Merz, *Mikrochim. Acta*, 456 (1959).
33. R. Puschel and H. Wittmann, *Mikrochim. Acta*, 670 (1960).
34. A. Dirscherl and F. Erne, *Mikrochim. Acta*, 775 (1960).
35. B. P. Kirk and H. C. Wilkinson, *Talanta*, **19**, 80 (1972).
36. J. Haslam, J. B. Hamilton and D. C. M. Squirrell, *Analyst*, **85**, 556 (1960).
37. B. Griepink, *Mikrochim. Acta*, 1151 (1964).
38. G. Hesse and V. Bockel, *Mikrochim. Acta*, 939 (1962).
39. J. Binkowski and P. Rutkowski, *Mickrochim. Acta*, **I**, 245 (1986).
40. F. G. Romer and B. Griepink, *Mikrochim. Acta*, 867 (1970).
41. W. Pfirter, *Mikrochim. Acta*, II, 515 (1975).
42. A. Campiglio, *Mikrochim. Acta*, **I,** 443 (1983).
43. H. Flaschka and A. Holasek, *Mikrochemie*, **39**, 101 (1952).
44. H. Buss, H. W. Kohlschutter and M. Preiss, *Fresenius' Z. Anal. Chem.*, **193**, 264 (1963).
45. H. Buss, H. W. Kohlschutter and M. Preiss, *Fresenius' Z. Anal. Chem.*, **193**, 326 (1963).
46. E. Lassner, R. Puschel and R. Scharf, *Fresenius' Z. Anal. Chem.*, **170**, 412 (1959).
47. R. Puschel, *Mikrochim. Acta*, 352 (1960).
48. J. Horacek, *Collect. Czech. Chem. Commun.*, **27**, 1811 (1962).
49. O. Budevsky, L. Pencheva, R. Russinova and E. Russeva, *Talanta*, **11**, 1225 (1964).
50. R. Flatt and G. Brunisholz, *Anal. Chim. Acta*, **1**, 124 (1947).
51. F. H. Firsching, *Anal. Chem.*, **33**, 873 (1961).
52. A. de Sousa, *Microchem. J.*, **18**, 137 (1973).
53. L. Maric, M. Siroki and Z. Stefanac, *Microchem. J.*, **21**, 129 (1976).
54. S. Shahine and S. El-Medany, *Microchem. J.*, **24**, 212 (1979).
55. G. D. Christian, E. C. Knoblock and W. C. Purdy, *Anal. Chem.*, **35**, 1869 (1963).
56. A. Pietrogrande, M. Zancato and G. Bontempelli, *Analyst*, **112**, 129 (1987).
57. W. Selig, *Mikrochim. Acta*, **II**, 9 (1976).
58. D. H. McColl and T. A. O'Donnell, *Anal. Chem.*, **36**, 848 (1964).
59. W. Selig, *Mikrochim. Acta*, 564 (1970).
60. R. P. A. Sims, *Analyst*, **86**, 584 (1961).
61. H. N. Wilson, *Analyst*, **76**, 65 (1951).
62. W. Selig, *Talanta*, **30**, 695 (1983).
63. W. S. Selig, *Mickrochim. Acta*, **II**, 133 (1984).
64. S. W. Bishara and F. M. El-Samman, *Microchem. J.*, **27**, 44 (1982).
65. D. B. Land and S. M. Edmonds, *Mikrochim. Acta*, 1013 (1966).
66. T. Imanari, S. Tanabe, T. Toida and T. Kawanishi, *J. Chromatogr.*, **250**, 55 (1982).
67. B. L. Griswold, F. L. Humoller and A. R. McIntyre, *Anal. Chem.*, **23**, 192 (1951).
68. F. K. Bell, C. J. Carr and J. C. Krantz, Jr, *Anal. Chem.*, **24**, 1184 (1952).
69. P. S. Chen, Jr, T. Y. Toribara and H. Warner, *Anal. Chem.*, **28**, 1756 (1956).
70. D. N. Fogg and N. T. Wilkinson, *Analyst*, **83**, 406 (1958).
71. I. Berenblum and E. Chain, *Biochem. J.*, **32**, 295 (1938).
72. N. S. Ging, *Anal. Chem.*, **28**, 1330 (1956).
73. F. L. Schaffer, J. Fong and P. L. Kirk, *Anal. Chem.*, **25**, 343 (1953).
74. M. Jean, *Anal. Chim. Acta*, **14**, 172 (1956).
75. J. B. Martin and D. M. Doty, *Anal. Chem.*, **21**, 965 (1949).
76. V. Djurkin, G. F. Kirkbright, and T. S. West, *Analyst*, **91**, 89 (1966).

77. C. H. Fiske and Y. Subarow, *J. Biol. Chem.*, **66**, 375 (1925).
78. B. B. Bauminger and G. Walters, *Analyst*, **91**, 205 (1966).
79. S. R. Benedict and R. C. Theis, *J. Biol. Chem.*, **61**, 63 (1924).
80. G. Telep and R. Ehrlich, *Anal. Chem.*, **30**, 1146 (1958).
81. T. Kuttner and L. Lichtenstein, *J. Biol. Chem.*, **86**, 671 (1930).
82. O. H. Lowry, N. R. Roberts, K. Y. Leiner, M.-L. Wu and A. L. Farr, *J. Biol. Chem.*, **207**, 1 (1954).
83. O. H. Lowry and J. A. Lopez, *J. Biol. Chem.*, **162**, 421 (1946).
84. Y. Hirai, N. Yoza and S. Ohashi, *J. Chromatogr.*, **206**, 501 (1981).
85. T. A. H. M. Janse, P. F. A. Van Der Wiel and G. Kateman, *Anal. Chim. Acta*, **155**, 89 (1983).
86. D. J. Malcolme-Lawes and K. H. Wong, *Analyst*, **115**, 65 (1990).
87. G. Schulze and A. Thiele, *Fresenius' Z. Anal. Chem.*, **329**, 711 (1988).
88. P. R. Freeman, I. D. McKelvie, B. T. Hart and T. J. Cardwell, *Anal. Chim. Acta*, **234**, 409 (1990).
89. S. Taguchi, E. Ito-Oka, K. Masuyama, I. Kasahara and K. Goto, *Talanta*, **32**, 391 (1985).
90. F. Umland and G. Wunsch, *Fresenius' Z. Anal. Chem.*, **213**, 186 (1965).
91. S. Sunita and V. K. Gupta, *J. Indian Chem. Soc.*, **64**, 114 (1987).
92. A. Chaube and V. K. Gupta, *Analyst*, **108**, 1141 (1983).
93. J. Raju and V. K. Gupta, *Microchem. J.*, **39**, 166 (1989).
94. I. Berenblum and E. Chain, *Biochem. J.*, **32**, 286 (1938).
95. G. Mission, *Chem. Ztg.*, **32**, 633 (1908).
96. R. A. Koenig and C. R. Johnson, *Ind. Eng. Chem., Anal. Ed.*, **14**, 155 (1942).
97. R. E. Kitson and M. G. Mellon, *Ind. Eng. Chem., Anal. Ed.*, **16**, 379 (1944).
98. T. S. Ma and J. D. McKinley, Jr, *Mikrochim. Acta*, 4 (1953).
99. A. G. Fogg, S. Soleymanloo and D. Thorburn Burns, *Anal. Chim. Acta*, **88**, 197 (1977).
100. L. Ducret and M. Drouillas, *Anal. Chim. Acta*, **21**, 86 (1959).
101. T. Matsuo, J. Shida and W. Kurihara, *Anal. Chim. Acta*, **91**, 385 (1977).
102. J. Shida and T. Matsuo, *Bull. Chem. Soc. Jpn.*, **53**, 2868 (1980).
103. A. K. Babko, Y. F. Shkaravskii and V. I. Kulik, *Zh. Anal. Khim.*, **21**, 196 (1966).
104. C. L. Penney, *Anal. Biochem.*, **75**, 201 (1976).
105. H. van Belle, *Anal. Biochem.*, **33**, 132 (1970).
106. S. Motomizu, T. Wakimoto and K. Toei, *Anal. Chim. Acta*, **138**, 329 (1982).
107. S. M. Hassan and F. Basyoni Salem, *Anal. Lett.*, **20**, 1 (1987).
108. S. Taguchi, I. Kasahara, Y. Fukushima and K. Goto, *Talanta*, **28**, 616 (1981).
109. M. Taga and M. Kan, *Bull. Chem. Soc. Jpn.*, **62**, 1482 (1989).
110. K. Itaya and M. Ui, *Clin. Chim. Acta*, **14**, 361 (1966).
111. S. Motomizu, T. Wakimoto and K. Toei, *Analyst*, **108**, 361 (1983).
112. S. Motomizu, T. Wakimoto and K. Toei, *Talanta*, **31**, 235 (1984).
113. T. Nasu and M. Kan, *Analyst*, **113**, 1683 (1988).
114. C. Matsubara, Y. Yamamoto and K. Takamura, *Analyst*, **112**, 1257 (1987).
115. H. Hashitani and M. Okumura, *Fresenius' Z. Anal. Chem.*, **328**, 251 (1987).
116. G. F. Kirkbright, R. Narayanaswamy and T. S. West, *Analyst*, **97**, 174 (1972).
117. G. F. Kirkbright, R. Narayanaswamy and T. S. West, *Anal. Chem.*, **43**, 1434 (1971).
118. T. Nasu and H. Minami, *Analyst*, **114**, 955 (1989).
119. M. Taga, M. Kan and T. Nasu, *Fresenius' Z. Anal. Chem.*, **334**, 45 (1989).
120. T. S. Prokopov, *Mikrochim. Acta*, 675 (1968).
121. Y. A. Gawargious and A. B. Farag, *Microchem. J.*, **16**, 333 (1971).
122. R. J. Bertolacini and J. E. Barney, II, *Anal. Chem.*, **30**, 202 (1958).
123. A. G. Fogg and N. K. Bsebsu, *Analyst*, **106**, 1288 (1981).
124. A. G. Fogg and N. K. Bsebsu, *Analyst*, **107**, 566 (1982).
125. M. Goto, Y. Miura, H. Yoshida and D. Ishii, *Mikrochim. Acta*, **I**, 121 (1983).
126. K. Matsunaga, I. Kudo, M. Yanada and K. Hasebe, *Anal. Chim. Acta*, **185**, 355 (1986).
127. W. S. Zaugg and R. J. Knox, *Anal. Chem.*, **38**, 1759 (1966).
128. G. F. Kirkbright, A. M. Smith and T. S. West, *Analyst*, **92**, 411 (1967).
129. O. J. de Blas, J. L. Pereda de Paz and J. H. Mendez, *Analyst*, **114**, 1675 (1989).
130. T. H. Handley and J. A. Dean, *Anal. Chem.*, **34**, 1312 (1962).
131. J. Nebot and E. Figuerola, *Int. J. Environ. Anal. Chem.*, **36**, 241 (1989).
132. W. A. Dippel, C. E. Bricker and N. H. Furman, *Anal. Chem.*, **26**, 553 (1954).
133. P. S. Chen, Jr, and T. Y. Toribara, *Anal. Chem.*, **25**, 1642 (1953).
134. C. Rocchiccioli and A. Townshend, *Anal. Chim. Acta*, **41**, 93 (1968).

135. E. Pungor, *Pure Appl. Chem.*, **23**, 51 (1970).
136. W. B. Barnett, *Anal. Chem.*, **44**, 695 (1972).
137. G. L. Long and C. B. Boss, *Anal. Chem.*, **54**, 624 (1982).
138. A. M. Maitra and E. Patsalides, *Anal. Chim. Acta*, **193**, 179 (1987).
139. A. Syty and J. A. Dean, *Appl. Opt.*, **7**, 1331 (1968).
140. D. W. Brite, *Anal. Chem.*, **27**, 1815 (1955).
141. R. M. Dagnall, K. C. Thompson and T. S. West, *Analyst*, **93**, 72 (1968).
142. G. L. Everett, T. S. West and R. W. Williams, *Anal. Chim. Acta*, **68**, 387 (1974).
143. S. S. Brody and J. E. Chaney, *J. Gas Chromatogr.*, **4**, 42 (1966).
144. W. N. Elliott, C. Heathcote and R. A. Mostyn, *Talanta*, **19**, 359 (1972).
145. R. Belcher, S. L. Bogdanski, O. Osibanjo and A. Townshend, *Anal. Chim. Acta*, **84**, 1 (1976).
146. G. F. Kirkbright, A. F. Ward and T. S. West, *Anal. Chim. Acta*, **62**, 241 (1972).
147. T. Ishizuka, K. Nakajima and H. Sunahara, *Anal. Chim. Acta*, **121**, 197 (1980).
148. D. R. Heine, M. B. Denton and T. D. Schlabach, *Anal. Chem.*, **54**, 81 (1982).
149. J. M. Cook and D. L. Miles, *Analyst*, **110**, 547 (1985).
150. J. L. Manzoori, A. Miyazaki and H. Tao, *Analyst*, **115**, 1055 (1990).
151. M. Siroki, G. Vujicic, V. Milun, Z. Hudovsky and L. Maric, *Anal. Chim. Acta*, **192**, 175 (1987).
152. D. R. Heine, M. B. Denton and T. D. Schlabach, *J. Chromatogr. Sci.*, **23**, 454 (1985)
153. D. L. Windsor and M. B. Denton, *J. Chromatogr. Sci.*, **17**, 492 (1979).
154. J. Ruzicka and E. H. Hansen, *Anal. Chim. Acta*, **78**, 145 (1975).
155. Y. Hirai, N. Yoza and S. Ohashi, *Anal. Chim. Acta*, **115**, 269 (1980).
156. N. Yoza, K. Ito, Y. Hirai and S. Ohashi, *J. Chromatogr.*, **196**, 471 (1980).
157. K. S. Johnson and R. L. Petty, *Anal. Chem.*, **54**, 1185 (1982).
158. S. Motomizu, T. Wakimoto and K. Toei. *Talanta*, **30**, 333 (1983).
159. S. Motomizu and M. Oshima, *Analyst*, **112**, 295 (1987).
160. J. L. Burguera, M. Burguera and D. Flores, *Anal. Chim. Acta*, **170**, 331 (1985).
161. I. D. McKelvie, B. T. Hart, T. J. Cardwell and R. W. Cattrall, *Analyst*, **114**, 1459 (1989).
162. A. Ambrus and H.-P. Thier, *Pure Appl. Chem.*, **58**, 1035 (1986).
163. R. S. Edmundson, *Dictionary of Organophosphorus Compounds*, Chapman and Hall, London, (1988).
164. P. A. Greve and C. E. Goewie, *Int. J. Environ. Anal. Chem.*, **20**, 29 (1985).
165. H. P. Burchfield and E. E. Storrs, *J. Chromatogr. Sci.*, **13**, 202 (1975).
166. A. Ambrus, J. Lantos, E. Visi, I. Csatlos and L. Sarvari, *J. Assoc. Off. Anal. Chem.*, **64**, 733 (1981).
167. G. J. Sharp, J. G. Brayan, S. Dilli, P. R. Haddad and J. M. Desmarchelier, *Analyst*, **113**, 1493 (1988).
168. D. C. G. Muir, N. P. Grift and J. Solomon, *J. Assoc. Off. Anal. Chem.*, **64**, 79 (1981).
169. A. H. Roos, A. J. van Munsteren, F. M. Nab and L. G. M. T. Tuinstra, *Anal. Chim. Acta*, **196**, 95 (1987).
170. I. S. Taylor and H.-P. Thier, *Pure Appl. Chem.*, **51**, 1603 (1979).
171. R. J. Hemingway, *Pure Appl. Chem.*, **56**, 1131 (1984).
172. L. R. Jones and J. A. Riddick, *Anal. Chem.*, **23**, 349 (1951).
173. L. R. Jones and J. A. Riddick, *Anal. Chem.*, **24**, 569 (1952).
174. M. E. Getz, *J. Assoc. Off. Anal. Chem.*, **45**, 393 (1962).
175. R. W. Storherr, M. E. Getz, R. R. Watts, S. J. Friedman, F. Erwin, L. Giuffrida and F. Ives, *J. Assoc. Off. Anal. Chem.*, **47**, 1087 (1964).
176. R. W. Storherr and R. R. Watts, *J. Assoc. Off. Anal. Chem.*, **48**, 1154 (1965).
177. R. R. Watts and R. W. Storherr, *J. Assoc. Off. Anal. Chem.*, **48**, 1158 (1965).
178. R. W. Storherr, P. Ott and R. R. Watts, *J. Assoc. Off. Anal. Chem.*, **54**, 513 (1971).
179. M. A. Luke, J. E. Froberg and H. T. Masumoto, *J, Assoc. Off. Anal. Chem.*, **58**, 1020 (1975).
180. M. A. Luke, J. E. Froberg, G. M. Doose and H. T. Masumoto, *J. Assoc. Off. Anal. Chem.*, **64**, 1187 (1981).
181. J. A. R. Bates, *Analyst*, **90**, 453 (1965).
182. L. J. Carson, *J. Assoc. Off. Anal. Chem.*, **64**, 714 (1981).
183. R. W. Martindale, *Analyst*, **113**, 1229 (1988).
184. J. J. Blaha and P. J. Jackson, *J. Assoc. Off. Anal. Chem.*, **68**, 1095 (1985).
185. J. Pflugmacher and W. Ebing, *J. Chromatogr.*, **160**, 213 (1978).
186. W. Specht and M. Tillkes, *Fresenius' Z. Anal. Chem.*, **301**, 300 (1980).
187. W. Specht and M. Tillkes, *Fresenius' Z. Anal. Chem.*, **322**, 443 (1985).

188. J. F. Lawrence, *Int. J. Environ. Anal. Chem.*, **29**, 289 (1987).
189. P. J. Wales, C. E. Mendoza, H. A. McLeod and W. P. McKinley, *Analyst*, **93**, 691 (1968).
190. M. Siewierski and K. Helrich, *J. Assoc. Off. Anal. Chem.*, **53**, 514 (1970).
191. A. Neicheva, E. Kovacheva and G. Marudov, *J. Chromatogr.*, **437**, 249 (1988).
192. A Consalter and V. Guzzo, *Fresenius' J. Anal. Chem.*, **339**, 390 (1991).
193. A. M. Wilson and R. J. Bushway, *J. Chromatogr.*, **214**, 140 (1981).
194. G. F. Ernst and M. J. P. T. Anderegg. *J. Assoc. Off. Anal. Chem.*, **59**, 1185 (1976).
195. C. Habig, A. Nomeir, R. T. DiGiulio and M. B. Abou-Donia, *J. Assoc. Off. Anal. Chem.*, **70**, 103. (1987).
196. D. C. Abbott, A. S. Burridge, J. Thomson and K. S. Webb, *Analyst*, **92**, 170 (1967).
197. R. G. Lehmann, L. M. Smith, R. H. Wiedmeyer and J. D. Petty, *J. Assoc. Off. Anal. Chem.*, **66**, 673 (1983).
198. J. S. Thornton and C. A. Anderson, *J. Agric. Food Chem.*, **14**, 143 (1966).
199. P. Cabras, P. Diana, M. Meloni and F. M. Pirisi, *J. Agric. Food Chem.*, **30**, 569 (1982).
200. M. C. Bowman and M. Beroza, *J. Agric. Food Chem.*, **16**, 399 (1968).
201. M. C. Bowman and M. Beroza, *J. Agric. Food Chem.*, **17**, 1054 (1969).
202. R. K. Keller, J. W. Tamkun and W. L. Adair, *Biochemistry*, **20**, 5831 (1981).
203. H. Egan, E. W. Hammond and J. Thomson, *Analyst*, **89**, 175 (1964).
204. H. Wan. *J. Chromatogr.*, **516**, 446 (1990).
205. H. V. Claborn and M. C. Ivey, *J. Agric. Food Chem.*, **13**, 353 (1965).
206. W. E. Dale and J. W. Miles, *J. Assoc. Off. Anal. Chem.*, **59**, 165 (1976).
207. J. Kjolholt, *J. Chromatogr.*, **325**, 231 (1985).
208. D. Barcelo, F. A. Maris, R. W. Frei, G. J. de Jong and U. A. T. Brinkman, *Int. J. Environ. Anal. Chem.*, **30**, 95 (1987).
209. C. Lentza-Rizos and E. J. Avramides, *Analyst*, **115**, 1037 (1990).
210. A. Di Muccio, A. Ausili, L. Vergori, I. Camoni, R. Dommarco, L. Gambetti, A Santilio and F. Vergori, *Analyst*, **115**, 1167 (1990).
211. L. D. Hutchins-Kumar, J. Wang and P. Tuzhi, *Anal. Chem.*, **58**, 1019 (1986).
212. R. R. Watts and R. W. Storherr, *J. Assoc. Off. Anal. Chem.*, **50**, 581 (1967).
213. W. A. Saner and J. Gilbert, *J. Liq. Chromatogr.*, **3**, 1753 (1980).
214. J. F. Lawrence, C. Renault and R. W. Frei, *J. Chromatogr.*, **121**, 343 (1976).
215. C. Mallet and V. N. Mallet, *J. Chromatogr.*, **481**, 37 (1989).
216. R. A. Henry, J. A. Schmit, J. F. Dieckman and F. J. Murphey, *Anal. Chem.*, **43**, 1053 (1971).
217. J. Askew, J. H. Ruzicka and B. B. Wheals, *Analyst*, **94**, 275 (1969).
218 V. N. Mallet, G. L. Brun, R. N. McDonald and K. Berkane, *J. Chromatogr.*, **160**, 81 (1978).
219. C. G. Daughton, D. G. Crosby, R. L. Garnas and D. P. H. Hsieh, *J. Agric. Food Chem.*, **24**, 236 (1976).
220. K. Berkane, G. E. Caissie and V. N. Mallet, *J. Chromatogr.*, **139**, 386 (1977).
221. G. G. Volpe and V. N. Mallet, *Int. J. Environ. Anal. Chem.*, **8**, 291 (1980).
222. G. L. LeBel, D. T. Williams, G. Griffith and F. M. Benoit, *J. Assoc. Off. Anal. Chem.*, **62**, 241 (1979).
223. A. Otsuki and T. Takaku, *Anal. Chem.*, **51**, 833 (1979).
224. R. Singh, *Analyst*, **114**, 425 (1989).
225. J. M. Vinuesa, J. C. M. Cortes, C. I. Canas and G. F. Perez, *J. Chromatogr.*, **472**, 365 (1989).
226. T. Lee, E. Barg and D. Lal, *Anal. Chim. Acta*, **260**, 113 (1990).
227. V. Drevenkar, Z. Frobe, B. Stengl and B. Tkalcevic, *Int. J. Environ. Anal. Chem.*, **22**, 235 (1985).
228. G. G. Jayson, T. A. Lawless and D. Fairhurst, *J. Colloid Interface Sci.*, **86**, 397 (1982).
229. J. G. Koen and J. F. K. Huber, *Anal. Chim. Acta*, **51**, 303 (1970).
230. C. Anglin and W. P. McKinley, *J. Agric. Food Chem.*, **8**, 186 (1960).
231. O. W. Grussendorf, A. J. McGinnis and J. Solomon, *J. Assoc. Off. Anal. Chem.*, **53**, 1048 (1970).
232. H. McLeod and P. J. Wales, *J. Agric. Food Chem.*, **20**, 624 (1972).
233. A. M. Gillespie and S. M. Walters, *J. Assoc. Off. Anal. Chem.*, **67**, 290 (1984).
234. J. D. Blake, M. L. Clarke and G. N. Richards, *J. Chromatogr.*, **398**, 265 (1987).
235. G. R. Bartlett, *Anal. Biochem.*, **124**, 425 (1982).
236. R. W. Storherr and R. R. Watts, *J. Assoc. Off. Anal. Chem.*, **52**, 511 (1969).
237. R. R. Watts, R. W. Storherr, J. R. Pardue and T. Osgood, *J. Assoc. Off. Anal. Chem.*, **52**, 522 (1969).
238. J. R. Pardue, *J. Assoc. Off. Anal. Chem.*, **54**, 359 (1971).

239. H. Beckman and D. Garber, *J. Assoc. Off. Anal. Chem.*, **52**, 286 (1969).
240. M. C. Ivey and H. V. Claborn, *J. Agric. Food Chem.*, **19**, 1256 (1971).
241. A. di Muccio, A. M. Cicero, I. Camoni, D. Pontecorvo and R. Dommarco, *J. Assoc. Off. Anal. Chem.*, **70**, 106 (1987).
242. A. Di Muccio, A. Ausili, I. Camoni, R. Dommarco, M. Rizzica and F. Vergori, *J. Chromatogr.*, **456**, 149 (1988).
243. B. G. Luke and J. C. Richards, *J. Assoc. Off. Anal. Chem.*, **67**, 902 (1984).
244. D. L. Stalling, R. C. Tindle and J. L. Johnson, *J. Assoc. Off. Anal. Chem.*, **55**, 32 (1972).
245. R. C. Tindle and D. L. Stalling, *Anal. Chem.*, **44**, 1768 (1972).
246. K. R. Griffitt and J. C. Craun, *J. Assoc. Off. Anal. Chem.*, **57**, 168 (1974).
247 L. D. Johnson, R. H. Waltz, J. P. Ussary and F. E. Kaiser, *J. Assoc. Off. Anal. Chem.*, **59**, 174 (1976).
248. M. L. Hopper, *J. Agric. Food Chem.*, **30**, 1038 (1982).
249. J. A. Ault, C. M. Schofield, L. D. Johnson and R. H. Waltz, *J. Agric. Food Chem.*, **27**, 825 (1979).
250. C. S. Hanes and F. A. Isherwood, *Nature (London)*, **164**, 1107 (1949).
251. L. C. Mitchell, *J. Assoc. Off. Anal. Chem.*, **43**, 810 (1960).
252. R. F. Irvine, A. J. Letcher, D. J. Lander and C. P. Downes, *Biochem. J.*, **223**, 237 (1984).
253. D. C. Abbott and H. Egan, *Analyst*, **97**, 475 (1967).
254. M. E. Getz and S. J. Friedman, *J. Assoc. Off. Anal. Chem.*, **46**, 707 (1963).
255. R. R. Watts, *J. Assoc. Off. Anal. Chem.*, **48**, 1161 (1965).
256. N. A. Smart and A. R. C. Hill, *J. Chromatogr.*, **30**, 626 (1967).
257. M. F. Kovacs, *J. Assoc. Off. Anal. Chem.*, **47**, 1097 (1964).
258. A. El-Rafai and T. L. Hopkins, *J. Agric. Food Chem.*, **13**, 477 (1965).
259. M. E. Getz and H. G. Wheeler, *J. Assoc. Off. Anal. Chem.*, **51**, 1101 (1968).
260. K. Nagasawa and H. Yoshidome, *J. Chromatogr.*, **39**, 282 (1969).
261. O. Antoine and G. Mees, *J. Chromatogr.*, **58**, 247 (1971).
262. A. Ambrus, E. Hargitai, G. Karoly, A. Fulop and J. Lantos, *J. Assoc. Off. Anal. Chem.*, **64**, 743 (1981).
263. K. C. Walker and M. Beroza, *J. Assoc. Off. Anal. Chem.*, **46**, 250 (1963).
264. B. Bush, R. Narang and C. Houck, *Anal. Lett.*, **10**, 187 (1977).
265. S. P. Srivastava and Reena, *Anal. Lett.*, **15**, 39 (1982).
266. J. Sherma and J. L. Boymel, *J. Chromatogr.*, **247**, 201 (1982).
267. L. H. Howe and C. F. Petty, *J. Agric. Food Chem.*, **17**, 401 (1969).
268. S. U. Bhaskar and N. V. Nanda Kumar, *J. Assoc. Off. Anal. Chem.*, **64**, 1312 (1981).
269. N. V. Nanda Kumar, K. Visweswariah and S. K. Majumder, *J. Assoc. Off. Anal. Chem.*, **59**, 641 (1976).
270. A. M. Gardner, *J. Assoc. Off. Anal. Chem.*, **54**, 517 (1971).
271. S. N. Tewari and S. P. Harpalani, *J. Chromatogr.*, **130**, 229 (1977).
272. K. Narayanaswami, B. Moitra, R. S. Kotangle and H. L. Bami, *J. Chromatogr.*, **95**, 181 (1974).
273. P. E. Belliveau and R. W. Frei, *Chromatographia*, **4**, 189 (1971).
274. C. E. Mendoza, P. J. Wales, H. A. McLeod and W. P. McKinley, *Analyst*, **93**, 173 (1968).
275. J. Sherma and W. Bretschneider, *J. Liq. Chromatogr.*, **13**, 1983 (1990).
276. J. D. McNeil, B. L. McLellan and R. W. Frei, *J. Assoc. Off. Anal. Chem.*, **57**, 165 (1974).
277. M. T. H. Ragab, *J. Assoc. Off. Anal. Chem.*, **50**, 1088 (1967).
278. R. L. Schutzmann and W. F. Barthel, *J. Assoc. Off. Anal. Chem.*, **52**, 151 (1969).
279. H. Breuer, *J. Chromatogr.*, **243**, 183 (1982).
280. J. A. Federici and J. Paul, *Microchem. J.*, **34**, 211 (1986).
281. M. Mazurek and Z. Witkiewicz, *J. Planar Chromatogr. Mod. TLC*, **4**, 416 (1991).
282. A. Nejtscheva, P. Wassileva-Alexandrova and E. Kovatscheva, *Mikrochim. Acta*, **I**, 393, (1984).
283. J. F. Lawrence and R. W. Frei, *J. Chromatogr.*, **98**, 253 (1974).
284. T. F. Bidleman, B. Nowlan and R. W. Frei, *Anal. Chim. Acta*, **60**, 13 (1972).
285. C. E. Mendoza, *J. Chromatogr.*, **78**, 29 (1973).
286. C. E. Mendoza, P. J. Wales, H. A. McLeod and W. P. McKinley, *Analyst*, **93**, 34 (1968).
287. R. Greenhalgh and J. Kovacicova, *J. Agric. Food Chem.*, **23**, 325 (1975).
288. J. A. Coburn and A. S. Y. Chau, *J. Assoc. Off. Anal. Chem.*, **57**, 1272 (1974).
289. I. H. Suffet, S. D. Faust and W. F. Carey, *Environ. Sci. Technol.*, **1**, 639 (1967).
290. J. Ruzicka, J. Thomson and B. B. Wheals, *J. Chromatogr.*, **30**, 92 (1967).
291. S. M. Prinsloo and P. R. de Beer, *J. Assoc. Off. Anal. Chem.*, **68**, 1100 (1985).

292. V. Bardarov and M. Mitewa, *J. Chromatogr.*, **462**, 233 (1989).
293. J. F. Lawrence and H. A. McLeod, *J. Assoc. Off. Anal. Chem.*, **59**, 637 (1976).
294. A. Ambrus, E. Visi, F. Zakar, E. Hargitai, L. Szabo and A. Papa, *J. Assoc. Off. Anal. Chem.*, **64**, 749 (1981).
295. R. R. Watts and R. W. Storherr, *J. Assoc. Off. Anal. Chem.*, **52**, 513 (1969).
296. G. Schomburg and H. Husmann, *Chromatographia*, **8**, 517 (1975).
297. L. Zenon-Roland, R. Agneessens, P. Nangniot and H. Jacobs, *J. High Resolut. Chromatogr. Chromatogr. Commun.*, **7**, 481 (1984).
298. H. -J. Stan and D. Mrowetz, *J. Chromatogr.*, **279**, 173 (1983).
299. C. Mallet and V. N. Mallet, *J. Chromatogr.*, **481**, 27 (1989).
300. L. Giuffrida, *J. Assoc. Off. Anal. Chem.*, **47**, 293 (1964).
301. H. -J. Stan and D. Mrowetz, *J. High Resolut. Chromatogr. Chromatogr. Commun.*, **6**, 255 (1983).
302. H. -J. Stan and H. Goebel, *J. Chromatogr.*, **268**, 55 (1983).
303. E. S. Kolesar, Jr. and R. M. Walser, *Anal. Chem.*, **60**, 1731 (1988).
304. J. W. Grate, M. Klusty, W. R. Barger and A. W. Snow, *Anal. Chem.*, **62**, 1927 (1990).
305. C. E. Cook, C. W. Stanley and J. E. Barney, II, *Anal. Chem.*, **36**, 2354 (1964).
306. B. Kolb and J. Bischoff, *J. Chromatogr. Sci.*, **12**, 625 (1974).
307. F. A. Maris, R. J. van Delft, R. W. Frei, R. B. Geergink and U. A. T. Brinkman, *Anal. Chem.*, **58**, 1634 (1986).
308. W. J. Trotter, *Int. J. Environ. Anal. Chem.*, **30**, 299 (1987).
309. S. Sass and G. A. Parker, *J. Chromatogr.*, **189**, 331 (1980).
310. S. Hashimoto, K. Fujiwara and K. Fuwa, *Anal. Chem.*, **57**, 1305 (1985).
311. P. Rodriguez, J. Permanyer, J. M. Grases and C. Gonzalez, *J. Chromatogr.*, **562**, 547 (1991).
312. C. V. Bowen and F. I. Edwards, Jr, *Anal. Chem.*, **22**, 706 (1950).
313. E. Sandi, *Fresenius' Z. Anal. Chem.*, **167**, 241 (1959).
314. H. B. Hanekamp, P. Bos, U. A. T. Brinkman and R. W. Frei, *Fresenius' Z. Anal. Chem.*, **297**, 404 (1979).
315. J. G. Koen, J. F. K. Huber, H. Poppe and G. den Boef, *J. Chromatogr. Sci.*, **8**, 192 (1970).
316. A. Verweij, C. E. Kientz and J. van den Berg, *Int. J. Environ. Anal. Chem.*, **34**, 191 (1988).
317. H. Small and T. E. Miller, Jr, *Anal. Chem.*, **54**, 462 (1982).
318. S. E. Meek and D. J. Pietrzyk, *Anal. Chem.*, **60**, 1397 (1988).
319. P. C. Bossle, D. J. Reutter and E. W. Sarver, *J. Chromatogr.*, **407**, 399 (1987).
320. P. Hoffmann, I. Schmidtke and K. H. Lieser, *Fresenius' Z. Anal. Chem.*, **335**, 402 (1989).
321. J. O. Bronstad and H. O. Friestad, *Analyst*, **101**, 820 (1976).
322. H. Yamaguchi, T. Nakamura, Y. Hirai and S. Ohashi, *J. Chromatogr.*, **172**, 131 (1979).
323. R. Stillman and T. S. Ma, *Mikrochim. Acta*, 491 (1973).
324. J. L. Meek and F. Nicoletti, *J. Chromatogr.*, **351**, 303 (1986).
325. R. W. Frei and J. F. Lawrence, *J. Chromatogr.*, **83**, 321 (1973).
326. H. Roseboom and C. J. Berkhoff, *Anal. Chim. Acta*, **135**, 373 (1982).
327. R. D. Voyksner and C. A. Haney, *Anal. Chem.*, **57**, 991 (1985).
328. C. E. Parker, C. A. Haney and J. R. Hass, *J. Chromatogr.*, **237**, 233 (1982).
329. V. L. McGuffin and M. Novotny, *Anal. Chem.*, **53**, 946 (1981).
330. V. L. McGuffin and M. Novotny, *Anal. Chem.*, **55**, 2296 (1983).
331. Y. Hirata, M. Novotny, T. Tsuda and D. Ishii, *Anal. Chem.*, **51**, 1807 (1979).
332. Y. Hirata and M. Novotny, *J. Chromatogr.*, **186**, 521 (1979).
333. Y. Hirata, P. Lin, M. Novotny and R. M. Wightman, *J. Chromatogr.*, **181**, 287 (1980).
334. J. C. Gluckman, D. Barcelo, G. J. de Jong, R. W. Frei, F. A. Maris and U. A. T. Brinkman, *J. Chromatogr.*, **367**, 35 (1986).
335. C. E. Kientz and A. Verweij, *Int. J. Environ. Anal. Chem.*, **30**, 255 (1987).
336. A. Farran, J. de Pablo and D. Barcelo, *J. Chromatogr.*, **455**, 163 (1988).
337. D. Barcelo and J. Albaiges, *J. Chromatogr.*, **474**, 163 (1989).
338. P. C. Bossle, J. J. Martin, E. W. Sarver and H. Z. Sommer, *J. Chromatogr.*, **267**, 209 (1983).
339. D. Barcelo, F. A. Maris, R. B. Geerdink, R. W. Frei, G. J. de Jong and U. A. T. Brinkman, *J. Chromatogr.*, **394**, 65 (1987).
340. L. D. Betowski and T. L. Jones, *Environ. Sci. Technol.*, **22**, 1430 (1988).
341. S. -N. Lin, C. -Y. Chen, S. D. Murphy and R. M. Caprioli, *J. Agric. Food Chem.*, **28**, 85 (1980).
342. B. Kennedy Keller, M. S. Fuller, G. D. Rottler and L. W. Connelly, *Anal. Biochem.*, **147**, 166 (1985).

343. K. Yamada, S. Abe, T. Suzuki, K. Katayama and T. Sato, *Anal. Biochem.*, **156**, 380 (1986).
344. C. Nanjundiah and T. L. Rose, *J. Electrochem. Soc.*, **133**, 955 (1986).
345. W. F. Smyth and M. R. Smyth, *Pure Appl. Chem.*, **59**, 245 (1987).
346. K. A. Ramsteiner and W. D. Hormann, *J. Chromatogr.*, **104**, 438 (1975).
347. I. Eggens, T. Chojnacki, L. Kenne and G. Dallner, *Biochim. Biophys. Acta.*, **751**, 355 (1983).
348. J. L. Burguera and M. Burguera, *Anal. Chim. Acta*, **179**, 497 (1986).
349. M. E. Leon-Gonzalez and A. Townshend, *Anal. Chim. Acta*, **236**, 267 (1990).
350. R. Kindervater, W. Kunnecke and R. D. Schmid, *Anal. Chim. Acta*, **234**, 113 (1990).
351. S. Kumaran and C. Tran-Minh, *Anal. Biochem.*, **200**, 187 (1992).
352. K. Levsen, *Org. Mass Spectrom.*, **23**, 406 (1988).
353. S. Sass and T. L. Fisher, *Org. Mass Spectrom.*, **14**, 257 (1979).
354. A. K. Singh, D. W. Hewetson, K. C. Jordon and M. Ashraf, *J. Chromatogr.*, **369**, 83 (1986).
355. D. Barcelo, M. Sole, G. Durand and J. Albaiges, *Fresenius' J. Anal. Chem.*, **339**, 676 (1991).
356. S. V. Hummel and R. A. Yost, *Org. Mass Spectrom.*, **21**, 785 (1986).
357. P. J. Arpino, *J. Chromatogr.*, **323**, 3 (1985).
358. A. P. Bruins, *J. Chromatogr.*, **233**, 99 (1985).
359. J. D. Rosen, *Applications of New Mass Spectrometry Techniques In Pesticide Chemistry* (*Chemical Analysis Series*, Vol. 91), Wiley, New York, (1987).
360. R. T. Rosen and J. E. Dziedzic, in *Applications of New Mass Spectrometry Techniques In Pesticide Chemistry* (*Chemical Analysis Series*, Vol. 91), (Ed. J. D. Rosen), Wiley, New York, 1987, p. 178.
361. C. R. Blakley and M. L. Vestal, *Anal. Chem.*, **55**, 750 (1983).
362. D. J. Liberato, C. C. Fenselau, M. L. Vestal and A. L. Yergey, *Anal. Chem.*, **55**, 1741 (1983).
363. J. J. Morelli and D. M. Hercules, *Anal. Chem.*, **58**, 1294 (1986).
364. J. N. Damico, *J. Assoc. Off. Anal. Chem.*, **49**, 1027 (1966).
365. H. -J. Stan, B. Abraham, J. Jung, M. Kellert and K. Steinland, *Fresenius' Z. Anal. Chem.*, **287**, 271 (1977).
366. K. L. Busch, M. M. Bursey, J. R. Hass and G. W. Sovocool, *Appl. Spectrosc.*, **32**, 388 (1978).
367. F. H. Field, *Acc. Chem. Res.*, **1**, 42 (1968).
368. D. F. Hunt, C. N. McEwen and R. A. Upham, *Tetrahedron Lett.*, **47**, 4539 (1971).
369. R. C. Dougherty, J. Dalton and F. J. Biros, *Org. Mass Spectrom.*, **6**, 1171 (1972).
370. D. F. Hunt and J. F. Ryan, *Anal. Chem.*, **44**, 1306 (1972).
371. D. F. Hunt and J. F. Ryan, *J. Chem. Soc., Chem. Commun.*, 620 (1972).
372. R. L. Holmstead and J. E. Casida, *J. Assoc. Off. Anal. Chem.*, **57**, 1050 (1974).
373. R. D. Voyskner, J. T. Bursey and E. D. Pellizzari, *Anal. Chem.*, **56**, 1507 (1984).
374. A. J. Alexander and P. Kebarle, *Anal. Chem.*, **58**, 471 (1986).
375. G. Schmelzeisen-Redeker, M. A. McDowall, U. Giessmann, K. Levsen and F. W. Rollgen, *J. Chromatogr.*, **323**, 127 (1985).
376. F. W. McLafferty and D. B. Stauffer, *Registry of Mass Spectral Data*, 2nd ed., Wiley, New York, (1989).
377. F. W. McLafferty, *Anal. Chem.*, **28**, 306 (1956).
378. J. L. Occolowitz and G. L. White, *Anal. Chem.*, **49**, 1179 (1963).
379. A. K. Singh, R. J. Zeleznikar, Jr, and L. R. Drewes, *J. Chromatogr.*, **324**, 163 (1985).
380. T. W. Sawyer, M. T. Weiss, P. A. D'Agostino, L. R. Provost and J. R. Hancock, *J. Appl. Toxicol.*, **12**, 1 (1992).
381. M. V. Norris, W. A. Vail and P. R. Averell, *Agric. Food Chem.*, **2**, 570 (1954).
382. K. Visweswriah and M. Jayaram, *Agric. Biol. Chem.*, **38**, 2031 (1974).
383. E. R. Clark and I. A. Qazi, *Analyst*, **104**, 1129 (1979).
384. E. R. Clark and I. A. Qazi, *Analyst*, **105**, 564 (1980).
385. E. R. Clark and I. A. Qazi, *Water Res.*, **14**, 1037 (1980).
386. A. C. Hill, M. Akhtar, M. Mumtaz and J. A. Osmani, *Analyst*, **92**, 496 (1967).
387. J. H. Mendez, O. J. De Blas, V. R. Martin and E. S. Lopez, *Anal. Lett.*, **18**, 2069 (1985).
388. R. T. Sane and S. S. Kamat, *J. Assoc. Off. Anal. Chem.*, **65**, 40 (1982).
389. S. U. Bhaskar and N. V. Nanda Kumar, *Talanta*, **27**, 757 (1980).
390. N. V. Nanda Kumar and M. Ramasundari, *J. Assoc. Off. Anal. Chem.*, **63**, 536 (1980).
391. S. U. Bhaskar and N. V. Nanda Kumar, *J. Assoc. Off. Anal. Chem.*, **65**, 1297 (1982).
392. S. Sass, T. L. Fisher, R. J. Steger and G. A. Parker, *J. Chromatogr.*, **238**, 445 (1982).
393. H. C. de Bisschop and E. Michiels, *Chromatographia*, **18**, 433 (1984).

394. H. P. Benschop, E. C. Bijleveld, M. F. Otto, C. E. A. M. Degenhardt, H. P. M. van Helden and L. P. A. de Jong, *Anal. Biochem.*, **151**, 242 (1985).
395. M. L. Shih and R. I. Ellin, *Anal. Lett.*, **19**, 2197 (1986).
396. S. J. Smith, *Talanta*, **10**, 725 (1983).
397. K. Alfthan, H. Kenttamaa and T. Zukale, *Anal. Chim. Acta.*, **217**, 43 (1989).
398. P. S. Hammond and J. S. Forster, *Anal. Biochem.*, **180**, 380 (1989).
399. H. O. Michel, E. C. Gordon and J. Epstein, *Environ. Sci. Technol.*, **7**, 1045 (1973).
400. K. B. Sipponen, *J. Chromatogr.*, **389**, 87 (1987).
401. A. Verweij, E. Burghardt and A. W. Koonings, *J. Chromatogr.*, **54**, 151 (1971).
402. H. P. Benschop, C. A. G. Konings and L. P. A. de Jong, *J. Am. Chem. Soc.*, **103**, 4260 (1981).
403. L. P. A. de Jong, E. C. Bijleveld, C. van Dijk and H. P. Benschop, *Int. J. Environ. Anal. Chem.*, **29**, 179 (1987).

Author index

This author index is designed to enable the reader to locate an author's name and work with the aid of the reference numbers appearing in the text. The page numbers are printed in normal type in ascending numerical order, followed by the reference numbers in parentheses. The numbers in *italics* refer to the pages on which the references are actually listed.

Index compiled by K. Raven

Subject index

Index compiled by P. Raven